ELECTRIC CIRCUIT ANALYSIS

THIS IS A WILEY ON-DEMAND PRODUCT

ELECTRIC CIRCUIT ANALYSIS

Robert A. Bartkowiak

The Pennsylvania State University

WILEY

JOHN WILEY & SONS, New York

Chichester, Brisbane, Toronto, Singapore

To Mickie and the Children

Electric Circuit Analysis

Copyright © 1985 by John Wiley & Sons, Inc.

Library of Congress Cataloging in Publication Data

Bartkowiak, Robert A.
     Electric circuit analysis.

     Includes index.
     1. Electric circuit analysis.   I. Title.
TK454.B362   1985   621.319′2   84–10815
ISBN 0-471-60355-4

     9

Contents

Preface

This textbook was written for use in electrical engineering technology programs. It can be used both in two-year programs and in the introductory courses in four-year programs. It can also be used in other technology disciplines in which a knowledge of electrical principles is considered essential. For this reason, descriptions of and problems pertaining to transducers, switches, transformers, and other devices that may be of interest to students in nonelectrical disciplines are included. Where appropriate, photographs are used to enhance device descriptions, so that students in nonelectrical disciplines who might not see these devices in a laboratory setting can obtain a better understanding of them.

The first few chapters require only basic physics and algebra. Although determinants should be covered in a prerequisite or concurrent algebra course, they are reviewed in the Appendix. This material can be introduced before the circuit analysis concepts of Chapter 6. Similarly, complex algebra is reviewed in Chapter 11, with particular emphasis on its application to ac circuits. The chapters on alternating current also require an understanding of trigonometry.

Since many technology programs include introductory calculus courses, the exactness of calculus is used in some formula derivations, but not until Chapter 7. Even then, calculus is used only as a tool in explaining certain electrical concepts. Because many students have access to a personal computer, computer programs in BASIC are included in the Appendix. These programs are useful for solving simultaneous equations and working with complex numbers, both of which generally require much student time.

Approximately the first half of the book is devoted to dc circuits. The basic quantities—current, voltage, and resistance—are extensively presented. Circuit analysis is considered, followed by discussions of capacitance, transients, magnetic circuits, and inductance. Dependent sources are discussed, as many electronic devices are modeled with such sources.

Next, ac circuits are studied, with emphasis on the fundamental ac relationships, phasor algebra, single-phase circuits, polyphase circuits, resonance, and nonsinusoidal waves. Network analysis and its associated theorems and applications are presented to demonstrate their usefulness when ac voltage and reactance are in a circuit.

Electrical measurements are often studied in associated laboratory course work, and therefore a chapter on measurement, Chapter 17, is included. The use of direct and alternating currents and of analog and digital instruments, such as ammeters, voltmeters, and ohmmeters, is described in this chapter.

The voltage versus current characteristics of nonlinear resistances and electric sources are first introduced in Chapters 4 and 5, respectively. These concepts are then combined in Chapter 18, which is devoted to dc graphical analysis. This chapter is particularly useful in the study of nonlinear circuits and can also serve as an introduction to electronic load line analysis.

Conventional current is used in the book. Also in keeping with the current recommendations of the Institute of Electrical and Electronics Engineers, SI units are used wherever possible.

Almost all sections contain drill problems and answers so that students can reinforce their understanding of the material before moving on. Additional questions and problems are provided at the end of each chapter. Most of these have been used in examinations and quizzes, and they have been chosen to reinforce the concepts presented within the chapters. Answers to odd-numbered problems are provided at the end of the book. These and the examples presented in the text have been specified to only two or three significant digits. A glossary is also included.

It has been said that a picture is worth a thousand words. Certainly, the photographs and device specifications provided by the many electrical device manufacturers will enhance the learning experience of the student using this textbook. My thanks are extended to the many marketing managers, sales administrators, and engineers for their willingness and assistance in providing this enhancement. My thanks are also extended to Rocky Bayer for his help with the author's own photographs—the ones without courtesy lines.

I wish to thank Professors Edward A. Dreisbach and Irving Engelson for their helpful suggestions regarding *Electric Circuits,* from which much of the material in this text comes. I also wish to thank Professors Clifford W. Cowan, Wesley G. Houser, and Lee Rosenthal, for reviewing the manuscript of this text and offering many helpful suggestions, and the staff of Harper & Row.

Finally, I wish to thank my wife, Michalene, for typing the manuscript and continually offering encouragement.

Robert A. Bartkowiak

Chapter 1

Introduction

1.1 UNITS

Throughout history, humans have developed methods of describing, measuring, and correlating various phenomena in the natural sciences. Records of ancient civilizations show that humans used parts of their bodies as measuring instruments. For example, the measure of length known as the yard can be traced back to the early Saxon word "gird," which refers to the circumference of a person's waist. However, because of the limited communication between early civilizations, it is not surprising that different systems were developed to measure the same quantities.

If we are interested in motion in physics, we wish to measure and relate the quantities length, mass, and time. Over the years measurement of these three quantities has led to at least three systems of measurement—the *centimeter-gram-second* (CGS), the *meter-kilogram-second* (MKS), and the *English* systems of units, which are shown in Table 1-1.

In October 1965 the Institute of Electrical and Electronics Engineers (IEEE) adopted the *International System of Units* (SI units) proposed by the 11th General Conference on Weights and Measures, held in France in 1960. The International System includes as subsystems the MKS system of units for mechanics and the MKSA system of units, which covers mechanics, electricity, and magnetism. The SI system of units has seven fundamental or base units plus two supplementary units. Each of the base units is precisely defined because it is used as a standard in itself and as a standard in secondary or derived units.

The SI units are shown in Table 1-2. The seven base quantities and two supplementary quantities are listed in the left column, the fundamental units are in the middle column, and their symbols are in the right column. Thus, the SI unit for mass is the kilogram, whose symbol is kg. It should be noted that the symbols never

TABLE 1-1 PHYSICAL UNITS

Unit system	Length	Mass	Time
CGS	centimeter (cm)	gram (g)	second (s)
MKS (MKSA)	meter (m)	kilogram (kg)	second
English	foot (ft)	slug (32 lbf)	second

have a distinct plural form, are not written with a period, and are upper- or lowercase only as defined.

Notice that in the SI system plane angles are measured in radians, for which 2π rad $= 360°$. However, degrees (°) are in such widespread use that they have been retained for general use even though they are not a defined SI unit. Similarly, degrees Celsius (°C), minutes (min), hour (h), and day (d) are so much a part of our everyday lives that they too are recognized units for general use.

The SI base units are indeed units upon which the measurement of other quantities is based. Figure 1-1 illustrates how a few secondary SI units are derived from the base units.

One of the simplest of derived units is that of velocity. Velocity is defined as a unit of length divided by a unit of time. Thus, as shown in Fig. 1-1, the SI unit of velocity is the meter per second (m/s). Futhermore, we recognize that acceleration is the change of velocity in unit time. Thus, the SI unit of acceleration is the meter per second squared (m/s²).

Most of us are somewhat aware of the terms "force," "energy," and "power" and of how they are used in physics and electrical studies. Figure 1-1 illustrates how these particular units relate to the SI base units. Note the following:

First, a force of 1 newton (N) is the force that causes a kilogram mass to have an acceleration of 1 m/s².

Second, energy is defined as force × distance and is described in the SI system in joules (J), where 1 J = 1 N of force acting through a distance of 1 m.

TABLE 1-2 THE INTERNATIONAL SYSTEM OF UNITS

Quantity	Unit	Symbol
Length	meter	m
Mass	kilogram	kg
Time	second	s
Electric current	ampere	A
Thermodynamic temperature	kelvin	K
Amount of substance	mole	mol
Luminous intensity	candela	cd
Supplementary units		
Plane angle	radian	rad
Solid angle	steradian	sr

Base units   Some derived units

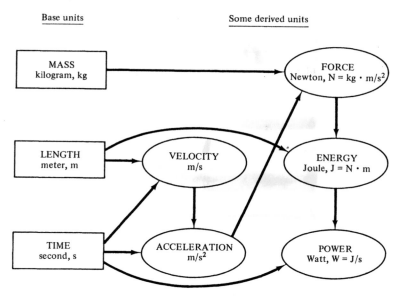

Figure 1-1   How a few derived units (ovals) relate to the SI base units (rectangles).

Third, power, defined as energy per unit time is described in watts (W), where 1 W = 1 J/s.

The secondary or derived units shown in Fig. 1-1 are only a few of the many you will encounter in your study of electric circuits. Fear not; this early introduction of the energy and power terms is only to illustrate that our electrical units have their derivation in the SI units. As these and other electrical units are introduced in the chapters that follow, they are defined and related to the SI base units. Although an illustration such as Fig. 1-1 might not be used, you are urged to think of the relationships in that fashion.

1.2 CONVERSION OF UNITS

With the introduction of the SI units and their acceptance by engineering societies has come the problem of recognition of and conversion to and from the other sets of units. The introduction of the SI units into our everyday lives has been gradual. In many instances a "soft" conversion has resulted in the simultaneous use of both sets of units. Thus, an electronic enclosure might have its volume listed in both cubic inches and liters.

In some instances it is advantageous or even necessary to convert from one set of units to another by use of a *conversion factor*, which is a constant that relates the two different units. Some common conversion factors are listed in Table 1-3. A systematic approach to their use is to divide either side of the equality by the other side, thus obtaining a ratio equal to 1. For example, 1 in. = 2.54 cm can also be

TABLE 1-3   SOME COMMON CONVERSION FACTORS

Length	1 inch (in.) = 2.54 centimeters (cm)
	1 m = 39.37 in.
	1 kilometer (km) = 0.6213 mile (mi)
Area	1 square inch (in.^{2}) = 6.452 cm^2
	1 square foot (ft^2) = 0.093 m^2
Volume	1 quart = 0.946 liter (l)
	1 l = 1000 cm^3
	1 cubic inch (in.^{3}) = 16.39 cm^3
Mass	1 pound (lb) = 453.6 g
	1 kg = 2.2046 lb
Force	1 newton (N) = 0.224 pound-force (lbf)
Energy	1 ft lbf = 1.3549 joules (J)
	1 Btu = 1055 J
	1 watt-hour (Wh) = 3600 J
Power	1 horsepower (hp) = 746 watt (W)
	1 hp = 550 ft lbf/s
Plane angle	1° = 0.01745 rad

expressed

$$\frac{1 \text{ in.}}{2.54 \text{ cm}} = 1 \quad \text{or} \quad \frac{2.54 \text{ cm}}{1 \text{ in.}} = 1$$

An equation can then be multiplied by either of the ratios without changing the value of the equation, since the net effect is a multiplication by 1. With the correct choice of ratio, cancellation of the desired unit results.

EXAMPLE 1-1

Convert 5 in. to an equivalent length in centimeters.

Solution

$$5 \text{ in.} \times \frac{2.54 \text{ cm}}{1 \text{ in.}} = 12.7 \text{ cm}$$

EXAMPLE 1-2

Convert 0.577 hp to its equivalent in watts.

Solution

$$0.577 \text{ hp} \times \frac{746 \text{ W}}{\text{hp}} = 430 \text{ W}$$

Drill Problem 1-1 ■

Perform the following conversions:
(a) 7.5 hp to watts;
(b) 250 in.^{3} to liters;
(c) 550 Btu to watt-hours.

Answer.   (a) 5595 W. (b) 4.1 l. (c) 161 Wh.   ■ ■

TABLE 1-4 INTERNATIONAL SYSTEM PREFIXES

Prefix	Pronunciation	Factor by which the unit is multiplied	Symbol
tera	ter′ȧ	10^{12}	T
giga	jĭ′gȧ	10^{9}	G
mega	mĕg′ȧ	10^{6}	M
kilo	kĭl′ō	10^{3}	k
milli	mĭl′ĭ	10^{-3}	m
micro	mĭ′krō	10^{-6}	μ
nano	năn′o	10^{-9}	n
pico	pē′kō	10^{-12}	p
femto	fĕm′tō	10^{-15}	f
atto	ăt′tō	10^{-18}	a

1.3 UNIT PREFIXES

Often the quantities being considered are much smaller or larger than the basic unit defined by the MKSA system. Rather than change the basic unit, a prefix is added to denote a multiple or submultiple of the basic unit. The prefixes used with the SI units are listed in Table 1-4.

The prefixes shown in Table 1-4 are those commonly used in the electrical and electronics field. Another prefix not listed but acceptable for SI use is the prefix "centi" (c), which is used in allied fields of study and means 10^{-2} units.

The multiplying factors of Table 1-4 are specified in powers of 10. In each case the power of 10 represents how many times the base (10) is multiplied by itself or how many decimal places the decimal point is shifted to the right or left when the number is written in its expanded form. If the exponent is positive, the decimal point is shifted to the right in its expanded form. For example,

$$2 \times 10^3 = 2 \times (10 \times 10 \times 10) = 2000$$

If the exponent is negative, the decimal point is shifted to the left. For example,

$$5 \times 10^{-3} = 5 \times (0.1 \times 0.1 \times 0.1) = 0.005$$

This exponential format for specifying large and small numbers is known as *scientific notation*. Calculators and pocket computers capable of handling entries directly in scientific notation have an *enter exponent* key $\boxed{\text{EE}}$ or its equivalent. To express a number in scientific notation, shift the decimal point so that a single non-zero digit remains to the left of the decimal point; the power of 10 is then changed to reflect the shift in the position of the decimal point.

EXAMPLE 1-3

Express the following numbers in scientific notation: **(a)** 51 300; **(b)** 0.082.

Solution. (a) $51\,300 = 5.13 \times 10^4$. (b) $0.082 = 8.2 \times 10^{-2}$.

Additionally, certain calculators and pocket computers have an $\boxed{\text{ENG}}$ key or its equivalent. This key refers to *engineering notation*, which is another form of exponential notation; the output (or answer) is shown with a power of 10, but the exponent used is divisible by 3 so that the answer corresponds directly to an SI prefix. When expressing a number in engineering notation, one shifts the decimal point so that one has a number with one, two, or three digits multiplied by a power of 10 (the exponent must be divisible by 3).

EXAMPLE 1-4

Express the following numbers in engineering notation: **(a)** 51 300; **(b)** 0.082.

Solution.   (a) $51\,300 = 51.3 \times 10^3$ (51.3 k-unit). (b) $0.082 = 82 \times 10^{-3}$ (82 m-unit).

Certain mathematical relationships apply to the use of exponents and powers of numbers. As applied to powers of 10, these relationships are defined by the following equations:

$$(10^a)(10^b) = 10^{a+b} \tag{1-1}$$

$$10^a/10^b = 10^{a-b} \tag{1-2}$$

$$(10^a)^b = 10^{ab} \tag{1-3}$$

In the preceding equations a and b may be either positive or negative, and their addition, subtraction, or multiplication is performed algebraically with attention to the signs.

EXAMPLE 1-5

Simplify

(a) $(10^{-2})(3 \times 10^4)$,

(b) $\dfrac{(10^2)}{5 \times 10^{-3}}$,

(c) $(2 \times 10^3)^2$.

Solution

(a) $(10^{-2})(3 \times 10^4) = 3 \times 10^{4-2} = 3 \times 10^2$.

(b) $\dfrac{10^2}{5 \times 10^{-3}} = 0.2 \times 10^{2-(-3)} = 0.2 \times 10^5 = 2 \times 10^4$.

(c) $(2 \times 10^3)^2 = 2^2 \times (10^3)^2 = 4 \times 10^6$.

Drill Problem 1-2 ■

Simplify and express in scientific notation

(a) $\dfrac{(10^3)^2}{2\pi(60)(6 \times 10^{-6})}$,

(b) $\dfrac{2\pi\,(1000)(3\,\times\,10^{-6})}{100}$.

Answer. (a) 4.42×10^8. (b) 1.885×10^{-4}. ■ ■

Drill Problem 1-3 ■

Express the answers of Drill Problem 1-2 with SI prefixes

Answer. (a) 442 M. (b) 188.5 μ. ■ ■

1.4 SIGNIFICANT DIGITS

As defined in Chapter 17 on measurements, *accuracy* is the amount a measured value agrees with its true value. When a measured value is free from error, the value is considered to be *correct*. In contrast, *precision* is the degree of detail with which a measurement is expressed. It is an indication of the smallest unit used on a measurement device or the smallest unit used in specifying a measurement or number. Any digit that is used in defining a number or measurement is called a *significant digit*.

Suppose one accurately measures the length and width of an electronic circuit board. If the length is measured as 6.25 mm, we say the measurement has a precision of three significant digits. If the width is measured as 3.15 mm, it likewise has a precision of three significant digits. If one determined the area from the two measurements with a calculator, one would probably see on the calculator display the following result:

$$\text{Area} = 6.25 \times 3.15 = 19.6875 \text{ mm}^2$$

The calculator produces a reading with a precision of six digits. However, it is unrealistic to specify such precision when the original data are specified only to three digits. It is correct to retain only those digits justified by the original measurements.

In the example, the length and width are each specified to three digits; hence, their product is correctly specified to only three significant digits by rounding to 19.7 mm². However, when an arithmetic operation involves quantities with differing numbers of significant digits, the quantity with the least number of digits determines the number of significant digits in the result. For example, the three-digit number 9.21 determines the significant digits in the calculation

$$9.21 \times 4.375 = 40.3$$

When a number has a zero as its last significant digit, that zero is normally used in determining the precision of the number. Thus, 8.2 is precise to two digits, whereas 8,200 is precise to four digits. However, it is correct to omit significant zeros if the intended precision is expressed in other ways. In this book we shall consider all values, even if expressed with only one or two digits, to be accurate to a precision of three digits. Thus, a current of 5 A will be understood to be 5.00 A. In the examples and problems in this book we shall generally limit ourselves to three significant digits, perhaps in a rare case extending the precision to four digits.

1.5 THE CONCEPT OF CHARGE

Although human interest in mechanics began in prehistoric times, most of our knowledge about electricity and magnetism has been developed during the last few centuries.

Thales, a Greek philosopher, made some brief observations on magnetism in 600 B.C., but it was not until 1600 that William Gilbert of England and some of his fellow experimenters observed that amber rubbed with fur acquired the ability to attract light objects to itself. The bodies were said to be electrified. It was also found that sometimes a force of *repulsion*, as well as one of *attraction*, could exist between two electrified bodies. In 1747 Benjamin Franklin introduced the nomenclature *positive* and *negative* to help explain the two types of electrification. About this time, the term *charge* was introduced to imply electrification or the presence of electric potential energy. Further experimentation in the eighteenth century led to the following statements:

1. When electrification is brought about, equal positive and negative charges are simultaneously produced.
2. Unlike charges attract each other; like charges repel.

In 1785, Charles Coulomb, a French physicist, introduced what is now known as Coulomb's law. This law states that when two point charges Q_1 and Q_2 are separated by a distance r, as in Fig. 1-2, the force of attraction or repulsion is given by

$$F = k \frac{Q_1 Q_2}{r^2} \tag{1-4}$$

The force given by Eq. 1-4 has units of newtons when r is in meters and Q_1 and Q_2 are in *coulombs*, the basic unit of charge. The constant of proportionality depends on the medium separating the charges; when air or free space is used, $k = 9 \times 10^9$ in SI units. Even if two charges do not touch, they exert a force on each other, and whenever a charge is acted on by an electric force, that charge is said to be in an *electric field of force* or simply *electric field*. The *electrostatic force* given by Eq. 1-4 is only one of the forces acting to hold atoms together.

EXAMPLE 1-6

A concentrated positive charge of 1×10^{-9} coulombs is located in free space 2 m from another positive charge of 4×10^{-6} C. What force exists between the charges?

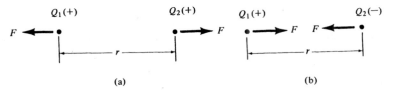

Figure 1-2 Two point charges: **(a)** like charges; **(b)** unlike charges.

Solution. Substituting into Eq. 1-4 yields

$$F = (9 \times 10^9) \frac{(1 \times 10^{-9})(4 \times 10^{-6})}{(2)^2}$$

$$= 9 \times 10^{-6} \text{ N}$$

$$= 9 \ \mu\text{N}$$

QUESTIONS

1. What is meant by the term "SI unit"?
2. What is scientific notation?
3. What is engineering notation; how does it differ from scientific notation?
4. What is a significant digit?
5. What is a conversion factor?
6. What is charge?
7. What is meant by the term "electric field"?

PROBLEMS

1. Make the following conversions:
   (a) 0.075 s to microseconds.
   (b) 0.075 s to nanoseconds.
   (c) 1.525 in. to meters.
   (d) 25 in.^{2} to square meters.
   (e) 0.57 in. to meters.
   (f) 946 W to horsepower.
   (g) 55 mi/h to meters per second (1 mi = 5280 ft).

2. Express the following in both milli- and microunits:
   (a) 0.002 52 A.
   (b) 0.0793 A.
   (c) 0.321 s.
   (d) 0.000 047 s.

3. Express the following in both kilo- and megaunits:
   (a) 1050 ohms (Ω).
   (b) 7252 W.
   (c) 75 000 m.
   (d) 7 252 000 W.

4. Express the speed 55 mi/h in kilometers per hour.

5. The SI unit for pressure is the pascal (Pa); 1 Pa = 1 N/m^2. Using conversion factors from Table 1-3, derive a factor to convert pounds per square inch (psi) to pascals.

6. Express the following in both scientific and engineering notation:
   (a) 0.452.
   (b) 157.
   (c) 52 400.
   (d) 0.000 58.

7. Perform the following calculations and express the answer in engineering notation:
   (a) $(5 \times 10^5)(4 \times 10^3)$.
   (b) $(25 \times 10^4) + (2 \times 10^5)$.
   (c) $(10)^2$.
   (d) $(0.01)^2(13.7 \times 10^{-2})$.
   (e) $(5 \times 10^{-6})/(9.2 \times 10^6)$.
   (f) $(-0.02)^3$.
   (g) $(52 \times 10^2)^2$.
   (h) $(25)^2(10^3)^2/10^4$.

8. Perform the following calculations and express the answers using SI prefixes (the derived electrical units will be explained in the course of this book):
   (a) $(75 \times 10^4) - (38 \times 10^3)$ W.
   (b) $(12 \times 10^{-12})(20 \times 10^{-12})/[(12 \times 10^{-12}) + (12 \times 10^{-12})]$ F.
   (c) $1/(37 \times 10^3)$ W.
   (d) $(11.25 \times 10^2)/[(50 \times 10^6) + (5 \times 10^7)]$ A.
   (e) $1/2(3.14)(10^3)(0.005 \times 10^{-6})$ Ω.
   (f) $2(3.14)(10 \times 10^3)(0.0016)$ Ω.

9. Find the force of attraction between the charges of Fig. 1-2 if $Q_1 = 6 \times 10^{-4}$ C, $Q_2 = -3 \times 10^{-4}$ C, and $r = 1$ m.

10. What will the force of attraction in Problem 9 become if the charge separation is increased by a factor of 10, that is, if $r = 10$ m.

11. Two electric charges, one twice the other, develop a mutually repulsive force of 12 N when located 0.08 m apart. Determine the magnitude of the charges.

12. Show that the SI unit for the constant k of Eq. 1-4 is newtons square meter per square coulomb. (Alternatively, as described in Chapter 7, the constant can be expressed in meters per farad.)

Chapter 2

Electrical Units

2.1 ATOMIC STRUCTURE

Matter is anything that has mass and occupies space. By subdividing matter into smaller and smaller pieces, one finally obtains the smallest particle of matter that still retains the properties of the original material (compound). This particle is the *molecule*; when subdivided it yields constituent chemical elements in the form of individual atoms.

Until the 1900s there was much speculation about the nature of the atom. In 1911 Rutherford, an English scientist, made particle-scattering experiments that led to his discovery of the atomic core or nucleus. He proposed a model in which the atom was primarily open space with all the mass concentrated in a small region called the *nucleus* (approximately 10^{-14} m in radius).

In 1913 the Danish physicist Niels Bohr extended Rutherford's model of the atom. Bohr theorized that the hydrogen atom consisted of a core or nucleus with a positive proton and a planetary negative electron revolving in a circular orbit (see Fig. 2-1). Bohr's theory allowed for only certain distinct orbits corresponding to distinct electron energies, the innermost orbit corresponding to the *ground state*. The electron could be *excited* or given enough kinetic energy to allow it to jump to other orbits.

The Bohr model can be expanded to the helium atom, which has two electrons, as in Fig. 2-2. Although the two electrons are shown to be in the same orbit, quantum theory tells us that the orbits, and therefore the electron energies, are just slightly different. This difference, due to a difference in the spin of the electrons on their own axes, also affects the magnetic properties of the materials (see Chapter 8).

More electrons are added to the atomic model by starting a second orbit or *shell*, which can contain eight electrons. A third shell can be added with a maximum

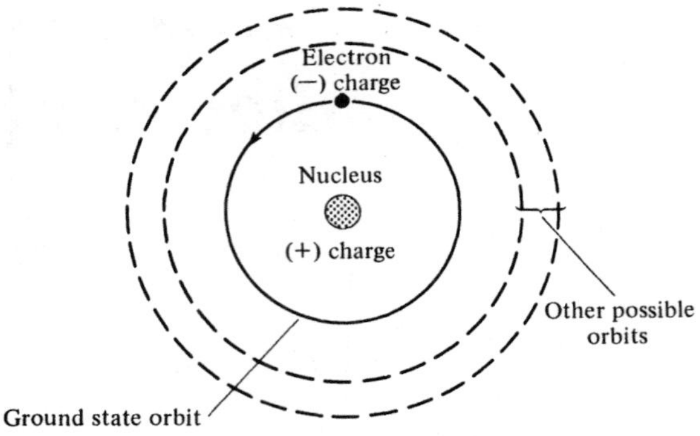

Figure 2-1 The Bohr model of the hydrogen atom.

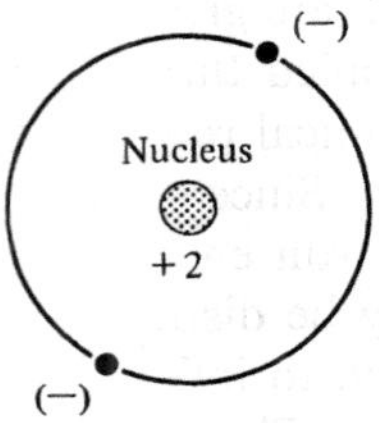

Figure 2-2 The helium atom model.

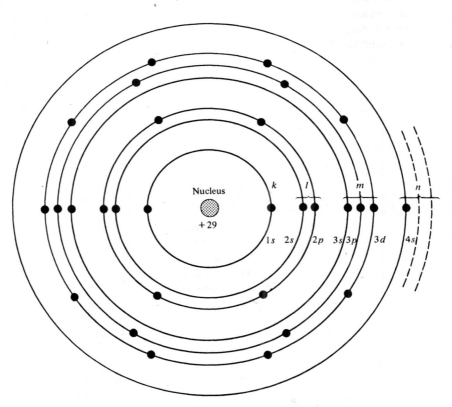

Figure 2-3 The copper atom model.

of 18 electrons. In the case of copper, which has 29 electrons (atomic number 29) a fourth shell containing one electron must be added, as in Fig. 2-3. The shells have been given identifying letters; from the innermost orbit they are k, l, m, n, and so forth, respectively. Subshells having slightly different energies within the principal shell are given the letter identifications s, p, d, and f. The outermost shell of electrons is called the *valence shell*; the electrons in this shell are the ones that take part in chemical reactions.

Since each electronic orbit corresponds to a particular kinetic energy, a plot of electron energy may be made as in Fig. 2-4. When an atom is isolated, the orbits may be distinct levels; however, as similar atoms are brought closer together, they begin to influence one another and force the electrons into slightly different energy levels. The energy levels then become densely packed bands, as indicated in Fig. 2-4. Those energy levels which are not considered to occur form "energy gaps" between the possible energy levels.

The electrons very close to the nucleus are held tightly in orbit; they are called *bound electrons*. On the other hand, the electrons in the valence orbit are farther from the nucleus, so that the forces attracting them toward the nucleus are not as great as for those electrons in the inner orbits. If energy is added to the atom, it is possible for a valence band electron to absorb this energy and move into an outer shell of a neighboring atom. An electron that is out of its valence state and free to move into an outer shell of a neighboring atom is called a *free electron*. In copper at room temperature, there are approximately 10^{24} free electrons per in.[3] An atom that has

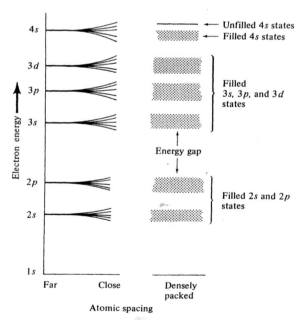

Figure 2-4 Energy level diagram.

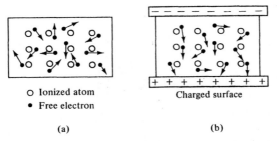

O Ionized atom
● Free electron

(a)   (b)

Figure 2-5 Instantaneous motion of free electrons: **(a)** no field, no net motion; **(b)** field applied, net motion.

lost one or more valence electrons has a net positive charge and is called a *positive ion*.

2.2 CURRENT

As pointed out in the previous section, thermal energy from the surroundings creates a number of free electrons. In the absence of any external electric field, these electrons move in random paths, as in the two-dimensional sketch of Fig. 2-5(a). The movement of any one electron is offset by the movement of another in an opposite direction so that there is no net movement of electrons. If an electric field is applied, as in Fig. 2-5(b), more of the electrons move toward the positive plate. The electrons then have a *net motion* or *drift* toward the positively charged plate. The drift can be thought of as being superimposed upon the random motion of free electrons.

If an imaginary plane is passed through a conductor, a material that allows current flow (see Fig. 2-6), and if an electric field is present, a net flow of charge flows

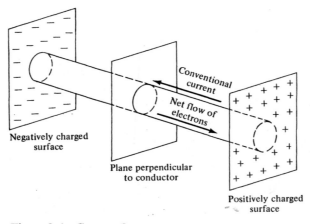

Conventional current

Net flow of electrons

Negatively charged surface

Plane perpendicular to conductor

Positively charged surface

Figure 2-6   Current flow.

past the plane. The net flow of electrons or charge past a given point per unit time is defined as the *electric current*.

As pointed out in Chapter 1, the basic unit of charge is the coulomb. A coulomb is equal to the charge carried by 6.24×10^{18} electrons. Conversely, the charge of one electron is

$$\text{electron charge} = \frac{1}{6.24 \times 10^{18}} = 1.6 \times 10^{-19} \text{ C}$$

Electric current, or the *rate of flow of charge*, is then measured in the units of coulombs per second. The symbol for current is I; in equation form the current is

$$I = \frac{Q}{t} \qquad \frac{\text{C}}{\text{s}} \tag{2-1}$$

In the MKS system the basic unit of current is the ampere, where 1 A is the rate of flow of electric current when 1 C of charge flows past a point in 1 s.

$$1 \text{ A} = \frac{1 \text{ C}}{1 \text{ s}} \tag{2-2}$$

Two methods are available for describing the direction of current flow (see Fig. 2-6).

Historically, current has been described as flowing from plus to minus, whereas the electron flow is from minus to plus. Either can be specified correctly because the electron is moving in one direction and the *electron vacancy* or *hole* is moving in the opposite direction. The flow of current from plus to minus is referred to as *conventional* or *positive current* and has been adopted by the IEEE. Conventional current will be used throughout this text.

Electric currents are often classified according to their direction of flow and variation with time. Current types are best described by graphs that show the instantaneous value of current i as a function of time t. When we consider quantities that may vary with time, we use small letters to indicate the variables; hence, we indicate current with a small i in Fig. 2-7.

A current is classified as *unidirectional* if it always maintains the same direction of flow and *bidirectional* if it changes direction. Both Figs. 2-7(a) and 2-7(b) represent unidirectional currents. Furthermore, if the unidirectional current is unchanging or changes negligibly as in Fig. 2-7(a), it is called *direct current* (dc).

Two types of bidirectional currents are illustrated in Figs. 2-7(c) and 2-7(d). In both cases, the currents change direction as indicated by the negative values of the illustrations. In particular, Fig. 2-7(c) depicts a current that changes direction while following the repetitive values of a sine wave. This sinusoidal type of current is referred to as *alternating current* (ac).

In the study of electric circuits it is convenient to consider separately the effects of dc and ac. However, we must remember that these represent only two of many current types. In the chapters that follow we shall see the significance of these classifications. We shall even study how any arbitrary repetitive current can be separated into dc and ac components. Incidently, alternating currents generally have a frequency of alternation, whereas direct currents do not. Thus, it is sometimes helpful to think of direct currents as zero frequency ac.

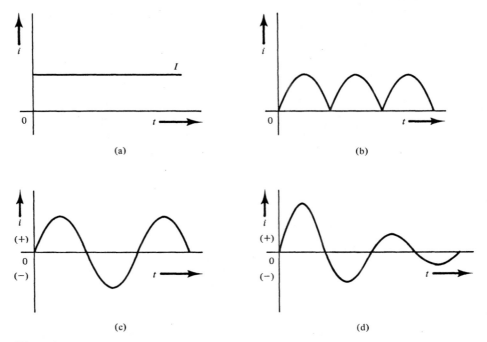

Figure 2-7 Some classes of currents: **(a)** direct (unidirectional, constant); **(b)** pulsating (unidirectional); **(c)** sinusoidal; **(d)** damped and oscillating.

Drill Problem 2-1 ■

(a) What amount of charge passes through a wire in 1 min if the wire carries 4 A of current? **(b)** How many electrons does the charge in (a) represent?

Answer. (a) 240 C. (b) 1.5×10^{21}.   ■ ■

2.3 VOLTAGE

In physics, work is described as a force moving an object through a distance, that is,

$$\text{work} = \text{force} \times \text{distance}$$

If a positive charge $+Q$ C is somewhere in space and another charge, a test charge of $+q_t$ C, is brought into the vicinity, a force of repulsion is given by Coulomb's law

$$F = k \frac{Qq_t}{r^2}$$

If the test charge is considered to be an infinite distance away from the $+Q$ charge, the force of repulsion is essentially zero. If the test charge is to be brought closer to the $+Q$ charge, force must be exerted in order to move the test charge against the repulsive force; that is, work must be done. This work increases the potential energy of the test charge, since if the test charge is left free, it will move under the repulsive force back to infinity.

The work required to move a unit charge in an electric field is defined as the *electric potential*; the rise or fall of potential energy involved in moving a unit charge from one point to another is the *potential difference*. Thus

$$\text{potential difference} = \frac{\text{work}}{\text{charge}} = \frac{W}{Q} \qquad \frac{\text{joule (J)}}{\text{coulomb (C)}} \qquad (2\text{-}3)$$

In the MKS system of units, a potential difference of 1 J/C is defined to be a *volt* (V) and the letter symbol V or E is used.

$$V \text{ or } E \quad 1 \text{ V} = \frac{1 \text{ J}}{1 \text{ C}} \qquad (2\text{-}4)$$

A potential developed in moving a charge from infinity to a point x is referred to as an *absolute potential*, whereas a potential developed in moving a charge from a point x to a point y is called a *relative potential*.

In Fig. 2-8 if q_t is a charge of 1 C, and 10 J of work are needed to move it from point x to point y, the relative potential is

$$V = \frac{10 \text{ J}}{1 \text{ C}} = 10 \text{ V}$$

Thus point y is at a potential of 10 V *relative* to point x.

It is often necessary to indicate precisely the two points used for a potential. We indicate this information by a set of double subscripts. The use of *double subscript notation* is described more fully in Section 5.8. For the time being, let us simply say that if point y is at a potential of 10 V *relative* to x, then

$$V_{yx} = 10 \text{ V}$$

Note that if y is greater in potential than x, then V_{yx} is positive; otherwise the potential is taken as negative.

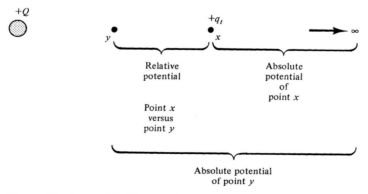

Figure 2-8 Potential difference between points.

Drill Problem 2-2 ■

What amount of charge is moved between two points if a potential of 1.5 V is developed and 0.25 J of energy is used?

Answer.   0.167 C.   ■ ■

2.4 POTENTIAL SOURCES

Devices that convert some other form of energy into electrical potential energy are known as sources of *electromotive force* (emf). We might characterize a source of electrical potential difference (or as it is more commonly called, a *voltage source*) as a shapeless object (or perhaps even a block) that has two terminals, one of which is positive relative to the other. Then, as indicated in Fig. 2-9(a), the application of another form of energy causes a charge to be transferred within the source so that one terminal becomes positively charged while the other becomes negatively charged. Note that if an electrical component is connected to the voltage source and takes current from the source, by our previous convention that current flows out of the positive terminal of the source.

A physical analogy to the voltage source is provided by Fig. 2-9(b). If a bowling ball rests on top of a hill (Why it is there and how it got there is immaterial!), it has

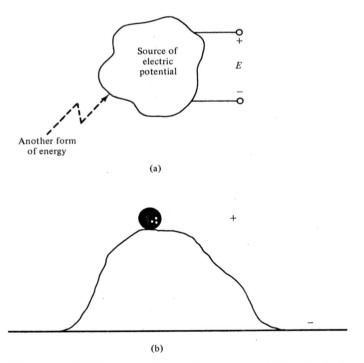

Figure 2-9   (a) The electrical potential source and (b) a physical analogy.

TABLE 2-1 SOURCES OF EMF

Energy source	Typical device
Chemical	Fuel cell, battery (voltaic cell)
Mechanical	Generator, alternator
Thermal	Thermocouple
Photoelectrical (light)	Solar cell, photo cell
Piezoelectrical (pressure)	Crystal

potential energy by virtue of its position relative to the bottom of the hill. If the ball is released, it will roll down the hill analogous to the flow of positive charges from the voltage source.

Depending on the type of energy conversion, the developed voltage may be fixed, varying, or even alternating, that is, such that the polarity at the terminals changes. Then the accompanying current would be anticipated as fixed, varying, or even alternating.

The voltage measured at an emf source is generally referred to by the symbol *E*. Some of the devices that convert other forms of energy into electrical energy are listed in Table 2-1.

2.5 CELLS AND BATTERIES

The *voltaic* or *chemical cell* is the basic unit for converting chemical energy into electrical energy. It consists of a pair of dissimilar metals immersed in a liquid or paste-type solution of ionic material called an *electrolyte*. (See Fig. 2-10.) The electrolyte ionizes or *dissociates* in solution. The positive ions enter into a chemical reaction at one metallic conductor, or *electrode*, the negative ions at the other electrode. The electrodes thus gain a net positive or negative charge. New compounds are formed at the electrode surfaces when the electrolyte ions combine chemically

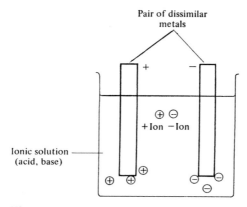

Figure 2-10 The simple chemical cell.

TABLE 2-2   NOMINAL CHEMICAL CELL VOLTAGES

Electrodes	Electrolyte	Nominal emf (V)
Zinc and copper	Sulfuric acid	1.0
Nickel and cadmium	Potassium hydroxide	1.2
Zinc and mercuric oxide	Potassium hydroxide	1.4
Zinc and manganese dioxide	Ammonium chloride	1.5 (flashlight cell)
Zinc and manganese dioxide	Potassium hydroxide	1.5 (alkaline cell)
Zinc and silver oxide	Potassium hydroxide	1.6
Lead and lead peroxide	Sulfuric acid	2.0 (auto/lead–acid cell)
Lithium and iodine	Lithium iodide	2.8

with the electrode materials. These new compounds eventually retard further chemical action.

The chemical activity at each electrode determines the overall voltage developed by the cell. The value of voltage normally available from a cell is called the *nominal voltage*. Nominal cell voltages for a few different electrode and electrolyte combinations are listed in Table 2-2.

You should note that the emf per cell is limited by the choice of materials and is relatively low (1–3 V). Thus, if the total emf or the current capacity is to be increased, similar cells must be interconnected. These collections of interconnected cells are called *batteries*.

The withdrawal of electrical energy from a cell is called *discharging*. During discharge, new compounds are formed at the electrodes and this retards further chemical activity. However, it is possible to reverse the chemical activity in some cells in order to restore the electrodes to their original condition. This process is called *charging* and is achieved by applying another source of emf so that current flows into the cell being *charged* in the opposite direction. A cell that can be restored in this fashion is called a *secondary cell*. A cell that cannot be restored or recharged is called a *primary cell*. If a secondary cell is overcharged, it is sometimes possible for gases to form as a result of the reverse chemical reaction. To relieve the pressure caused by these gases some secondary cells may be vented. The venting is generally accomplished with the use of plugs that are normally sealed but allow the controlled escape of gases.

Cells and batteries are available in a wide variety of sizes and shapes. Certain cell types, such as AA, C, and D, are somewhat standard and can be obtained in a variety of chemical types. Of course, the selection of size, capacity, and chemical type of cell depends on how and where the cell is to be used.

Perhaps the most familiar cell is the ordinary flashlight or dry cell, also known as the *carbon–zinc cell*. In this type of cell the positive electrode or *cathode* is manganese dioxide (MnO_2) and the negative electrode or *anode* is zinc. Although a carbon rod is used in the construction of the battery, it is not the actual electrode; it is simply a collector area that provides a large contact area for the manganese dioxide. In an ordinary dry cell, a paste mixture contains the electrolyte that is a mixture of ammonium chloride and zinc chloride. A modification of this type of cell is the *zinc–chloride cell*, depicted in Fig. 2-11. In this heavy duty cell the electrolyte is only zinc chloride. This difference in electrolytes provides the zinc–chloride cell with a higher efficiency

(a)

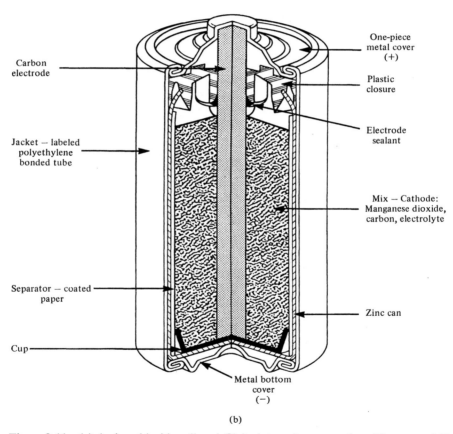

(b)

Figure 2-11 **(a)** A zinc-chloride cell and **(b)** its internal construction. (Courtesy of Union Carbide Corporation.)

than the traditional dry cell. Both types of carbon–zinc cells are designed as primary cells with a nominal voltage rating of 1.5 V.

The *alkaline cell* is similar to the carbon–zinc cell in that the two electrodes are zinc and manganese dioxide. In the alkaline cell, however, the electrolyte is a strongly alkaline solution of potassium hydroxide (KOH). The alkaline cell has a voltage characteristic similar to the carbon–zinc cell so that the two are often used interchangeably. The advantages of the alkaline cell are its longer shelf life and operating life compared to the carbon–zinc cell.

Three other popular primary cells are the mercuric oxide, silver oxide, and manganese dioxide cells. In particular, the mercuric oxide and silver oxide types are used where high-energy density is important. Thus, these cells are widely used in cameras, hearing aids, and watches. A cutaway view of a miniature version of one of these cells is shown in Fig. 2-12. The anode of any of these cells is a gelled mixture of amalgamated zinc powder and electrolyte. In turn, the cathodes are silver oxide (AgO_2), mercuric oxide (HgO), or manganese dioxide (MnO_2), and the electrolytes are aqueous solutions of potassium or sodium hydroxide.

Another important primary cell is the lithium cell which has both an exceptionally long shelf life and a long operating life. A number of different electrodes are used in the cells that make up the "generic" lithium cell.

In a lithium cell, the very chemically reactive metal, lithium is used as the negative electrode. Some of the materials used for the positive electrode are manganese dioxide, carbon fluoride, sulfur dioxide, and iodine. In turn, the electrolyte is either a solid salt or a liquid organic compound. The characteristic feature of a lithium cell is its relatively high nominal voltage of 2.6–3.6 V, depending on the materials used.

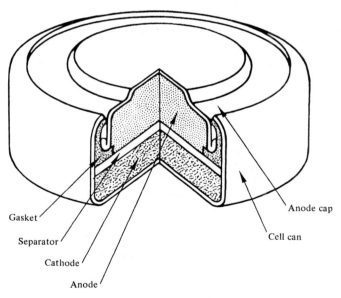

Gasket

Separator

Cathode

Anode

Anode cap

Cell can

Figure 2-12   The internal construction of a miniature aqueous solution cell. (Courtesy of Union Carbide Corporation.)

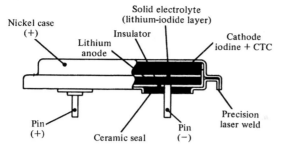

Figure 2-13 Cross-section of the typical lithium–iodine cell. (Courtesy of Catalyst Research Corporation.)

Of particular importance is the *solid-state lithium cell*, composed of solid electrodes and a solid electrolyte. Shelf lives of 10–20 years and longer are characteristic of these cells. Initially, these solid-state cells were used in cardiac pacemakers; they are now used as backup power sources for low-current (μA) microcomputer memory circuits. A cross section of a typical lithium–iodine cell is shown in Fig. 2-13.

You will notice in Fig. 2-13 that lithium is the negative electrode and iodine plus an organic compound that forms a conductive charge transfer complex (CTC) serves as the positive electrode. An initial chemical reaction between these two components forms a third component, a lithium–iodide layer that acts as a solid electrolyte. The pin construction allows for printed circuit board mounting. Three such lithium–iodine solid-state cells are shown in Fig. 2-14. Also shown is a 24-pin integrated-circuit memory chip for which the cells might be used as a backup supply to insure that the stored data will not be lost if a break in the main power source occurs.

An indication of the long life of lithium–iodine cells is shown in Fig. 2-15. Such a graph, which shows terminal voltage as a function of time (in this case years) at various current loads, is called a *discharge curve*. Note that an increase in discharge current results in a decreased life.

Figure 2-14 Three LITHIODE™ lithium–iodide cells and a memory chip. (Courtesy of Catalyst Research Corporation.)

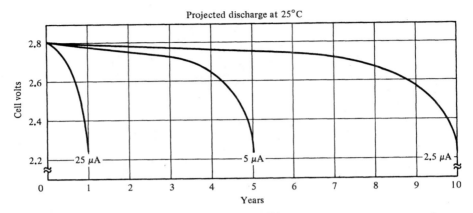

Figure 2-15 The discharge curve for a LITHIODE™ model 1935 cell. (Courtesy of Catalyst Research Corporation.)

As was pointed out earlier, the difference between primary and secondary cells is that chemical reactions at the electrodes of a secondary cell are readily reversible so the cell can be restored to a fully charged state. It follows then that primary cells and batteries are more useful for short duration discharge at low-power levels, whereas secondary cells and batteries are useful for long-term, high-discharge levels. Two of the many types of secondary cells are the nickel–cadmium cell and the lead–acid cell, both of which are listed in Table 2-2.

The *nickel–cadmium cell* consists of positive plates containing nickel salts and negative plates containing cadmium salts. An absorbent material separates the positive and negative plates and holds the potassium hydroxide electrolyte. For relatively small capacities the nickel–cadmium cell is assembled as a nonvented cell. The two plates and the separator are rolled into a tight roll that is then placed into the cell container along with the electrolyte. For larger capacities the nickel–cadmium cell is constructed as a vented cell, as indicated by the construction process and cutaway view of Figs. 2-16 and 2-17, respectively.

By placing the proper number of cells in a battery container and interconnecting the cells, a battery of suitable voltage and current can be constructed. For example, a vented nickel–cadmium cell, as shown in Fig. 2-17, is just one of a number of cells used in the aircraft battery of Fig. 2-18.

Another secondary cell worth noting is the lead–acid cell. The positive electrode consists of lead peroxide (PbO_2) supported on a grid and the negative electrode is pure "spongy" lead (Pb), also supported on a grid. The electrolyte is sulfuric acid (H_2SO_4) in water. During discharge the sulfuric acid is consumed and water is formed according to the chemical equation

$$Pb + 2H_2SO_4 + PbO_2 \longrightarrow 2PbSO_4 + 2H_2O$$

Thus, a unique feature of this type of cell is that the state of charge can be monitored by measuring the amount of sulfuric-acid electrolyte. This is done by a *specific gravity* measurement.

Specific gravity is the ratio of the density of a substance to the density of a standard substance (water). For a fully charged lead–acid cell the specific gravity should

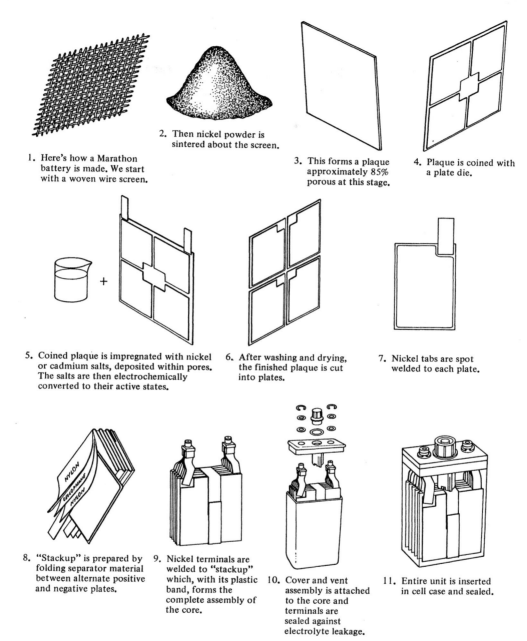

1. Here's how a Marathon battery is made. We start with a woven wire screen.

2. Then nickel powder is sintered about the screen.

3. This forms a plaque approximately 85% porous at this stage.

4. Plaque is coined with a plate die.

5. Coined plaque is impregnated with nickel or cadmium salts, deposited within pores. The salts are then electrochemically converted to their active states.

6. After washing and drying, the finished plaque is cut into plates.

7. Nickel tabs are spot welded to each plate.

8. "Stackup" is prepared by folding separator material between alternate positive and negative plates.

9. Nickel terminals are welded to "stackup" which, with its plastic band, forms the complete assembly of the core.

10. Cover and vent assembly is attached to the core and terminals are sealed against electrolyte leakage.

11. Entire unit is inserted in cell case and sealed.

Figure 2-16 Construction process of a vented nickel–cadmium cell. (Courtesy of Marathon Battery Company.)

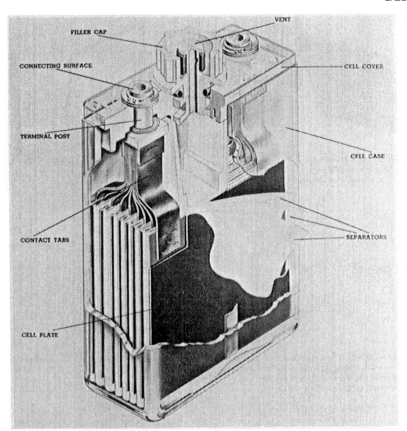

Figure 2-17 A cutaway view of the vented nickel-cadmium cell. (Courtesy of Marathon Battery Company.)

be near 1.3; as discharge occurs the specific gravity drops toward 1.0.

Most of us are familiar with the automobile battery as an example of the use of lead–acid cells. In a 12 V version of this type of battery, six cells are connected together. Another version of the lead–acid cell is the sealed type cell where a gelatin electrolyte solution is used.

Cells and batteries are not only rated for nominal voltage but also for capacity when fully charged. *Cell* or *battery capacity* is simply a product of current and time. It is expressed in ampere–hours (Ah). Capacity is dependent on a number of factors: temperature of cell during discharge, rate of discharge, final cutoff voltage, and cell history. Ideally, a cell or battery with a given capacity rating will have a life calculated by

$$\text{life} = \frac{\text{capacity}}{\text{current}} \tag{2-5}$$

EXAMPLE 2-1

A certain cell has a capacity of 24 Ah. If the cell is discharged at a rate of 1.5 A, what is its ideal life?

Figure 2-18 An aircraft battery assembled from individual nickel-cadmium cells. (Courtesy of Marathon Battery Company.)

Solution. Equation 2-5 yields

$$\text{life} = \frac{24 \text{ Ah}}{1.5 \text{ A}} = 16 \text{ h}$$

A term often used in the comparison of cell performance is *C rate*. The *C rate* is the rate of current at which a battery will discharge its capacity in one hour. Thus, a 30 Ah battery has a *C* rate of 30 A. Fractions or multiples of a *C* rate are also used.

EXAMPLE 2-2

Determine the *C*, *2C*, and *C/5* rates of a 3 Ah battery.

Solution. *C* rate = 3 Ah/1 h = 3 A. *2C* rate = 2(3 A) = 6 A. *C/5* rate = $\frac{1}{5}$(3 A) = 0.6 A.

The effects that temperature and discharge rate have on the capacity of a cell are typified by Figs. 2-19 and 2-20, respectively. Although these graphs are for nickel–cadmium cells, similar effects occur for all types of cells. Thus, it is no wonder that those of us who live in regions with cold winters sometimes experience trouble starting our automobiles!

EXAMPLE 2-3

A 30 Ah nickel–cadmium battery is to be used at $-7°C$. What is its capacity at this temperature?

Solution. From Fig. 2-19, the capacity at $-7°C$ is approximately 80% of its room-temperature capacity.

$$\text{capacity} = 0.8 \times 30 \text{ Ah} = 24 \text{ Ah}$$

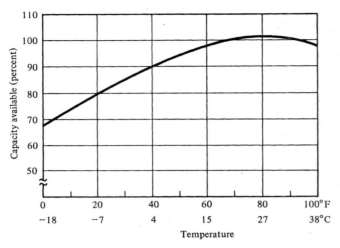

Figure 2-19 Relationship between temperature and available capacity for a nickel–cadmium cell discharging at the C rate. (Courtesy of Marathon Battery Company.)

Drill Problem 2-3 ■

A 40 Ah, 36H120 cell is discharged at a rate of 8 A. **(a)** What percentage of the capacity is removed when the voltage drops to 1.1 V? **(b)** How many hours does it take to discharge to 1.1 V? (Use Fig. 2-20.)

Answer. (a) 92%. (b) 4.6 h. ■ ■

Drill Problem 2-4 ■

A 36H120 nickel–cadmium cell is discharged to a 1 V level at a rate of 300 A. **(a)** What capacity (in Ah) is removed? **(b)** What C rate is used?

Answer. (a) 20 Ah. (b) 7.5C. ■ ■

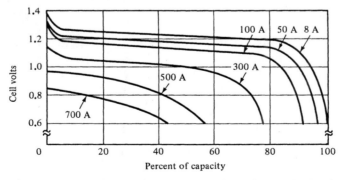

Figure 2-20 Typical discharge characteristics for a 36H120, 40 Ah nickel–cadmium cell. (Courtesy of Marathon Battery Company.)

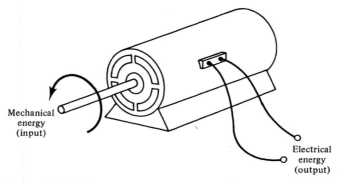

Figure 2-21   The generator as an energy conversion device.

2.6 OTHER POTENTIAL SOURCES

Sources of emf other than batteries are also important. For instance, most of our electrical energy is obtained at the expense of mechanical energy.

If wires capable of carrying current are moved in a magnetic field, an electrical potential is generated. Such a device, which provides for the rotation of a shaft with conductors through a magnetic field is called a *generator*. The effect by which the conversion occurs will be studied in Chapter 9. It is sufficient at this point to consider only the pictorial representation of this conversion as illustrated in Fig. 2-21.

Typically, generators have a greater voltage and current capability than the other sources of emf listed in Table 2-1 (up to hundreds of volts and amperes). Depending on its design, a generator may produce either unidirectional or alternating voltage. If the generator produces ac, it is called an *alternator*.

One device that converts thermal energy into electrical energy is the thermo-couple shown in Fig. 2-22. Two metallic junctions are formed, and if t_1 is kept cold while t_2 is heated, a small emf of the order 10^{-4} V will be developed for every 1°C difference in temperature between the junctions. Since the generated emf is so small, thermocouples are primarily used for temperature measurements.

One type of photoelectric generation of an emf is based on the principle that when light rays strike a semiconductor junction, the number of free electrons and

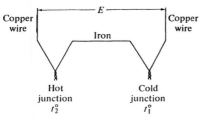

Figure 2-22   The copper–iron thermocouple.

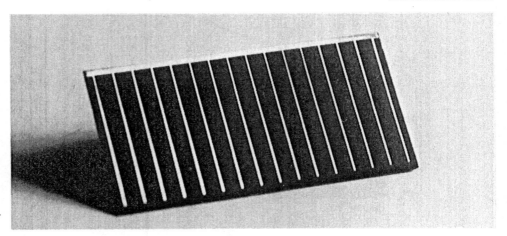

Figure 2-23   A solar cell.

holes at the junction is increased. The illuminated semiconductor region then acts like a source of voltage, similar to a battery. Such a device is called a *photovoltaic cell*.

The photovoltaic cell develops a relatively low voltage (about 0.5–0.7 V dc) and can be used for general light sensing. For "power generation" a larger light collection area is used. When the cell is designed for direct conversion of sunlight into electrical energy it is called a *solar cell*. A solar cell measuring approximately 2 cm × 4 cm is shown in Fig. 2-23. In bright sunlight it generates about 0.5 V and can supply a current of approximately 100 mA.

When certain crystalline materials such as quartz and special ceramics are placed under mechanical strains, a voltage can be measured across the crystalline structure. This effect is known as the *piezoelectric effect*. The effect is also reversible; that is, if a voltage is applied to the piezoelectric crystal, the crystal will vibrate mechanically. Piezoelectric circuit devices are available for each case.

As a voltage source the piezoelectric device is generally used as a pressure or stress sensor. Figure 2-24 depicts such a sensor. The crystal is mounted between two

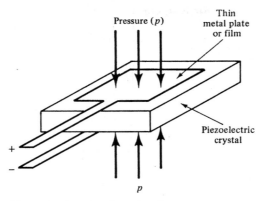

Figure 2-24   The piezoelectric sensor.

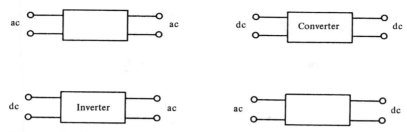

Figure 2-25 Four types of power supplies.

conducting plates to which external connections are made. Pressure on the crystal causes a charge to be generated. The voltage, in turn, depends on this developed charge and other circuit aspects.

A *power supply* is a device that converts one type of electrical potential or current into another type. Four types of conversion may be made and these are depicted in Fig. 2-25. The actual conversion process may be electromechanical or electronic. You should note, in particular, that a dc-to-dc power supply that changes dc levels is called a *converter* and a dc-to-ac power supply is called an *inverter*.

One power supply encountered frequently is the benchtop ac-to-dc laboratory power supply. This power supply uses electronic conversion. A benchtop laboratory power supply is shown in Fig. 2-26. Notice that it is a dual voltage supply with separate voltage and current controls and separate metering.

Figure 2-26 A laboratory dc power supply. (Courtesy of Acopian Corporation.)

Figure 2-27   An open-frame power supply.

Another ac-to-dc power supply is shown in Fig. 2-27. This particular power supply is an open-frame power supply that is intended for enclosure within a larger piece of equipment. The supply shown develops 5 V and might be used to power a microcomputer system.

2.7 CONDUCTORS, SEMICONDUCTORS, AND INSULATORS

Based on the number of free electrons available for conduction, different materials require different magnitudes of applied electric force in order to provide current flow. Materials that allow current to flow with only a small applied voltage are called *con-*

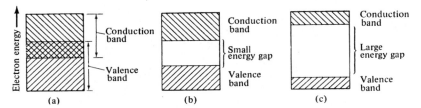

Figure 2-28   Energy level diagrams: **(a)** conductor; **(b)** semiconductor; **(c)** insulator.

TABLE 2-3 AVERAGE DIELECTRIC
STRENGTHS OF SOME
COMMON INSULATORS

Material	Average dielectric strength (V/mm)
Air	3,000
Bakelite	21,000
Glass	35,000
Mica	60,000
Oil	10,000
Paper	20,000
Rubber	25,000
Teflon	60,000

ductors. These types of materials such as copper and silver have approximately 10^{24} free electrons per in.^{3} at room temperature.

On the other hand, materials that allow very little current flow are called *insulators* or *dielectrics*. Insulators such as air, teflon, and porcelain have approximately 10^{10} free electrons per in.^{3} at room temperature.

A third group of materials has approximately 10^{17} free electrons per in.^{3} at room temperature. Materials such as carbon, silicon, and germanium fall in this category and can be rated as poor insulators or *semiconductors*. Semiconductor materials having four valence electrons and needing four additional electrons to fill a subshell combine to form a crystalline structure.

In terms of an energy level diagram (Fig. 2-28), most metals have a valence band that is only partially filled with electrons and overlaps the conduction band. In a semiconductor material, there is a small gap between the valence band and first available conduction band, but it is not difficult for electrons to acquire the energy needed to jump across the gap. In an insulator, this energy gap is wider and is not easily bridged by an electron. If a high enough electric field is applied to an insulator, it is possible to free a large number of electrons. The point at which substantial current begins to flow in an insulator is called *breakdown*.

Since the potential difference needed for breakdown depends on the thickness of the insulator across which the voltage is applied, it is convenient to define a term involving a unit thickness. *Dielectric strength* or *breakdown strength* is the voltage per unit thickness at which breakdown occurs. Some dielectric strengths are given in Table 2-3.

2.8 RESISTANCE

Resistance is the opposition of a material (or circuit) to the flow of electrical current. The letter symbol is *R* and the basic unit is the ohm. The Greek capital letter omega

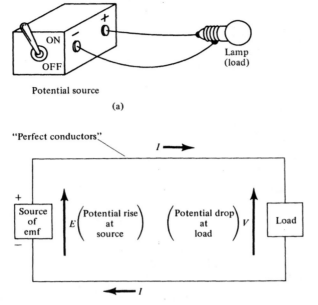

Figure 2-29   The simple electric circuit: **(a)** pictorial sketch; **(b)** block diagram.

(Ω) is used as the unit abbreviation. Relatively speaking, a conductor has low resistance, whereas an insulator has high resistance.

Resistance properties are discussed more fully in the following chapter, but first we consider the complete electric circuit that is formed when a source of emf is connected to a *load* or device that converts electrical energy back into some other form of energy (see Fig. 2-29). The current in Fig. 2-29 travels via a path formed by "perfect conductors" or at least conductors with very little resistance compared to that presented by the load. As noted, the current passes through a rise in potential E in passing through the source but must pass through a potential drop at the load. It is the potential drop at the load that is denoted V.

2.9 CIRCUIT SYMBOLS

A drawing that shows symbolically or schematically the interconnection of circuit components is called a *circuit diagram*. Although it is possible to draw circuit diagrams in block form as in Fig. 2-29, it is customary to use more specialized graphic symbols. The graphic symbols used in this text are shown in Fig. 2-30. As a circuit element or device is introduced, the reader should refer to Fig. 2-30 for its graphic symbol.

Figure 2-30   Graphic symbols used in the text.

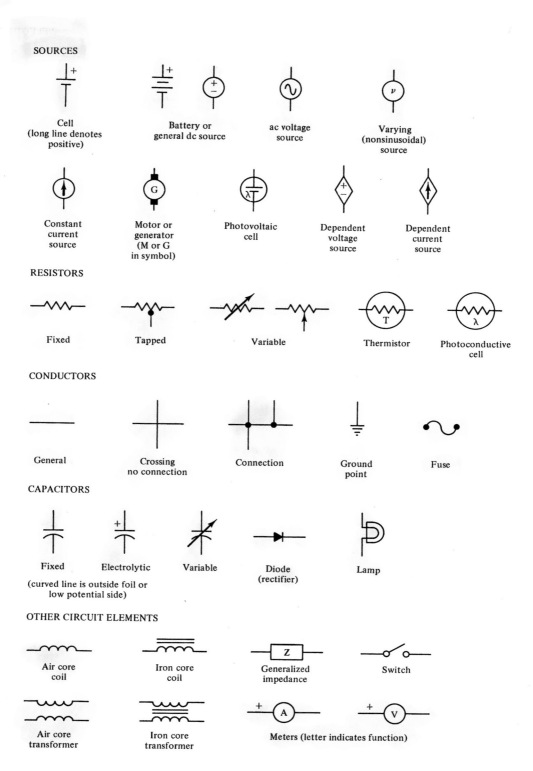

SOURCES

Cell
(long line denotes positive)

Battery or general dc source

ac voltage source

Varying (nonsinusoidal) source

Constant current source

Motor or generator (M or G in symbol)

Photovoltaic cell

Dependent voltage source

Dependent current source

RESISTORS

Fixed

Tapped

Variable

Thermistor

Photoconductive cell

CONDUCTORS

General

Crossing no connection

Connection

Ground point

Fuse

CAPACITORS

Fixed

Electrolytic

Variable

Diode (rectifier)

Lamp

(curved line is outside foil or low potential side)

OTHER CIRCUIT ELEMENTS

Air core coil

Iron core coil

Generalized impedance

Switch

Air core transformer

Iron core transformer

Meters (letter indicates function)

QUESTIONS

1. What are valence electrons?
2. What is meant by energy gaps?
3. What are free electrons?
4. What is an ion?
5. What is current?
6. State the unit for current.
7. What is meant by a hole in the atomic structure?
8. Differentiate between direct and alternating current.
9. Define electric potential.
10. Distinguish between a primary cell and a secondary cell.
11. Explain the operation of a cell.
12. What is charging?
13. State five ways of developing an emf.
14. Describe the following: conductors, insulators, and semiconductors.
15. Define dielectric strength.
16. What is resistance?

PROBLEMS

1. Find the current in amperes if 725 C of charge pass through a wire in 40 s.
2. How many coulombs of charge pass through a lamp in 10 min if the current is 250 mA?
3. What current flows if 2×10^{21} electrons pass through a wire in 50 ms?
4. How many electrons pass through a conductor in 1 min if the current is 75 mA?
5. How long will it take 70 mC of charge to pass by a point in a conductor if the conductor current is 8 A?
6. How many electrons must move from the positive terminal inside a battery to the negative terminal if 0.5 J of energy is converted to 1.5 V?
7. The emf of a battery is 1.5 V; how much charge moves if the energy converted is 15 J?
8. How much work is done in moving an electron from a point where the voltage is 5 V to a point where the potential is 8 V?
9. What potential is applied to a conductor if 25 J of energy are dissipated as heat during the 1 min interval during which 250 mA flows?
10. A charge of 200 μC is moved against an electric field by applying 72 μJ. Through what potential difference did the charge move?
11. If a LITHIODE™ model 1935 cell is discharged at 5 μA to a voltage of 2.6 V, how long can it be expected to last? (Use Fig. 2-15.)
12. If a cell has a rating of 400 mAh and is discharged at 100 μA; how many days can it be expected to last?
13. A 12 V battery is discharged completely at a rate of 2.5 A for 24 h. What is its capacity?

14. A 12 V auto battery is rated at 60 Ah. **(a)** What is its C rate? **(b)** What is its $C/5$ rate?

15. What fraction of a C rate is a discharge of 3.5 A for a 70 Ah battery?

16. A nickel–cadmium battery has a capacity of 70 Ah at room temperature. What is its capacity at 40°F?

17. A fully charged 40 Ah, 36H120 cell is discharged at a rate of 100 A to 1 V. **(a)** What capacity is removed? **(b)** How many hours does it take to discharge?

18. Two 40 Ah, 36H120 cells are discharged to 1 V. One is discharged at a rate of 100 A, the other at a rate of 300 A (discharge rate factor of 3). By what factor do the discharge times differ?

19. Two parallel plates are separated by mica dielectric. If a potential of 9 kV is to be safely applied to the plates, what should be the minimum thickness of the dielectric?

20. A set of charged plates is separated by a 0.08 mm layer of rubber. What maximum voltage can the charge accumulation represent if the insulator is not to fail?

Chapter 3

The Nature of Resistance

3.1 INTRODUCTION

By acquiring enough kinetic energy, an electron becomes a free electron and moves about until it *collides,* that is, makes contact, with an atom. Upon colliding with an atom, the free electron loses some or all of its kinetic energy, some of which goes into the excitation of other electrons. Some is also absorbed as heat energy by the atoms, which are thought to be moving in a vibratory manner. If an electric potential is applied to a conductor, the electrons are given increased kinetic energy and then collide more frequently with atoms, thereby increasing the temperature of the conductor. Thus, when electric current flows in a conductor, some of the electrical potential energy is converted to heat energy, so that resistance is associated not only with opposition to the current flow but also with the development of heat energy in a conductor.

If an electric appliance is connected to an electrical receptacle, the connecting wires should not dissipate heat. In this instance, resistance is clearly undesirable. In fact, if the resistance of the connecting wires was not negligible, there might even be a chance that heat would develop faster than the conductors could lose it to the surroundings. The wire temperature could increase and create a fire hazard.

Just as there are times when we wish to produce heat in a lamp or heater element, there are times when we must purposely use up some potential energy. You will see how this is accomplished when we study the series-type circuit connection. At the same time we shall purposely design some resistance into the electric circuit.

A device specifically designed to have resistance is called a *resistor.* Depending on the materials used and their physical construction, resistors can be categorized as *carbon-composition, wire-wound, thick* or *thin film,* or *semiconductor.* Additionally, they can be classified into the two groups: *fixed* or *variable.* The various resistors are discussed in this chapter, along with the various factors that affect resistance.

3.2 RESISTIVITY

The resistance that a conductor offers to current flow depends not only on the material used, but on the length, cross-sectional area, and temperature of the conductor. If the effects of temperature are neglected, resistance depends directly on the length of the conductor and inversely on the cross-sectional area; that is,

$$R \propto (l/A)$$

where R is the resistance, l the length, and A the cross-sectional area.

If a constant of proportionality is used in the relationship, Eq. 3-1 results.

$$R = \rho(l/A) \tag{3-1}$$

The constant of proportionality in Eq. 3-1 is called the *resistivity* and has the symbol ρ (Greek lowercase rho). The resistivity or *specific resistance* depends on the conducting material and is defined as the resistance per unit length and cross-sectional area of the material. Since the SI units of length and cross section are meters and meters squared, respectively, the resistivity of a material is the resistance between the opposite faces of a cube, 1 m on an edge, of the material. For this reason resistivity is sometimes called the *resistance per meter cube*.

If we solve Eq. 3-1 for ρ and the units associated with ρ, we obtain

$$\rho = \frac{RA}{l} = \frac{(\Omega)(m^2)}{m} = \Omega \cdot m$$

The SI unit for resistivity is thus the ohm-meter and is usually specified at 20°C (Celsius or centigrade), although it is a simple matter to find the resistivity at another temperature. The resistivities of some common conducting materials are given in Table 3-1.

TABLE 3-1 THE RESISTIVITIES OF COMMON CONDUCTING MATERIALS AT 20°C

Conducting material	Resistivity $(\Omega \cdot m)$	Resistivity $(\Omega \cdot \text{cir mil/ft})$
Aluminum	2.83×10^{-8}	17.0
Brass	7×10^{-8}	42.0
Copper, annealed	1.72×10^{-8}	10.37
Copper, hard-drawn	1.78×10^{-8}	10.7
Gold	2.45×10^{-8}	14.7
Lead	22.1×10^{-8}	132
Nichrome	100×10^{-8}	600
Platinum	10×10^{-8}	60.2
Silver	1.64×10^{-8}	9.9
Tin	11.5×10^{-8}	69.2
Tungsten	5.52×10^{-8}	33.2
Zinc	6.23×10^{-8}	37.4

most common (handwritten annotation)

EXAMPLE 3-1

An annealed copper bus bar measures 1×3 cm and is $\frac{1}{2}$ m long. What is the resistance between the ends of the bar?

Solution

$$R = \rho \frac{l}{A} = (1.724 \times 10^{-8}) \frac{(5 \times 10^{-1})}{(10^{-2} \times 3 \times 10^{-2})} = 2.85 \times 10^{-5} \; \Omega$$

Because electrical conductors are generally circular in cross section, with wire diameters measured in thousandths of an inch (mils), it is convenient to define an area called the *circular mil*. The circular mil is the area of a circular cross section having a diameter of 1 mil. The area actually represented by a circular mil is

$$A = \frac{\pi d^2}{4} = \frac{\pi (1 \text{ mil})^2}{4} = \frac{\pi}{4} \text{ mil}^2$$

Thus 1 cir mil contains $\pi/4$ mil^2, and conversely, 1 mil^2 contains $4/\pi$ circular mils. This is seen in Fig. 3-1, which clearly shows that there is more than 1 cir mil (actually $4/\pi$) in 1 mil^2.

Since the area depends on the square of the diameter, then, for any arbitrary diameter d in mils, the area in circular mils is given by

$$A \text{ (cir mil)} = d^2 \tag{3-2}$$

It is also appropriate to define the resistivity in the English system, as the resistance of a 1 ft length of conductor with an area of 1 circular mil. Table 3-1 lists resistivity values not only in ohm·meters but also in the English unit of ohm·cir mil per foot.

EXAMPLE 3-2

Find the resistance of an annealed copper wire 200 ft long, with a 100 mil diameter.

Solution

$$A \text{ (cir mil)} = d^2 = (100)^2 = 10^4 \text{ cir mil}$$

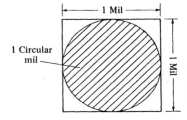

Figure 3-1   The relationship between a square mil and a circular mil.

Then $$R = \rho \frac{l}{A} = (10.37) \frac{(200)}{10^4} = 20.7 \times 10^{-2} = 0.207 \ \Omega$$

Often the conductor being considered is rectangular in cross section, with the physical dimensions given in inches and feet. The cross-sectional area is then calculated in square mils and converted to circular mils by the $4/\pi$ relationship relating the two; that is, we multiply the number of square mils by $4/\pi$ to get circular mils.

EXAMPLE 3-3

Find the resistance of an aluminum strip 5 ft long with the cross-section dimensions $\frac{1}{2}$ in. by 0.005 in.

Solution

$$A \ (\text{mil}^2) = (500)(5) = 2500 \ \text{mils}^2$$

$$A \ (\text{cir mil}) = 2500 \times \frac{4}{\pi} = 3180 \ \text{cir mil}$$

Then $$R = \rho \frac{l}{A} = \frac{(17)(5)}{3.18 \times 10^3} = 26.7 \times 10^{-3} \ \Omega$$

Drill Problem 3-1 ■

Calculate the area in cir mil for the following diameters: **(a)** 0.020 in.; **(b)** 0.125 in.; **(c)** 0.01 ft.

Answer. (a) 400. (b) 15 625. (c) 14 400.   ■ ■

Drill Problem 3-2 ■

Calculate the diameter in inches for the following areas: **(a)** 850 cir mil; **(b)** 6530 cir mil.

Answer. (a) 0.029 in. (b) 0.081 in.   ■ ■

Drill Problem 3-3 ■

Calculate the resistance of a 2 ft annealed copper strap having a cross section of $\frac{1}{4} \times \frac{1}{2}$ in.

Answer. $1.3 \times 10^{-4} \ \Omega$.   ■ ■

3.3 WIRE TABLES

Circular cross-section wire is designated by a gauge size or number representing a certain diameter wire. The American Wire Gauge (AWG) sizes are given in Table 3-2. Notice that the larger the gauge number, the smaller the diameter of the conductor. The AWG wire table was set up in such a manner that each decrease in one gauge number represents a 25% increase in cross-sectional area. Examination of the table reveals that a change in 10 gauge numbers represents a change of 10 times the

TABLE 3-2 AMERICAN WIRE GAUGE, STANDARD ANNEALED COPPER WIRE AT 20°C
Excerpted from National Bureau of Standards H100

	English units				Metric units		
		Cross section				Cross section	
Gauge	Diameter (mil)	(cir mil)	(in.^{2})	Ω/1000 ft.	Diameter (mm)	(mm^2)	Ω/km
0000	460.0	211 600	0.166 2	0.049 01	11.68	107.2	0.160 8
000	409.6	167 800	0.131 8	0.061 82	10.40	85.01	0.202 8
00	364.8	133 100	0.104 5	0.077 93	9.266	67.43	0.255 7
0	324.9	105 600	0.082 91	0.098 25	8.252	53.49	0.322 3
1	289.3	83 690	0.065 73	0.123 9	7.348	42.41	0.406 5
2	257.6	66 360	0.052 12	0.156 3	6.543	33.62	0.512 8
3	229.4	52 620	0.041 33	0.197 1	5.827	26.67	0.646 6
4	204.3	41 740	0.032 78	0.248 5	5.189	21.15	0.815 2
5	181.9	33 090	0.025 99	0.313 4	4.620	16.77	1.028
6	162.0	26 240	0.020 61	0.395 2	4.115	13.30	1.297
7	144.3	20 820	0.016 35	0.498 1	3.665	10.55	1.634
8	128.5	16 510	0.012 97	0.628 1	3.264	8.367	2.061
9	114.4	13 090	0.010 28	0.792 5	2.906	6.631	2.600
10	101.9	10 380	0.008 155	0.998 8	2.588	5.261	3.277
11	90.7	8230	0.006 46	1.26	2.30	4.17	4.14
12	80.8	6530	0.005 13	1.59	2.05	3.31	5.21
13	72.0	5180	0.004 07	2.00	1.83	2.63	6.56
14	64.1	4110	0.003 23	2.52	1.63	2.08	8.28
15	57.1	3260	0.002 56	3.18	1.45	1.65	10.4
16	50.8	2580	0.002 03	4.02	1.29	1.31	13.2
17	45.3	2050	0.001 61	5.05	1.15	1.04	16.6
18	40.3	1620	0.001 28	6.39	1.02	0.823	21.0
19	35.9	1200	0.001 01	8.05	0.912	0.653	26.4
20	32.0	1020	0.000 804	10.1	0.813	0.519	33.2
21	28.5	812	0.000 638	12.8	0.724	0.412	41.9
22	25.3	640	0.000 503	16.2	0.643	0.324	53.2
23	22.6	511	0.000 401	20.3	0.574	0.259	66.6
24	20.1	404	0.000 317	25.7	0.511	0.205	84.2
25	17.9	320	0.000 252	32.4	0.455	0.162	106
26	15.9	253	0.000 199	41.0	0.404	0.128	135
27	14.2	202	0.000 158	51.4	0.361	0.102	169
28	12.6	159	0.000 125	65.3	0.320	0.0804	214
29	11.3	128	0.000 100	81.2	0.287	0.0647	266
30	10.0	100	0.000 078 5	104	0.254	0.0507	340
31	8.9	79.2	0.000 062 2	131	0.226	0.0401	430
32	8.0	64.0	0.000 050 3	162	0.203	0.0324	532
33	7.1	50.4	0.000 039 6	206	0.180	0.0255	675
34	6.3	39.7	0.000 031 2	261	0.160	0.0201	857
35	5.6	31.4	0.000 024 6	331	0.142	0.0159	1090
36	5.0	25.0	0.000 019 6	415	0.127	0.0127	1360
37	4.5	20.2	0.000 015 9	512	0.114	0.0103	1680
38	4.0	16.0	0.000 012 6	648	0.102	0.008 11	2130
39	3.5	12.2	0.000 009 62	847	0.089	0.006 21	2780
40	3.1	9.61	0.000 007 55	1080	0.079	0.004 87	3540

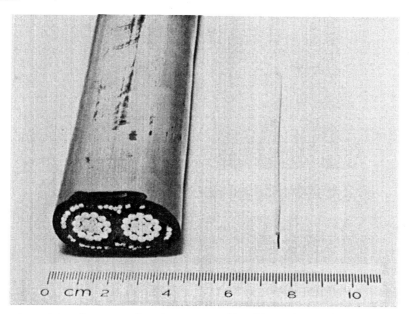

Figure 3-2 Size comparison between a cable with one 2/0 and two 4/0 conductors and a 22-gauge wire.

area; for example, in going from AWG 10 to AWG 20 wire, the area changes from 10 380 cir mils to 1022 cir mils. If follows that a change in 10 gauge numbers represents a change in resistance by approximately a factor of 10. Thus AWG 10 has a resistance of approximately 1 Ω/1000 ft, whereas AWG 20 has a resistance of approximately 10 Ω/1000 ft. Any wire larger than 0000 is specified not by a gauge number but by its cross-sectional area in circular mils.

The resistance values in Table 3-2 are for standard annealed copper. Resistance values for other conductors can be found in standard handbooks.

Often a wire is stranded or made up of smaller wires for increased flexibility. If, in addition, high tensile strength is required, a steel core may be used.

A relative comparison of wire size is shown in Fig. 3-2. On the right side of Fig. 3-2 is a single conductor of size AWG 22. This type of wire is used in electronic circuits. On the left is a "three-wire" cable in which each of the three wires is actually a group of stranded conductors. Each of the two inner wires consists of 19 strands that form a composite area equivalent to 0000 (4/0) wire size. The third wire is a group of 33 strands that are spread around the two inner wires. Even though there are more conductors for this third wire, the strands have a smaller cross section so that the net equivalent area is equal to that of a 00 (2/0) wire size. Surrounding each of the three wires is insulation. The three-wire cable of Fig. 3-2 is used for power distribution.

Microcomputer signals are often transferred over ribbon cables such as those depicted in Fig. 3-3. Here the conducting wires and insulating material are molded into a flat ribbon. Both cables contain stranded AWG 28 wires.

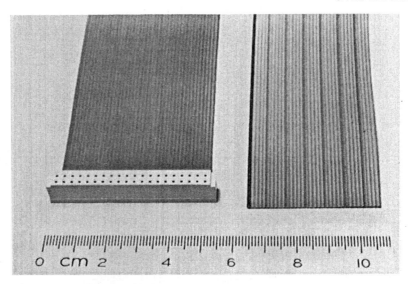

Figure 3-3   Ribbon cables.

Limited by their capacity to transfer heat to their surroundings, wires are rated according to the maximum current they can safely handle. These ratings are highly dependent on insulation type and environment and are listed in standard handbook tables.

Drill Problem 3-4 ■

What is the resistance of 400 ft of AWG 6 wire?

Answer.   0.158 Ω.   ■ ■

Drill Problem 3-5 ■

The total distance from a power source to a motor and back to the source is 250 ft. What is the minimum copper wire size needed for a power line that is to connect to the motor and have a total resistance of less than 0.04 Ω?

Answer.   Size AWG 2.   ■ ■

3.4 TEMPERATURE EFFECTS

As a conductor increases in temperature, either from current flowing through it or by heat absorption from the surrounding medium, increased atomic motion results in the conductor. Because of this increased atomic motion, most conductors experience an increase in resistance with an increase in temperature. If we plot a curve of resistance versus temperature for a conductor such as copper, we obtain the relationship shown in Fig. 3-4. Notice that a linear relationship between resistance and temperature exists in the temperature range over which the resistance is normally

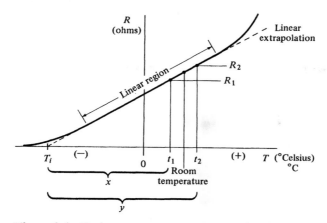

Figure 3-4 Resistance versus temperature, for a conducting metal.

used. Even though the curve becomes nonlinear as the resistance approaches zero, a straight line extrapolation can be made as a continuation of the linear portion of the curve. The extrapolated curve intercepts the temperature axis at a point T_i called the *inferred zero-resistance temperature* or *inferred absolute zero* ($T_i = -234.5°C$ for annealed copper).

Considering two resistance values R_1 and R_2 at temperatures t_1 and t_2, respectively (Fig. 3-4), we see that the linear extrapolation provides a similar triangle relationship relating R_1 and R_2. Making a ratio of the sides of the similar right triangles, we find that

$$\frac{R_1}{x} = \frac{R_2}{y}$$

Since sides x and y have lengths $|T_i| + t_1$ and $|T_i| + t_2$, respectively,

$$\frac{R_1}{|T_i| + t_1} = \frac{R_2}{|T_i| + t_2} \tag{3-3}$$

EXAMPLE 3-4

What is the resistance of an annealed copper wire at 10°C if the resistance is 50 Ω at 60°C?

Solution

$$\frac{R_1}{234.5° + t_1} = \frac{R_2}{234.5° + t_2}$$

$$\frac{R_1}{234.5° + 10°} = \frac{50\ \Omega}{234.5° + 60°}$$

or
$$R_1 = 50\left(\frac{244.5}{294.5}\right) = 41.5\ \Omega$$

The temperature intercepts of some other materials are listed in Table 3-3.

TABLE 3-3   THE TEMPERATURE INTERCEPTS AND
COEFFICIENTS FOR COMMON
CONDUCTING MATERIALS

Conducting material	Inferred absolute zero (°C)	Temperature coefficient ($\Omega \cdot °C/\Omega$ at 0°C)
Aluminum	−236	0.004 24
Brass	−480	0.002 08
Copper, annealed	−234.5	0.004 27
Copper, hard drawn	−242	0.004 13
Gold	−274	0.003 65
Lead	−224	0.004 66
Nichrome	−2270	0.000 44
Platinum	−310	0.003 23
Silver	−243	0.004 12
Tin	−218	0.004 58
Tungsten	−202	0.004 95
Zinc	−250	0.004 00

Because of the linear relationship between the resistance and the temperature, the slope $\Delta R/\Delta T$ is constant and a change of 1°C results in the same change in resistance ΔR (see Fig. 3-5). The per unit change of resistance per °C change in temperature referred to any point n on the R versus T curve is defined as the *temperature coefficient of resistance*, α (Greek alpha); that is,

$$\alpha_n = \frac{\Delta R}{\Delta T} \frac{1}{R_n} \qquad \frac{\Omega}{°C \cdot \Omega} \tag{3-4}$$

The subscript of α denotes the reference temperature; it should be apparent that α varies with temperature. In Fig. 3-5, $\alpha_1 = \Delta R/R_1$ and $\alpha_3 = \Delta R/R_3$, and since $R_3 > R_1$, then $\alpha_1 > \alpha_3$.

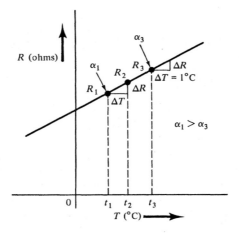

Figure 3-5   Temperature coefficient.

It is possible to calculate the temperature coefficient of resistance from the inferred zero-resistance temperature. If we refer to Fig. 3-4 and substitute $\Delta T = |T_i| + t_n$ and $\Delta R = 0 + R_n$ into Eq. 3-4, we obtain the expression

$$\alpha_n = \frac{R_n}{|T_i| + t_n}\left(\frac{1}{R_n}\right) = \frac{1}{|T_i| + t_n} \tag{3-5}$$

From this last relationship, it is seen that if $t_n = 0°C$, then α_0, the temperature coefficient at $0°C$, is the reciprocal of T_i.

$$\alpha_0 = 1/|T_i| \tag{3-6}$$

Table 3-3 also contains the temperature coefficients at $0°C$.

The resistance value R_2 of Fig. 3-5 can be expressed in terms of R_1 as

$$R_2 = R_1 + \Delta R \tag{3-7}$$

Then if the change in resistance ΔR (obtained from Eq. 3-4 as $\Delta R = \alpha_1 R_1 \Delta T$) is substituted into Eq. 3-7, the following equation results:

$$R_2 = R_1(1 + \alpha_1 \Delta T) \tag{3-8}$$

where $$\Delta T = t_2 - t_1$$

EXAMPLE 3-5

What is the resistance of a copper wire at $10°C$ if the resistance at $20°C$ is $4.33 \ \Omega$ and if $\alpha_{20°C} = 0.003\,93$?

Solution

$$R_2 = R_1(1 + \alpha_1 \Delta T) = 4.33[1 + 0.003\,93(10° - 20°)]$$
$$= 4.33(1 - 0.0393) = 4.33(0.961)$$
$$= 4.16 \ \Omega$$

Most materials exhibit an increase in resistance with an increase of temperature and are said to have a *positive temperature coefficient of resistance*. However, some materials, such as semiconductor materials, exhibit a decrease in resistance with an increase of temperature and are said to have a *negative temperature coefficient of resistance*. Resistor manufacturers generally specify the temperature coefficient as the change in resistance in parts per million per degree centigrade (ppm/°C).

The carbon-composition resistor discussed in the next section has the resistance temperature characteristic of Fig. 3-6. It is interesting to note from this characteristic that above room temperature a positive temperature coefficient exists, but below room temperature a negative temperature coefficient exists.

One might wonder whether the resistance actually becomes zero as the temperature of a material approaches the inferred zero-resistance temperature. Experimental information indicates that the resistance does approach zero. Although not equal to zero as suggested by the linear extrapolation, the resistivity of these super-

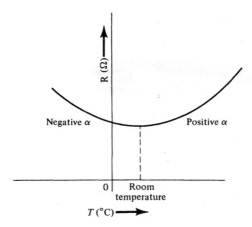

Figure 3-6 R versus T curve for a carbon composition resistor.

conductors is of the order of 10^{-25} $\Omega\cdot$m, which is approximately 17 orders of magnitude lower than that of copper.

Drill Problem 3-6 ■

What is the resistance of an aluminum wire at 60°C if the resistance is 2.56 Ω at 10°C?

Answer.   3.08 Ω.   ■ ■

Drill Problem 3-7 ■

What is the temperature coefficient of resistance of aluminum at 60°C?

Answer.   0.003 38 Ω/°CΩ.   ■ ■

3.5 WIRE-WOUND AND CARBON-COMPOSITION RESISTORS

A resistor of fixed value can be made simply by winding a wire of suitable length and cross section on a suitable core or form. Such *wire-wound resistors* are available commercially; they usually consist of Nichrome or copper–nickel wire space-wound on a ceramic tube and protected from mechanical or environmental hazards by a protective coating of vitreous enamel or silicone (Fig. 3-7). Wire-wound resistors are used whenever large amounts of power (watts) must be dissipated. Although power is not discussed until the next chapter, it is sufficient to say that power dissipation refers to the heat energy developed in a resistor.

A second type of commercially available resistor is the *carbon-composition resistor*, which has been used extensively in electronic applications. A mixture of carbon and binding material is either applied as a coating to the outside surface of a glass tube or molded into a dense structure, as shown in Fig. 3-8.

Carbon-composition resistors are relatively inexpensive and are available in power ratings of $\frac{1}{10}$ to 5 W. The resistance value of a carbon-composition resistor

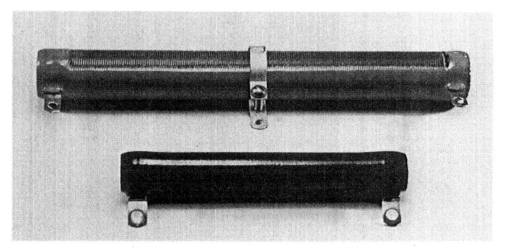

Figure 3-7 Wirewound resistors.

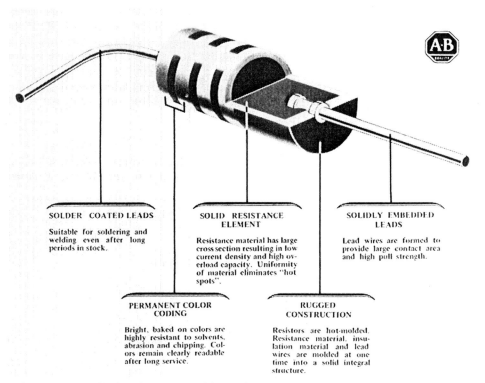

SOLDER COATED LEADS	SOLID RESISTANCE ELEMENT	SOLIDLY EMBEDDED LEADS
Suitable for soldering and welding even after long periods in stock.	Resistance material has large cross section resulting in low current density and high overload capacity. Uniformity of material eliminates "hot spots".	Lead wires are formed to provide large contact area and high pull strength.

PERMANENT COLOR CODING	RUGGED CONSTRUCTION
Bright, baked on colors are highly resistant to solvents, abrasion and chipping. Colors remain clearly readable after long service.	Resistors are hot-molded. Resistance material, insulation material and lead wires are molded at one time into a solid integral structure.

Figure 3-8 Molded carbon composition resistor. (Courtesy of Allen-Bradley.)

TABLE 3-4 COLOR CODE FOR CARBON
COMPOSITION RESISTORS

Color	Digit or number of zeros
Black	0
Brown	1
Red	2
Orange	3
Yellow	4
Green	5
Blue	6
Violet	7
Gray	8
White	9
Gold	0.1 multiplier or $\pm 5\%$ tolerance
Silver	0.01 multiplier or $\pm 10\%$ tolerence

is specified by a set of color-coded bands appearing on the resistor housing. Each color represents a digit in accordance with Table 3-4. As noted in Fig. 3-9, the color bands are read starting from the end with a band closest to it. The first and second bands indicate the first and second digits, respectively. The third band indicates the number of zeros that follow the first two digits, except that when gold and silver are used, they represent multiplying factors of 0.1 and 0.01, respectively. A fourth color band indicates the manufacturing tolerance. The absence of this fourth band means that the resistance value is within $\pm 20\%$ of the stated value. A fifth color band indicates that the resistor meets a certain reliability specification. For example, if a resistor has color bands of blue, gray, silver, and gold, the resistance value is $0.68 \ \Omega \ \pm 5\%$.

Drill Problem 3-8 ■

Determine the range of resistance for the carbon composition resistors having color bands of **(a)** green, blue, red, and gold and **(b)** yellow, violet, orange, and silver.

Answer. (a) $5.32 - 5.88 \ \mathrm{k}\Omega$. (b) $42.3 - 51.7 \ \mathrm{k}\Omega$.   ■ ■

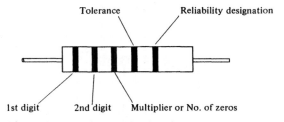

Figure 3-9 Resistor color bands.

3.6 FILM AND SEMICONDUCTOR RESISTORS

A *thick-film resistor* is defined by the electronics industry as a resistor whose resistance element is in the form of a film with a thickness of greater than 0.000 001 in. (1 millionth of an inch). On the other hand, a *thin-film resistor* has a resistance element less than 1 millionth of an inch thick. The film is generally applied to a ceramic core or a thin flat ceramic surface, called a *substrate*, by vacuum evaporation, cathode sputtering, or printing with resistive inks. Depending on the material used, the resistors are classified as carbon film, metal-film, or metal-oxide film resistors.

It is convenient to discuss film resistors in terms of *sheet resistance*, which is the resistance of a square of the film material. Consider the film resistance of Fig. 3-10.

The basic equation for bulk resistance (from Eq. 3-1) is

$$R = \rho \frac{l}{A}$$

The cross-sectional area of the film resistor of Fig. 3-10 is $A = Wt$, where t is the thickness of the film and W is the width of the film, so that the resistance formula is given by

$$R = \left(\frac{\rho}{t}\right)\left(\frac{l}{w}\right)$$

The ratio l/w is called N, the number of squares, and ρ/t is called ρ_s, the *sheet resistivity* in ohms per square. Thus

$$R = \rho_s N \tag{3-9}$$

For example, if in Fig. 3-10 the length is 0.09 in. and the width is $W = 0.03$ in., there are three squares of film material. Then if the sheet resistivity is 12.5 kΩ/sq, the resistance is

$$R = 12.5 \text{ k}\Omega/\text{sq} \times 3 \text{ sq} = 37.5 \text{ k}\Omega$$

Sheet resistivities vary from approximately 10 to 5000 Ω/sq, depending on the type of material used and the method of depositing the film.

The temperature characteristics of film resistors tend to depend on the film thickness, with thicker films tending to have positive temperature coefficients

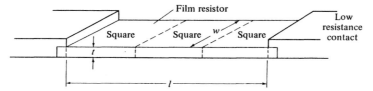

Figure 3-10 Film resistor.

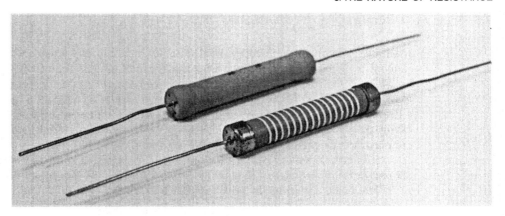

Figure 3-11 A film resistor having a ceramic core.

like the bulk material and thinner films tending toward negative temperature coefficients.

Figure 3-11 shows a film resistor that consists of a helix pattern cut into the resistance film that has been applied to a ceramic core. Figure 3-12 presents a cross section and a size comparison of a thick-film fixed chip resistor. Figure 3-12(a) is a cross section of the chip, and Fig. 3-12(b) is a photo of some chips and a pencil for size comparison. The chip resistors shown in Fig. 3-12 measure 1.6 × 3.2 × 0.6 mm and are designed for soldering on to printed circuit boards.

Film resistors are also fabricated in single in-line packages (SIP) and dual in-line packages (DIP) that are similar to the chip enclosures used for integrated circuits. Some DIP thick-film resistor networks and their circuit configurations are shown in Fig. 3-13. Resistor networks such as DIPs are useful where repetitive circuit configurations or values are required. Those shown in Fig. 3-13 are used with logic circuits.

Adjustable or variable resistors are also available as films. Two such resistors are shown in Fig. 3-14. Inside both resistors are a resistance film and an adjustable wiper arm that makes contact with that film. The adjustable wiper arm and the two ends of the film are terminated with contacts for printed circuit board soldering. Any type of variable resistor with three terminals, one for the movable contact and two for the resistor ends, is called a *potentiometer*.

Although a thorough discussion of semiconductor resistances is beyond the scope of this book, a brief comment follows. Semiconductors have electrical characteristics that classify them between conductors and insulators. If certain other materials called *impurities* are added to semiconductor material, the electrical resistance changes markedly and it is possible to obtain desired resistances. The reader may be aware that diodes and transistors are also fabricated from semiconductor materials. Resistors, diodes, and transistors may then be fabricated simultaneously and may be *integrated* into a small electrical circuit having no connecting wires. On the other hand, resistors or any other circuit elements that are separate or individually distinct are said to be *discrete components*.

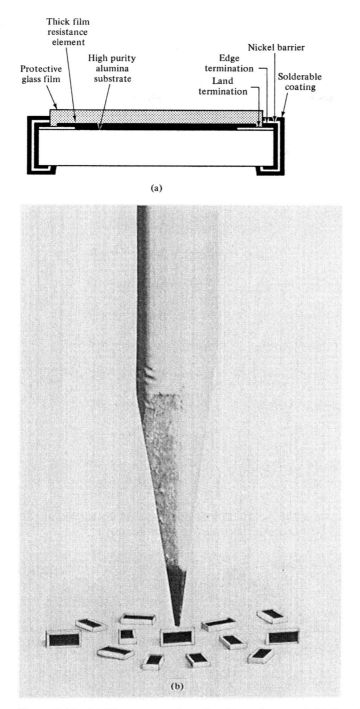

Thick film
resistance
element

High purity
alumina
substrate

Protective
glass film

Edge
termination

Land
termination

Nickel barrier

Solderable
coating

(a)

(b)

Figure 3-12 **(a)** The cross section of a chip resistor, and **(b)** the relative size of chip resistors. (Courtesy of Allen–Bradley.)

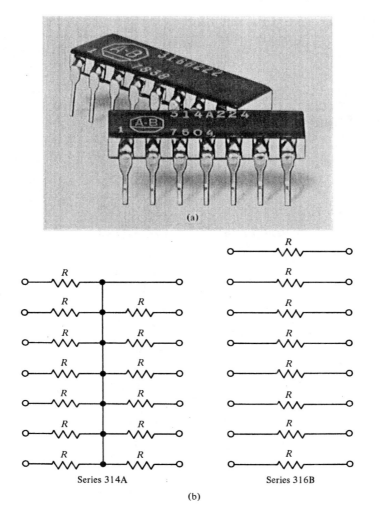

(a)

Series 314A   Series 316B

(b)

Figure 3-13 **(a)** Type 314 and 316 DIP resistors, and **(b)** the configurations for the type 314 and 316 DIP networks. (Courtesy of Allen–Bradley.)

Figure 3-14 Miniature potentiometers.

3.7 SOME RESISTIVE TRANSDUCERS

Because the resistance of a conducting film varies with length, cross section, and temperature according to the equations we have studied, it is possible to use resistance measurements to sense changes in any of those parameters. A device that responds to one physical effect by producing another physical effect is called a *transducer*. Electrical transducers are often used in instrumentation for nonelectrical quantities. Thus it is possible to measure changes in voltage, current, or resistance as a result of changes in length, temperature, light intensity, and so on. In fact, some of the voltage conversion devices, such as the piezoelectric generator are really transducers.

The *thermistor* is a two-terminal device designed specifically to exhibit a change in electrical resistance with a change in its body temperature. Most thermistors have

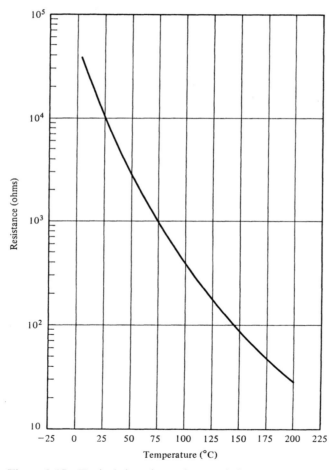

Figure 3-15 Typical thermistor characteristic.

negative temperature coefficients of resistance so that a graph of resistance versus temperature would show decreasing resistance with increasing temperature. Hence, the typical characteristic graph of Fig. 3-15. Depending on construction, thermistors may be of the thick film or semiconductor type.

You will notice that Fig. 3-15 is a semilogarithmic graph; that is, one scale has linear values, whereas the second has exponential or power of 10 values. A change in one power of 10 along the logarithmic scale is called a *cycle* or *decade*. When interpolating values on a logarithmic scale you must remember that within a cycle the scale is nonuniform, and beginning with large spaces and ending with small spaces. This nonlinearity is indicated in Fig. 3-16.

Although a complete logarithmic scale is supplied by commercially available graph paper, publishers usually simplify the scales to save artwork time. Hence, in reading logarithmic scales in this text you should utilize Fig. 3-16 "to read between the lines."

The *strain gauge* is a transducer used to convert linear dimension changes into resistive changes. One type of strain gauge consists of a resistive film applied in a serpentine configuration to a thin flexible sheet of nonconductive material. A general-purpose strain gauge is shown in Fig. 3-17. It has a constantan metal alloy grid that is fully encapsulated in polyimide and is available with nominal resistance values of 120 or 350 Ω.

In order to measure changes in length, a strain gauge is applied with a proper adhesive to the material that is to be measured. As the length of the measured material changes, the gauge moves, which causes changes in length and cross section and, in turn, changes in resistance.

Another useful resistive transducer is the *photoconductive cell* which is a two-terminal device whose resistance depends upon the intensity of light impinging on the cell. There are two types of photoconductive cells. One uses a semiconductor junction whose resistance changes with the light falling on the junction, and the

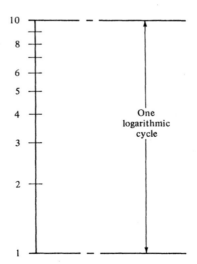

Figure 3-16 The logarithmic scale.

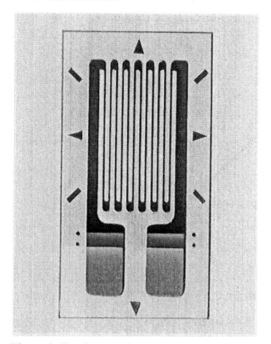

Figure 3-17 A general purpose type CEA-series strain gauge. (Courtesy of Micro-Measurements Division, Measurements Group, Inc., Raleigh, North Carolina.)

second type, depicted in Fig. 3-18, consists of a film of resistive material, often placed in serpentine fashion within a protective housing. The resistive elements are often made of cadmium sulfide or cadmium selenide.

The photoconductive cell exhibits its maximum resistance in total darkness. With an increase in the light level, the resistance decreases in a fashion typified by

Figure 3-18 A photoconductive cell.

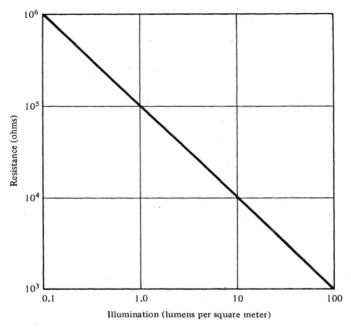

Figure 3-19   The resistance characteristic for the photoconductive cell in Fig. 3-18.

Fig. 3-19. For example, the photoconductive cell shown in Fig. 3-18 changes in resistance from 3 MΩ in total darkness to 40 Ω in bright sunlight.

You should note that both scales in Fig. 3-19 are logarithmic. Along the horizontal axis is the SI unit for illumination, lm/m². A 100 W light bulb emits about 1700 lumen (lm).

3.8 CONDUCTANCE AND CONDUCTIVITY

Since resistance is the opposition to current flow, the inverse or reciprocal of resistance is the ease with which current flows in a resistance. The reciprocal of resistance is called *conductance* and is given the symbol *G*. It is measured in siemens:

$$G = 1/R \qquad S \tag{3-10}$$

Prior to the adoption of SI units by the IEEE, the unit of conductance was recognized as the reciprocal of ohm and was given the name mho (℧—inverted Greek omega). You may still encounter the use of this latter unit. Notice that, since resistance and conductance are reciprocal quantities, a large resistance is associated with a small conductance and vice versa.

Drill Problem 3-9 ■

Determine the change in conductance of the photoconductive cell discussed in Section 3-7 if the cell illumination changes from total darkness to bright sunlight.

Answer. 0.333 μS–0.025 S. ∎ ∎

In a similar manner, the reciprocal of resistivity is *conductivity*, which is the *specific conductance*, or conductance per unit length and cross section. The symbol for conductivity is the Greek letter sigma (σ), and the unit in the SI system is siemens per meter:

$$\sigma = 1/\rho \qquad \text{S/m} \qquad (3\text{-}11)$$

Sometimes the conductivities of different materials are compared. In making a comparison, the conductivity of standard annealed copper is taken as 100%. Then the *percent conductivity* of a given material is the ratio of the conductivity of that material to the conductivity of standard annealed copper, expressed as a percentage.

EXAMPLE 3-6

Determine the percent conductivity of aluminum.

Solution

$$\frac{\sigma_{Al}}{\sigma_{Cu}} = \frac{1/\rho_{Al}}{1/\rho_{Cu}} = \frac{\rho_{Cu}}{\rho_{Al}} = \frac{10.37}{17} = 0.61$$
$$= 61\%$$

QUESTIONS

1. Resistance depends on what four quantities?
2. Define resistivity and give its units.
3. What is a circular mil?
4. What is the relationship between a square mil and a circular mil?
5. Explain how the wire size, cross section, and resistance change in going from AWG 10 wire to AWG 30 wire.
6. Explain what is meant by inferred absolute zero.
7. Why is wire stranded?
8. What is the temperature coefficient of resistance?
9. Explain the difference between positive and negative temperature coefficients.
10. What is the significance of the color bands on a carbon-composition resistor?
11. Describe the construction of a wire-wound resistor.
12. What is the difference between a thick and a thin-film resistor?
13. What is sheet resistance and sheet resistivity?
14. What is a transducer?
15. Describe a few resistive transducers.
16. What is the difference between conductance and conductivity?
17. What is the difference between resistivity and conductivity?
18. What is meant by percent conductivity?

PROBLEMS

1. Determine the resistance of a copper bar at 20°C that has cross-sectional dimensions of 3 × 0.5 cm and a length of 5 m.

2. A certain piece of copper wire has a resistance of 50 Ω. If an aluminum wire of the same size is used, what is the resistance?

3. Calculate the resistance of an aluminum bus bar measuring 0.25 cm × 5 cm × 1 m long.

4. A piece of copper wire is 50 ft long. How long must a Nichrome wire of the same cross section be in order to have the same resistance?

5. Calculate the area in circular mils for the following wire diameters:
   (a) 0.025 in.   (b) 1.125 in.
   (c) 0.272 in.   (d) 114 mil.

6. Given the following areas in circular mils, calculate the corresponding wire diameters:
   (a) 31 831 cir mil.   (b) 420 cir mil.
   (c) 3260 cir mil.   (d) 63.2 cir mil.

7. Calculate the resistance per mile for both the inner and outer copper conductors of the coaxial cable in Fig. 3-20.

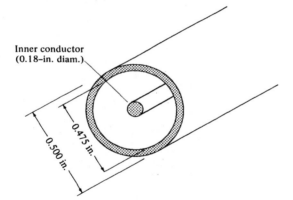

Inner conductor
(0.18-in. diam.)

0.475 in.

0.500 in.

Figure 3-20   Problem 3-7.

8. A resistance of 10 Ω is to be constructed from Nichrome AWG 29 wire. What length wire must be used?

9. A circular wire 300 ft long has a diameter of 0.020 in. and a resistance of 7.85 Ω. What is the resistivity of the wire?

10. A coppper strip $\frac{1}{10}$ × $\frac{1}{2}$ × 3 in. long is used as an ammeter shunt. What is the resistance of the copper strip?

11. A coil of copper wire has a resistance of 11.7 Ω at 20°C. What will its resistance be at 50°C?

12. Determine the resistance of the following lengths of standard annealed copper wire:
    (a) 200 ft of AWG 2.   (b) 100 ft of AWG 12.   (c) 25 m of AWG 14.

13. A motor is located 300 ft away from a power source. If the total resistance of the copper conductors to the motor and back to the source must be no greater than 1.0 Ω, can AWG 12 wire be used?

14. A large spool contains an unknown length of insulated AWG 2 aluminum wire. The two ends of the wire are accessible so a resistance measurement is performed. If the resistance is 0.13 Ω, how many feet of wire are on the spool?

15. A certain copper cable is made of seven strands, each 0.123 in. in diameter.
    (a) What is the closest AWG equivalent?
    (b) What is the resistance per mile of the cable?

16. Resistivity is usually given in tables for 20°C only. At what temperature does annealed copper have a resistivity of 11.5 Ω cir mil/ft, assuming no change in length or diameter?

17. What will be the resistance of a silver wire at 50°C if the resistance at 0°C is 20 Ω and the temperature coefficient at 0°C is 0.004 11?

18. The resistance of a copper wire is 8.2 Ω at 40°C. What is its resistance at −40°C?

19. Calculate the temperature coefficient for annealed copper at 20° and at 60°C.

20. The copper field winding of a motor has a resistance of 480 Ω at 25°C. After the motor is used for some time, the resistance is measured to be 554 Ω. What is the temperature rise of the winding?

21. A coil of a transformer (wound with copper wire) has a resistance of 40 Ω at 80°C. What is the temperature of the coil when the resistance is 36 Ω?

22. The temperature of the filament of a tungsten lamp is to be determined. The lamp resistance is measured as 25 Ω at 20°C and with the rated potential across the lamp is measured as 240 Ω. What temperature is the filament with the rated potential applied?

23. Determine the color code for the following resistors:
    (a) 91 kΩ ± 10%.   (b) 620 Ω ± 20%.
    (c) 2.2 kΩ ± 5%.   (d) 0.18 Ω ± 5%.
    (e) 0.56 Ω ± 10%.   (f) 390 kΩ ± 20%.

24. What are the ohmic values and tolerance of the resistors with the following color bands:
    (a) brown−gray−red−gold.   (b) orange−orange−gold−none.
    (c) red−violet−orange−silver.   (d) brown−green−yellow−gold.
    (e) green−gray−red−none.   (f) violet−green−black−silver.

25. A material of 200 Ω/sq sheet resistivity is to be used in making a thin-film resistor of 10 kΩ. If the width is to be 5 mil, what length must the film be?

26. Determine the conductance of the copper bar in Problem 1.

27. Using the typical thermistor characteristic of Fig. 3-15, determine the resistance values at 25°, 50°, and 75°C.

28. Determine the change in conductance as the illumination on a photoconductive cell with the characteristic of Fig. 3-19 changes from 1 to 100 lm/m^2.

29. Determine the conductances for the following resistances:
    (a) 470 Ω.   (b) 5.2 MΩ.   (c) 0.25 Ω.

30. What is the conductance of a strain gauge with a resistance of 120 Ω?

31. What is the conductivity of Nichrome in the metric system?

32. Using Eqs. 3-1 and 3-10, obtain an expression for conductance in terms of the conductivity, length, and cross-sectional area.

33. Determine the percent conductivity of silver. Explain the significance of the determined value.

Chapter 4

Ohm's Law, Power, and Energy

4.1 OHM'S LAW

In the 1800s, in Germany, Georg Simon Ohm investigated the relationship between the voltage that existed across a simple electric circuit and the current through that circuit. He found that provided the circuit resistance did not change in temperature as the voltage was increased, the current changed in direct proportion; that is, the ratio of voltage to current was constant. Since the plot of voltage versus current is a straight line (see Fig. 4-1), Ohm was able to define a constant of proportionality k such that

$$V = kI \qquad (4\text{-}1)$$

The constant of proportionality is known as resistance, and Eq. 4-1 (Ohm's law) is rewritten

$$V = IR \qquad (4\text{-}2)$$

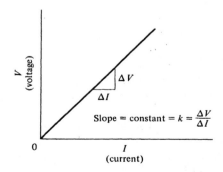

Figure 4-1   Voltage versus current curve (a linear relationship).

Figure 4-2 A resistance creates a voltage drop when current flows through it.

where V is in volts, I in amperes, and R in ohms.

EXAMPLE 4-1

A resistance of 300 Ω has a current of 0.6 A flowing through it. What voltage is developed across the resistance?

Solution

$$V = IR = (0.6)(300) = 180 \text{ V}$$

As current flows through a resistance, the resistance absorbs energy from the source supplying the current. Then there is a drop in potential across the resistance. As indicated in Fig. 4-2, *the voltage drop across a resistor is taken in the direction of the flow of conventional current.*

Drill Problem 4-1 ∎

An incandescent lamp is connected to 120 V and allows a current flow of 0.88 A. What is the resistance?

Answer. 136 Ω. ∎ ∎

Drill Problem 4-2 ∎

A carbon-composition resistor of 120 ± 5% Ω has a current of 15 μA flowing through it. What range of voltage can be expected across the resistor?

Answer. 1.71–1.89 mV. ∎ ∎

4.2 LINEAR AND NONLINEAR RESISTANCES

Ohm's law is based on a linear relationship between the voltage and the current. A resistance whose voltage versus current relationship is a straight line, as in Fig. 4-1, is a *linear resistance*. A linear resistance obeys Ohm's law because the resistance is always constant. However, a resistance whose ohmic value does not remain constant is defined as a *nonlinear resistance*. Because of temperature effects, all resistances are basically nonlinear, but those exhibiting only a small resistance change over a range of operating voltage and current are often called linear. Voltage versus current characteristics are shown in Fig. 4-3 for linear and nonlinear resistances. Curve (a) of Fig.

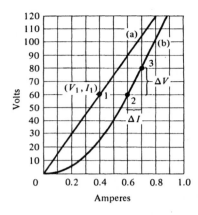

Figure 4-3 $V-I$ curves: **(a)** linear; **(b)** nonlinear.

4-3 has a constant slope corresponding to the resistance value that can be calculated at any point along the curve. At point 1,

$$R = \frac{V_1}{I_1} = \frac{60 \text{ V}}{0.4 \text{ A}} = 150 \ \Omega$$

Curve (b), on the other hand, has a slope that changes, but it is possible to determine a small change in voltage (ΔV) as a result of a small change in current (ΔI). Then the *dynamic resistance* in ohms is defined as

$$R = \frac{\Delta V}{\Delta I} = \frac{V_x - V_y}{I_x - I_y} \qquad \frac{\text{volts}}{\text{ampere}} \tag{4-3}$$

where V_x, I_x and V_y, I_y are the respective coordinates of a set of chosen points x and y.

EXAMPLE 4-2

What is the dynamic resistance of the nonlinear curve of Fig. 4-3 in the range between points 2 and 3?

Solution

$$R = \frac{\Delta V}{\Delta I} = \frac{V_3 - V_2}{I_3 - I_2} = \frac{82 - 60}{0.1} = \frac{22}{0.1} = 220 \ \Omega$$

The smaller the change in voltage and current, the more representative is the dynamic resistance in a particular range.

In addition to the linear and nonlinear classifications, resistances can be classified as *bilateral* or *unilateral*. A *bilateral resistance* is a device that exhibits the same voltage–current characteristics regardless of the direction of current through the device. The nonlinear characteristic curve of Fig. 4-3(b) is that of a 100 W incandescent lamp. Current flowing through the lamp in one direction produces the same

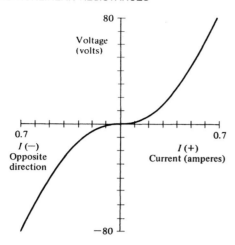

Figure 4-4 *V–I* curve for a nonlinear bilateral device (incandescent lamp).

characteristic as the current flowing in the opposite direction. The voltage versus current characteristics can then be shown in two quadrants, as in Fig. 4-4.

A *unilateral resistance* is one whose resistance or voltage–current curve changes markedly with the direction of the current through the device. A semiconductor diode exhibits the typical characteristic shown in Fig. 4-5. With voltage applied so that current flows in a *forward direction*, a relatively large current flows, whereas with voltage applied so that the current flows in a *reverse direction*, relatively low current flows. Notice, for example, that the forward voltage axis divisions are in tenths of a volt, and the current axis divisions are in hundreds of milliamperes. However, for the reverse direction the voltage divisions are in hundreds of volts and the current divisions are in microamperes. Figure 4-5 is a current versus voltage curve, and the actual slope of the curve is given in terms of dynamic conductance, where

$$G = \frac{\Delta I}{\Delta V} \qquad (4\text{-}4)$$

Drill Problem 4-3 ■

Calculate the approximate resistance for a diode with the typical characteristic of Fig. 4-5 for currents of **(a)** 5 μA reverse; **(b)** 50 mA forward.

Answer.   (a) 80 MΩ. (b) 14 Ω.   ■ ■

Drill Problem 4-4 ■

Determine the dynamic conductance and resistance of a lamp with a characteristic of Fig. 4-3(b) in the range 20–30 V. Compare the resistance value to that obtained in Example 4-2.

Answer.   0.007 S; 143 Ω.

The resistance is smaller at this lower voltage because the slope is smaller.   ■ ■

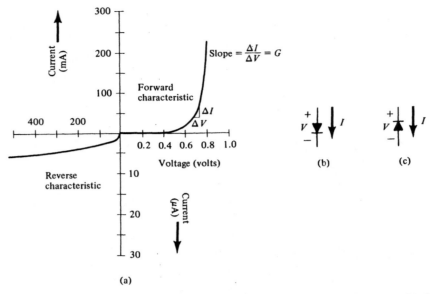

(a)

Figure 4-5 **(a)** $I-V$ curve for diode; **(b)** diode connection, forward current; **(c)** diode connection, reverse current.

4.3 POWER AND ENERGY

Power is *the rate of doing work*, or alternatively, *the work per unit time*. The MKS unit of power is the joule per second or watt,

$$P = W/t \quad \text{W} \tag{4-5}$$

where P is the power in watts, W the work in joules, and t the time in seconds.

If one recalls the units for current and voltage presented in Chapter 2,

$$V = \frac{\text{joules}}{\text{coulomb}} \quad \text{and} \quad I = \frac{\text{coulombs}}{\text{second}}$$

it is seen that a product of voltage and current results in power.

$$VI = \left(\frac{\text{joules}}{\text{coulomb}}\right)\left(\frac{\text{coulombs}}{\text{second}}\right) = \frac{\text{joules}}{\text{second}} \tag{4-6}$$

Thus, $P = VI$, where P is the electrical power in watts, V the voltage in volts, and I the current in amperes.

If Ohm's law is substituted into Eq. 4-6, the result is

$$P = I^2 R \tag{4-7}$$

and if $I = V/R$ is substituted in Eq. 4-6, the result is

$$P = \frac{V^2}{R} \tag{4-8}$$

These last two power equations allow one to solve for power if the resistance and either the current or voltage is known.

EXAMPLE 4-3

What is the maximum current that a 40 kΩ, 10 W resistor can handle without overheating?

Solution. Solving for I^2 in Eq. 4-7, we obtain

$$I^2 = \frac{P}{R} = \frac{10}{40 \times 10^3} = 2.5 \times 10^{-4} \text{ A}^2$$

Then, taking the square root, we find that

$$I = \sqrt{2.5 \times 10^{-4}} = 1.58 \times 10^{-2} = 15.8 \text{ mA}$$

Electric appliances usually have the input power requirements marked on them. Some typical power ratings for household appliances are listed in Table 4-1.

Since *energy* is work that is either stored or expended, it can be calculated from the power and the period of time over which the power is used. Rearrangement of

TABLE 4.1 TYPICAL POWER REQUIREMENTS FOR HOUSEHOLD APPLIANCES

Appliance	Power rating (W)
Blender	200
Can opener	200
Clothes dryer	6000
Clothes washer	525
Coffee maker	900
Floor polisher	125
Hair dryer	1500
Iron	1100
Microwave oven	1500
Oil-burner motor	270
Radio	15
Refrigerator	200
Stereo	125
Television set	200
Toaster	1200
Vacuum cleaner	630
Window fan	170

Eq. 4-5 provides the expression for energy,

$$\text{energy} = W = Pt \tag{4-9}$$

where, in electrical units, energy = watt-second. It is generally more convenient to use a larger unit of electrical energy; therefore, a *watt-hour* or *kilowatt-hour* is used, where

$$\text{kilowatt-hour} = \text{kWh} = \frac{\text{watts} \times \text{hours}}{1000} \tag{4-10}$$

Power companies charge for electrical energy by the kilowatt-hours. A sliding scale or step down in rates is generally used, so that an increase in the use of kilowatt-hours results in a lower cost per kilowatt-hour.

EXAMPLE 4-4

A student used a 100 W desk lamp while studying for 6 hr. If the average cost of energy is 3.5 cents/kWh, what is the energy cost?

Solution

$$\text{energy} = Pt = (100 \text{ W})(6 \text{ h}) = 600 \text{ Wh} = 0.6 \text{ kWh}$$
$$\text{cost} = 0.6 \text{ kWh} \times 3.5 \text{ cents/kWh} = 2.1 \text{ cents}$$

You should note that in practice electrical energy is measured in watt-hours or kilowatt-hours not joules. As can be seen in Table 1-3, there are 3600 J in every watt-hour.

Power is measured by a wattmeter, and energy is measured by a watt-hour meter. Power companies generally use a kilowatt-hour meter to measure the energy supplied to a customer. Watt-hour (or kilowatt-hour) meters can be obtained either with a digital readout or with a group of four or five dials and pointers as indicated in Fig. 4-6.

The dial-and-pointer-type meter is read from left to right so that the left dial and pointer indicate the most significant digit. When reading the meter one takes the number just passed by the pointer on each scale. As noted in Fig. 4-6, because of the internal gearing the dials have alternate clockwise and counterclockwise scales. It follows that the amount of energy used over a period of time is found by subtracting a previous reading from the current reading. Both the watt-hour meter and watt-meter are discussed more fully in Chapter 17, on measurements.

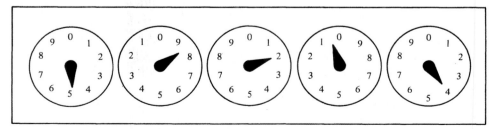

Figure 4-6   The typical dial and pointer arrangement of a kilowatt-hour meter.

Drill Problem 4-5 ■

What is the maximum current and voltage that can be applied to a 10 kΩ, 2 W resistor?

Answer.   14.1 mA; 141 V.   ■ ■

Drill Problem 4-6 ■

What power is supplied to a 12 V dc motor that has a current of 0.9 A flowing to it?

Answer.   10.8 W.   ■ ■

Drill Problem 4-7 ■

Determine the power taken by the lamp with the characteristic of Fig. 4-3(b) at currents of 0.6, 0.7, and 0.8 A. (The power changes nonlinearly.)

Answer.   36, 57.4, and 81.6 W.   ■ ■

Drill Problem 4-8 ■

What is the conductance of a 60 W lamp that is connected to its rated voltage of 120 V?

Answer.   4.17 mS.   ■ ■

Drill Problem 4-9 ■

Using the kilowatt-hour reading of Fig. 4-6 and a previous reading of 46 952, calculate the total energy cost if the cost per kilowatt-hour is 6 cents.

Answer. $75.12.   ■ ■

4.4 EFFICIENCY

When any system or device converts one form of energy to another, some energy is lost or wasted, usually in the form of heat. Thus the energy or power put into a system must be greater than that received from the system. *Efficiency* is the ratio of output energy to input energy over the same time period. Since the same period of time is used, efficiency may be expressed as a ratio of output to input power. The Greek letter eta is used for efficiency:

$$\text{efficiency} = \eta = \frac{\text{output}}{\text{input}} \tag{4-11}$$

$$= \frac{\text{output}}{\text{output} + \text{losses}} \tag{4-12}$$

EXAMPLE 4-5

What is the efficiency of a transistor stereo amplifier that draws 145 W of power while supplying each of two speakers with 30 W of power?

Solution

$$\eta = \frac{\text{output power}}{\text{input power}} = \frac{60 \text{ W}}{145 \text{ W}} = 0.414 = 41.4\%$$

As noted in Example 4-5, efficiency may be expressed as a decimal or a *percent*.

Often a larger system is considered to be made up of smaller systems in *cascade*, that is, the output from one system goes directly into a succeeding system or stage (see Fig. 4-7). In Fig. 4-7 the stage 1, stage 2, and overall efficiencies are η_1, η_2, and η_0, respectively, where

$$\eta_1 = \frac{P_{\text{output 1}}}{P_{\text{input 1}}}$$

$$\eta_2 = \frac{P_{\text{output 2}}}{P_{\text{input 2}}}$$

and

$$\eta_0 = \frac{P_{\text{output 2}}}{P_{\text{input 1}}}$$

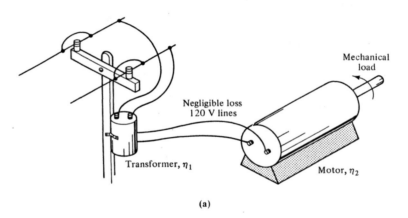

(a)

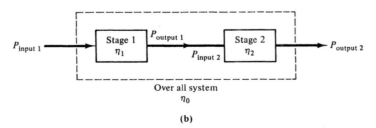

(b)

Figure 4-7 A cascaded system: **(a)** transformer input, motor output; **(b)** the block diagram for efficiency calculations.

Substituting $P_{\text{output 2}} = \eta_2 P_{\text{input 2}}$ into the equation for η_0 yields

$$\eta_0 = \frac{P_{\text{output 2}}}{P_{\text{input 1}}} = \frac{\eta_2 P_{\text{input 2}}}{P_{\text{input 1}}}$$

But $P_{\text{input 2}} = P_{\text{output 1}}$, so that

$$\eta_0 = \frac{\eta_2 P_{\text{output 1}}}{P_{\text{input 1}}}$$

or

$$\eta_0 = \eta_2 \eta_1$$

Thus the overall efficiency of a cascaded system equals the product of the stage efficiencies. In general, this last statement is expressed

$$\eta_0 = \eta_1 \eta_2 \eta_3 \cdots \eta_n \tag{4-13}$$

where η_0 is the overall efficiency and $\eta_1, \eta_2, \eta_3, \ldots, \eta_n$ are the individual stage efficiencies.

Many devices have ratings specified for their output power. Good examples are electromechanical devices such as motors and generators. Motors, of course, have mechanical power for their output, whereas, generators have electrical power for their output. Even though the SI unit of power is the watt, the horsepower is the prevailing unit in describing mechanical output. As pointed out in Chapter 1, the equation for power conversion is 1 hp = 746 W.

EXAMPLE 4-6

A motor has an output rating of 2.5 hp and an efficiency of 85%. If the motor is operated from 240 V, what current does it take to supply the rated output?

Solution

$$P_{\text{out}} = 2.5 \text{ hp} \times 746 \text{ W/hp} = 1865 \text{ W}$$

$$P_{\text{in}} = \frac{P_{\text{out}}}{\eta} = \frac{1865}{0.85} = 2194 \text{ W}$$

then

$$I = \frac{P_{\text{in}}}{V} = \frac{2194 \text{ W}}{240 \text{ V}} = 0.914 \text{ A}$$

Drill Problem 4-10 ■

A solar cell has an output of 0.5 V at 0.1 A. If the efficiency is 15%, what is the radiant input power?

Answer. 0.333 W. ■ ■

Drill Problem 4-11 ■

A transformer supplies 250 W of power to a load at an efficiency of 98%. What are the transformer losses?

Answer. 5.1 W. ■ ■

Drill Problem 4-12 ■

With regard to Fig. 4-7, if the transformer efficiency is 96%, what motor efficiency is needed to obtain an overall efficiency of at least 78%?

Answer. 81.25%.   ■ ■

QUESTIONS

1. What is meant by linear and nonlinear resistances?
2. What is dynamic resistance?
3. What is dynamic conductance?
4. State the difference between bilateral and unilateral resistance.
5. What is power?
6. Define energy.
7. What is meant by efficiency?
8. What is meant by a cascaded system?

PROBLEMS

1. A current of 54 μA flows through a 2.7 MΩ resistor connected across a potential source of E volts. What is E?
2. A 2200 Ω resistor causes a voltage drop of 1.5 V; how much current is flowing through the resistor?
3. Which resistance causes the greater voltage drop:
   (a) a 1 MΩ resistance having a current flow of 75 mA;
   (b) a 150 Ω resistor having a current of 5.5 A.
4. A fuse wire with 0.012 Ω resistance has a voltage drop of 250 μV. What current is flowing through the fuse?
5. A relay coil having a resistance of 48 Ω must carry 0.18 A in order to close. What voltage must be supplied to it?
6. How much voltage appears across a 1.5 MΩ resistor when 1.2 μA flows through it?
7. A variable resistor has a voltage of 12 V across it. If the resistance is varied in a range of 500–1.2 kΩ, what range of current flows in the resistor?
8. A resistive element has a value of 4.7 kΩ $\pm$ 5%. If a voltage of 15 V is applied, what current range is expected?
9. A nonlinear resistor has the V versus I characteristic shown in Fig. 4-8. What is the dynamic resistance **(a)** in the range of current 0.1–0.2 A and **(b)** in the range of current 0.3–0.4 A?
10. Referring to Fig. 4-8, what are the dynamic conductances in the current ranges **(a)** 0.1–0.2 A and **(b)** 0.3–0.4 A?
11. Given the I versus V curve of Fig. 4-9, what is the dynamic resistance and the dynamic conductance in the range of 0.6–0.65 V?

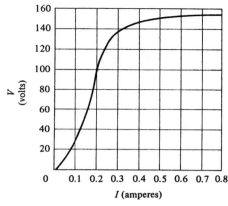

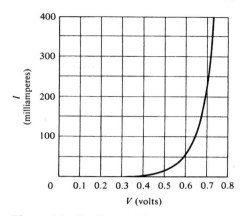

Figure 4-8 Problems 4-9 and 4-10.   **Figure 4-9** Problem 4-11.

12. A 2700 Ω resistor is connected to a 10 V supply. How much power does it dissipate?

13. A 120 V appliance having a resistance of 18 Ω operates for 1 h. How much energy is used?

14. How much current must be flowing in a 7500 Ω resistor if the power consumed is 20 W?

15. What is the maximum voltage that should be applied to a $\frac{1}{2}$ W, 2.7 kΩ resistor?

16. What current will a 550 W heater draw from a 120 V line?

17. A voltage of 12 V is connected to a 100 Ω resistor. What power is absorbed by the resistor? What value would the resistor have to have to absorb twice as much power?

18. A number of 60 W lamps are connected in parallel to a 120 V house-wiring circuit. If the circuit is rated at 15 A, how many lamps can be used?

19. What current should not be exceeded in a $\frac{1}{2}$ W, 1.5 MΩ resistor?

20. A solar cell has an output of 0.45 V at 0.3 A. If a group of these cells is to supply power for an electronic circuit that draws 0.5 W, how many cells are needed?

21. A resistive load is supplied with 10 A of current by a cable of AWG 10 copper wire, as shown in Fig. 4-10. How much power is lost in the cable?

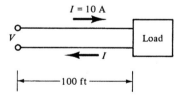

Figure 4-10 Problem 4-21.

22. A 12 V battery is discharged at 0.5 A for 6 h. How much energy is removed? Express the energy in both watt-hours and joules.

23. How many joules of energy does a 1500 W microwave oven use in 45 min?

24. At 6 cents/kWh, how much does it cost to run a typical clothes dryer (6 kW) for 90 min?

25. A portable television receiver takes 180 W from a power line. What is the cost of operation per day if the receiver is used for 6 h and the cost per kilowatt-hour is 3 cents?

TABLE 4-2 RATE SCHEDULE

Kilowatt-hours	Cost ($)
First 24	4.00
Next 36	0.10 each
Next 80	0.09 each
Next 230	0.052 each
All additional	0.032 each

26. A power company utilizes the sliding rate scale shown in Table 4-2. What would a consumer pay for 400 kWh of energy?

27. A watt-hour meter has dial indications as in Fig. 4-11. If the previous reading was 25 925 and the energy costs are as given in Table 4-2, what is the energy charge?

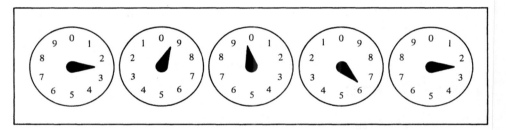

Figure 4-11   Problem 4-27.

28. What input power does a transformer require if it supplies 1.2 kW at an efficiency of 92%?

29. A 10 kW generator of 78% efficiency has to be driven by an engine of what horsepower? (Remember that 1 hp = 746 W.)

30. A 120 V motor produces an output of 2 hp at an efficiency of 72%. What is the current that flows to the motor under these conditions?

31. A $\frac{1}{4}$ hp motor with an efficiency of 80% is rated for 120 V operation. What current must be supplied for a rated output of $\frac{1}{4}$ hp?

32. The overall efficiency of a two-stage system is 45%. One of the stages is 80% efficient. How efficient is the second stage?

33. The input power to an audio cable is 2 W and the output is 1.5 W. What is the efficiency of the cable?

34. A transmission line supplied with 2700 W delivers 2500 W at the load end. What is the efficiency of the line?

35. A motor supplies 1.5 hp to a mechanical load and is 82% efficient. Determine the motor losses in watts.

36. A four-stage system has an overall efficiency of 42%. If the first three stages have efficiencies of 98%, 72%, and 85%, what efficiency must the fourth stage possess?

Chapter 5

Kirchhoff's Laws
and Simple Circuits

5.1 NODES, BRANCHES, AND LOOPS

A basic electric circuit consists of a source connected to a load. In the simple circuit arrangement depicted in Fig. 5-1, it is easy to recognize the Ohm's law relationship between voltage and current. With circuits containing more sources and loads, it becomes more difficult to relate the various voltages and currents. In this chapter we expand on the simple, single-source, single-load circuit and consider the more complicated circuits formed by connecting many circuit elements.

The connection point between two or more elements is called a *node*. In Fig. 5-1 the top connection point of the source and load is one node and the bottom connection is a second node. Each node in Fig. 5-1 is identified by a large dot that can physically represent a solder joint, twisted wires, a connecting lug, or some other form of connection.

Most of the time connecting wires extending from a device or element to the actual point of connection have such a small resistance that they are at the same

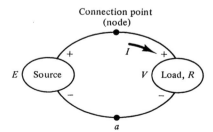

Figure 5-1   A simple electric circuit with two nodes.

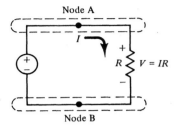

Figure 5-2 A node containing connecting wires.

potential as the actual point of connection. Hence, it is convenient to think of a node as containing the connecting wires as well. This is illustrated in Fig. 5-2, where dashed or broken lines are used to show the expanse of the nodes.

The idea of a node extending beyond the actual solder joint or connection point to include the connecting wires is well illustrated with a printed circuit board. A drawing of a printed circuit board is shown in Fig. 5-3. Most of us have seen such circuit boards and recognize that circuit elements are soldered to conducting pads that, in turn, are connected by conducting foil. A combination of connecting foil and solder pads is considered one node as long as it does not pass through a circuit element. Such a node is identified by dashed lines in Fig. 5-3.

When a node connects *only* two elements it is called a *minor node*. The nodes in Figs. 5-1 and 5-2 are minor nodes. However, when a node connects three or more elements it is called a *major node*. Node A of Fig. 5-3 is a major node because it connects three elements plus a pin connection that, in turn, is connected to at least one other element.

A *branch* is a single element or device connected between two nodes. Notice that two branches are present in both Fig. 5-1 and Fig. 5-2.

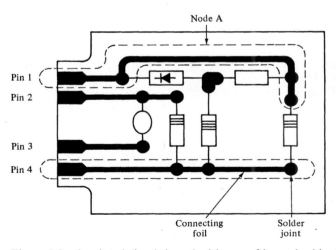

Figure 5-3 A printed circuit board with one of its nodes identified.

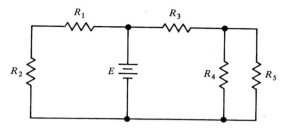

Figure 5-4 Circuit for Example 5-1.

EXAMPLE 5-1

Determine the number of branches and nodes for the circuit in Fig. 5-4.

Solution. As shown in Fig. 5-5, there are six branches (labeled b_1 through b_6) and four nodes (dashed lines).

Notice in Fig. 5-4 that the node connecting R_1 and R_2 is a minor node. Any current passing through R_1 must also pass through R_2 via that minor node. Because it uses the same current flow, a string of elements such as R_1 and R_2 of Fig. 5-4 is sometimes considered, collectively, a single branch. Thus, in a larger context, the *branch* is defined as an element or string of elements connected between major nodes.

When two or more simple circuit elements are connected together an *electric network* is formed. If the network contains at least one closed path, the network is called an *electric circuit*. Notice that each of the networks in Fig. 5-1 through Fig. 5-5 is an electric circuit since each has at least one closed path. Furthermore, a closed path in an electric circuit is called a *loop*. To define any loop, begin at a node, trace a closed path without passing through any element or node more than once, and end at the first node. Thus, in Fig. 5-5, branches b_1, b_2, and b_6 form just one of the many loops in that circuit.

When an electric network has a path that is not closed, we say that the circuit is an *open circuit* (∞ ohms). Two such circuits are shown in Fig. 5-6. As in Fig. 5-6, each circuit lacks an element between the two points *a* and *b* so that no current flows to or from the open circuit.

If a circuit element is connected to points *a* and *b* of either of the open circuits of Fig. 5-6, that circuit forms a closed loop and current flows. A special case arises

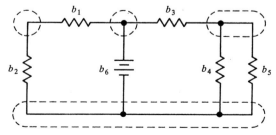

Figure 5-5 The nodes and branches of Fig. 5-4.

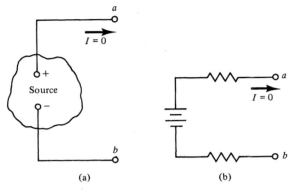

Figure 5-6   Two examples of an open circuit.

when a resistance of 0 Ω is used to close a circuit, as in Fig. 5-7. That circuit condition is called a *short circuit*.

The current that flows in a short circuit, I_{sc}, can be very high and sometimes dangerous. Before we can say how high and how dangerous, we must know something about the circuit being shorted. Finally, just as the short circuit is associated with a resistance of 0 Ω across the circuit, the open circuit is associated with a resistance of infinite ohms across the circuit. Both of these conditions will be explored further in this chapter.

Drill Problem 5-1 ■

Determine the number of branches and nodes for the circuit board in Fig. 5-3. Remember that each pin is connected to at least one other circuit element off the board.

Answer.   6; 5.   ■ ■

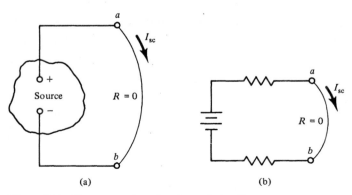

Figure 5-7   Two examples of a shorted circuit.

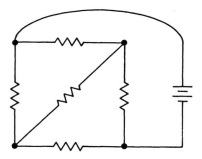

Figure 5-8 Circuit for Drill Problem 5-2.

Drill Problem 5-2 ■

Determine the number of branches and nodes for the circuit of Fig. 5-8.

Answer.   6; 4.   ■ ■

Drill Problem 5-3 ■

Referring to Fig. 5-5, which of the following sets of branches form loops: **(a)** b_3, b_4, b_6; **(b)** b_1, b_6, b_4; **(c)** b_1, b_2, b_4, b_3?

Answer.   (a) and (c).   ■ ■

5.2 KIRCHHOFF'S LAWS

The simplest electric circuit is one with a single source of electromotive force supplying energy to a resistive load, as in Fig. 5-1. With ideal conductors, all the energy supplied by the source is converted as the current moves through the resistive load; that is, the conservation of energy requires that around any closed electrical path, the algebraic sum of electromotive forces must equal the algebraic sum of the potential differences. This idea was developed in 1845 by a German physicist named Gustav Robert Kirchhoff and is known as *Kirchhoff's voltage law*. Let us state this formally:

In any closed loop the algebraic sum of the voltage drops and rises equals zero.

Kirchhoff's voltage law is stated mathematically as

$$\sum V = 0 \tag{5-1}$$

where it is understood that the summation of voltages is an algebraic one; that is, we use one algebraic sign for the drops in potential and the opposite sign for rises in potential. For example, in Fig. 5-1, if we start at point a and traverse the electrical loop in a clockwise direction, a voltage rise is first encountered (E). Next, a voltage drop ($V = IR$) is encountered, after which we return to point a, thereby closing the

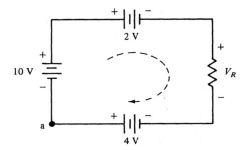

Figure 5-9   Circuit for Example 5-2.

path. If we algebraically call the potential rise positive (+), it follows that a potential drop is negative (−), so that the Kirchhoff voltage equation for Fig. 5-1 is

$$E - V = 0$$

For convenience, Kirchhoff's voltage law is sometimes abbreviated KVL.

EXAMPLE 5-2

Write the KVL equation for the circuit of Fig. 5-9 and use that equation to solve for the resistor voltage V_R.

Solution.   Starting at point a, traversing the loop clockwise, and algebraically labeling the voltage drops positive (+) and the rises negative (−), we obtain

$$-10 + 2 + V_R - 4 = 0$$

Solving for V_R, we obtain

$$V_R = 4 + 10 - 2 = 12 \text{ V}$$

The current in Fig. 5-1 has only one path, although many paths may be available in a typical electric circuit. Figure 5-10 represents a node of a more complicated circuit. The currents coming into a node must equal the currents leaving the same node, which is analogous to the water flow in a hydraulic system in which all water coming into a junction of pipes must leave the junction. This may be stated in a slightly

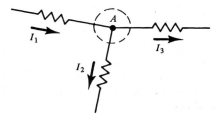

Figure 5-10   Currents at a node.

different form:

> **At any node the algebraic sum of the currents entering and leaving the node equals zero.**

This statement is known as *Kirchhoff's current law* (sometimes abbreviated KCL) and is based on the conservation of charge; that is, there is no accumulation or disappearance of charge at a node.

Kirchhoff's current law is stated as

$$\sum I = 0 \qquad (5\text{-}2)$$

where currents coming into a node are arbitrarily assigned one algebraic sign (+ or −) and those leaving are assigned the opposite algebraic sign (− or +), respectively. A node equation is then a statement of Kirchhoff's current law at a node. For example, if the currents into the node are positive and the currents leaving the node are negative, the equation for node A of Fig. 5-10 is

$$I_1 - I_2 - I_3 = 0$$

EXAMPLE 5-3

Solve for the current I_4 in Fig. 5-11.

Solution. With currents into the node as positive,

$$\sum I = I_1 - I_2 + I_3 - I_4 = 0$$

Substituting values for I_1, I_2, and I_3 and then solving for I_4, we obtain

$$5 - 2 + 3 - I_4 = 0$$

or $I_4 = 6$ A.

As noted by the preceding examples, KVL and KCL are generally used to solve for unknown voltages and currents of an electric circuit. Although the polarities of the voltages and directions of the currents were given in the examples, there are times when they will not be specified. In that case we simply choose the polarities or directions and solve the necessary KVL and KCL equations. Should a mathematical solution indicate an unknown with a minus sign, the unknown has a polarity or direction opposite to that initially chosen.

Kirchhoff's two laws provide the basic tools for circuit analysis, and their usefulness will become more apparent as we proceed.

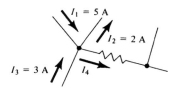

Figure 5-11 Currents at a node for Example 5-3.

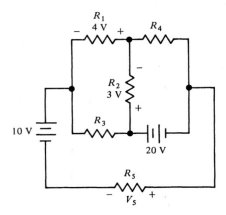

Figure 5-12   Circuit for Example 5-4.

EXAMPLE 5-4

Determine the voltage V_5 as specified in Fig. 5-12.

Solution.   A KVL equation can be written around a loop containing R_1, R_2, the 20 V source, R_5, and the 10 V source. With drops algebraically positive and proceeding clockwise, we obtain

$$-4 - 3 + 20 + V_5 - 10 = 0$$

or $$V_5 = 10 - 20 + 4 + 3 = -3\ \text{V}$$

The negative sign indicates that the voltage is reversed in polarity from the specified polarity.

Drill Problem 5-4 ■

Determine the unknown voltages for each of the circuits shown in Fig. 5-13.

Answer.   (a) 9 V. (b) −8 V; 3 V.   ■ ■

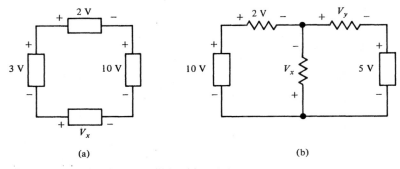

(a)   (b)

Figure 5-13   Circuits for Drill Problem 5-4.

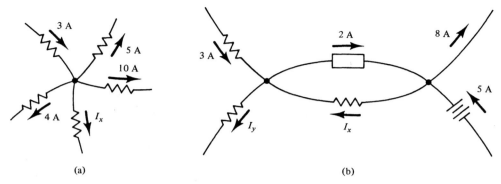

Figure 5-14 Networks for Drill Problem 5-5.

Drill Problem 5-5 ■

Determine the unknown currents for each of the circuits of Fig. 5-14.

Answer. (a) -16 A. (b) -1 and 0 A. ■ ■

5.3 THE SERIES CIRCUIT

When resistances or other circuit elements are connected end to end or in tandem, as are the resistances of Fig. 5-15(a), the elements are said to be in *series*. From the principle of conservation of charge and an inspection of Fig. 5-15(a), it should be evident that the current leaving the battery (I) is the same current that flows in each of the resistances (I_1, I_2, and I_3, respectively); that is, $I = I_1 = I_2 = I_3$. Thus a *series circuit* may alternatively be defined as a circuit having only one path for the current.

Applying Kirchhoff's voltage law to the circuit by starting at the positive terminal of the battery, traversing the loop clockwise, and taking a voltage drop as positive,

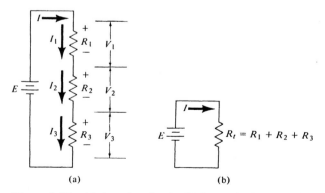

Figure 5-15 **(a)** A series circuit; **(b)** its equivalent circuit.

we obtain

$$V_1 + V_2 + V_3 - E = 0$$

or $$E = V_1 + V_2 + V_3 \tag{5-3}$$

Equation 5-3 states that in a series circuit the applied emf equals the sum of the individual voltage drops. If the individual voltage drops ($V_1 = IR_1$, $V_2 = IR_2$, and $V_3 = IR_3$) are substituted in Eq. 5-3 and the equation subsequently solved for I, we find

$$E = IR_1 + IR_2 + IR_3$$
$$E = I(R_1 + R_2 + R_3)$$

or $$I = \frac{E}{R_1 + R_2 + R_3}$$

The preceding expression indicates that in a series circuit, the total or equivalent resistance R_t equals the sum of the resistances; hence, we have the equivalent circuit of Fig. 5-15(b). In general,

$$R_t = R_1 + R_2 + R_3 + \cdots + R_n \tag{5-4}$$

and with N resistors of the same value R in series

$$R_t = NR \tag{5-5}$$

Multiplying Eq. 5-3 by the current I, we find that

$$EI = IV_1 + IV_2 + IV_3$$

or $$P_t = P_1 + P_2 + P_3 \tag{5-6}$$

That is, in accordance with the law of conservation of energy, the total power supplied by the battery equals the sum of the power losses in the resistive load.

EXAMPLE 5-5

A 10, a 15, and a 30 Ω resistor are connected in series across a 120 V source. **(a)** What is the total series resistance? **(b)** What current flows? **(c)** What power is dissipated by the resistances?

Solution

(a) $R_t = R_1 + R_2 + R_3 = 10 + 15 + 30 = 55\ \Omega$.

(b) $I = \dfrac{E}{R_t} = \dfrac{120\ \text{V}}{55\ \Omega} = 2.18\ \text{A}$.

(c) $P_t = EI = (120\ \text{V})(2.18) = 262\ \text{W}$.

A characteristic of a series circuit is that the current must pass through each part; if a break in the circuit occurs, that is, the circuit becomes *open*, the entire circuit ceases to function. *Fuses* and *circuit breakers* are protective devices that are designed

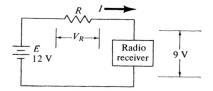

Figure 5-16   Example 5-6.

to open a circuit when too high a current flows in the circuit. They are then placed in series with the circuit being protected.

A series resistance is often used to reduce the voltage across a device, as in Example 5-6.

EXAMPLE 5-6

A radio receiver requiring 9 V is operated from a 12 V source, as indicated in Fig. 5-16. Calculate the resistance and power rating of the series limiting resistor if the normal current required by the receiver is 120 mA.

Solution.   By Kirchhoff's voltage law, the voltage across resistance R is

$$V_R = E - 9 \text{ V} = 3 \text{ V}$$

Then

$$R = \frac{V_R}{I} = \frac{3 \text{ V}}{120 \text{ mA}} = 25 \text{ } \Omega$$

and

$$P_R = V_R I = 3 \text{ V}(120 \text{ mA}) = 0.36 \text{ W}$$

Voltage sources in series are often encountered when series elements are considered. Two series voltage sources are depicted in Fig. 5-17(a). Although the network is not really a closed circuit, for KVL considerations the network is closed by a resistance of infinite ohms across which the voltage V_x is measured. Then a KVL equation allows us to say that $V_x = 8$ V with an equivalent as shown in Fig. 5-17(b).

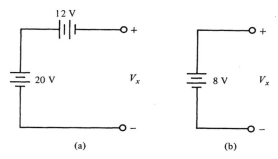

Figure 5-17   **(a)** Two series voltage sources and **(b)** their equivalent.

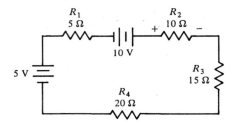

Figure 5-18 Circuit for Example 5-7.

EXAMPLE 5-7

Determine the voltage drop across a resistor R_2 for the circuit in Fig. 5-18.

Solution. The equivalent source voltage is

$$E = 5 + 10 = 15\,\text{V}$$

Then

$$R_t = R_1 + R_2 + R_3 + R_4$$
$$= 5 + 10 + 15 + 20 = 50\,\Omega$$

$$I = \frac{E}{R_t} = \frac{15\,\text{V}}{50\,\Omega} = 0.3\,\text{A}$$

and

$$V_2 = IR_2 = 0.3(10) = 3\,\text{V}$$

Drill Problem 5-6 ■

A series connection of three resistors (2, 5, and 13 kΩ) and a 12 V source will result in what power dissipation in the 2 kΩ resistor?

Answer. 0.72 mW.   ■ ■

Drill Problem 5-7 ■

What is the value of the unknown resistances for the networks of Fig. 5-19?

Answer. (a) 15.9 kΩ. (b) 2.5 Ω.   ■ ■

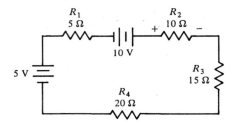

(a)   (b)

Figure 5-19 Networks for Drill Problem 5-7.

Drill Problem 5-8 ■

An electrical device with a resistance of 150 Ω is connected in series with an unknown resistor R_x to a 120 V source. What is the value of R_x in order for the current in the device to be limited to 0.5 A?

Answer.   90 Ω.   ■ ■

5.4 THE PARALLEL CIRCUIT

When resistances or other circuit elements are connected so that they have the same pair of terminal points or nodes as the resistances of Fig. 5-20(a), the elements are said to be in *parallel*. In the parallel circuit, the voltage drops across the parallel elements are the same, so that in Fig. 5-20(a),

$$E = V_1 = V_2 = V_3$$

Thus a *parallel circuit* may also be defined as a circuit having a common voltage across its elements. The total current I flowing from the battery divides into the individual *branch currents* I_1, I_2, and I_3 at node a; then the branch currents recombine at node b to form the current I flowing back to the battery. By Kirchhoff's current law,

$$I = I_1 + I_2 + I_3 \tag{5-7}$$

If the individual branch currents,

$$I_1 = \frac{E}{R_1} \qquad I_2 = \frac{E}{R_2} \quad \text{and} \quad I_3 = \frac{E}{R_3}$$

are substituted into Eq. 5-7, we obtain

$$I = \frac{E}{R_1} + \frac{E}{R_2} + \frac{E}{R_3}$$

$$I = E\left(\frac{1}{R_1} + \frac{1}{R_2} + \frac{1}{R_3}\right)$$

or

$$I = E(G_1 + G_2 + G_3)$$

The preceding expression indicates that in a parallel circuit, the total or equivalent conductance G_t equals the sum of the conductances, and hence the equivalent

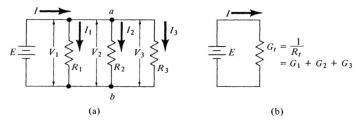

(a)   (b)

Figure 5-20   **(a)** A parallel circuit; **(b)** its equivalent circuit.

circuit of Fig. 5-20(b). In general,

$$G_t = G_1 + G_2 + G_3 + \cdots + G_n \qquad (5\text{-}8)$$

or

$$\frac{1}{R_t} = \frac{1}{R_1} + \frac{1}{R_2} + \frac{1}{R_3} + \cdots + \frac{1}{R_n} \qquad (5\text{-}9)$$

When N resistors of the same value are in parallel,

$$G_t = NG \qquad (5\text{-}10)$$

and

$$R_t = \frac{R}{N} \qquad (5\text{-}11)$$

Notice the advantage of a parallel circuit: if any one of the branches becomes open, the other branches are not affected. Thus, in house wiring, the use of parallel circuits makes it possible to switch any light or appliance on or off without affecting the other circuits.

EXAMPLE 5-8

What total current is supplied by the battery in Fig. 5-21?

Solution. We find the total resistance

$$\frac{1}{R_t} = \frac{1}{10} + \frac{1}{50} + \frac{1}{25} + \frac{1}{12.5} = \frac{12}{50}$$

so that

$$R_t = \frac{50}{12} = 4.16 \ \Omega$$

Then

$$I_t = \frac{E}{R_t} = \frac{100}{4.16} = 24 \ \text{A}$$

Alternatively, the individual currents can be found:

$$I_1 = \frac{100}{10} = 10 \ \text{A} \qquad I_2 = \frac{100}{50} = 2 \ \text{A}$$

$$I_3 = \frac{100}{25} = 4 \ \text{A} \qquad I_4 = \frac{100}{12.5} = 8 \ \text{A}$$

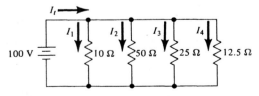

Figure 5-21   Example 5-8.

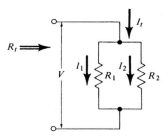

Figure 5-22 Two parallel resistors.

so that

$$I_t = I_1 + I_2 + I_3 + I_4 = 24 \text{ A}$$

It is evident from Example 5-8 that the total resistance (sometimes called the equivalent resistance) of a parallel circuit R_t is less than the resistance of any individual branch. As another example of this, consider five 150 Ω resistors in parallel. By Eq. 5-11,

$$R_t = \frac{R}{N} = \frac{150}{5} = 30 \ \Omega$$

Frequently, a parallel circuit will consist of only two resistors, as in Fig. 5-22. The total resistance in this case may be found from Eq. 5-9,

$$\frac{1}{R_t} = \frac{1}{R_1} + \frac{1}{R_2} = \frac{R_2 + R_1}{R_1 R_2}$$

Using the reciprocal,

$$R_t = \frac{R_1 R_2}{R_1 + R_2} \tag{5-12}$$

we find that Eq. 5-12 states simply that the *total resistance of two resistances in parallel is their product over their sum.*

Equation 5-12 can also be used to simplify a parallel network containing three or more resistances by systematically simplifying just two resistances at a time. This is illustrated in Example 5-9.

EXAMPLE 5-9

Determine the total resistance for the network in Fig. 5-23(a).

Solution. The steps are illustrated by Figs. 5-23(b), 5-23(c), and 5-23(d). First, the 10 Ω resistances are simplified by Eq. 5-11:

$$R_{t1} = \frac{10}{2} = 5 \ \Omega$$

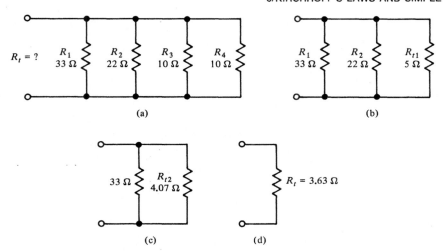

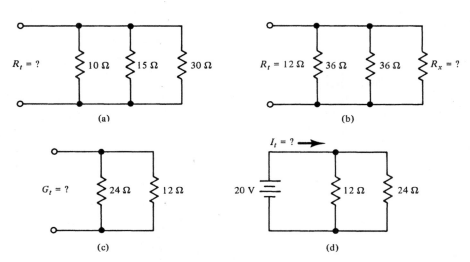

Figure 5-23   Network simplification for Example 5-9.

Then

$$R_{t2} = 5\,\Omega \,\|\, 22\,\Omega = \frac{(5)(22)}{27} = 4.07\,\Omega$$

and

$$R_t = 33\,\Omega \,\|\, R_{t2} = \frac{(33)(4.07)}{37.07} = 3.63\,\Omega$$

where the symbol $\|$ means "in parallel with."

Figure 5-24   Networks for Drill Problem 5-9.

Drill Problem 5-9 ■

Determine the unknown quantities for the networks in Fig. 5-24.

Answer. (a) 5 Ω. (b) 36 Ω. (c) 0.125 S. (d) 2.5 A. ■ ■

5.5 VOLTAGE AND CURRENT DIVISION

Often in analyzing a series circuit it becomes necessary to find the voltage drops across one or more of the resistances. A simple relationship for the voltage drops may be obtained by referring to Fig. 5-15(a). The total current is given by

$$I = \frac{E}{R_1 + R_2 + R_3}$$

and the voltage drops are given by

$$V_1 = IR_1 = E\frac{R_1}{R_1 + R_2 + R_3}$$

$$V_2 = IR_2 = E\frac{R_2}{R_1 + R_2 + R_3}$$

and

$$V_3 = IR_3 = E\frac{R_3}{R_1 + R_2 + R_3}$$

Notice in these relationships that the voltage drop across a resistor is proportional to the ratio of that resistance to the total resistance of the circuit. A general statement called the voltage divider rule can be made as follows:

In a series circuit, the voltage drop across a particular resistor R_n is the line voltage times the fraction R_n/R_t.

$$V_n = E\frac{R_n}{R_t} \tag{5-13}$$

EXAMPLE 5-10

Determine the voltage drops V_1, V_2, and V_3 in the circuit of Fig. 5-25.

Solution.   The total resistance $R_t = 60 + 85 + 55 = 200$ Ω. Then by Eq. 5-13

$$V_1 = E\frac{R_1}{R_t} = 10\left(\frac{60}{200}\right) = 3.0 \text{ V}$$

$$V_2 = E\frac{R_2}{R_t} = 10\left(\frac{85}{200}\right) = 4.25 \text{ V}$$

$$V_3 = E\frac{R_3}{R_t} = 10\left(\frac{55}{200}\right) = 2.75 \text{ V}$$

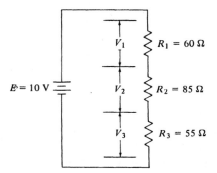

Figure 5-25   Voltage division.

A check by Kirchhoff's voltage law indicates that

$$V_1 + V_2 + V_3 = 3 + 4.25 + 2.75 = 10 \text{ V} = E$$

The special case of two resistors in parallel lends itself to a current division rule. In Fig. 5-22, the total current coming to the parallel combination of R_1 and R_2 divides into the currents I_1 and I_2, respectively. These branch currents are

$$I_1 = \frac{V}{R_1} \quad \text{and} \quad I_2 = \frac{V}{R_2}$$

However, the voltage drop V equals $I_t R_t$, where R_t is given by Eq. 5-12.

$$V = I_t R_t = I_t \frac{R_1 R_2}{R_1 + R_2}$$

If the expression for V is then substituted in the branch current, the expressions I_1 and I_2 are found to be

$$I_1 = I_t \frac{R_2}{R_1 + R_2} \tag{5-14}$$

and

$$I_2 = I_t \frac{R_1}{R_1 + R_2} \tag{5-15}$$

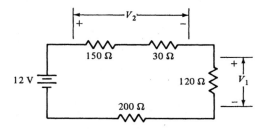

Figure 5-26   Circuit for Example 5-11.

Equations 5-14 and 5-15 are the mathematical statements of the *current divider rule*, which is stated as follows:

With two resistors in parallel, the current in either resistor is the total current times the ratio of the opposite resistor over the sum of the two resistors.

EXAMPLE 5-11

Using voltage division, determine voltages V_1 and V_2 for the circuit of Fig. 5-26.

Solution. $R_t = 150 + 30 + 120 + 200 = 500 \; \Omega$. Then

$$V_1 = E \frac{R_1}{R_t} = 12\left(\frac{120}{500}\right) = 2.88 \text{ V}$$

and

$$V_2 = E \frac{R_2}{R_t} = 12\left(\frac{180}{500}\right) = 4.32 \text{ V}$$

EXAMPLE 5-12

Find the branch currents I_1 and I_2 for the parallel circuit of Fig. 5-27.

Solution. By current division,

$$I_1 = 20\left(\frac{5 \text{ k}\Omega}{5 \text{ k}\Omega + 10 \text{ k}\Omega}\right) = 20\left(\frac{1}{3}\right) = 6.7 \text{ mA}$$

$$I_2 = 20\left(\frac{5 \text{ k}\Omega}{15 \text{ k}\Omega}\right) = 20\left(\frac{2}{3}\right) = 13.3 \text{ mA}$$

In Example 5-12, notice that since R_1 is twice the resistance of R_2, one-half as much current will flow in it as in R_2. Thus in any parallel circuit the ratio between any two branch currents equals the inverse of their resistance ratio.

Drill Problem 5-10 ■

Using voltage or current division determine the unknown quantities in the circuits of Fig. 5-28.

Answer. (a) 4 V; 6 V. (b) 1 A; 4 A; 2 A.   ■ ■

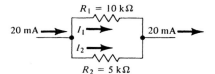

Figure 5-27 Current division.

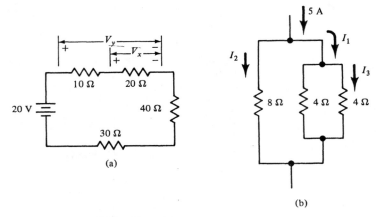

Figure 5-28 Drill Problem 5-10.

5.6 SERIES–PARALLEL CIRCUITS

Circuits that contain both series and parallel combinations are called *series–parallel circuits*. In a series–parallel circuit parts of the circuit contain series elements in which the same current flows, whereas other parts contain parallel elements across which the same voltage occurs.

Some examples of networks containing series–parallel combinations are shown in Fig. 5-29. As with any circuit, we wish to determine the various voltages and currents. Before doing so, we should study the circuit configuration to determine what parts are in series and what parts are in parallel.

EXAMPLE 5-13

Referring to Fig. 5-29, determine which elements are in series and which are in parallel.

Solution. (a) First, notice that elements B and C are connected to the same nodes, have the same voltage, and are in parallel. In turn, that parallel combination is in series with element A.

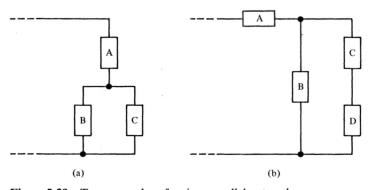

Figure 5-29 Two examples of series–parallel networks.

(b) Notice that elements C and D form a series branch and have the same current. Element B is in parallel with the series branch containing C and D. In turn, that parallel combination is in series with A.

One way of analyzing series–parallel resistance circuits is to begin with the innermost parts and then progressively simplify the circuit to one equivalent resistance. Then the desired voltages and currents are determined by the use of equations including the voltage and current division equations.

Unfortunately, the "starting point" for analysis of series–parallel circuits is as varied as the circuits themselves. However, the following examples provide some suggestions.

EXAMPLE 5-14

Find the total resistance and the currents that flow in the circuit in Fig. 5-30(a).

Solution. Inspection of Fig. 5-30(a) leads one to replace the parallel combination of R_2 and R_3 by R_{t1}, as in Fig. 5-30(b).

$$R_{t1} = R_2 \parallel R_3 = \frac{(8)(12)}{8 + 12} = 4.8 \ \Omega$$

The total resistance presented to the source is found by a second reduction, as in Fig. 5-30(c). Since R_{t1} is in series with R_1,

$$R_{t2} = R_{t1} + 8 \ \Omega = 12.8 \ \Omega$$

The total current I_1 equals

$$I_1 = \frac{E}{R_{t2}} = \frac{24}{12.8} = 1.875 \ \text{A}$$

and by current division

$$I_2 = I_1 \left(\frac{R_3}{R_2 + R_3} \right) = 1.875 \left(\frac{8}{20} \right) = 0.75 \ \text{A}$$

$$I_3 = I_1 \left(\frac{R_2}{R_2 + R_3} \right) = 1.875 \left(\frac{12}{20} \right) = 1.125 \ \text{A}$$

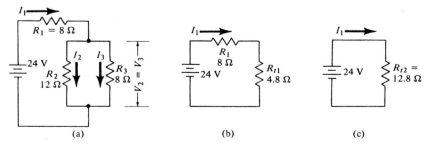

(a) (b) (c)

Figure 5-30 (a) A series–parallel circuit. (b) After the first reduction. (c) After the second reduction.

In the solution of any circuit problem, one will find many alternate ways to the desired results. For instance, having found the total current I_1 in Example 5-14, we do not need to use the current divider rule but can find V_2 and V_3.

$$V_2 = V_3 = I_1 R_{t1} = 1.875(4.8) = 9 \text{ V}$$

Then

$$I_2 = \frac{V_2}{R_2} = \frac{9}{12} = 0.75 \text{ A}$$

and

$$I_3 = \frac{V_3}{R_3} = \frac{9}{8} = 1.125 \text{ A}$$

EXAMPLE 5-15

Find the voltages and currents specified in the circuit of Fig. 5-31(a).

Solution.   Following the reductions shown in Fig. 5-31(b)–(d)

$$R_{t1} = R_4 \| R_5 = \frac{(4)(12)}{16} = 3 \Omega$$

$$R_{t2} = R_{t1} + R_3 = 3 + 6 = 9 \Omega$$

$$R_{t3} = R_2 \| R_{t2} = \frac{(9)(18)}{27} = 6 \Omega$$

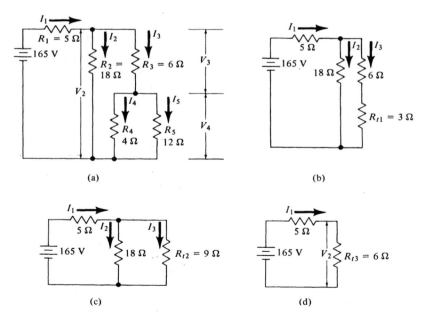

(a)   (b)

(c)   (d)

Figure 5-31 **(a)** Circuit for Example 5-15. **(b)** The circuit after the first reduction. **(c)** The circuit after the second reduction. **(d)** The circuit after the third reduction.

Then

$$I_1 = \frac{E}{R_1 + R_{t3}} = \frac{165}{5 + 6} = \frac{165}{11} = 15 \text{ A}$$

and

$$V_2 = I_1 R_{t3} = 15(6) = 90 \text{ V}$$

so that

$$I_2 = \frac{V_2}{R_2} = \frac{90}{18} = 5 \text{ A}$$

$$I_3 = \frac{V_2}{R_{t2}} = \frac{90}{9} = 10 \text{ A}$$

By voltage division,

$$V_3 = V_2 \left(\frac{R_3}{R_3 + R_{t1}} \right) = 90 \left(\frac{6}{6 + 3} \right) = 60 \text{ V}$$

$$V_4 = V_2 - V_3 = 90 - 60 = 30 \text{ V}$$

Finally,

$$I_4 = \frac{V_4}{R_4} = \frac{30}{4} = 7.5 \text{ A}$$

and

$$I_5 = \frac{V_5}{R_5} = \frac{30}{12} = 2.5 \text{ A}$$

Not all electric circuits can be reduced by series and parallel reductions to a simple circuit of one equivalent resistance connected to a single source of power. Consider the circuit in Fig. 5-32. This circuit, called a *bridge circuit* or configuration, cannot be reduced directly by series and parallel combinations. The bridge circuit and other complex circuits require more elaborate circuit techniques, which are developed in the next chapter.

Drill Problem 5-11 ■

Referring to Fig. 5-33, describe the series–parallel connections of resistances for the circuits shown.

Answer. (a) $(R_1 + R_2) \| R_3$. (b) $R_1 + R_2 + (R_3 \| R_4)$.   ■ ■

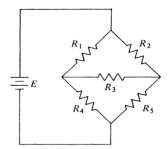

Figure 5-32   Bridge circuit.

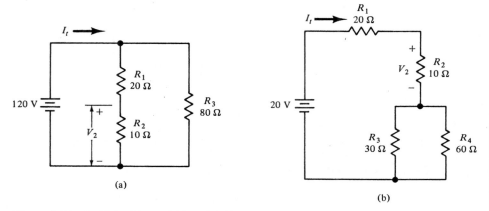

(a)

(b)

Figure 5-33 Drill Problems 5-11 and 5-12.

Drill Problem 5-12 ■

Referring to Fig. 5-33, determine the total current I_t and the voltage V_2 for each of the circuits.

Answer. (a) 3.5 A; 80 V. (b) 0.4 A; 4 V.   ■ ■

5.7 SOURCES AND INTERNAL RESISTANCE

In Chapter 2, the voltage source was presented as an *ideal voltage source*, that is, a source that could supply its rated voltage regardless of the amount of current drawn from the source. This is not the case in reality, because it is found that the terminal voltage of a source, E_t, decreases as the load current drawn from the source increases.

Consider the simple chemical cell that generates an emf E. When the cell is connected to an external circuit, the current to complete its path, must flow through the electrodes and the electrolyte, both of which have a particular electrical resistance. This type of resistance found within a source is called *internal resistance* and is denoted by the symbol R_i. Then the simple cell, and in fact any electrical energy source, is more accurately represented by an ideal voltage source that develops a constant emf E at any current in series with the source internal resistance, as shown in Fig. 5-34.

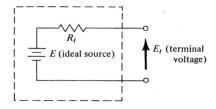

Figure 5-34 A practical source of emf.

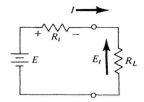

Figure 5-35   A practical source supplying current.

With no current supplied by the source, the voltage drop across R_i is zero and the terminal voltage E_t equals the source emf E. This value of the terminal voltage E is sometimes called the open circuit voltage E_{oc} or *no-load voltage* E_{nl}.

Voltage sources generally have relatively low internal resistances. For example, "fresh" C and D flashlight cells have internal resistances of about 0.47 and 0.27 Ω, respectively. The cell resistance increases with age, since the cell dries out and, with use, the MnO_2 is converted to Mn_2O_3, which has a higher resistance.

If one were to measure the voltage of a practical source, the voltmeter, which normally has a high resistance ($k\Omega$ or $M\Omega$), would draw negligible current so that the voltage measured would essentially be the open-circuit voltage. On the other hand, if a load draws considerable current the internal resistance causes a noticeable voltage drop within the source. Consider then the circuit in Fig. 5-35.

A voltage equation written for the circuit in Fig. 5-35 shows that the terminal voltage $E_t = IR_L$ is less than the emf by the internal voltage drop; that is,

$$E_t = E - IR_i \qquad (5\text{-}16)$$

The voltage versus current characteristic for the ideal and practical voltage sources are shown in Fig. 5-36. The E versus I characteristics of the practical source are shown to be linear in Fig. 5-36, indicative of a linear internal resistance in which the internal resistance equals the slope of the curve. Since the internal resistance may be nonlinear, some sources display a nonlinear E versus I curve.

As an increasing amount of current is taken from a practical source, the voltage drops toward zero until, as in Fig. 5-36, the terminal voltage actually becomes zero and the current is a maximum value I_{sc}. This condition corresponds to that shown

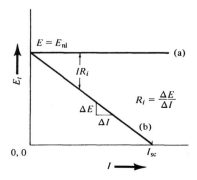

Figure 5-36 E versus I. **(a)** Ideal voltage source. **(b)** Practical voltage source.

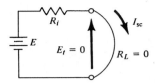

Figure 5-37   A short-circuited voltage source.

in Fig. 5-37, where a very short conductor of essentially zero resistance is connected to the source terminals. The source that is said to be *short-circuited* has zero terminal voltage and a current limited only by the internal resistance, which is called the *short-circuit current* I_{sc}. From Eq. 5-16, the short-circuit current is related to the open-circuit voltage E by

$$I_{sc} = E/R_i \tag{5-17}$$

The voltage variation that exists in going from no-load to a point called full-load is specified by a unitless ratio called the *voltage regulation*, which is defined by

$$\text{V.R.} = \frac{E_{nl} - E_{fl}}{E_{fl}} \tag{5-18}$$

where E_{nl} is the no-load voltage and E_{fl} the full-load voltage.

Notice from Eq. 5-18 that a source approaching the ideal has zero regulation: no change between no load and full load. Voltage regulation is frequently expressed as a percent,

$$\%\text{V.R.} = (\text{V.R.}) \times 100 \tag{5-19}$$

EXAMPLE 5-16

The emf of a source is 45 V. With a load of 140 mA, the terminal voltage falls to 40 V. What is the **(a)** source internal resistance and **(b)** voltage regulation?

Solution

(a) $R_i = \dfrac{E - E_t}{I} = \dfrac{(45 - 40)}{0.14} = \dfrac{5}{0.14} = 35.7 \ \Omega$

(b) $\text{V.R.} = \dfrac{E_{nl} - E_{fl}}{E_{fl}} = \dfrac{(45 - 40)}{40} = \dfrac{5}{40} = 0.125 = 12.5\%$

A primary battery cannot be charged, but a secondary battery *can* be charged. When the battery is charged, current flows into the battery, as indicated in Fig. 5-38(a). Then the terminal voltage must exceed the cell emf, since the internal drop now adds to the emf:

$$E_t = E + IR_i \tag{5-20}$$

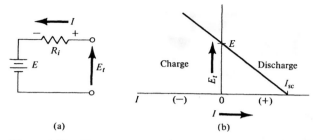

Figure 5-38 The secondary battery **(a)** with charging current. **(b)** E versus I characteristic.

If one considers the discharge current of the battery positive $(+)$ and the charge current negative $(-)$, the $E-I$ characteristic extends into the second quadrant, as indicated in Fig. 5-38(b). Here again, a linear internal resistance is indicated.

A second type of electrical source is the *ideal current source*, which supplies a constant value of current regardless of the load resistance connected to its terminals. The schematic symbol for the ideal current source is shown in Fig. 5-39(a), and its E versus I characteristic in Fig. 5-39(c). The arrow in the current source schematic indicates the direction of conventional current I supplied by the source.

Just as the ideal voltage source supplies a particular current that depends upon the external circuitry, the ideal current source develops a voltage across its terminals that also depends on the external circuitry. For example, a 1-A ideal current source develops 10 V across its terminals when connected to a 10 Ω resistance, and 200 V when connected to a 200 Ω resistance. The polarity of these voltages is such that a voltage drop is developed in the direction of positive current flow in the external resistance.

The inverse or interchanged role that a current source has with respect to a voltage source is known as *duality*. The practical voltage source has a *series internal resistance* that is ideally zero. In a dual manner, the practical current source has a

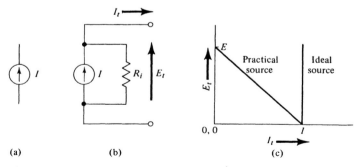

Figure 5-39 **(a)** The ideal current source. **(b)** The practical current source. **(c)** Their E versus I characteristics.

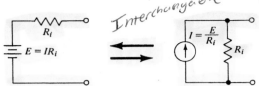

Figure 5-40   Voltage and current source equivalents.

parallel internal conductance G_i that is ideally zero. Alternately, the practical current source is described as having an internal resistance $R_i = 1/G_i$ (ideally it is infinite) that shunts part of the developed current so that part does not reach the source terminals [see Fig. 5-39(b)]. The terminal characteristic E_t versus I_t for the practical current source is shown in Fig. 5-39(c); notice that it resembles that of the practical voltage source.

Two sources are *equivalent* if they produce identical currents in the same resistive load. If the terminals of the practical current source are shorted, the current flowing from the terminals is

$$I_t = I = I_{sc} \tag{5-21}$$

With an open-circuit condition, the current produced by the current source flows through R_i, producing the open-circuit voltage,

$$E_{oc} = IR_i \tag{5-22}$$

Combining Eqs. 5-21 and 5-22, we obtain Eq. 5-17. Thus a current source can be converted to a voltage source and a voltage source to a current source, provided that the proper circuit configuration is used, as noted by Fig. 5-40. This source equivalence is considered further when Thevenin's and Norton's theorems are discussed in Chapter 6.

EXAMPLE 5-17

Convert the sources shown in Figs. 5-41(a) and 5-41(b) to their equivalents.

Solution

(a) $\qquad\qquad I_{sc} = E_{oc}/R_i = 12 \text{ V}/20 = 0.6 \text{ A}$

(b) $\qquad\qquad E_{oc} = I_{sc}R_i = 20 \text{ mA}(3 \text{ k}\Omega) = 60 \text{ V}$

hence, the circuits of Figs. 5-41(c) and 5-41(d).

Earlier it was said that series voltage sources can be combined by simply recognizing that KVL applies. In a dual fashion, we can say that parallel current sources can be combined, in this latter case, by the use of KCL.

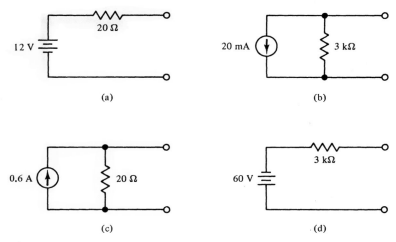

Figure 5-41 **(a), (b)** Sources for Example 5-17; **(c), (d)** Sources for their equivalents.

EXAMPLE 5-18

Determine the voltage equivalent source for the circuit in Fig. 5-42(a).

Solution. First, KCL is applied to the top node:

$$I_t = 5 + 5 - 3 = 7 \text{ A}$$

hence, Fig. 5-42(b). Then,

$$E_{oc} = I_{sc}R_i = 7 \text{ A}(6\Omega) = 42 \text{ V}$$

hence, Fig. 5-42(c).

Drill Problem 5-13 ■

A source with an open-circuit voltage of 13.2 V has a terminal voltage of 9.2 V when a full-load current of 30 A is drawn from the source. Determine **(a)** the source internal resistance, **(b)** the power lost within the source, and **(c)** the voltage regulation.

Answer. (a) 0.133 Ω. (b) 120 W. (c) 43.5%. ■ ■

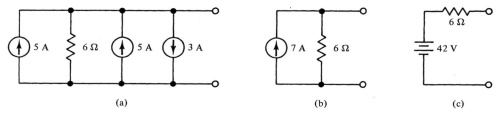

Figure 5-42 **(a)** The network for Example 5-18. **(b), (c)** Two equivalents.

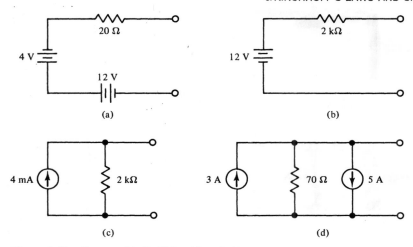

Figure 5-43   Sources for Drill Problem 5-14.

Drill Problem 5-14 ■

Determine the current and voltage equivalent sources for the sources in Fig. 5-43.

Answer. (a) 0.4 A; 20 Ω. (b) 6 mA; 2 kΩ. (c) 8 V; 2 kΩ. (d) 140 V; 70 Ω.   ■ ■

5.8 VOLTAGE DIFFERENCES AND DOUBLE SUBSCRIPTS

It is often necessary to specify the potential difference between two different points or nodes of a circuit. A convenient way of doing this is to use *double-subscript notation.* In accordance with the IEEE standards for semiconductors, the *first subscript designates the point at which the voltage is measured with respect to the reference point designated by the second subscript.* When the reference node is understood, the second subscript may be omitted. For example, in Fig. 5-44 the voltages across the 40 and 20 Ω resistors are by voltage division 20 and 10 V, respectively. Using the double-

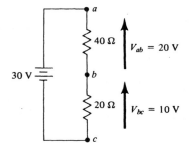

Figure 5-44   Use of double subscripts.

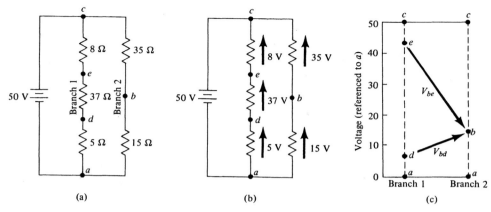

Figure 5-45 (a) A series–parallel circuit. (b) Voltage drops in the circuit. (c) The potential diagram.

subscript notation, the voltage at a referenced to b is $V_{ab} = +20$ V. It follows that if the reference or negative lead of a voltmeter is connected to point b and the positive lead connected to point a, a positive or *upscale* voltage reading results. Conversely, point b is negative with respect to point a so that $V_{ba} = -20$ V; that is,

$$V_{ab} = -V_{ba} \qquad (5\text{-}23)$$

An arrow is sometimes used to indicate the voltage under consideration. Then, as in Fig. 5-44, the head of the arrow is the point of measurement and the tail is the point of reference. Another way of considering the algebraic sign associated with the double-subscript voltage is to consider the voltage V_{ab} to be a voltage drop *from a to b*, so that for Fig. 5-44 $V_{ab} = +20$ V. Then V_{ba} defines a voltage drop *from b to a*, and since the voltage from b to a is a rise or negative drop, $V_{ba} = -20$ V.

Often the potential between two points becomes more apparent if one sketches a graph of the potentials as in the following example:

EXAMPLE 5-19

Determine voltages V_{bd} and V_{be} for the circuit in Fig. 5-45(a).

Solution. By voltage division, the voltage drops are specified as in Fig. 5-45(b). Then with a as a reference point, that is, zero potential, the voltages at points b, c, d, and e are 15, 50, 5, and 42 V, respectively; hence we have the potential diagram of Fig. 5-45(c). Then

$$V_{bd} = +10\,\text{V} \quad \text{and} \quad V_{be} = -27\,\text{V}$$

Drill Problem 5-15 ■

Determine the specified voltages in the circuits of Fig. 5-46.

Answer. (a) $V_{ab} = 7$ V; $V_{ac} = 4$ V; $V_{bc} = -3$ V. (b) 56 V.   ■ ■

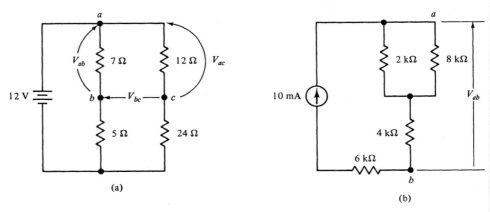

Figure 5-46   Circuits for Drill Problem 5-15.

5.9 VOLTAGE DIVIDERS

A voltage divider is a series combination of resistors chosen so that a single voltage source can supply one or more reduced voltages. A voltage divider that will provide a supply voltage E and two reduced voltages is shown in Fig. 5-47. In practice, resistive loads are connected between the reference point or ground (taken as the negative terminal in Fig. 5-47) and the voltage terminals or *taps A, B,* and *C*. If these loads or circuits require negligible current, that is, the voltage taps supply effectively zero current, the divider is said to be *unloaded.* Then the only current present is the *bleeder current I,* the current confined to the divider network, which in this case equals the supply voltage divided by the total resistance.

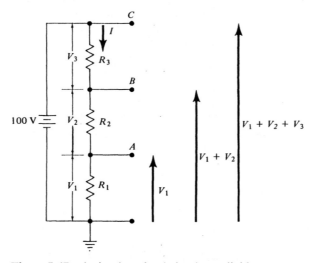

Figure 5-47   A simple unloaded voltage divider.

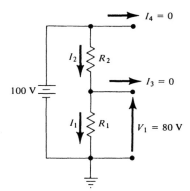

Figure 5-48 An example of a voltage divider.

As an example, suppose that voltages of 100 and 80 V are to be provided by a 100 V source. The divider network requires two resistors as in Fig. 5-48. Then if the bleeder current is arbitrarily chosen as 20 mA, the resistances R_1 and R_2 are calculated by

$$R_1 = \frac{80\ \text{V}}{20\ \text{mA}} = 4\ \text{k}\Omega$$

$$R_2 = \frac{100 - 80}{20\ \text{mA}} = \frac{20\ \text{V}}{20\ \text{mA}} = 1\ \text{k}\Omega$$

But consider what happens if the load current I_3 supplied by the divider in Fig. 5-48 changes so that it is no longer negligible. The current I_2 flowing in R_2 previously was only the bleeder current but now equals the sum of I_1 and I_3. The voltage drop across R_2 increases, thereby decreasing the voltage drop V_1 from the desired 80 V.

If one has an idea of the load currents to be supplied, one can account for these currents and hence design a *loaded voltage divider*. Again, the bleeder current is that minimum current that is confined to the divider network. If not specified, it is usually taken as 10% of the total load currents to be supplied. This rule of thumb for bleeder current provides a compromise; a larger value of bleeder current results in an excessive power loss in the divider and a smaller value results in poorer voltage regulation should the load currents change, as they invariably do. The design of the loaded voltage divider proceeds as in the following example.

EXAMPLE 5-20

A 14 V source supplies 20 mA at 3 V, 30 mA at 9 V, and 50 mA at 12 V. Design the voltage divider using a bleeder current equal to 10% of the load currents.

Solution. The design requires four resistors, as shown in Fig. 5-49. The total load current is $20 + 30 + 50 = 100$ mA. Then the bleeder current is 10% (100 mA) = 10 mA. Since I_1 is the minimum current confined to the divider net-

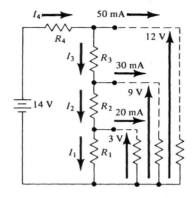

Figure 5-49   Example 5-20.

network, $I_1 = 10$ mA so that

$$R_1 = \frac{V_1}{I_1} = \frac{3 \text{ V}}{10 \text{ mA}} = 300 \text{ }\Omega$$

Solving for R_2, we find that

$$I_2 = I_1 + 20 \text{ mA} = 30 \text{ mA}$$
$$R_2 = \frac{9 \text{ V} - V_1}{I_2} = \frac{6 \text{ V}}{30 \text{ mA}} = 200 \text{ }\Omega$$

Solving for R_3, we obtain

$$I_3 = I_2 + 30 \text{ mA} = 60 \text{ mA}$$
$$R_3 = \frac{12 \text{ V} - 9 \text{ V}}{60 \text{ mA}} = \frac{3 \text{ V}}{60 \text{ mA}} = 50 \text{ }\Omega$$

and finally

$$I_4 = I_3 + 50 \text{ mA} = 110 \text{ mA}$$
$$R_4 = \frac{14 \text{ V} - 12 \text{ V}}{110 \text{ mA}} = \frac{2 \text{ V}}{110 \text{ mA}} = 18.2 \text{ }\Omega$$

Calculation of the power requirements of the divider resistors completes the solution.

$$P_1 = V_1 I_1 = (3 \text{ V}) (10 \text{ mA}) = 0.030 \text{ W}$$
$$P_2 = V_2 I_2 = (6 \text{ V}) (30 \text{ mA}) = 0.180 \text{ W}$$
$$P_3 = V_3 I_3 = (3 \text{ V}) (60 \text{ mA}) = 0.180 \text{ W}$$
$$P_4 = V_4 I_4 = (2 \text{ V}) (110 \text{ mA}) = 0.220 \text{ W}$$

In the preceding example, the zero potential point or ground was taken as the negative terminal of the voltage source. In some electronic applications, both negative and positive voltages with respect to ground are desired. These dual polarity voltage

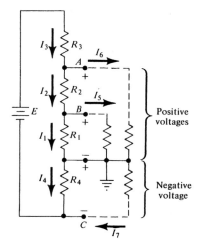

Figure 5-50 A dual polarity voltage divider.

dividers are obtained using a ground or zero potential tap (see Fig. 5-50). The positive voltages are measured at points A and B referenced to the ground point, and a negative voltage is measured at point C referenced to ground. Notice that the currents in the positive portion of the divider (I_1, I_5, and I_6) combine at the ground node, and their sum equals the sum of the currents that leave the ground node to flow in the negative portion of the network. The bleeder current is then calculated from the positive load currents only.

5.10 SWITCHES AND PROTECTIVE DEVICES

Up to this point, the circuits have been somehow connected prior to our investigation of them. As a practical matter, however, we need to be able to complete a circuit, break a circuit, or, in some cases, change the connections.

A device to make, break, or change electrical connections is called a *switch*. Switches can be operated manually, electronically, or automatically using pressure, temperature, or other effects.

Many classifications are used in describing switches. One method of classification is by the mechanics of the switching mechanism, for which we have descriptive terms such as pushbutton, rocker-button, toggle, rotary, slide, thumbwheel, and membrane.

Switches are also classified according to the connections that are made or broken. The schematic symbols for a few different types of make/break arrangements are shown in Fig. 5-51. The switch nomenclature used to describe the switch arrangements in Fig. 5-51 is

SPST—single-pole, single-throw

SPDT—single-pole, double-throw

DPST—double-pole, single-throw

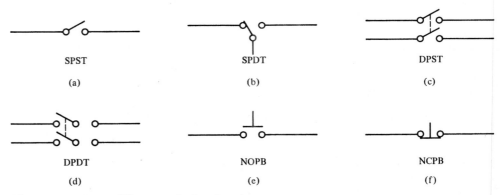

SPST	SPDT	DPST
(a)	(b)	(c)
DPDT	NOPB	NCPB
(d)	(e)	(f)

Figure 5-51 Some different make/break switch arrangements.

DPDT—double-pole, double-throw

NOPB—normally opened, pushbutton

NCPB—normally closed, pushbutton

A few switch types from your author's collection are shown in Fig. 5-52.

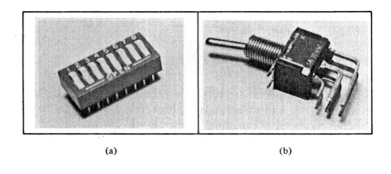

(a)	(b)

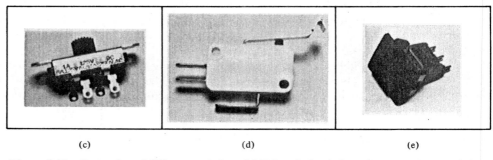

(c)	(d)	(e)

Figure 5-52 Examples of different switches. **(a)** Printed circuit board mounted DIP, eight position, SPST; **(b)** Printed circuit board mounted toggle, DPDT; **(c)** slide, DPST; **(d)** momentary lever, SPDT; **(e)** momentary pushbutton, SPDT.

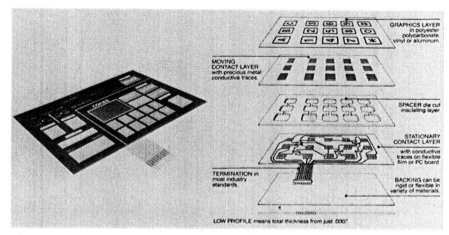

Figure 5-53 A membrane keyboard and typical membrane keyboard construction. (Courtesy of Cherry Electrical Products Corporation.)

A particularly thin and lightweight switch found on home appliances, office equipment, and computers is the membrane switch. Groups of these switches are usually combined into a unit called a membrane keyboard. One such keyboard and the typical membrane keyboard construction are shown in Fig. 5-53. Notice that as proper pressure is applied to the top layer, the moving contact layer makes contact with the stationary contact layer and closes the appropriate circuit.

An electric circuit should be provided with some type of protection from excessive currents due to device malfunction or accidental short circuit. Two devices that offer protection against excessive currents are fuses and circuit breakers.

As the name implies, a *fuse* consists of a wire or strip of fusible metal that melts when the current becomes excessive. In contrast, a *circuit breaker* is actually a manual switch that can open itself automatically under overload conditions. In a circuit breaker the overload condition is generally sensed by a thermal element. On overload the thermal element bends or deforms and by this action or in combination with other means, such as magnetic means, opens the protected circuit. As noted in an earlier part of this chapter, fuses or circuit breakers are inserted in series with the circuit being protected. The symbols for a fuse and a circuit breaker are shown in Fig. 5-54. Examples of ferrule fuses and a circuit breaker are shown in Fig. 5-55. Although the two fuses are in similar glass tubes, the lower fuse has a single fusible element and the upper fuse has a dual element. The dual element design provides a time delay to allow for the high initial currents that characterize motors, inductors, and transformers.

Fuse   Circuit breaker

Figure 5-54 Fuse and circuit-breaker symbols.

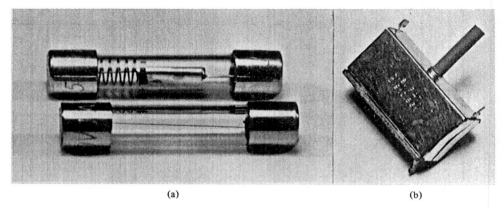

Figure 5-55 **(a)** Two ferrule fuses and **(b)** a circuit-breaker designed for chassis mounting.

QUESTIONS

1. What is a node?
2. Distinguish between a minor and major node.
3. What is a branch? What is a series branch?
4. What is the difference between an electric network and an electric circuit?
5. What is meant by the term "loop"?
6. What do we mean by the terms "open circuit" and "short circuit"?
7. What is Kirchhoff's voltage law?
8. State Kirchhoff's current law.
9. What is meant by a series circuit?
10. What is meant by a parallel circuit?
11. What is an advantage of a parallel circuit?
12. Explain the principles of voltage and current division.
13. What is a series–parallel circuit?
14. Sketch the bridge configuration.
15. Distinguish between an ideal and a practical voltage source.
16. Sketch the V–I characteristics for the ideal and practical voltage sources.
17. Define internal resistance.
18. What is meant by open-circuit voltage and short-circuit current?
19. Define voltage regulation.
20. Distinguish between an ideal and a practical current source.
21. Sketch the V–I characteristics for the ideal and practical current sources.
22. Discuss the use of double-subscript notation.
23. What is meant by bleeder current?
24. How can positive and negative voltages be developed simultaneously by a voltage divider?

25. What is a switch?

26. Explain the difference between a fuse and a circuit breaker.

PROBLEMS

Draw the circuit diagram for those problems where none is given.

1. Determine the number of branches and nodes for the circuits in Fig. 5-56.

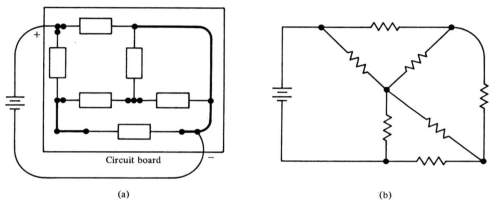

(a)   (b)

Figure 5-56   Problem 5-1.

2. Identify all the possible loops in the circuit in Fig. 5-57.

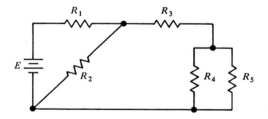

Figure 5-57   Problem 5-2.

3. Determine the unknown voltages for the circuits in Fig. 5-58.

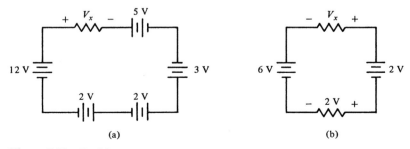

(a)   (b)

Figure 5-58   Problem 5-3.

4. Using KVL equations, solve for the unknown voltages V_x and V_y for each circuit in Fig. 5-59.

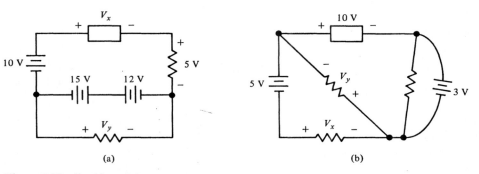

(a)   (b)

Figure 5-59   Problem 5-4.

5. Find the unknown currents for the networks in Fig. 5-60.

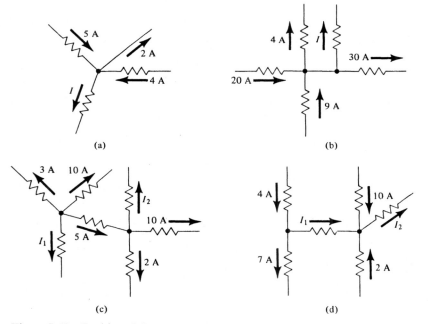

(a)   (b)

(c)   (d)

Figure 5-60   Problem 5-5.

6. What is the total resistance and current flow in each of the networks in Fig. 5-61.

7. Determine the unknown quantities for the circuits in Fig. 5-62.

8. A light-emitting diode is to be connected in series with a resistance so that when the combination is connected to 5 V the diode voltage and current will be 1.6 V and 20 mA, respectively. Determine the unknown resistance and its power dissipation.

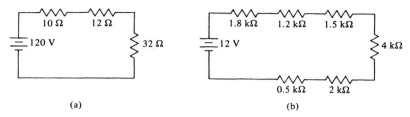

Figure 5-61 Problem 5-6.

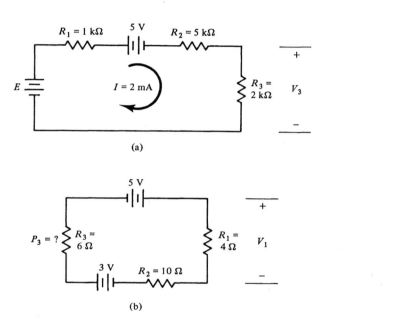

Figure 5-62 Problem 5-7.

9. A 20 kΩ resistor and a 35 kΩ resistor are connected in series to a 12 V source. Determine the total current, the total power dissipated, and the voltage across each resistor.

10. Determine the resistance and the power dissipation of a resistor that must be placed in series with a 100 Ω resistor and a 120 V source in order to limit the power dissipation in the 100 Ω resistor to 90 W?

11. A series string of Christmas tree lights consists of eight 6 W lamps. If the string of lights is designed for use on a 120 V source, what current flows and what is each lamp resistance?

12. Two heaters are each rated at 1 kW and 220 V. If it is assumed that the heater resistance remains constant, what is the total dissipated power when the two are connected in series across 220 V?

13. An electric meter has a resistance of 20 Ω and produces a maximum needle deflection with 10 mA flowing through the meter. What resistance must be connected in series with the meter so that maximum needle deflection occurs when the series combination is connected to 150 V?

14. Two resistors are in series: a 9.2 kΩ resistor rated at 1 W and a 5.1 kΩ resistor rated at $\frac{1}{2}$ W. What maximum current can safely flow in the circuit? What maximum voltage can safely be supplied to the combination?

15. An appliance outlet is connected to a 120 V source by 50 ft of two-conductor aluminum cable of 12 gauge wire. If a load of 20 A is connected to the outlet, what is the voltage at the outlet?

16. The shunt field of a motor takes 1.8 A when connected to a 120 V line. What resistance must be added in series to limit the field current to 1.2 A?

17. A load of 10 kW is connected to a 230 V source by a two-conductor cable of number 8 gauge copper wire. If the voltage at the load terminals is only 215 V, how far is the load from the source?

18. Find the total resistance and the total current flow in each of the circuits of Fig. 5-63.

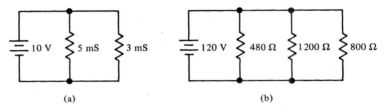

(a)   (b)

Figure 5-63   Problem 5-18.

19. Resistors of 1, 2, 5, 10, and 20 Ω are connected in parallel across a 100 V source. What is the current in each resistor, the total current, and the total resistance of the parallel combination?

20. The equivalent resistance of two parallel resistors is 400 Ω. If one of the resistors is 1000 Ω, what is the value of the second resistor?

21. Five 1.5 MΩ resistors are in parallel. What resistance value is a sixth resistor if it is placed in parallel with the others to make a total of 200 Ω?

22. A total current of 10 mA flows to three resistors, of 47, 56, and 82 kΩ, connected in parallel. What is the branch current and the voltage drop across the resistors?

23. If three lamps (of 60, 40, and 25 W) are connected in parallel to 120 V, what is their total resistance and what total current flows?

24. An electric meter has a resistance of 20 Ω and produces a maximum needle deflection with 10 mA flowing through the meter. What resistance must be connected in parallel with the meter so that maximum needle deflection occurs when 100 mA flows into the parallel combination?

25. An electric circuit in a house has a 100 W lamp, a 1100 W toaster, and a 240 W refrigerator connected in parallel across a 110 V line. Find (a) the current drawn by each device, (b) the total resistance, and (c) the total current.

26. A 12 V source delivers 9.2 A to three resistors connected in parallel. One resistor passes 2.4 A. If the two other resistors each pass the same current, what is their value?

27. Three resistances of 4, 6, and R Ω are connected in parallel and have a total conductance of 0.6 S. What is the value of R?

28. Use voltage division or the current division relationship to solve for the indicated voltages or currents for the circuits in Fig. 5-64.

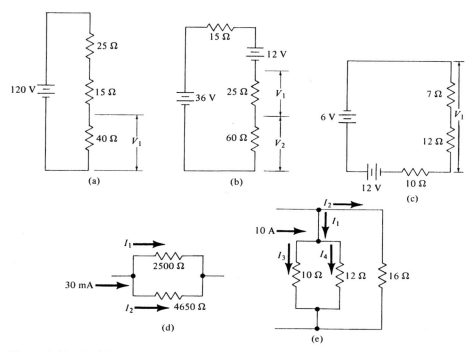

Figure 5-64   Problem 5-28.

29. Find the equivalent resistance for each of the networks in Fig. 5-65.

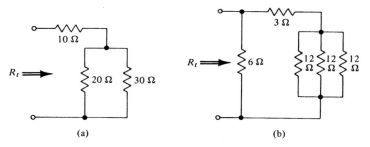

Figure 5-65   Problem 5-29.

30. Determine the unknown resistances for the resistance networks in Fig. 5-66. [Network (b) looks like a bridge network, but with nodes *a* and *b* connected together, the circuit is really a series–parallel combination.]

31. Solve for the voltages V_x and V_y in Fig. 5-67.

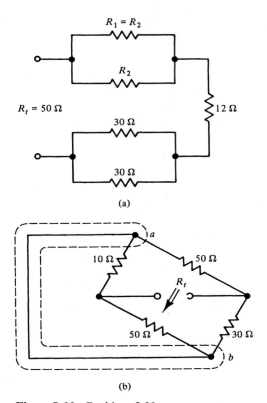

(a)

(b)

Figure 5-66   Problem 5-30.

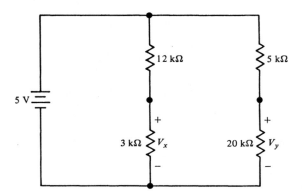

Figure 5-67   Problem 5-31.

32. Solve for the indicated voltages and currents for the circuits in Fig. 5-68.

33. The lamp shown in Fig. 5-69 is rated at 12 V, 0.30 A. What must the supply voltage E be for the lamp to be operated at the rated value?

34. If the voltage across the 10 Ω resistor in Fig. 5-70 is 24 V, what is the value of R?

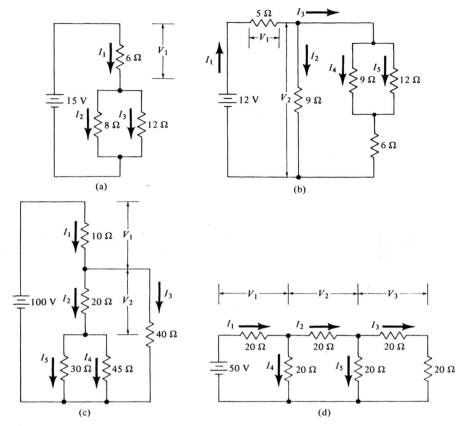

Figure 5-68   Problem 5-32.

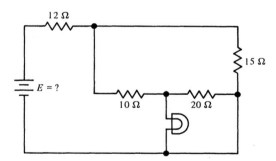

Figure 5-69   Problem 5-33.

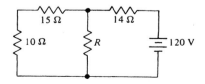

Figure 5-70   Problem 5-34.

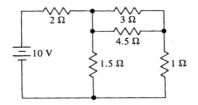

Figure 5-71   Problem 5-35.

35. How much power is dissipated by the 1 Ω resistor in Fig. 5-71?

36. Multiply Eq. 5-16 by the current I and explain the significance of this new equation.

37. A 30 V source has an internal resistance of 2 Ω. Find the terminal voltage when a 10 Ω load resistor is connected.

38. A battery with an open-circuit voltage of 1.58 V develops a 30 A short-circuit current. **(a)** What is the battery internal resistance? **(b)** What is the terminal voltage when 5 A is supplied by the battery?

39. If the efficiency of a practical voltage source is the power developed by the ideal portion of the source divided into the power delivered to the load, what is the efficiency for the battery of Problem 38 when 5 A is supplied?

40. A certain voltage source delivers 12 V at 3 A. When connected to a different load, it delivers 11.2 V at 8 A. What is the internal resistance?

41. A 12 V supply is used to charge a 6 V battery having a 0.8 Ω internal resistance. What series resistance is necessary to limit the charging current to 600 mA?

42. A 15 A current source has an internal resistance of 2 Ω. Find the terminal voltage and voltage regulation when a 10 Ω resistor is connected.

43. Convert the practical sources of Fig. 5-72 to their current or voltage equivalent.

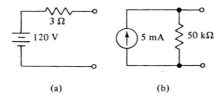

(a)   (b)   **Figure 5-72**   Problem 5-43.

44. Convert the voltage source networks of Fig. 5-73 to their current source equivalents.

45. Convert the current source networks of Fig. 5-74 to their voltage source equivalents.

46. A practical source delivers 5 A at 6 V and 10 A at 4 V. Determine the practical voltage and current sources that have these characteristics.

47. Determine the voltage V_{ab} in Fig. 5-75.

48. Determine voltages V_{ab} and V_{bc} in Fig. 5-76.

49. A source of 24 V is used in a transistor power amplifier to supply the following loads: 12 V at 3 A and 6 V at 1 A. By using a bleeder current of $\frac{1}{2}$ A, design the required voltage divider network. Include the power ratings of the resistors.

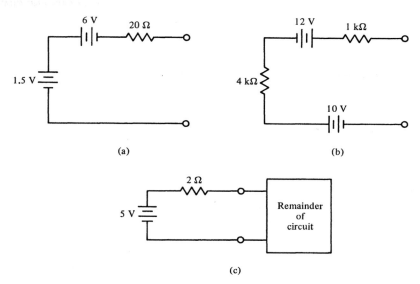

(a)

(b)

(c)

Figure 5-73 Problem 5-44.

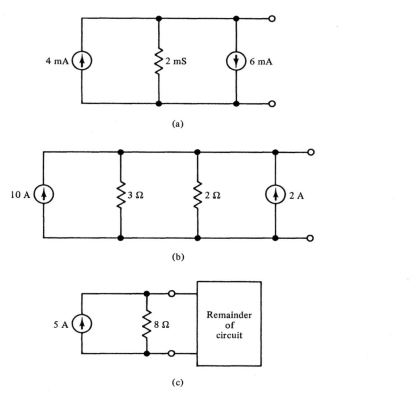

(a)

(b)

(c)

Figure 5-74 Problem 5-45.

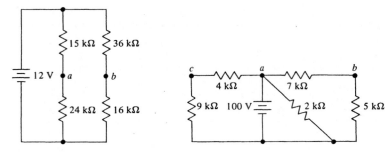

Figure 5-75   Problem 5-47.   **Figure 5-76**   Problem 5-48.

50. Design a loaded voltage divider to be used with a 300 V source to supply the following loads: 75 V at 20 mA, 120 V at 30 mA, and 300 V at 100 mA. Use the 10% rule for bleeder current. Find the required power ratings for the resistors.

51. Given a 24 V source, use the 10% rule for bleeder current and design a voltage divider to supply the following loads: + 18 V at 4 mA; + 12 V at 10 mA; − 6 V at 10 mA.

Chapter 6

Circuit Analysis Techniques

6.1 INTRODUCTION

It became obvious in the previous chapter that some networks, such as the bridge circuit, might require more sophisticated techniques of analysis than those applied to ordinary series–parallel circuits. This need for other methods of analyzing networks becomes more evident as one studies the various electronic and electric-power systems that contain more than one source of energy. In an earlier chapter, a node was defined as a junction point for two or more circuit elements. Thus, as a reminder, points *A*, *B*, and *C* of Fig. 6-1 are nodes. However, we previously defined a node such as node *B* as a *minor* or *trivial node* because the KCL equation relates that the two elements connected to that node have the same current. Of course, that relationship should be already obvious from inspection of the network.

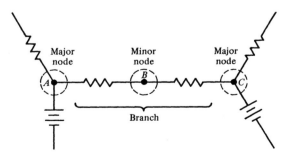

Figure 6-1   An electric network.

A *major node*, such as node A or node C of Fig. 6-1, connects at least three elements and thus relates at least three currents. The current relationship is not obvious, and a KCL equation is most valuable.

In the last chapter we defined a branch as the path between two nodes and expanded that definition to include the series connection of elements. In this larger context, a *branch* defines the path, whether a single element or series string of elements, between two major nodes. Some circuit analysts refer to a branch containing a group of series elements as a *major branch*.

We call a circuit *complex* if there are two or more energy sources in different branches of the circuit, or if the circuit cannot be reduced entirely by series–parallel combinations. In the study of complex circuits, we wish to determine all the branch currents and node voltages. This requires that we apply Ohm's law, KVL, and KCL.

In general, a circuit will possess b branches and n major nodes. For example, the circuit of Fig. 6-2 has three branches (b_1, b_2, and b_3) and two major nodes (n_1 and n_2).

If we obtain the branch current values for a circuit such as the one in Fig. 6-2, we can apply Ohm's law to obtain the node voltage values. Comparably, if we obtain the node voltage values, we can apply Ohm's law to obtain the branch currents. In either case we have enough information about the circuit to describe it electrically.

We begin an analysis of an electric circuit by counting the unknown branch currents and unknown node voltages in the circuit. Then we write enough simultaneous equations based on KVL and KCL to obtain a solution to the circuit. Mathematically, each of the equations must be independent, that is, not obtainable from any other equation or from any combination of the other equations. However, you need not concern yourself with this requirement in this chapter. The procedures presented here will provide you with a set of simultaneous equations that you will solve by substitution, determinants, or Gaussian elimination. The determinant solution of a set of equations is discussed in the Appendix. Some computer solutions are also presented there.

In this chapter three methods of analysis involving the use of simultaneous equations are presented. These three methods are generalized or branch analysis, loop analysis, and nodal analysis. In addition, a number of other circuit analysis procedures and theorems are presented.

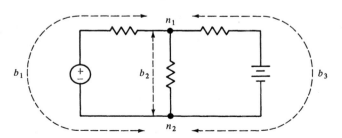

Figure 6-2   Branches and nodes of a circuit.

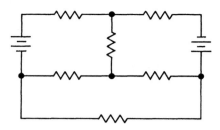

Figure 6-3   Circuit for Drill Problem 6-1.

Drill Problem 6-1 ■

Determine the number of branches and major nodes for the circuit of Fig. 6-3.

Answer.   6, 4.   ■ ■

6.2 GENERALIZED KIRCHHOFF ANALYSIS

We begin our analysis of complex circuits with the *generalized Kirchhoff* or *branch current method*. We use both KVL and KCL to obtain enough independent equations to solve for the unknown branch currents. If there are b branches, there are b unknown currents, and hence b independent equations to be determined.

The steps one takes in solving a circuit by the generalized Kirchhoff or branch current method are the following:

1. Assume unknown currents in each branch and indicate their directions by arrows. (The choice of directions is arbitrary.)
2. Assign polarity marks to indicate the voltage drop or rise that a particular current causes in passing through a circuit element.
3. Write node equations at $n - 1$ nodes, where n equals the number of major nodes. (Equations are written at all but one node, which is considered a reference node. Although the choice is arbitrary, the reference node is often chosen as the one at the bottom of the circuit or the one connecting the largest number of branches.)
4. Write voltage equations for $b - n + 1$ closed loops, where b is the total number of branches. (The loops are successively chosen such that each new loop includes at least one new branch not previously included in the path of a loop. Here it is sufficient to choose all those loops forming topological "windows," that is, those loops forming closed contours and having no lines within.)
5. Solve the simultaneous equations for the unknown currents.

EXAMPLE 6-1

Solve for the branch currents of the circuit of Fig. 6-4(a).

Solution.   First notice that the circuit contains three branches so that three unknown currents are assigned, as indicated in Fig. 6-4(b).

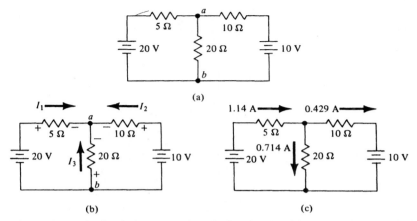

Figure 6-4 (a) Circuit for Example 6-1. (b) Circuit with assumed currents. (c) Actual currents.

Since there are two major nodes, one node, node a, is chosen to supply an equation. Calling currents into the node positive, we find that

$$I_1 + I_2 + I_3 = 0 \qquad (6\text{-}1)$$

Two voltage equations are then written, first for the left loop. Starting at the positive side of the 20 V battery, proceeding clockwise, and calling voltage drops positive, we obtain

$$5I_1 - 20I_3 - 20 = 0$$

Starting at the positive side of the 10 V battery in the right loop, proceeding clockwise, and again calling voltage drops positive, we get

$$10 + 20I_3 - 10I_2 = 0$$

Each of the voltage equations can be divided by a common factor and rewritten as

$$I_1 - 4I_3 = 4 \qquad (6\text{-}2)$$

$$-I_2 + 2I_3 = -1 \qquad (6\text{-}3)$$

Equations 6-1–6-3 can be solved by determinants with the results

$$D = \begin{vmatrix} 1 & 1 & 1 \\ 1 & 0 & -4 \\ 0 & -1 & 2 \end{vmatrix} = -1 - 4 - 2 = -7$$

$$D_1 = \begin{vmatrix} 0 & 1 & 1 \\ 4 & 0 & -4 \\ -1 & -1 & 2 \end{vmatrix} = -4 + 4 - 8 = -8$$

$$D_2 = \begin{vmatrix} 1 & 0 & 1 \\ 1 & 4 & -4 \\ 0 & -1 & 2 \end{vmatrix} = 8 - 1 - 4 = 3$$

$$D_3 = \begin{vmatrix} 1 & 1 & 0 \\ 1 & 0 & 4 \\ 0 & -1 & -1 \end{vmatrix} = 1 + 4 = 5$$

$$I_1 = \frac{D_1}{D} = \frac{-8}{-7} = 1.144 \text{ A}$$

$$I_2 = \frac{D_2}{D} = \frac{3}{-7} = -0.429 \text{ A}$$

$$I_3 = \frac{D_3}{D} = \frac{5}{-7} = -0.714 \text{ A}$$

The minus signs for I_2 and I_3 only indicate that those currents actually flow in directions opposite to the directions initially chosen. The actual currents are as shown in Fig. 6-4(c).

6.3 LOOP ANALYSIS

For a given network, the number of simultaneous equations to be solved may be reduced by *loop analysis*. The loop method of analysis is based on writing Kirchhoff's voltage equations only. In the last chapter a *loop* was defined as any closed electrical path. Thus, in the circuit of Fig. 6-5 it is possible to identify three loops: *a*, *b*, and *c*. With loop analysis the currents chosen as unknowns are circulating currents rather than branch currents, where a *circulating* or *loop current* flows throughout a closed loop. Since each circulating current leaves each node it enters, the Kirchhoff

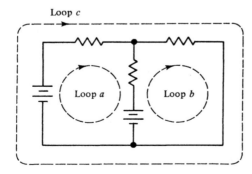

Figure 6-5 A circuit with the possible loops for which KVL equations may be written.

current equations are automatically satisfied; therefore, only the voltage equations are necessary.

The question then arises, How many and which loops must be used in writing KVL equations? The answer, of course, is just enough loops to develop a complete set of independent voltage equations. *We can be sure that the proper number of independent KVL equations are chosen if the loops are successively chosen so that each new loop contains at least one new element not contained in a previously chosen loop.* Applying this rule to the circuit of Fig. 6-5, we recognize three possible sets of loops for a necessary and sufficient set of KVL equations. The sets are loops *a* and *b*, loops *a* and *c*, and loops *b* and *c*. These are indicated in Fig. 6-6.

A *mesh* is a closed loop that contains no other closed loops. Thus, in Fig. 6-6 loops *a* and *b* are meshes. Meshes are the topological "windows" of a circuit, and for the circuits we shall consider it is sufficient to use the equations for the meshes as our loop equations.

Circuit analysts often refer to loop analysis involving only mesh equations as *mesh analysis*. With mesh analysis, the unknown mesh currents are generally selected all in one direction, that is, clockwise; and all current sources are converted to their voltage equivalents. However, for the time being we shall not be so restrictive.

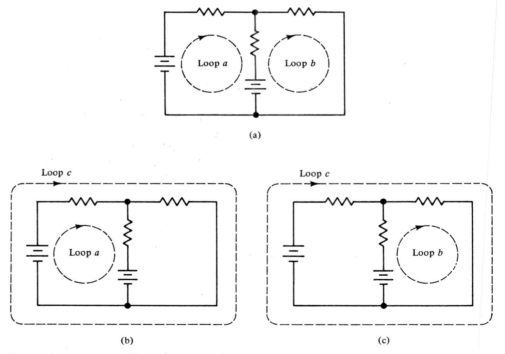

Figure 6-6 The sets of loops that will yield a sufficient set of KVL equations for the circuit of Fig. 6-5.

The steps used in solving a circuit by loop analysis are as follows:

1. Determine the independent loops to be used. The meshes constitute a set of independent loops.
2. Assume a positive current flow in each closed loop. The currents can be arbitrarily chosen to flow in either a clockwise or a counterclockwise direction.
3. Assign polarity marks to indicate the voltage drop or rise that a particular loop current causes in passing through a circuit element. If a circuit element is shared between two loops, it will have two loop currents flowing through it and two individual voltage rises or drops.
4. Start at one point of a loop and follow the loop current around that particular loop while algebraically adding the voltage drops and rises as the loop is traversed.
5. Solve the simultaneous equations for the unknown loop currents.

EXAMPLE 6-2
Solve for the currents shown in the circuit of Figure 6-7(a).

Solution. The loop currents and resultant polarity marks are shown in Fig. 6-7(b). Considering the rises to be positive and following in the same direction as the loop currents, we obtain two equations,

$$\text{loop } a: \quad 10 - 10I_a - 30I_a + 30I_b = 0$$
$$\text{loop } b: \quad -10 + 30I_a - 30I_b - 20I_b = 0$$

Simplified, the two equations are

$$4I_a - 3I_b = 1$$
$$3I_a - 5I_b = 1$$

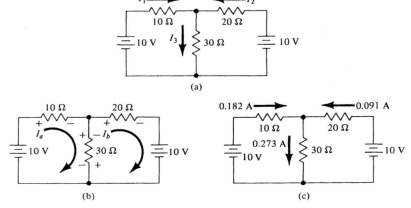

(a)

(b)

(c)

Figure 6-7 (a) Circuit for Example 6-2. (b) Circuit with assumed currents. (c) Actual currents.

The two equations can then be solved for the unknown loop currents I_a and I_b.

$$I_a = \begin{vmatrix} 1 & -3 \\ 1 & -5 \end{vmatrix} \bigg/ \begin{vmatrix} 4 & -3 \\ 3 & -5 \end{vmatrix} = -2/-11 = 0.182 \text{ A}$$

$$I_b = \begin{vmatrix} 4 & 1 \\ 3 & 1 \end{vmatrix} \bigg/ D = 1/-11 = -0.091 \text{ A}$$

Solving for the branch currents, we find that

$$I_1 = I_a = 0.182 \text{ A}$$
$$I_2 = -I_b = -(-0.091) = 0.091 \text{ A}$$
$$I_3 = I_a - I_b = 0.182 - (-0.091) = 0.273 \text{ A}$$

Notice that if the branch current method had been used in Example 6-2, three equations, one for each branch, would have been required. However, only two were required with the loop analysis method. The smaller number of equations needed to solve a given problem by loop analysis becomes more apparent with increasing circuit complexity.

EXAMPLE 6-3

Determine the power dissipated in the 3 Ω resistance of Fig. 6-8(a). Use loop analysis to solve for the current in that resistance.

Solution. With loop currents I_a and I_b chosen as in Fig. 6-8(b) we obtain the equations,

$$\text{loop } a: \qquad 7I_a - 5I_b = 2$$
$$\text{loop } b: \qquad -5I_a + 8I_b = 5$$

The equations can be solved simultaneously with the result,

$$I_b = \begin{vmatrix} 7 & 2 \\ -5 & 5 \end{vmatrix} \bigg/ \begin{vmatrix} 7 & -5 \\ -5 & 8 \end{vmatrix} = 45/31 = 1.45 \text{ A}$$

Since $I = I_b = 1.45$ A, then $P = I^2R = (1.45)^2(3) = 6.31$ W.

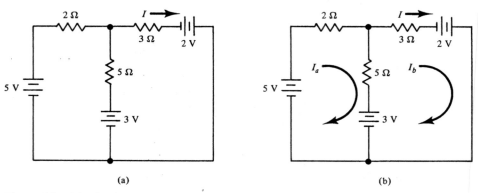

(a)   (b)

Figure 6-8   (a) Circuit for Example 6-3. (b) Circuit with assumed loop currents.

EXAMPLE 6-4

Determine the loop equations when the loops are meshes with currents chosen as in Fig. 6-9.

Solution. With currents chosen as in the figure, we obtain the KVL equations.

$$\text{loop } a: \quad 30I_a - 10I_b + 0I_c = 10$$
$$\text{loop } b: \quad -10I_a + 55I_b - 20I_c = 0$$
$$\text{loop } c: \quad 0I_a - 20I_b + 35I_c = -20$$

Notice in Example 6-4 that the mesh currents were all in the same direction. The system determinant for this set of equations is

$$\Delta = \begin{vmatrix} 30 & -10 & 0 \\ -10 & 55 & -20 \\ 0 & -20 & 35 \end{vmatrix}$$

principal diagonal

Note that in the system determinant the terms along the principal diagonal are positive and each equals the sum of the resistances in the appropriate mesh. Also, there is symmetry about the principal axis, with the symmetrical terms being negative. For bilateral resistances, the system determinant will always be symmetrical about the principal diagonal. Moreover, if we use mesh currents, all of which are defined in the same direction, the terms on either side of the principal diagonal will all be minus (or possibly zero as in Example 6-4). We can use this relationship for either checking or developing the system determinant. Thus,

$$\Delta = \begin{vmatrix} R_{11} & -R_{12} & -R_{13} \\ -R_{21} & R_{22} & -R_{23} \\ -R_{31} & -R_{32} & R_{33} \end{vmatrix}$$

where R_{ii} = total "self resistance" through which I_i flows,

R_{ij} = the "mutual resistance," that is, the total resistance shared by the meshes i and j.

Of course, the symmetry of the system determinant cannot be used as a check if current sources are in the network.

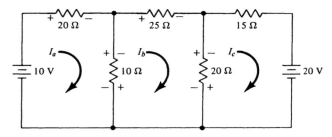

Figure 6-9   Three-loop problem from Example 6-4.

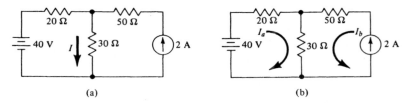

(a)   (b)

Figure 6-10   **(a)** Circuit for Example 6-5. **(b)** Circuit with loop currents.

If a branch has a current source determining its current as in Fig. 6-10(a), the loop current in the loop containing that branch is determined by the branch current.

EXAMPLE 6-5

Find the current I that flows through the 30 Ω resistor of Fig. 6-10(a).

Solution.   The loop currents I_a and I_b are chosen as in Fig. 6-10(b). Since the current I_b is the only loop current flowing through the branch containing the current source, and since it is specified in the same direction, it is equal to 2 A. Then, with the equation from loop a, the two circuit equations are

$$I_b = 2 \text{ A}$$
$$50I_a + 30I_b - 40 = 0$$

Substituting I_b from the first equation into the second equation and solving for I_a, we get

$$I_a = \frac{-20}{50} = \frac{-2}{5} = -0.4 \text{ A}$$

The current I is then found.

$$I = I_a + I_b = -0.4 + 2 = 1.6 \text{ A}$$

Drill Problem 6-2 ■

Refer to Fig. 6-11. Using loop analysis, determine the current in the 4 Ω resistances.

Answer.   0.05 A counterclockwise.   ■ ■

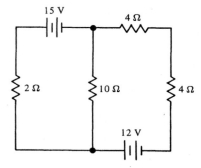

Figure 6-11   Drill Problem 6-2.

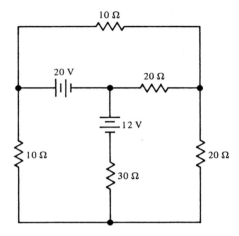

Figure 6-12 Drill Problem 6-3.

Drill Problem 6-3 ■

Refer to Fig. 6-12. Using loop analysis, determine the current in the 10 Ω resistance.

Answer. 0.693 A clockwise. ■ ■

6.4 NODAL ANALYSIS

The *nodal method* of circuit analysis is based on writing Kirchhoff's current equations in terms of the node potentials, which are taken as the unknowns for a set of simultaneous equations. In an electric circuit there are *n* major nodes; one of these nodes is selected as a reference node and arbitrarily assigned a potential of 0 V. The other major nodes are then each assigned different symbolic potentials.

The steps one takes in solving a circuit by nodal analysis are as follows:

1. Select one major node as the reference node and assign each of the $n - 1$ remaining nodes its own unknown potential (with respect to the reference node).
2. Assign branch currents to each branch. (The choice of direction is arbitrary.)
3. Express the branch currents in terms of the node potentials.
4. Write a current equation at each of the $n - 1$ unknown nodes.
5. Substitute the current expressions (step 3) into the current equations (step 4), which then become a set of simultaneous equations in the unknown node voltages.
6. Solve for the unknown voltages and ultimately the branch currents.

In step 1 it is sometimes convenient to choose the node at the bottom of a network or the node that has the largest number of branches connected to it. This node can be identified by adding a ground symbol to indicate the reference point.

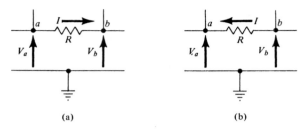

(a)   (b)

Figure 6-13 (a) Current assumption. (b) Opposite current assumption.

The node having the largest number of branches connected to it produces a node equation with the largest number of terms, so that choosing it as the reference point eliminates it from further consideration.

When expressing the branch currents in terms of the unknown potentials, one must be consistent with the idea that positive current flows from a higher potential to a lower potential. Consider the part of a network shown in Fig. 6-13. The two nodes a and b have the respective potentials V_a and V_b relative to the reference node. If the current is assumed to flow from a to b, then it follows that the mathematical expression for the current must state that $V_a > V_b$. Thus, for Fig. 6-13(a),

$$I = \frac{V_a - V_b}{R}$$

On the other hand, if the current is assumed to flow from b to a, as in Fig. 6-13(b),

$$I = \frac{V_b - V_a}{R}$$

EXAMPLE 6-6

Use nodal analysis to solve for the branch currents of the circuit of Fig. 6-14(a).

Solution. The reference node and branch currents are arbitrarily chosen as in Fig. 6-14(b).

With node b selected as reference, it is seen that the voltage on the one side of the 5 Ω resistor is fixed at 20 V relative to the reference. Then if current

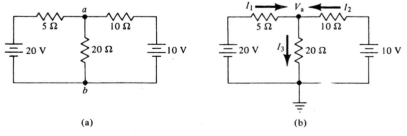

(a)   (b)

Figure 6-14 (a) Circuit for Example 6-6. (b) Circuit with assumed currents and reference node.

I_1 flows from that node to node a, the current I_1 is

$$I_1 = \frac{20 - V_a}{5}$$

Similarly, the voltage on the one side of the 10 Ω resistor is fixed at 10 V, so that if I_2 leaves that 10 V point,

$$I_2 = \frac{10 - V_a}{10}$$

Then we obtain the expression for I_3,

$$I_3 = \frac{V_a - 0}{20}$$

If the current equation is written at node a with the currents into the node algebraically assumed positive,

$$I_1 + I_2 - I_3 = 0$$

Substituting the expressions for I_1, I_2, and I_3 into the node equation, we get

$$\frac{20 - V_a}{5} + \frac{10 - V_a}{10} - \frac{V_a - 0}{20} = 0 \qquad (6\text{-}4)$$

It can be seen that Eq. 6-4 is an equation that can be solved for the unknown voltage V_a. Clearing fractions and combining terms, we obtain

$$100 - 7V_a = 0$$

or

$$V_a = \frac{100}{7} = 14.28 \text{ V}$$

Returning to the branch current expressions and substituting for V_a, we find that

$$I_1 = \frac{20 - V_a}{5} = \frac{20 - 14.28}{5} = \frac{5.72}{5} = 1.145 \text{ A}$$

$$I_2 = \frac{10 - V_a}{10} = \frac{10 - 14.28}{10} = \frac{-4.28}{10} = -0.428 \text{ A}$$

$$I_3 = \frac{V_a - 0}{20} = \frac{14.28}{20} = 0.714 \text{ A}$$

EXAMPLE 6-7

Find the current I that flows through the 30 Ω resistor of Fig. 6-15(a).

Solution. The assumptions for currents shown in Fig. 6-15(b) lead to the expressions

$$I_1 = \frac{40 - V_a}{20} \qquad I_2 = \frac{V_a}{30} \qquad I_3 = 2$$

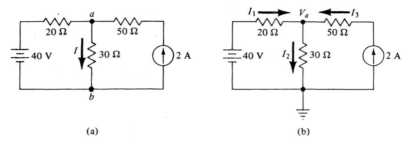

(a)   (b)

Figure 6-15 **(a)** Circuit for Example 6-7. **(b)** Circuit with reference node.

The resulting node equation is

$$I_1 - I_2 + I_3 = 0$$

or

$$\frac{40 - V_a}{20} - \frac{V_a}{30} + 2 = 0$$

Simplifying this last equation,

$$240 - 5V_a = 0$$

$$V_b = \frac{240}{5} = 48 \text{ V}$$

$$I = I_2 = \frac{V_a}{30} = \frac{48}{30} = 1.6 \text{ A}$$

EXAMPLE 6-8

Obtain the simultaneous equations necessary to solve the circuit of Fig. 6-9 by the node method.

Solution. The circuit is redrawn in Fig. 6-16. In accordance with the assumptions shown,

$$I_1 = \frac{10 - V_a}{20} \qquad I_2 = \frac{V_b - V_a}{25}$$

$$I_3 = \frac{V_a - 0}{10} \qquad I_4 = \frac{20 - V_b}{15}$$

$$I_5 = \frac{V_b - 0}{20}$$

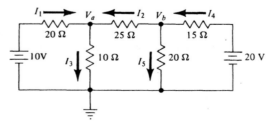

Figure 6-16 Example 6-8.

The node equations are as follows:

$$\text{node } a: \quad I_1 + I_2 - I_3 = 0$$
$$\text{node } b: \quad I_4 - I_2 - I_5 = 0$$

Substituting the current expressions into the node equations, we get

$$\text{node } a: \quad \frac{10 - V_a}{20} + \frac{V_b - V_a}{25} - \frac{V_a - 0}{10} = 0$$

$$\text{node } b: \quad \frac{20 - V_b}{15} - \frac{V_b - V_a}{25} - \frac{V_b - 0}{20} = 0$$

Clearing fractions and combining terms, we finally obtain the two equations:

$$\text{node } a: \quad 19V_a - 4V_b = 50$$
$$\text{node } b: \quad 12V_a - 47V_b = -400$$

An alternative approach to nodal analysis involves changing all series branches containing voltage sources to their current source equivalents. In particular, notice in Fig. 6-17(a) that the voltage source equivalent has the current source equivalent of Fig. 6-17(b). This transformation is performed using the techniques of the previous chapter.

Although it is easier to determine the current entering node b with the current source configuration of Fig. 6-17(b), we should note that the minor node a is lost in the transformation. It is important to keep this in mind since the branch current I of the original circuit is

$$I = \frac{V_a - V_b}{R_1}$$

Many unwary students of electric circuit analysis will perform nodal analysis using current source equivalents as in Fig. 6-17(b) and, after solving for V_b, will state

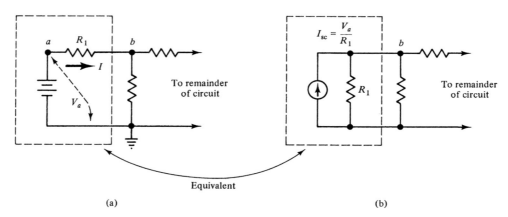

Equivalent

(a)   (b)

Figure 6-17   Conversion of voltage source equivalent for nodal analysis.

erroneously that $I = V_b/R_1$. One way of eliminating this error is to *return to the original circuit with its voltage sources after you find the unknown node voltages.*

Another way of approaching nodal analysis is to consider all nodes, even minor ones, to have an unknown potential. Of course, this results in a few more simultaneous equations, some of which simply state trivial information. With computer-based nodal analysis the extra equations are not a problem. The following problems illustrate how source conversions are used.

EXAMPLE 6-9

Determine the currents specified in Fig. 6-18(a). Use source conversions and nodal analysis.

Solution. The current source equivalents are determined first. For the right source,

$$I_{sc} = \frac{10\ V}{10\ \Omega} = 1\ A$$

and for the left source,

$$I_{sc} = \frac{20\ V}{5\ \Omega} = 4\ A$$

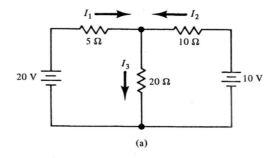

(a)

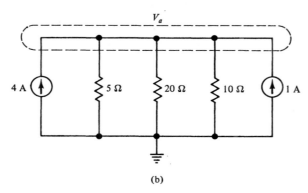

(b)

Figure 6-18 Example 6-9.

The source equivalents are replaced and a reference node selected as in Fig. 6-18(b). Then the single KCL equation is

$$4 + 1 - \frac{V_a}{5} - \frac{V_a}{20} - \frac{V_a}{10} = 0$$

$$5 = (\tfrac{1}{5} + \tfrac{1}{20} + \tfrac{1}{10})V_a$$

$$= 0.35 V_a$$

$$V_a = 14.28 \text{ V}$$

Then,

$$I_1 = \frac{20 - V_a}{5} = \frac{5.72}{5} = 1.145 \text{ A}$$

$$I_2 = \frac{10 - V_a}{10} = \frac{-4.28}{10} = -0.428 \text{ A}$$

and

$$I_3 = \frac{V_a - 0}{20} = \frac{14.28}{20} = 0.714 \text{ A}$$

EXAMPLE 6-10

Using source conversions, determine the simultaneous nodal equations necessary to solve the circuit of Fig. 6-19(a).

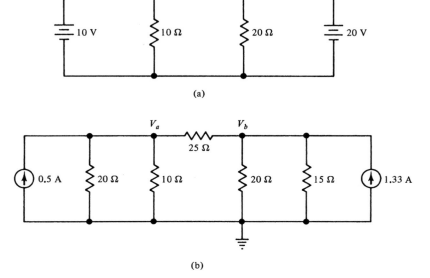

(a)

(b)

Figure 6-19   Example 6-10. **(a)** Original circuit and **(b)** circuit after source conversion.

Solution. First, as shown in Fig. 6-19(b), the sources are converted to their equivalents.

For this problem write each node equation by considering that all resistor currents leave a particular node when we consider the node equation for that node. At node *a*.

$$-0.5 + \frac{V_a}{20} + \frac{V_a}{10} + \frac{V_a - V_b}{25} = 0$$

and at node *b*,

$$-1.33 + \frac{V_b}{20} + \frac{V_b}{15} + \frac{V_b - V_a}{25} = 0$$

Collecting terms, we obtain

node *a*: $\left(\frac{1}{20} + \frac{1}{10} + \frac{1}{25}\right) V_a - \left(\frac{1}{25}\right) V_b = 0.5$

node *b*: $-\left(\frac{1}{25}\right) V_a + \left(\frac{1}{20} + \frac{1}{15} + \frac{1}{25}\right) V_b = 1.33$

Rewritten, this becomes

node *a*: $0.19 V_a - 0.04 V_b = 0.5$

node *b*: $-0.04 V_a + 0.157 V_b = 1.33$

Notice that symmetry exists with respect to the coefficients of V_a and V_b. This leads to the following statement: *Whenever we use all current source equivalents and take all resistor currents away from each node as we write the current equation at each node, we obtain a symmetrical system determinant of conductances.*

Thus, for a system with three unknown voltages,

$$\Delta = \begin{vmatrix} G_{11} & -G_{12} & -G_{13} \\ -G_{21} & G_{22} & -G_{23} \\ -G_{31} & -G_{32} & G_{33} \end{vmatrix}$$

where G_{ii} = "self conductance" for a node,

G_{ij} = "shared conductance," that is, the conductance between nodes *i* and *j*, each having unknown potentials.

This symmetry relationship can be used for checking or developing nodal equations.

Drill Problem 6-4 ■

Use nodal analysis to determine the node voltage V_a of Fig. 6-20.

Answer.   6.25 V.   ■ ■

Drill Problem 6-5 ■

Using nodal analysis, determine the current in the 10 Ω resistor of Fig. 6-21.

Answer.   0.2 A.   ■ ■

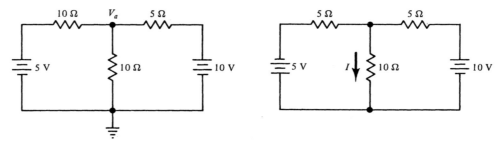

Figure 6-20   Drill Problem 6-4.   **Figure 6-21**   Drill Problem 6-5.

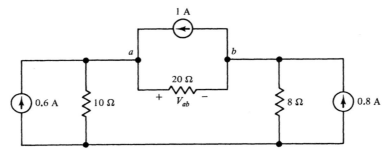

Figure 6-22   Drill Problem 6-6.

Drill Problem 6-6 ■

Determine the voltage across the 20 Ω resistor of Fig. 6-22.

Answer.   9.26 V.   ■ ■

6.5 SUPERPOSITION

Simultaneous equations may be avoided in the solution of a complex network by the application of the *superposition theorem*, which is stated as follows:

> **The current in any circuit element or voltage across any element of a linear, bilateral network is the algebraic sum of the currents or voltages separately produced by each source of energy.**

As the effect of each source is considered, the other sources are removed from the circuit, but their internal resistances must remain. Thus when a voltage source is removed, a very low internal resistance, ideally a short circuit, remains. On the other hand, when a current source is removed, a very high internal resistance remains and, ideally, an open circuit results from the removal of that current source. Finally, after each source is considered, the total current or voltage for an element is found by superimposing each of the component currents or voltages.

EXAMPLE 6-11

Find the currents flowing in the circuit of Fig. 6-23(a) by superposition.

Solution. If we consider the currents due to the left-hand source, as in Fig. 6-23(b), we obtain

$$R_T = 2 + \frac{(3)(6)}{9} = 4 \ \Omega$$

and

$$I_4 = \frac{4 \ V}{4 \ \Omega} = 1 \ A$$

By current division,

$$I_5 = \left(\frac{6}{9}\right) I_4 = 0.667 \ A$$

and

$$I_6 = 0.333 \ A$$

Next, by considering the effect of the right-hand source in Fig. 6-23(c),

$$R_T = 6 + \frac{(2)(3)}{5} = 6 + 1.2 = 7.2 \ \Omega$$

and

$$I_7 = \frac{6 \ V}{7.2 \ \Omega} = 0.833 \ A$$

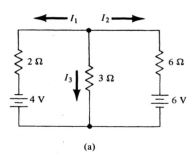

(a)

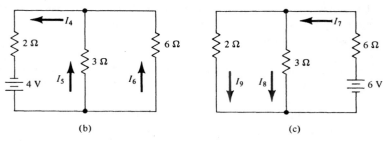

(b)   (c)

Figure 6-23 (a) Circuit for Example 6-11. (b) Left source considered. (c) Right source considered.

By current division,

$$I_8 = \left(\frac{2}{5}\right) I_7 = 0.333 \text{ A}$$

and
$$I_9 = 0.5 \text{ A}$$

If we consider the current flowing in the 2 Ω resistor, we obtain

$$I_1 = I_4 + I_9 = 1.0 + 0.5 = 1.5 \text{ A}$$

Similarly,

$$I_2 = -I_6 - I_7 = -0.333 - 0.833 = -1.166 \text{ A}$$
$$I_3 = -I_5 + I_8 = -0.667 + 0.333 = -0.334 \text{ A}$$

It must be stressed that superposition holds only for the linear bilateral circuit, where there is a linear relationship between voltage, current, and resistance. Since the power loss in a resistor results from a squared relationship, the *power cannot be found by superimposing power losses but must be calculated from a total voltage or current value.* That is, the power in the 2 Ω resistor of Example 6-11 is not given by

$$P = I_4^2(2) + I_9^2(2) = (1)^2(2) + (0.5)^2(2)$$
$$= 2 + 0.5 = 2.5 \text{ W} \qquad \text{(incorrect)}$$

but by

$$P = I_1^2(2) = (1.5)^2(2) = 4.5 \text{ W}$$

It should be realized that the three methods of analysis, loop, nodal, and superposition, perform the same task; namely, they enable us to solve a circuit problem. Often personal preference dictates the final choice of method, yet a preliminary examination of the circuit problem might also suggest the shortest or most direct method. Thus, we might count the number of loops and major nodes and choose between the loop and nodal methods by selecting the method that requires the smaller number of simultaneous equations.

Certain network theorems also simplify the analysis of electric circuits by providing procedures for reducing a given network to a simpler but equivalent network. In the next few sections we shall explore some of these theorems and some special concepts used in circuit analysis.

Drill Problem 6-7 ■

Use superposition to determine the current in the 10 Ω resistor of Fig. 6-24.

Answer.   2.25 A, upward.   ■ ■

Drill Problem 6-8

Determine by superposition the power loss in the 5 Ω resistor of Fig. 6-25.

Answer.   20 W   ■ ■

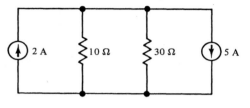

Figure 6-24 Drill Problem 6-7.

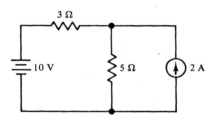

Figure 6-25 Drill Problem 6-8.

6.6 THEVENIN'S THEOREM

In Section 5.7, an equivalent electric circuit was developed for a source of emf—an ideal emf in series with an internal resistance. Certainly, since the source has a distributed rather than a lumped resistance, the two circuits are not the same physically; yet, they are the same electrically. This concept is extended to any two-terminal network by Thevenin's theorem, proposed by M. L. Thevenin (France, 1883):

> **Any two-terminal network containing voltage or current sources can be replaced by an equivalent circuit consisting of a voltage equal to the open-circuit voltage of the original circuit in series with the resistance measured back into the original circuit.**

Thus, the two-terminal network in Fig. 6-26(a) can be represented electrically by the Thevenin equivalent of Fig. 6-26(b). The voltage measured under open conditions E_{oc} is sometimes called the Thevenin voltage E_{th}.

Thevenin's theorem may be applied to any of the networks considered so far by treating one branch of the network as a load and the remaining part of the net-

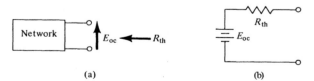

(a)   (b)

Figure 6-26   **(a)** Two-terminal network. **(b)** Thevenin equivalent.

work as a two-terminal network supplying that load. The steps one takes in applying the theorem are then as follows:

1. That portion of the original network considered as the load is removed or imagined to be removed. The terminals are then identified for later polarity determination.
2. The open-circuit voltage is calculated.
3. The Thevenin resistance is calculated *looking back* into the network. The sources are removed, but their internal resistances remain.
4. The equivalent circuit is drawn, the load reconnected, and the load current determined.

EXAMPLE 6-12

Determine the Thevenin circuit for the circuit of Fig. 6-27(a).

Solution. First, notice that the terminals *a* and *b* are already opened and identified; hence, whatever is considered a load to the circuit is removed.

The voltage across the 8 Ω resistor is the open circuit voltage and by voltage division is

$$E_{oc} = 9 \text{ V} \left(\frac{8 \, \Omega}{8 \, \Omega + 10 \, \Omega} \right) = 4 \text{ V}$$

R_{th} is calculated after the source is removed and a short circuit is inserted as its internal resistance. From Fig. 6-27(b) it is seen that

$$R_{th} = 8 \, \Omega \,\|\, 10 \, \Omega = \frac{(8)(10)}{18} = 4.44 \, \Omega$$

hence, the Thevenin equivalent of Fig. 6-27(c).

When we determine an equivalent network we have a new network that behaves the same electrically as the one it replaces. Obviously, Fig. 6-27(c) is not the same as Fig. 6-27(a), yet each network allows the same current flow and will have the same voltage drop across its terminals when the same load is connected to each. What we

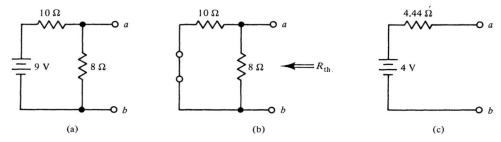

Figure 6-27 **(a)** Circuit for Example 6-12. **(b)** Determination of R_{th}. **(c)** The equivalent network.

gain for our efforts is a simpler way of viewing the electrical behavior of the original network.

EXAMPLE 6-13

Use Thevenin's theorem to solve for the current in the resistance R_L of Fig. 6-28(a), when R_L is **(a)** 10 Ω and **(b)** 50 Ω.

Solution. The load is removed as in Fig. 6-28(b) and with no current flowing from terminal a, E_{oc} is given by voltage division as

$$E_{oc} = E' = 40 \text{ V} \left(\frac{20}{40}\right) = 20 \text{ V}$$

(a is at higher potential than b). Then the source is removed as in Fig. 6-28(c) and R_{th} is calculated as

$$R_{th} = 20 + (20) \| (20) = 20 + 10 = 30 \text{ Ω}$$

The load is then connected to the Thevenin equivalent as in Fig. 6-28(d) and the load currents calculated as

(a) $$R_L = 10 \text{ Ω}$$

$$I = \frac{E_{oc}}{R_{th} + 10} = \frac{20}{30 + 10} = \frac{20}{40} = 0.5 \text{ A}$$

(b) $$R_L = 50 \text{ Ω}$$

$$I = \frac{E_{oc}}{R_{th} + 50} = \frac{20}{80} = 0.25 \text{ A}$$

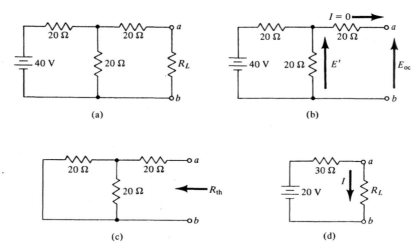

Figure 6-28 **(a)** Circuit for Example 6-13. **(b)** Determination of E_{oc}. **(c)** Determination of R_{th}. **(d)** Equivalent network.

Thus Thevenin's theorem allows one to observe a specific portion of a network, thereby making it easy to see the effect accompanying a change in a single resistance while all other resistances and emfs remain unchanged.

EXAMPLE 6-14

Use Thevenin's theorem to solve for the specified current in Fig. 6-29(a)

Solution.   Removing the 20 Ω resistor considered a load and redrawing the circuit, we obtain Fig. 6-29(b). Then $E_{oc} = 10$ V plus the rise across the 10 Ω resistor. Writing a loop equation for circuit b, we get

$$5I' + 10I' = 20 - 10 = 10$$

Thus

$$I' = \frac{10}{15} = 0.667 \text{ A}$$

and $$E_{oc} = 10 + I'(10) = 10 + 6.67 = 16.67 \text{ V}$$

Solving for R_{th} by Fig. 6-29(c), we obtain

$$R_{th} = (5) \| (10) = 3.33 \ \Omega$$

Then the equivalent circuit is the same as in circuit d, and the current I is

$$I = \frac{E_{oc}}{R_{th} + 20} = \frac{16.67}{23.33} = 0.714 \text{ A}$$

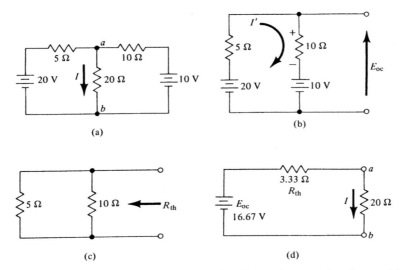

Figure 6-29 **(a)** Circuit for Example 6-14. **(b)** Load removed, redrawn. **(c)** Finding R_{th}. **(d)** Equivalent circuit.

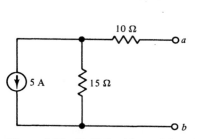

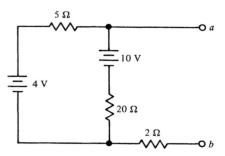

Figure 6-30   Drill Problems 6-9 and 6-11.   **Figure 6-31**   Drill Problems 6-10 and 6-12.

Drill Problem 6-9 ■

Determine the Thevenin network for Fig. 6-30.

Answer.   75 V (*a* negative); 25 Ω.   ■ ■

Drill Problem 6-10 ■

Determine the Thevenin network for Fig. 6-31.

Answer.   5.2 V (*a* positive); 6 Ω.   ■ ■

6.7 NORTON'S THEOREM

In Chapter 5, the practical voltage source, an ideal voltage source in series with an internal resistance, was shown to have an equivalent current source representation. It is not surprising, then, that a companion theorem exists to Thevenin's theorem that enables one to replace a complicated network or its Thevenin equivalent by a practical current source representation. This companion theorem is known as Norton's theorem.

> **Any two-terminal network containing voltage or current sources can be replaced by an equivalent circuit consisting of a current source equal to the short-circuit current from the original network in parallel with the resistance measured back into the original circuit.**

Thus the two-terminal network of Fig. 6-32(a) is represented electrically by the Norton equivalent of Fig. 6-32(b).

As noted in Fig. 6-32, the current source has a current value called the Norton current I_n, or simply the short-circuit current, I_{sc}. Notice that the parallel resistance obtained as the Norton resistance R_n equals the Thevenin resistance R_{th}. Sometimes the Norton equivalent conductance is specified for a two-terminal network; it is simply the reciprocal of R_n or R_{th}.

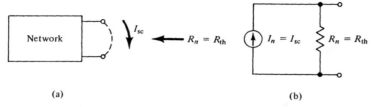

(a) (b)

Figure 6-32 **(a)** Two-terminal network. **(b)** Norton equivalent.

To apply the theorem, one follows the general approach of Thevenin's theorem, except that when the load is removed the two terminals of the network are shorted so that I_{sc} can be calculated.

EXAMPLE 6-15

Determine the Norton equivalent network for the circuit of Example 6-12.

Solution. The original circuit appears in Fig. 6-33(a). When the output terminals are shorted as in Fig. 6-33(b), the short-circuit current is found to be

$$I_{sc} = \frac{9\text{ V}}{10\ \Omega} = 0.9\text{ A}$$

From Example 6-12, $R_{th} = 4.44\ \Omega = R_n$; hence, we have the equivalent network of Fig. 6-33(c).

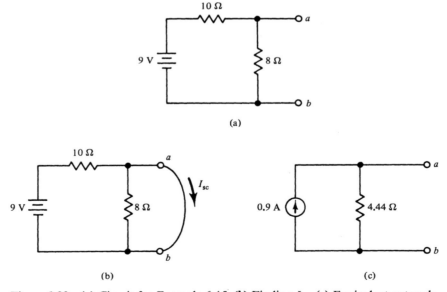

Figure 6-33 **(a)** Circuit for Example 6-15. **(b)** Finding I_{sc}. **(c)** Equivalent network.

EXAMPLE 6-16

Solve Example 6-13 by Norton's theorem.

Solution. The original circuit appears in Fig. 6-34(a). When the load is removed and the output terminals are shorted as in Fig. 6-34(b), the short-circuit current is found.

$$I' = \frac{40}{20 + (20 \parallel 20)} = \frac{40}{30} = \frac{4}{3} \text{ A}$$

By current division,

$$I_{sc} = \frac{1}{2} I' = \frac{2}{3} \text{ A}$$

As in Example 6-13, $R_{th} = 30 \ \Omega$; thus the equivalent circuit of Fig. 6-34(c). Finally, the load currents are found by current division.

(a) $R_L = 10 \ \Omega$

$$I = I_{sc} \left(\frac{30}{40} \right) = \frac{2}{3} \left(\frac{3}{4} \right) = 0.5 \text{ A}$$

(b) $R_L = 50 \ \Omega$

$$I = I_{sc} \left(\frac{30}{80} \right) = \frac{2}{3} \left(\frac{3}{8} \right) = 0.25 \text{ A}$$

Both of these agree with the currents obtained by the Thevenin equivalent circuit.

A simple equation relates the Thevenin and Norton circuits. If the Thevenin circuit is equivalent to some complicated network, then the short-circuit current of

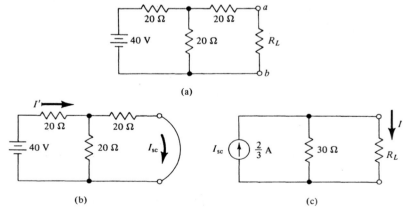

(a)

(b)   (c)

Figure 6-34   (a) Circuit for Example 6-16. (b) Finding I_{sc}. (c) Equivalent network.

the Thevenin equivalent is the short-circuit current for the original network. By shorting the cirucit of Fig 6-26(b), one obtains

$$I_{sc} = \frac{E_{oc}}{R_{th}} \qquad (6-5)$$

Whether we wish to obtain a Thevenin or Norton equivalent network, we must always determine R_{th}, which equals R_n. Yet Eq. 6-5 provides us with a choice. If we desire a Thevenin equivalent, we can solve directly for the Thevenin voltage, or we can first determine the Norton current and then determine the Thevenin voltage by Eq. 6-5. Similarly, if we desire a Norton equivalent, we have the choice of directly solving for I_{sc}, or solving for the Thevenin voltage and then using Eq. 6-5. The choice frequently depends on which appears easier to obtain in a given situation.

Finally, Thevenin's and Norton's theorems are for linear bilateral networks. Certainly not all circuit devices are linear and bilateral. Even so, there may be a certain range of operation over which a device or circuit is linear. A Thevenin or Norton equivalent may be used over that range. Thus, even complicated electronic devices such as transistors are modeled by a Thevenin or Norton equivalent for a linear region of operation.

When a Thevenin or Norton equivalent circuit is used to represent electronic devices such as a transistor, the resistance that the load "sees" as it "looks back" into the network is often called the *output resistance* of the network.

Drill Problem 6-11 ■

Determine the Norton equivalent for the circuit of Fig. 6-30, first by solving directly for I_{sc} and then by converting from the Thevenin equivalent of Drill Problem 6-9.

Answer.   3 A (leaves *b*); 25 Ω.   ■ ■

Drill Problem 6-12 ■

Determine the Norton equivalent for the circuit of Fig. 6-31, first by solving directly for I_{sc} and then by converting from the Thevenin equivalent of Drill Problem 6-10.

Answer.   0.866 A (leaves *a*); 6 Ω.   ■ ■

6.8 MAXIMUM POWER TRANSFER THEOREM

Any circuit or network may be represented, insofar as load current, voltage, and power are concerned by a Thevenin equivalent circuit. The Thevenin resistance is comparable to a source internal resistance that, in turn, absorbs some of the power available from the ideal voltage source. In Fig. 6-35 a variable load resistance R_L is connected to a Thevenin circuit. The current for any value of load resistance is

$$I = \frac{E}{R_i + R_L} \qquad (6-6)$$

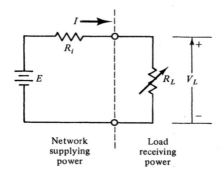

Figure 6-35 shows the network supplying power, the load receiving power, with labels:

Network supplying power | Load receiving power

Figure 6-35 Thevenin circuit with a load resistor.

Then by $I^2 R_L$, the power delivered to the load is

$$P_L = \frac{E^2 R_L}{(R_i + R_L)^2} \tag{6-7}$$

The load power depends on both R_i and R_L; however, R_i is considered constant for any particular network. Then one might get an idea of how P_L varies with a change in R_L by assuming values for the Thevenin circuit of Fig. 6-35 and, in turn, calculating P_L for different values of R_L.

For example, with $E = 10$ V and $R_i = 5\ \Omega$, one can substitute various values for R_L into Eq. 6-7 and obtain the values for P_L listed in Table 6-1. The results of Table 6-1 are plotted in Fig. 6-36.

Whenever a curve has a maximum or minimum point, the slope is zero. By applying calculus (differentiation) to Eq. 6-7, we find that the maximum point of the P_L versus R_L relationship occurs at $R_L = R_i$. Notice that this condition is empirically indicated by Fig. 6-36. Thus *maximum power is transferred from a source when the load resistance equals the internal resistance of the source.* Also, notice that under the condition of maximum power transfer, the load voltage is, by voltage division, one-half of the open-circuit emf.

TABLE 6-1 MAXIMUM POWER TRANSFER

R_L (Ω)	P_L (W)	Efficiency (%)
0	0	0
1	2.78	16.6
2	4.08	28.5
3	4.70	37.6
4	4.95	43.0
5	5.00	50.0
6	4.96	54.5
7	4.85	58.2
8	4.75	61.8
10	4.45	67.0

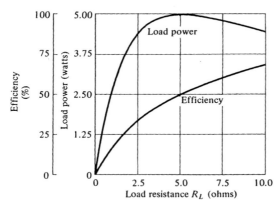

Figure 6-36 Maximum power transfer curves.

Another item of importance is efficiency. The efficiency for the system under consideration is

$$\eta = \frac{P_L}{P_D} \tag{6-8}$$

where P_L is the load power and P_D is the power developed by the source (the ideal emf × the current). Under maximum power transfer conditions, the current in the example is 1 A so that P_D is 10 W. Since P_L is 5 W, the efficiency is only 50%. The efficiencies for other values of R_L appear in Table 6-1 and are summarized by the efficiency versus load curve of Fig. 6-36.

The problem of obtaining maximum power at the highest efficiency is resolved by compromise. In electronics and communications systems, it is usually important to obtain maximum power from low power sources, such as an antenna. Then the source and load resistances are *matched* for maximum power at the expense of efficiency. However, electric power companies try to keep their losses low by operating at high efficiency.

EXAMPLE 6-17

A battery has an open-circuit voltage of 6 V and a short-circuit current of 30 A. What is the maximum power available from the battery?

Solution. The internal resistance is obtained,

$$R_i = \frac{\Delta V}{\Delta I} = \frac{E_{oc}}{I_{sc}} = \frac{6}{30} = 0.2 \ \Omega$$

For maximum power,

$$R_L = R_i = 0.2 \ \Omega$$

$$I = \frac{E}{R_i + R_L} = \frac{6}{0.4} = 15 \ A$$

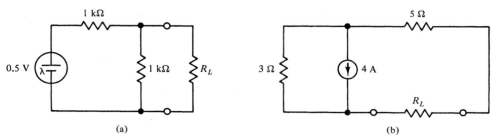

(a)   (b)

Figure 6-37 Circuits for Drill Problem 6-13.

By voltage division,

$$V_L = \tfrac{1}{2}(6) = 3 \text{ V}$$

Then

$$P_L = V_L I_L = 3(15) = 45 \text{ W}$$

Drill Problem 6-13

Determine each load resistance R_L for maximum power and the maximum power delivered to R_L for the circuits of Fig. 6-37.

Answer.   (a) 500 Ω; 31.3 μW. (b) 8 Ω; 4.5 W.   ■ ■

6.9 DEPENDENT SOURCES

Up to this point we have considered each voltage source or current source to have an electrical operation that depends solely on its own electrical characteristic. These sources are said to be *independent*. We now study the use of another type of source, the *dependent source*.

A *dependent* or *controlled source* is a source such that the voltage or current depends on the voltage or current in some other part of the circuit. Dependent sources appear in the electric circuit models of many electronic devices, such as the transistor and the operational amplifier, abbreviated *op amp*. Although the operation of electronic devices such as the transistor and the op amp are certainly exciting, we shall not study them in depth. Rather, we shall consider a few of the circuit models and how we may apply circuit analysis techniques to them.

The circuit symbols used for dependent sources are usually distinguished from the independent source symbols. The symbols used in this text are characterized by a diamond-shaped symbol such as that shown in Fig. 6-38. The voltage polarity and current direction are specified within the diamond shapes, and the voltage V and current I are functions of other voltages or currents.

Consider as an example, the circuit of Fig. 6-39. Notice the dependent voltage source with a voltage value of $2.5V_a$. Thus, the dependent source voltage is 2.5 times whatever value V_a is at any time. The quantity on which a dependent source depends, in this case V_a, is called the *control variable*.

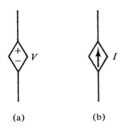

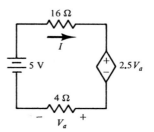

Figure 6-38   Schematic symbols for the dependent (a) voltage source and (b) current source.

Figure 6-39   A circuit with a dependent voltage source; Example 6-18.

EXAMPLE 6-18

Calculate the current I that flows in the circuit of Fig. 6-39.

Solution.   First, a KVL equation is written:

$$16I + 2.5V_a + 4I = 5$$

or

$$20I + 2.5V_a = 5$$

Two unknowns, I and V_a, are present; a second equation is needed. The second equation comes from the control variable V_a and Ohm's law:

$$V_a = 4I$$

Substituting this last equation into the KVL equation and solving for I, we obtain

$$20I + 2.5(4I) = 5$$
$$30I = 5$$
$$I = 5/30 = 0.167 \text{ A}$$

EXAMPLE 6-19

Determine the voltage V_L for the circuit of Fig. 6-40.

Solution.   In this circuit two Norton equivalent circuits share a common node. Notice that the right source depends on the control current I_x that flows in the left part of the circuit.

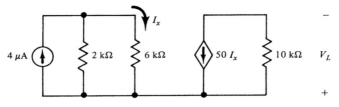

Figure 6-40   Circuit for Example 6-19.

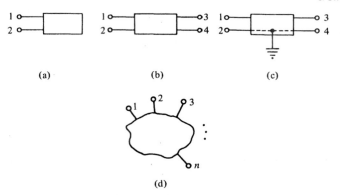

(a)   (b)   (c)

(d)

Figure 6-41   Networks: **(a)** two terminal; **(b)** four terminal; **(c)** three terminal; **(d)** n terminal.

Using current division for the left part, we obtain

$$I_x = 4\ \mu A\left(\frac{2\ k\Omega}{2\ k\Omega + 6\ k\Omega}\right) = 1\ \mu A$$

Then, for the right part

$$V_L = 50I_x(10\ k\Omega)$$
$$= 50(1 \times 10^{-6}\ A)(10 \times 10^{3}) = 0.5\ V$$

Networks are termed *active* if they contain voltage or current sources and *passive* if they contain no sources. In addition, they may be classified as *two-terminal*, *three-terminal*, *four-terminal*, or *n-terminal*, depending on the number of electrical terminals externally available. (See Fig. 6-41.)

So far we have been able to describe the electrical behavior of two-terminal networks ranging from a resistor to the Thevenin or Norton equivalent of a more complicated network. Characterizing the electrical behavior of the four-terminal network of Fig. 6-41 requires a bit more effort. Two of the terminals serve as an *input pair* to which excitation is applied, and the remaining two terminals serve as the response or *output pair*.

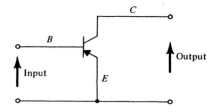

Figure 6-42   A three-terminal device (transistor).

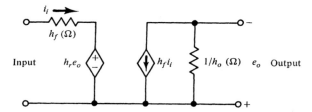

Figure 6-43 One of the many transistor circuit models (*h*-parameter circuit).

If an internal connection exists between two of the terminals of a four-terminal network as in Fig. 6-41(c), the network degenerates to a three-terminal network. An example of a three-terminal device is the transistor, which is shown in Fig. 6-42 with a common or grounded terminal between the input and output.

One of the many models used to describe the transistor electrically is shown in Fig. 6-43. That circuit configuration is known as the *h-parameter model*, where the *h*s are constants. We discuss *h* parameters in more detail in Chapter 20. Notice that on the input side a Thevenin circuit with a dependent voltage source is present, whereas, on the output side a Norton circuit with a dependent current source is present.

Another example of an electronic device that is modeled with a dependent source is the operational amplifier or op amp. The op amp is a high-gain, directly coupled voltage amplifier that amplifies a voltage e_d called the differential voltage that appears at the input. Its schematic symbol and an equivalent circuit are shown in Fig. 6-44.

When electronic devices are placed in circuits with other elements, the device models or equivalent circuits are used for analysis. As simple as these models may be, they are frequently approximated for simplicity of analysis. For example, the resistance in the transistor circuit, $1/h_o$ is generally high and is approximated as an open circuit. Similarly, the term R_i represents the input resistance to the op amp and is extremely high. It follows that unless the input current is the quantity being

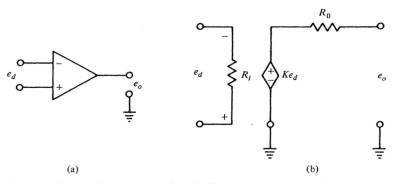

(a)   (b)

Figure 6-44 **(a)** The op amp symbol; **(b)** an equivalent circuit.

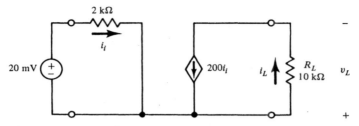

Figure 6-45 Example 6-20.

investigated, R_i is approximated as an open circuit. But let us not worry too much about how or why the electronic equivalent circuits are simplified. Instead, let us consider how the circuit analyst applies KVL and KCL equations in solving for the various voltages and currents in a given circuit shown to contain electronic models.

EXAMPLE 6-20

The circuit of Fig. 6-45 contains a simplified model of a transistor. Determine the voltage across R_L.

Solution.   The KVL equation for the left loop is

$$(2 \text{ k}\Omega)i_i = 20 \text{ mV}$$

$$i_i = \frac{20 \times 10^{-3}}{2 \times 10^3} = 10 \times 10^{-6} \text{ A}$$

Then
$$v_L = i_L R_L = (200 i_i)(10 \text{ k}\Omega)$$
$$= 200(10 \times 10^{-6})(10 \times 10^3) = 20 \text{ V}$$

Drill Problem 6-14 ■

Determine i_i and the voltage v_L for the circuit of Fig. 6-46 that contains the model of an op amp.

Answer.   1 pA, -98 mV.   ■ ■

Drill Problem 6-15 ■

Determine v_L for the circuit of Fig. 6-47.

Answer.   2.4 V.   ■ ■

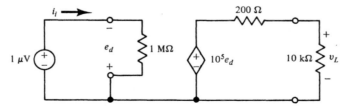

Figure 6-46 Drill Problem 6-14.

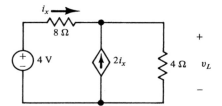

Figure 6-47 Drill Problem 6-15.

6.10 WYE AND DELTA CONVERSIONS

Two networks that occur frequently in electric circuit analysis are the wye (Y) and delta (Δ). These two circuits identified in Fig. 6-48 are sometimes part of a larger circuit and obtain their names from their configurations. Notice that Fig. 6-48(b) forms an inverted Greek letter Δ.

Although the networks of Fig. 6-48 are three-terminal networks, they are sometimes redrawn as four-terminal networks as shown in Fig. 6-49. In comparing Figs. 6-48 and 6-49, notice that when the two arms R_1 and R_2 are moved into the horizontal position they form a T configuration, and when the two sides of the inverted delta R_A and R_B are spread apart, a pi (π) configuration is obtained. The T and π terminology is generally used in communications and electronics, whereas, the Y and Δ terminology is generally used for power circuits.

If a network has mirror-image symmetry with respect to some centerline, that is, if a line can be found to divide the network into two symmetrical halves, the network is a *symmetrical network*. The T network is symmetrical when $R_1 = R_2$, and the π network is symmetrical when $R_A = R_B$. Furthermore, if all the elements in either the T or π are equal, the T or π is said to be *balanced*.

We can obtain equations for direct transformation or conversion from a T to a π or from a π to a T by considering that for equivalence the two networks must

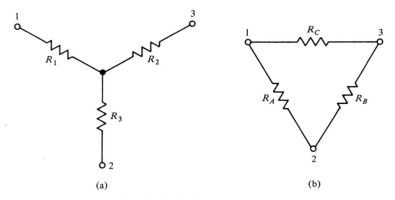

Figure 6-48 Networks: **(a)** Y; **(b)** Δ.

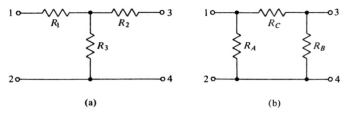

Figure 6-49 Networks: **(a)** T or Y; **(b)** π or Δ.

have the same resistance when measured between similar pairs of terminals. That is, we can "look into" a pair of terminals like 1 and 2 to determine R_{12} while the other terminals are left open. For the networks of Fig. 6-49 we can write equations of equivalence for the resistances R_{12}, R_{13}, and R_{34} with the results

$$R_1 + R_3 = \frac{R_A(R_B + R_C)}{R_A + R_B + R_C} \tag{6-9}$$

$$R_1 + R_2 = \frac{R_C(R_A + R_B)}{R_A + R_B + R_C} \tag{6-10}$$

$$R_2 + R_3 = \frac{R_B(R_A + R_C)}{R_A + R_B + R_C} \tag{6-11}$$

To convert a π to a T, relationships for R_1, R_2, and R_3 must be obtained in terms of the π resistances R_A, R_B, and R_C. Letting $S = R_A + R_B + R_C$ and solving Eqs. 6-9, 6-10, and 6-11 by determinants, we obtain

$$D = \begin{vmatrix} 1 & 0 & 1 \\ 1 & 1 & 0 \\ 0 & 1 & 1 \end{vmatrix} = 1 + 1 = 2$$

$$D_1 = \begin{vmatrix} \dfrac{R_A(R_B + R_C)}{S} & 0 & 1 \\[2mm] \dfrac{R_C(R_A + R_B)}{S} & 1 & 0 \\[2mm] \dfrac{R_B(R_A + R_C)}{S} & 1 & 1 \end{vmatrix}$$

$$= \frac{R_A(R_B + R_C)}{S} + \frac{R_C(R_A + R_B)}{S} - \frac{R_B(R_A + R_C)}{S}$$

$$= \frac{R_A R_B + R_A R_C + R_A R_C + R_B R_C - R_A R_B - R_B R_C}{S}$$

$$= \frac{2R_A R_C}{S}$$

Thus,

$$R_1 = \frac{D_1}{D} = \frac{2R_A R_C}{2S} = \frac{R_A R_C}{R_A + R_B + R_C} \tag{6-12}$$

In a similar fashion, R_2 and R_3 are found to be

$$R_2 = \frac{R_B R_C}{R_A + R_B + R_C} \tag{6-13}$$

and

$$R_3 = \frac{R_A R_B}{R_A + R_B + R_C} \tag{6-14}$$

To convert a T to a π, one obtains relationships for R_A, R_B, and R_C in terms of the T resistances R_1, R_2, and R_3. The resulting relationships are

$$R_A = \frac{R_1 R_2 + R_2 R_3 + R_3 R_1}{R_2} \tag{6-15}$$

$$R_B = \frac{R_1 R_2 + R_2 R_3 + R_3 R_1}{R_1} \tag{6-16}$$

and

$$R_C = \frac{R_1 R_2 + R_2 R_3 + R_3 R_1}{R_3} \tag{6-17}$$

In order to note the symmetry of the transformation equations, the Y (T) and Δ (π) networks have been superimposed on each other in Figure 6-50. Notice that each Y resistor equals the product of the two adjacent legs of the Δ divided by the sum of the three legs of the Δ. On the other hand, each leg of the Δ equals the sum of the possible products of the Y resistances divided by the opposite Y resistance.

When a Y or Δ is balanced, the conversion equations reduce to

$$R_Y = \frac{R_\Delta}{3} \tag{6-18}$$

and

$$R_\Delta = 3R_Y \tag{6-19}$$

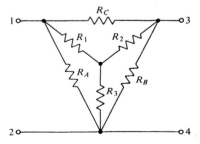

Figure 6-50 Superimposed Y and Δ networks.

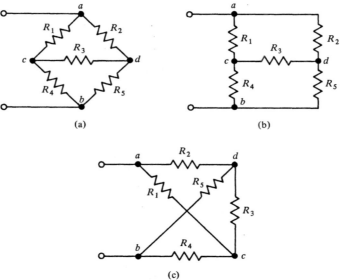

(a)   (b)

(c)

Figure 6-51 (a) The bridge circuit. (b), (c) Two other ways of representing the bridge circuit.

where R_Y is the value of one of the equal-valued Y resistors and R_Δ is the value of one of the equal-valued Δ resistors.

The bridge circuit was introduced in Chapter 5 as one type of resistance circuit that is not reducible by series–parallel combinations. The bridge circuit and two ways of portraying it are shown in Fig. 6-51. Notice that the bridge circuit consists of two Δs sharing a common resistance.

Although resistances are used here, other devices, such as rectifying diodes, are used in the bridge configuration. The bridge is also used for resistance and impedance measurements as noted in Chapter 17. In this latter case, the resistance R_3 is replaced by a sensitive instrument that can detect small differences in voltage between points c and d.

When no current flows in the detection or bridged portion as indicated in Fig. 6-52, the bridge is said to be *balanced*. For this to be the case,

$$V_{ac} = V_{ad}$$

and

$$V_{cb} = V_{db}$$

With substitutions, the relationships become

$$I_1 R_1 = I_2 R_2$$

and

$$I_4 R_4 = I_5 R_5$$

Since $I_1 = I_4$ and $I_2 = I_5$, we can make these substitutions into the preceding equa-

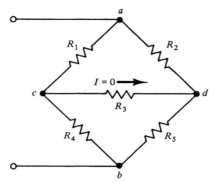

Figure 6-52 The balanced bridge has no current in the "bridged" portion.

tions with the result

$$\frac{R_1}{R_4} = \frac{R_2}{R_5} \tag{6-20}$$

Equation 6-20 defines the conditions for a balanced bridge.

If a bridge circuit is not balanced, we may, of course, apply loop or nodal analysis to determine the various currents and voltages. However, if we are interested in only the external behavior of the bridge, we can use Δ to Y conversion, as indicated by the following example.

EXAMPLE 6-21

What current I flows to the bridge circuit of Fig. 6-53(a)?

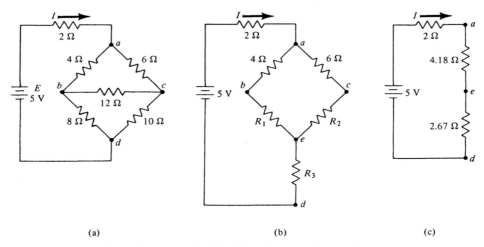

(a)   (b)   (c)

Figure 6-53 **(a)** Circuit for Example 6-21. **(b)** Δ-Y conversion. **(c)** Equivalent series circuit.

Solution. A Y is substituted for the Δ between points *b*, *c*, and *d*, as shown in Figure 6-53(b); then

$$R_1 = \frac{(12)(8)}{8 + 12 + 10} = \frac{(12)(8)}{30} = 3.2 \ \Omega$$

$$R_2 = \frac{(10)(12)}{8 + 12 + 10} = 4 \ \Omega$$

$$R_3 = \frac{(8)(10)}{8 + 12 + 10} = 2.67 \ \Omega$$

The resistance between points *a* and *e* is obtained as

$$R_{ae} = (4 + 3.2) \| (6 + 4) = \frac{(7.2)(10)}{17.2} = 4.18 \ \Omega$$

Then, from Fig. 6-53(c),

$$R_t = 2 + 4.18 + 2.67 = 8.85 \ \Omega$$

and

$$I = \frac{E}{R_t} = \frac{5}{8.85} = 0.565 \ \text{A}$$

Drill Problem 6-16 ■

What is the equivalent Δ for a Y having the values $R_1 = 500 \ \Omega$, $R_2 = 300 \ \Omega$, and $R_3 = 300 \ \Omega$?

Answer. 1.3 kΩ; 780 Ω; 1.3 kΩ.   ■ ■

Drill Problem 6-17 ■

Determine the total resistance between terminals *a* and *b* for Fig. 6-54.

Answer. 8 kΩ.   ■ ■

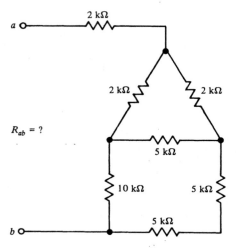

Figure 6-54   Drill Problem 6-17.

QUESTIONS

1. What is the difference between a major and a minor node?
2. A network of b branches requires how many unknown branch currents?
3. What is the difference between a branch current and a loop current?
4. A particular network has l loops and n major nodes. How many loop equations are necessary to solve the network? How many node equations are needed?
5. What are the steps needed to solve a network problem by either loop or nodal analysis?
6. How does a mesh differ from a loop?
7. What is mesh analysis?
8. State the conditions for which the system determinant for a set of loop equations will be symmetrical about the principal diagonal.
9. State the conditions for which the system determinant for a set of node equations will be symmetrical about the principal diagonal.
10. State the superposition theorem.
11. Does the superposition theorem apply when nonlinear elements are in a circuit? Explain.
12. What is Thevenin's theorem?
13. What is the value of Thevenin's theorem?
14. What is Norton's theorem?
15. What is the difference between the Thevenin and Norton resistances?
16. State the maximum power transfer theorem.
17. Under maximum power transfer conditions, what is the system efficiency?
18. Describe the difference between an independent and a dependent source.
19. Describe the different types of dependent sources.
20. What is meant by the term *"symmetrical network"*?
21. What is meant by the descriptions *balanced* Y and *balanced* Δ?
22. When is a bridge circuit balanced?
23. How are Y–Δ and Δ–Y conversions useful?

PROBLEMS

1. Determine the number of major nodes and branches in the circuit of Fig. 6-55.
2. Using the branch current method of solution, determine the current in each branch of the circuit of Fig. 6-56.
3. Using loop analysis, determine the voltage V_{ab} for the circuit of Fig. 6-57.
4. Using loop analysis, solve for the current I for the circuit of Fig. 6-58.
5. Write the loop equations for the circuit of Fig. 6-56. Solve for the loop currents and the related branch currents.
6. Write the loop equations for the circuit of Fig. 6-59. Solve for the loop currents and the related branch currents.

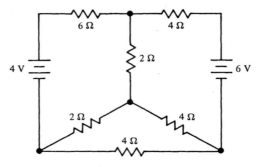

Figure 6-55   Problems 6-1 and 6-11.

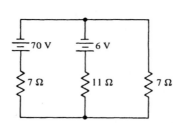

Figure 6-56   Problems 6-2 and 6-5.

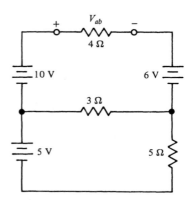

Figure 6-57   Problem 6-3.

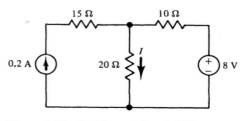

Figure 6-58   Problems 6-4 and 6-13.

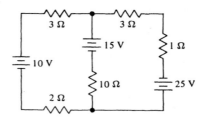

Figure 6-59   Problem 6-6.

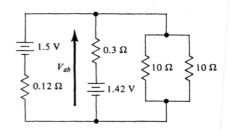

Figure 6-60   Problem 6-7.

7. Write the loop equations for the circuit of Fig. 6-60. Solve for the loop currents and the related branch currents.

8. By using loop analysis, solve for the branch currents of Fig. 6-61.

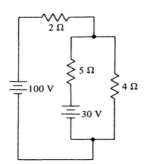

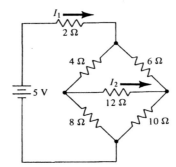

Figure 6-61   Problems 6-8 and 6-14.   **Figure 6-62**   Problems 6-9 and 6-15.

9. By using loop analysis, find the currents I_1 and I_2 for the unbalanced bridge circuit of Fig. 6-62.

10. By using loop analysis, find the current I in the 4 Ω resistor of Fig. 6-63.

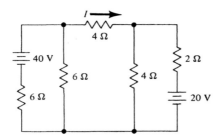

Figure 6-63   Problems 6-10 and 6-16.

11. Using loop analysis, solve for the current in the 6 Ω resistor of the circuit of Fig. 6-55.

12. Using loop analysis, solve for the current I and the voltage V_{ab} for the circuit of Fig. 6-64.

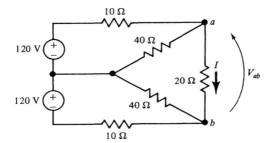

Figure 6-64   Problems 6-12, 6-17, and 6-24.

13. Convert the voltage source in the circuit of Fig. 6-58 into a current source. Use nodal analysis in solving for the current I.

14. Use nodal analysis to solve the circuit of Fig. 6-61.

15. Use nodal analysis to solve for the currents I_1 and I_2 in the bridge circuit of Fig. 6-62.

16. Use nodal analysis to solve for the current I in the circuit of Fig. 6-63.

17. Convert the voltage sources of Fig. 6-64 to current source equivalents and using nodal analysis determine I and V_{ab}.

18. Using nodal analysis, determine V_{ab} and the power dissipated in the 2 kΩ resistor of Fig. 6-65.

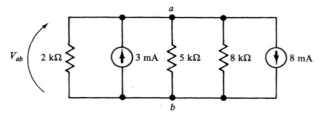

Figure 6-65   Problems 6-18 and 6-25.

19. Using nodal analysis, solve for the voltages V_a, V_b, and V_c (all referenced to the ground point) for the circuit of Fig. 6-66.

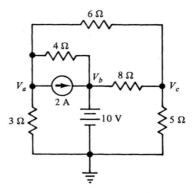

Figure 6-66   Problem 6-19.

20. Using nodal analysis, solve for the voltage V_{ab} for the circuit of Fig. 6-67. Notice that conductance values are specified.

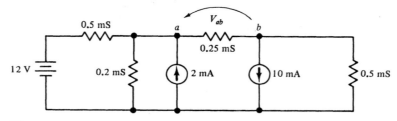

Figure 6-67   Problem 6-20.

21. Solve by superposition for the current I in the circuit of Fig. 6-10(a).

22. Find the current in the 40 Ω resistor of Fig. 6-68 using the superposition theorem.

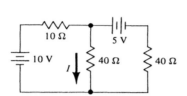

Figure 6-68   Problems 6-22 and 6-28.   Figure 6-69   Problem 6-23.

23. Use superposition to find the branch currents for Fig. 6-69.

24. Use superposition to determine I and V_{ab} for the circuit of Fig. 6-64.

25. Use superposition to determine V_{ab} for the circuit of Fig. 6-65.

26. Solve for the Thevenin circuits that could replace the networks of Fig. 6-70.

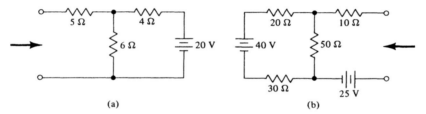

(a)   (b)

Figure 6-70   Problems 6-26 and 6-32.

27. Use Thevenin's theorem to find the range of current that flows in the load resistor R_L of Fig. 6-71 if R_L varies between 1 and 5 Ω.

28. Referring to Fig. 6-68, find the specified current in the 40 Ω resistor by replacing the rest of the network by a Thevenin equivalent.

29. Obtain the Thevenin equivalent circuit looking back into terminals a and b of Fig. 6-72.

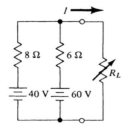

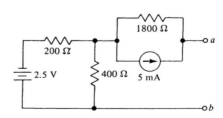

Figure 6-71   Problem 6-27.   Figure 6-72   Problems 6-29 and 6-33.

30. By using Thevenin's theorem, solve for the current in the branch *a–b* in the circuit of Fig. 6-73.

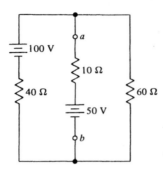

Figure 6-73   Problems 6-30 and 6-34.

31. A battery has an open-circuit voltage of 12.6 V and a short-circuit current of 40 A. What is **(a)** its Thevenin equivalent and **(b)** its Norton equivalent?

32. Find the Norton equivalent circuits for the circuits of Fig. 6-70.

33. Obtain the Norton equivalent circuit for Fig. 6-72.

34. Referring to terminals *a* and *b* of Fig. 6-73, find the equivalent Norton circuit.

35. Refer to Fig. 6-74. How many equations are necessary to solve for the current *I* by **(a)** loop analysis and **(b)** node analysis? Replace everything to the left of the 5 Ω resistor by a Thevenin equivalent and everything to the right of the resistor by a Thevenin equivalent; then solve for the current *I*.

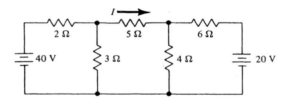

Figure 6-74   Problem 6-35.

36. What load resistance if connected to terminals 1 and 2 of Fig. 6-75 would receive maximum power from the network?

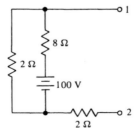

Figure 6-75   Problem 6-36.

37. Referring to Fig. 6-76, what value must R_L be for maximum power transfer? What is the maximum power deliverable to R_L?

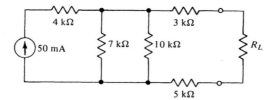

Figure 6-76   Problem 6-37.

38. Referring to Fig. 6-77, determine the value of R_L for maximum power transfer and evaluate this power.

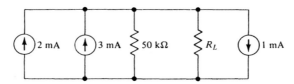

Figure 6-77   Problem 6-38.

39. Given the circuits of Fig. 6-78, each of which contains a dependent source, determine the voltage across the 5 Ω resistor in each case.

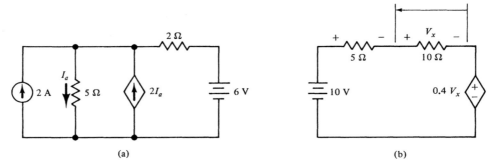

(a)   (b)

Figure 6-78   Problem 6-39.

40. Determine the voltage across R_L for the simplified transistor circuit of Fig. 6-79.

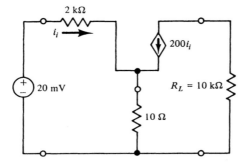

Figure 6-79   Problem 6-40.

41. Determine v_L and i_L for the simplified op amp circuit of Fig. 6-80.

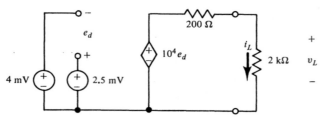

Figure 6-80   Problem 6-41.

42. Determine v_L for the op amp circuit of Fig. 6-81.

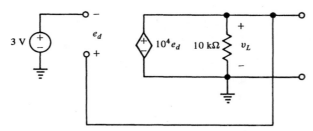

Figure 6-81   Problem 6-42.

43. Determine the equivalent Δ or Y networks for Fig. 6-82.

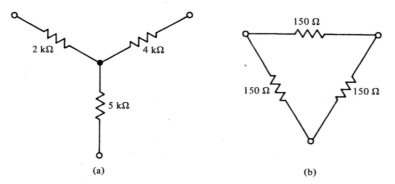

Figure 6-82   Problem 6-43.

44. Show the progressive simplifications necessary in calculating the resistance between terminals a and b in Fig. 6-83.

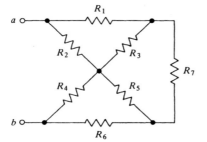

Figure 6-83   Problem 6-44.

45. Determine the resistances between terminals a and b for the circuits in Fig. 6-84.

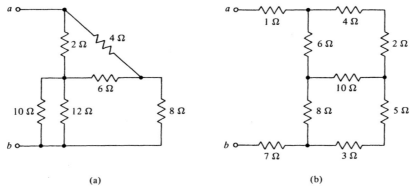

(a)   (b)

Figure 6-84   Problem 6-45.

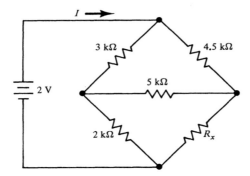

Figure 6-85   Problem 6-46.

46. Given the bridge circuit of Fig. 6-85, determine **(a)** the value of R_x for balance, **(b)** the voltage across R_x at balance, and **(c)** the current I at balance.

Chapter 7

The Electric Field and Capacitance

7.1 THE ELECTROSTATIC FIELD

The previous chapters were concerned with the resistive properties of electric circuits. Resistance, which is the opposition to current flow, is associated with the dissipation of energy. In addition to resistance, an electric circuit may also possess the properties of inductance and capacitance, both of which are associated with the storage of energy. Inductance, which is the property of an electric circuit to oppose any change in current through the circuit, is considered in Chapter 9.

On the other hand, *capacitance is the property of an electric circuit to oppose any change in voltage across the circuit.* Alternatively, capacitance is the ability of an electric circuit to store energy in an electrostatic field. *Electrostatics*, the study of electricity at rest, was introduced in Chapter 1. The concepts of electrostatics may be summarized as follows: When electrification is brought about, equal positive and negative charges result; like charges repel, and unlike charges attract; and the force exerted by two charges is given by Coulomb's law,

$$F = k\frac{Q_1 Q_2}{r^2} = \left(\frac{1}{4\pi\epsilon}\right)\frac{Q_1 Q_2}{r^2} \tag{7-1}$$

The force given by Eq. 7-1 is in newtons when Q_1 and Q_2 are in coulombs, r is in meters, and k is in terms of a new constant ϵ (Greek epsilon), called the *absolute permittivity*, a term defined in the next section. The value of k given in Chapter 1 (9×10^9) holds only for the case in which the charges are separated by air or free space. For any other medium, the k must be calculated using the proper value of ϵ.

From the discussion of Coulomb's law, it can be said that the region surrounding a charge is characterized by the fact that forces act on another charge brought

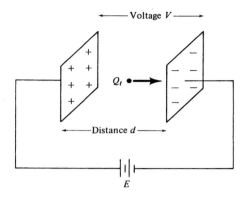

Figure 7-1 Path of a test charge in an electric field.

into the region. This leads to the term *electric* or *electrostatic field*, which is a region in space wherein a charge, a test charge, experiences an electric force. The test charge Q_t is taken as a positive charge sufficiently small so that it does not disturb the existing field. Consider the two charged plates of Fig. 7-1. An insulating material, such as air, separates the plates and it is assumed that the voltage E is low enough not to cause a breakdown. As indicated by Fig. 7-1, the left-hand plate becomes positively charged, since the positive terminal of the voltage source removes enough electrons to equalize the charge in that portion of the circuit. Similarly, the right-hand plate becomes negatively charged, since the negative terminal of the battery forces excess electrons onto it.

It is seen that an electric field exists between the plates in Fig. 7-1 because a positive test charge Q_t placed between the plates will be simultaneously repelled from the positive plate and attracted to the negative plate. A test charge moves or drifts along a certain path within an electric field. To help explain the action of the electric field, the test particle is said to move along an invisible line of force. Specifically, an *electric line of force* represents the path along which a test charge moves when attracted by another charge.

Electric lines of force are considered by convention to possess the following characteristics: (1) they originate on a positive charge and terminate on a negative charge, and (2) they enter or leave a charged body normal (perpendicular) to its surface. The first characteristic is illustrated by Fig. 7-2, which shows the lines of force representing the electric field between two unlike charges and between two like charges. The perpendicular emanation of force lines from a surface is illustrated in Fig. 7-3, which shows the force lines leaving radially from a sphere and a current-carrying conductor, which forms a cylindrical charged body.

If a charge Q_2 is brought into the field of another charge Q_1, a force will be exerted on Q_2 in accordance with Eq. 7-1. But if Q_2 is relatively small compared to Q_1 so that it does not disturb the field of Q_1, then the intensity of the electric field at the point occupied by Q_2 can be measured in terms of the force exerted on the charge Q_2. The electric force per unit charge at a particular point in space is defined as the

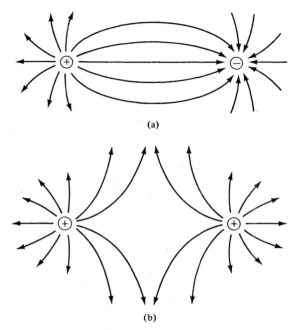

(a)

(b)

Figure 7-2 Electric field between **(a)** two unlike charges and **(b)** two like charges.

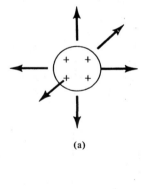

(a)

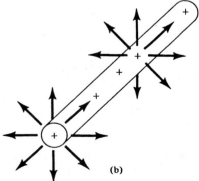

(b)

Figure 7-3 Flux lines leaving **(a)** a sphere and **(b)** a conductor.

electric field intensity. The symbol for electric field intensity is $\mathscr{E}$ and by definition

$$\mathscr{E} = F/Q \qquad \text{N/C} \qquad (7\text{-}2)$$

In the special case of two charged plates, as in Fig. 7-1, the electric field intensity is uniform between the plates. Consider that the plates are separated by a distance d and that a positive test charge Q_t is moved from the negative plate to the positive plate. Work is done in moving the test charge against the electric field; this work is the product of the average electric force F and the distance d over which it acts:

$$W = Fd$$

In Chapter 2, a volt is defined as a joule per coulomb. Alternatively, a charge Q_t moving against a potential V requires work given by

$$W = Q_t V$$

Equating the two expressions for work, we obtain

$$Fd = Q_t V$$

Then, dividing both sides by Q_t and comparing the results with Eq. 7-2, we see that the electric field intensity is

$$\mathscr{E} = V/d \qquad \text{V/m} \qquad (7\text{-}3)$$

Equation 7-3 indicates that for a set of parallel plates having a voltage difference V, the potential is distributed uniformly along a line of force. Even though the two units for field intensity are equivalent, volts per meter is preferred to newtons per coulomb.

The question arises as to how many electric lines of force are used to represent the field around a charge. The *electric flux* or total number of lines of electric force is indicated by the Greek letter ψ (psi) and is given by Gauss's law (Germany, 1777–1855). *Gauss's law* states that the electric flux passing through any closed surface is equal to the total charge enclosed:

$$\psi = Q \qquad (7\text{-}4)$$

Thus the flux is given in coulombs or lines of force because by Gauss's law there is one force line per unit charge. The number of flux lines per unit area measured in coulombs per square meter or lines per square meter is the *electric flux density*. Mathematically, the flux density D is specified by

$$D = \psi/A \qquad (7\text{-}5)$$

EXAMPLE 7-1

An isolated point charge of 60×10^{-6} C is suspended in air. Determine **(a)** the field intensity at a point 0.1 m from the charge, and **(b)** the flux density at 0.1 m from the charge.

Solution. (a) From Eqs. 7-1 and 7-2, the field intensity is

$$\mathscr{E} = k\frac{Q}{r^2} = 9 \times 10^9 \frac{(60 \times 10^{-6})}{(0.1)^2} = 5.4 \times 10^7 \text{ V/m}$$

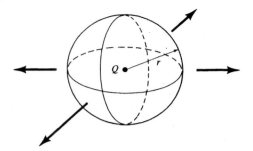

Figure 7-4 Example 7-1.

(b) The electric lines of force leave the point charge radially and pass through a spherical surface area, as shown by Fig. 7-4. Then

$$A = 4\pi r^2 = 4\pi(10^{-2})\,\text{m}^2$$

and

$$D = \frac{\psi}{A} = \frac{60 \times 10^{-6}}{4\pi(10^{-2})} = 4.77 \times 10^{-4}\,\text{C/m}^2.$$

Drill Problem 7-1 ■

What is the field intensity between the plates of Fig. 7-1 if 20 V is applied and the spacing of the plates is 20 mm.

Answer. 1 kV/m.   ■ ■

7.2 DIELECTRIC MATERIALS

According to Section 2.7, insulators or dielectrics are characterized by the fact that there are very few free electrons present, that is, the electrons are tightly bound to their nuclei. When a dielectric is placed in an electric field, free charges are not available; yet an electrical stress is created in the dielectric. The electric force acts in such a way that the positive and negative charges of the atom are pulled slightly away from each other, as shown in Fig. 7-5.

Without an electric field applied, a dielectric atom is symmetrical, as shown by Fig. 7-5(a), but in the presence of an electric field the atom becomes unsymmetrical, as in Fig. 7-5(b). The electrons shift so that they are closer to the positive charge causing the field, that is, toward the origin of the flux lines. Atoms so disturbed are said to be *polarized* and are sometimes called *induced dipoles*. When the field is removed, the atoms revert back to their normal states, although some residual polarization may remain. The consequence of this residual polarization is that even though the charged surfaces that originally created the field are momentarily connected, a voltage may later be measured between those surfaces as those atoms with residual polarization return to their normal states.

A measure of how well electric lines of force are established in a dielectric is provided by the term *permittivity*. The *absolute permittivity* ϵ of a dielectric medium

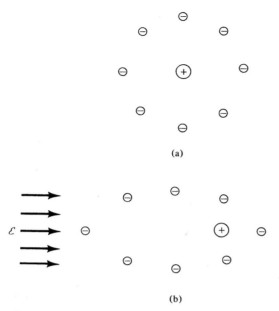

Figure 7-5 (a) A dielectric atom (no field). (b) A polarized atom (in a field).

is the ratio of the flux density in the dielectric to the electric field across the dielectric:

$$\epsilon = D/\mathscr{E} \tag{7-6}$$

Coulomb's law (Eq. 7-1) shows that the unit for ϵ is square coulombs per newton-square meter. An alternative unit is the farad per meter, where a *farad* (F) is defined as a square coulomb per newton meter.

The permittivity of a vacuum of free space is a constant ϵ_0, given by

$$\epsilon_0 = \frac{1}{36\pi \times 10^9} = 8.85 \times 10^{-12} \quad \text{F/m}$$

The *relative permittivity* ϵ_r is the ratio of absolute permittivity of a material to that for a vacuum and is thus defined as

$$\epsilon_r = \epsilon/\epsilon_0 \quad \text{(dimensionless)} \tag{7-7}$$

The relative permittivity, sometimes called the *dielectric constant*, is a ratio of similar quantities and thus dimensionless. The dielectric constant for air is 1.0006, so that the absolute permittivity for air is usually taken as ϵ_0. Other dielectric constants are listed in Table 7-1.

EXAMPLE 7-2

In Example 7-1, the field intensity was found to be 5.4×10^7 V/m at a point 1.0 m from the given charge. Using the formula relating the field intensity to flux density, calculate D and compare it to the value obtained in Example 7-1.

TABLE 7-1   AVERAGE DIELECTRIC CONSTANTS
AND STRENGTHS

Dielectric	Dielectric constant (ϵ_r)	Dielectric strength (V/mm)
Air	1.0006	3 000
Bakelite	5	21 000
Glass	6	35 000
Mica	5	60 000
Oil	4	10 000
Paper	2.5	20 000
Rubber	3	25 000
Teflon	2	60 000

Solution.   By Eq. 7-6,

$$D = \epsilon\mathscr{E} = 8.85 \times 10^{-12} \times 5.4 \times 10^{7}$$
$$= 4.77 \times 10^{-4} \text{ C/m}^2$$

This is the same result as in Example 7-1.

In Chapter 2, *dielectric strength* was defined as the voltage per unit thickness at which breakdown occurs. The dielectric strength thus corresponds to the field intensity required for breakdown. As the field intensity is increased, the polarization of the dielectric atoms becomes more pronounced. Finally, a value of field intensity may be reached at which so much force is exerted on the orbital electrons that they are torn free from their orbits. The dielectric strengths of some dielectrics are listed in Table 7-1. Notice that no relationship exists between the dielectric constant and the dielectric strength. The dielectric constant and dielectric strength vary with impurities, temperature, frequency, and other factors so that those values appearing in Table 7-1 are average or typical values.

7.3 CAPACITANCE

A certain relationship exists between the voltage applied to a pair of plates separated by a dielectric and the charge appearing on those plates. Consider the pair of plates in Fig. 7-6, which are initially uncharged; that is, $q = 0$, $v = 0$.

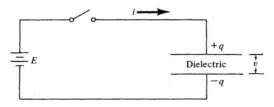

Figure 7-6   The simple capacitor.

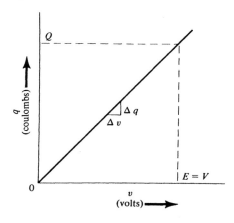

Figure 7-7 Charge versus voltage relationship.

When the switch is closed, charges from the source will distribute themselves on the plates, that is, a current will flow. Initially, this current i is large, but as more charge is accumulated, and hence more voltage developed across the plates, the accumulated charge tends to oppose the further flow of charge. Finally, when enough charge has been transferred from one plate to the other, a voltage $v = E$ will have been developed across the plates. The plates are then charged to a maximum, and since the voltage across the plates equals the source voltage, the current i must be zero. In the ideal situation, the transfer of charge occurs in zero time, but in the practical situation the charging process requires a very short but finite time.

Notice in Fig. 7-6 that the current, voltage, and charge are represented by small letters. The IEEE recommends the use of small letters to symbolize quantities that change with respect to time, *instantaneous quantities*.

If one plots the accumulated charge versus the voltage developed across the plates, one finds a linear relationship, as in Fig. 7-7. The constant of proportionality relating the charge and voltage, that is, the slope, is called the capacitance:

$$C = Q/V$$

or
$$Q = CV \tag{7-8}$$

The unit of capacitance is the coulomb per volt, called the *farad* in honor of the English physicist Michael Faraday (1791–1867). The farad is too large a unit for practical circuits; therefore, one finds capacitance values expressed in microfarads (μF, 10^{-6} farad) or picofarads (pF, 10^{-12} farad).

It is seen from the preceding discussion that capacitance results whenever charges are separated by a dielectric. It is possible to obtain an expression for capacitance in terms of the dielectric and other physical factors by considering the parallel plates of Fig. 7-8. A charge is placed on the plates, each having a surface area A. Then the flux density between the plates is given by Eqs. 7-4 and 7-5 as

$$D = Q/A$$

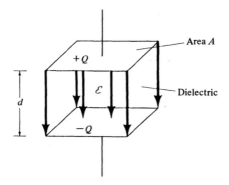

Figure 7-8 Parallel-plate capacitor.

As a consequence of the charge, a voltage V occurs across the capacitor and the field intensity given by Eq. 7-3 is

$$\mathscr{E} = V/d$$

Substituting the preceding expressions for D and $\mathscr{E}$ into Eq. 7-6, we find that

$$\epsilon = \frac{D}{\mathscr{E}} = \left(\frac{Q}{V}\right)\left(\frac{d}{A}\right)$$

but since Q/V is the capacitance,

$$C = \epsilon A/d \qquad\qquad (7\text{-}9)$$

Equation 7-9 indicates that the capacitance is determined by the geometric factors A and d and by the type of dielectric separating the plates. When the plate area is increased, the capacitance is increased. Similarly, when the separation between plates is decreased, the capacitance is increased. Substituting the expression for ϵ (Eq. 7-7) into Eq. 7-9, we obtain

$$C = \frac{\epsilon_r \epsilon_0 A}{d} = \frac{\epsilon_r (8.85 \times 10^{-12})A}{d} \qquad\qquad (7\text{-}10)$$

where C is in farads and A and d are in square meters and meters, respectively. In turn, a *capacitor* is a device specifically designed to have capacitance.

EXAMPLE 7-3

A 300 pF capacitor is constructed using a mica dielectric between two plates, each having an area of 1×10^{-4} m^2. What is the thickness of the dielectric?

Solution. From Table 7-1, $\epsilon_r = 5$. Then, using Eq. 7-10,

$$C = \frac{\epsilon_r \epsilon_0 A}{d} = 300 \text{ pF}$$

or $$d = \frac{5(8.85 \times 10^{-12})(1 \times 10^{-4})}{300 \times 10^{-12}} = \frac{43.3 \times 10^{-4}}{300} = 1.47 \times 10^{-5} \text{ m}$$

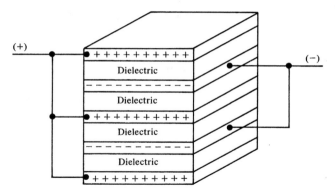

Figure 7-9 Multiple-plate capacitor.

In order to obtain a large plate area with reasonable overall dimensions, a capacitor is sometimes constructed by stacking a number of plates as in Fig. 7-9. In multiple-plate capacitors, both sides of the plates are used so that alternate plates carry the same sign of charge. Usually, an odd number of plates is used. Then the two outside plates that are connected can be joined to the negative charge (usually the ground point in electronic circuits). In this case the outside plates tend to form a *shield*, eliminating the effects of external fields.

It is seen in Fig. 7-9 that the total effective plate area equals $(n - 1) \times$ the area of one plate, where n is the total number of plates. The equation for a multiple-plate capacitor is thus

$$C = \frac{(n - 1)\epsilon_r\epsilon_0 A}{d} \tag{7-11}$$

Substituting the constant ϵ_0, we obtain

$$C = \frac{(n - 1)(\epsilon_r)(8.85 \times 10^{-12})A}{d} \tag{7-12}$$

It is assumed in the derivation of the capacitance relationship (Eq. 7-9) that all of the flux lines pass within the area bounded by the body of the capacitor. However, leakage of flux lines, called *fringing*, does take place at the edges of the plates (see Fig. 7-10). Since the same number of flux lines passes through a larger than

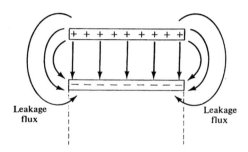

Leakage flux

Leakage flux

Figure 7-10 Fringing.

expected area, the capacitance is slightly larger than expected. The plate separation is usually small enough that the effects of fringing can be neglected.

It must be pointed out that capacitance is present whenever two conductors are separated by a dielectric. Equation 7-10 holds only for the parallel-plate configuration. However, the capacitance for any particular situation is found to depend on the same parameters: dielectric type, charge surface, and charge separation. In fact, there are times when we are "stuck" with capacitive effects not purposely designed by us. Some of these effects are discussed later in this chapter.

Drill Problem 7-2 ■

An air dielectric capacitor is constructed with 7 square plates each measuring 1 cm on an edge and separated by 1 mm. What is its value of capacitance?

Answer.   5.31 pF.   ■ ■

Drill Problem 7-3 ■

What is the maximum voltage that can be applied to the capacitor of Drill Problem 7-2 if the dielectric strength is not to be exceeded?

Answer.   3 kV.   ■ ■

Drill Problem 7-4 ■

What charge is present on a 10 μF capacitor that has a voltage of 5 V across it?

Answer.   50 μC.   ■ ■

7.4 TYPES OF CAPACITORS

In the preceding section the dielectric separating the charged plates of a capacitor was considered ideal; that is, it was assumed that no current flows in the dielectric. However, as noted in Chapter 2, dielectrics do have some free electrons, and the free electrons that flow in the dielectric when an electric field is applied constitute a *leakage current*. The leakage current is usually so small that it is considered negligible; in fact, its effect can be represented by a very high resistance (approximately 1000 MΩ) shunting the capacitor as indicated in Fig. 7-11. This shunting resistance is called *leakage resistance*. A consequence of the leakage resistance is that if a charge is placed on a capacitor and not "refreshed," it will eventually leak through the dielectric. However, for purposes of circuit analysis the leakage resistance is usually high enough for the circuit to be considered open.

Commercially available capacitors are described by the dielectric used and by whether the capacitor is fixed or variable. The symbols for fixed and variable capacitors are shown in Fig. 2-30 and are repeated in Fig. 7-12 for convenience. The curved line of the capacitor symbol represents that plate of the capacitor that is connected to the point of lower potential. It also corresponds to the outside foil of a rolled

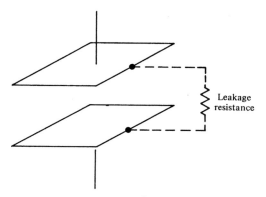

Figure 7-11 Dielectrics offer some leakage resistance.

capacitor, which in turn is identified by a stripe on the body of the capacitor. Cutaway views illustrating the construction features of some typical fixed capacitors appear in Fig. 7-13.

One type of *ceramic capacitor* consists of a ceramic disk or tube with a dielectric constant ranging from 10 to 10 000. A thin layer of silver is applied to each side of the dielectric, leads are attached, and an outer insulated coating is applied. Another type of ceramic capacitor, the multilayer type, consists of metal plates stacked alternately with ceramic dielectrics and then molded or heated into a single block. This construction is called monolithic (from the Greek for "single stone"). Two monolithic ceramic capacitors are shown in Fig. 7-13. The radial dipped ceramic capacitor in Fig. 7-13(a) has lead wires, whereas the ceramic chip capacitor in Fig. 7-13(b) has solder pads on the ends to facilitate the soldering of the chip to a circuit board. Ceramic capacitors are characterized by low loss and small size. They range in values from a few picofarads to about 2 μF.

The *paper capacitor* consists of aluminum foil and Kraft paper (usually impregnated with wax or resin) rolled and molded to form a compact unit. Paper capacitors range from about 0.0005 μF to about 2 μF.

The *plastic film capacitor* is very similar in construction to the paper capacitor. Plastic film dielectrics such as polystyrene and polyester separate the metal foil used for the plates. The capacitor is rolled and housed in a metal or plastic case.

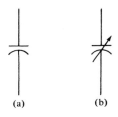

(a)   (b)

Figure 7-12 Capacitor symbols: **(a)** Fixed; **(b)** variable.

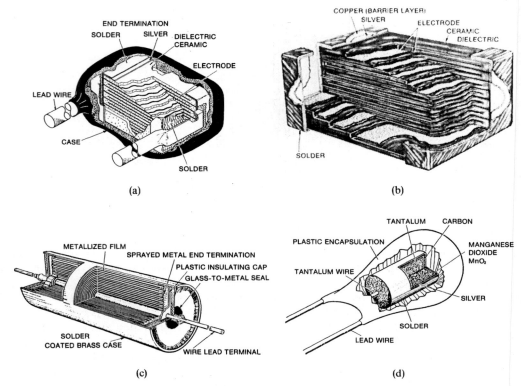

Figure 7-13 The construction features of some typical fixed capacitors. **(a)** KEMET® radial dipped monolithic ceramic capacitor. **(b)** KEMET® monolithic ceramic chip capacitor. **(c)** Hermetically sealed metalized film capacitor. **(d)** KEMET® Ultradip solid tantalum capacitor. (Courtesy of Union Carbide Corporation, Electronics Divisions.)

The *mica capacitor* consists of a stack of alternate layers of mica dielectric and conducting foil. The basic unit is then enclosed in a molded case of phenolic resin or in a "dip-coated" case. Silvered mica capacitors differ from the foil type in that the conducting plate is a thin layer of silver that is fired onto the mica surface. The silvered mica units have closer tolerances and a particular capacitance change versus temperature relationship.

An *electrolytic capacitor* consists of two plates separated by an electrolyte and a dielectric as in Fig. 7-14. As shown in Fig. 7-14(a), one of the plates is oxidized and it is this oxide that forms the dielectric. The electrolyte is the true negative plate, and it has the ability to oxidize the positive plate where any imperfections in the oxide dielectric may exist. The second metallic plate serves as the negative terminal and provides the contact surface for the electrolyte. Generally, electrolytic capacitors utilize aluminum or tantalum as the base metal. Tantalum oxide has almost twice the dielectric constant of aluminum oxide, thereby resulting in a higher capacitance per volume. The electrolyte may be in a liquid, paste, or solid form.

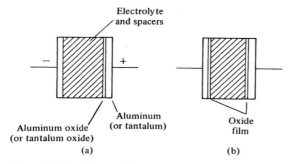

Figure 7-14 Electrolytic capacitors: **(a)** Polarized; **(b)** nonpolarized.

The oxide dielectric continues to act as an insulator only as long as the same polarity is used that was originally used in forming the oxide. Thus a capacitor with a configuration such as in Fig. 7-14(a) requires that the polarity of the applied voltage never be reversed. A voltage of reversed polarity causes a substantial current to flow, with subsequent damage to the capacitor. This electrolytic is of the *polarized* type.

A *nonpolarized* electrolytic is constructed by forming an equal oxide film on both terminal plates, as in Fig. 7-14(b). This type of capacitor is used where the voltage is expected to reverse.

Electrolytic capacitors have high values of capacitance, ranging from about one microfarad to on the order of thousands of microfarads. The leakage currents are somewhat higher than those of other capacitors.

Variable capacitors generally use air as the dielectric and have one set of plates meshing between a second set of plates. The fixed plate assembly is called the *stator* and the movable set is called the *rotor*. Because of fringing, the minimum value of capacitance occuring when the plates are unmeshed is not zero but a finite value.

Another type of variable capacitor is the *trimmer* or *padder*, consisting of two or more plates separated by a dielectric of mica. A screw is mounted so that tightening the screw compresses the plates more tightly against the dielectric, thereby reducing the dielectric thickness and increasing the capacitance.

It should also be pointed out that fixed capacitors may be made by thin-film techniques. In this process a conducting material is deposited on a ceramic substrate. Next, a layer of dielectric is applied; this is followed by another conducting layer. Thin-film capacitors are generally used in microcircuits and although there is no upper limit on the size, capacitors much over 0.01 μF become so large that one is better off attaching separate discrete capacitors to the microcircuit assembly. The voltage ratings of thin-film capacitors are also limited, but capacitors having breakdown voltages of 10–20 V are easily produced.

Capacitance values may be stamped on the capacitor or may be indicated by color-coded bands or dots. The color code introduced for resistors in Section 3.5 is basically the same one used for capacitors; however, the band or dot significance varies for different types of capacitors. Additionally, a voltage rating may be stamped or color-coded on the capacitor. Figure 7-15 shows some external features of capacitors.

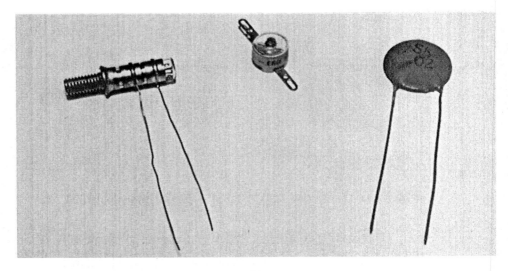

(a)

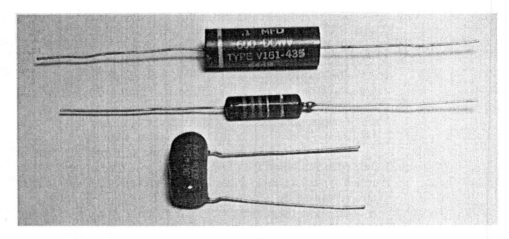

(b)

(c)

Figure 7-15 The external features of some capacitors. **(a)** Ceramic (left to right): variable, trimmer, disc; **(b)** (top to bottom) paper, plastic film, mica; **(c)** Air variable.

7.5 SERIES AND PARALLEL CAPACITORS

When capacitors are connected in series, the impressed emf E is divided among the capacitors and the equivalent or total capacitance C_t is less than that of the smallest unit. This is indicated in the analysis that follows.

Consider three capacitors connected in series, as in Fig. 7-16. The capacitors are initially uncharged so that when a potential V_t is applied with the indicated polarity, plate a becomes positively charged and plate c' negatively charged. As a result of this charging action, plate a' takes on an equal negative charge, while plate c takes on an equal positive charge; in similar fashion, plates b and b' become charged. Thus all plates in a series capacitor circuit acquire exactly the same charge, $Q_t = Q_1 = Q_2 = Q_3$, and so on.

In accordance with Kirchhoff's voltage law, the total V_t is

$$V_t = V_1 + V_2 + V_3$$

where by Eq. 7-8

$$V_1 = \frac{Q_1}{C_1} \qquad V_2 = \frac{Q_2}{C_2} \qquad V_3 = \frac{Q_3}{C_3}$$

Then

$$V_t = \frac{Q_t}{C_t} = \frac{Q_t}{C_1} + \frac{Q_t}{C_2} + \frac{Q_t}{C_3}$$

Dividing this last equation by Q_t and solving for C_t, one obtains

$$C_t = \frac{1}{1/C_1 + 1/C_2 + 1/C_3} \tag{7-13}$$

One recognizes that Eq. 7-13 for series capacitors is similar to that for the total resistance of parallel resistors. For two capacitors in series, the total capacitance is found by the product over the sum relationship,

$$C_t = \frac{C_1 C_2}{C_1 + C_2} \tag{7-14}$$

When capacitors are connected in parallel, the total charge flowing to the combination divides among the capacitors and the total capacitance is the sum of the individual capacitances. Consider the parallel capacitor circuit of Fig. 7-17. Each

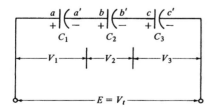

Figure 7-16   Capacitors in series.

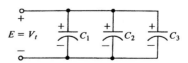

Figure 7-17   Capacitors in parallel.

capacitor takes a charge given by

$$Q_1 = C_1 V_1 \qquad Q_2 = C_2 V_2 \qquad Q_3 = C_3 V_3$$

where

$$V_t = V_1 = V_2 = V_3$$

The total charge transferred is then

$$Q_t = C_t V_t = Q_1 + Q_2 + Q_3 = C_1 V_t + C_2 V_t + C_3 V_t$$

Dividing the last equation by V_t, we obtain an expression for the total parallel capacitance. This expression is given in general by Eq. 7-15.

$$C_t = C_1 + C_2 + C_3 + \cdots + C_n \tag{7-15}$$

EXAMPLE 7-4

Three series capacitors are connected to a 120 V source as depicted in the series connection of Fig. 7-16. If $C_1 = 10 \ \mu\text{F}$, $C_2 = 20 \ \mu\text{F}$, and $C_3 = 30 \ \mu\text{F}$, what is (a) the total capacitance and (b) the voltage across each capacitor?

Solution. (a) The total capacitance is calculated from Eq. 7-13:

$$\frac{1}{C_t} = \frac{1}{C_1} + \frac{1}{C_2} + \frac{1}{C_3}$$

$$= \frac{1}{10 \ \mu\text{F}} + \frac{1}{20 \ \mu\text{F}} + \frac{1}{30 \ \mu\text{F}}$$

$$= 0.1 \times 10^6 + 0.05 \times 10^6 + 0.0333 \times 10^6$$

$$= 0.1833 \times 10^6 \qquad 1/\text{F}$$

Then

$$C_t = \frac{1}{0.1833 \times 10^6} = 5.45 \times 10^{-6} = 5.45 \ \mu\text{F}$$

(b) For the series combination,

$$Q_1 = Q_2 = Q_3 = Q_t = C_t V_t$$

$$= 120(5.45 \times 10^{-6}) = 6.54 \times 10^{-4} \ \text{C}$$

Then

$$V_1 = \frac{Q_1}{C_1} = \frac{6.54 \times 10^{-4}}{10 \times 10^{-6}} = 65.4 \ \text{V}$$

$$V_2 = \frac{Q_2}{C_2} = \frac{6.54 \times 10^{-4}}{20 \times 10^{-6}} = 32.7 \ \text{V}$$

$$V_3 = \frac{Q_3}{C_3} = \frac{6.54 \times 10^{-4}}{30 \times 10^{-6}} = 21.8 \ \text{V}$$

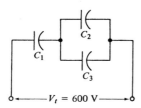

Figure 7-18 Example 7-5 and Drill Problem 7-6.

EXAMPLE 7-5

Find **(a)** the total capacitance, **(b)** the charge on each capacitor, and **(c)** the voltage across each capacitor when 600 V is applied across the capacitor configuration of Fig. 7-18. Here $C_1 = 12\ \mu F$, $C_2 = 6\ \mu F$, and $C_3 = 30\ \mu F$.

Solution. $C_2 \| C_3 = 6 + 30 = 36\ \mu F$.

(a) $$C_t = \frac{(12)(36)}{12 + 36} = \frac{36}{4} = 9\ \mu F$$

(b) $$Q_t = C_t V_t = (9 \times 10^{-6})(600) = 54 \times 10^{-4}\ C = 5.4 \times 10^{-3}\ C$$

But for the series combination,

$$Q_1 = Q_t = Q_2 + Q_3 = 5.4 \times 10^{-3}\ C$$

Then the capacitor voltages are

(c) $$V_1 = \frac{Q_1}{C_1} = \frac{5.4 \times 10^{-3}}{12 \times 10^{-6}} = 4.5 \times 10^2 = 450\ V$$

$$V_2 = V_3 = 600 - 450 = 150\ V$$

Then

$$Q_2 = C_2 V_2 = (6 \times 10^{-6})(150) = 0.9 \times 10^{-3}\ C$$
$$Q_3 = C_3 V_3 = (30 \times 10^{-6})(150) = 4.5 \times 10^{-3}\ C$$

Notice that when capacitors are connected in series, the voltage ratio between any two capacitors is the inverse of their capacitance ratio.

Drill Problem 7-5 ■

What is the total capacitance for four capacitors, each 0.1 μF, connected **(a)** in series and **(b)** in parallel?

Answer.   (a) 0.025 μF. (b) 0.4 μF.   ■ ■

Drill Problem 7-6 ■

Refer to Fig. 7-18. If $C_1 = 10$ pF, $C_2 = 4$ pF, $C_3 = 2$ pF, and $V_t = 20$ V, determine **(a)** the total capacitance and **(b)** the voltages across C_1 and C_2.

Answer.   (a) 3.75 pF. (b) 7.5 and 12.5 V.   ■ ■

7.6 TRANSIENTS

The voltages and currents considered so far have been unidirectional quantities of constant value. These constant-valued quantities are called *steady state* quantities.

Consider the action of charging a capacitor. The potential of the capacitor cannot change instantaneously; that is, a definite time is required for the transition of the capacitor from the uncharged to the charged state. Similarly, a circuit having inductance requires a finite time for any change in current to occur in the inductance. In electric circuits, these unsettled or temporary states are called *transient states*. Transient states are generally associated with the opening and closing of switches, but a circuit may be in a state of change for other reasons, such as a change in applied voltage.

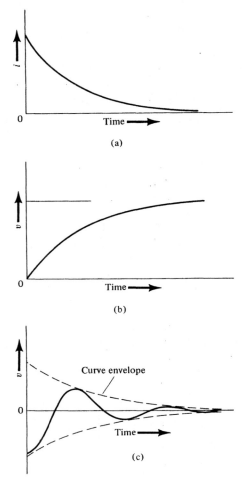

Figure 7-19 (a) Exponential decay of current; (b) exponential increase of voltage; (c) exponential envelope of a decaying oscillating voltage.

The majority of transient states are characterized by voltages and currents that increase or decrease in an exponential manner, as indicated in Fig. 7-19. The exponential family of curves used to mathematically describe electrical transients is that family given by

$$y = e^{-x} \tag{7-16}$$

where e, the base of natural logarithms, $= 2.71\,828$, x is a function of time, and y is the voltage or current. Equation 7-16 is plotted in Fig. 7-20(a). The complete curve is shown in Fig. 7-20(a), but since x is a function of time, that part of the curve where $x \geq 0\,(t \geq 0)$ is more important and is shown as a continuous line.

On the other hand, the basic equation of exponential growth is given by

$$y = 1 - e^{-x} \tag{7-17}$$

and is shown in Fig. 7-20(b) for $x \geq 0$.

Although both curves begin at a definite initial point, they never reach a final value. However, they do approach a final value closely enough so that one can

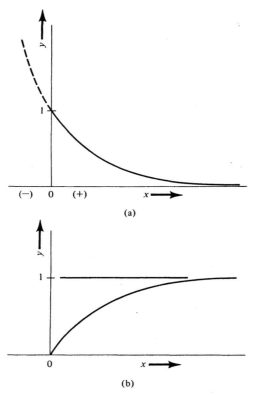

(a)

(b)

Figure 7-20 (a) Graph of $y = e^{-x}$ and graph of $y = 1 - e^{-x}$ for $x \geq 0$.

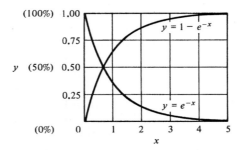

Figure 7-21   Universal exponential curves.

consider for all practical purposes that a final value has been reached. For instance, for $x \geq 5$, the value of y is within 1% of its final value.

Tables of the exponential functions of Eqs. 7-16 and 7-17 for different values of x and y are found in most mathematics handbooks. Just as conveniently, scientific hand calculators and computers also provide the exponential function and its "inverse" or antilogarithm.

We can compare the increasing and decreasing exponential functions by plotting on common x and y axes as in Fig. 7-21. Such curves are called *universal exponential curves*. Notice that the y axis has a normalized scale of zero to unity. When we actually use the exponential curves for plotting transient currents or voltages, the maximum value is determined by the circuit and may be any finite value, not necessarily of magnitude 1. Hence, it is convenient to think of the y axis scale as a percentage scale, as also indicated in Fig. 7-21. The universal exponential curves are useful for solving graphically for charging and discharging transient values.

7.7 THE CHARGING TRANSIENT

The charging concept was introduced in Section 7-3, where it was pointed out that when an uncharged capacitor is connected to a source, a current flows until the charge stored on the capacitor produces a potential exactly equal to the source potential. In turn, the change from an initial uncharged state to a final charged state implies a transient state during which the current i varies in some manner as a function of time.

During the transient period, the charge and voltage across the capacitor are constantly changing. Using instantaneous quantities in Eq. 7-8, one has

$$q = Cv$$

and for a small change in voltage Δv, the change in charge is

$$\Delta q = C \, \Delta v \tag{7-18}$$

In calculus, infinitesimal changes are studied and the symbol Δ is replaced by a d, so that Eq. 7-18 in terms of infinitesimal changes is expressed

$$dq = C \, dv_c \tag{7-19}$$

A subscript c has been added to the voltage term to denote a capacitor voltage, and although it is redundant now, it becomes necessary later.

Furthermore, since the charge and voltage are changing with time, it is appropriate to express their infinitesimal changes with respect to time.

$$\frac{dq}{dt} = C\frac{dv_c}{dt} \tag{7-20}$$

The quantities dq/dt and dv_c/dt are the respective changes in charge and voltage occurring in the infinitesimal time dt; that is, they are rates of change of q and v_c. The rate of change of charge with respect to time, however, is the instantaneous current. Thus

$$i = C\frac{dv_c}{dt} \tag{7-21}$$

The charging circuit introduced in Section 7.3 is an ideal situation, since no resistance is shown in series with the capacitor. Some resistance is always present in the charging circuit, even if only in conductors; therefore, a more accurate circuit diagram is Fig. 7-22.

The charging transient of Fig. 7-22 begins when the switch is closed, at a time called $t = 0$ s. If the switch is closed, the voltage equation is

$$E = v_R + v_C$$
$$= iR + v_C \tag{7-22}$$

At the instant the switch is closed, the capacitor, which is considered initially uncharged, has zero voltage ($v_C = 0$) so that Eq. 7-22 becomes

$$E = iR + 0$$

or
$$i = I_0 = E/R \tag{7-23}$$

The initial charging current I_0 is thus limited by the resistance of the circuit and is determined by the simple application of Ohm's law. On the other hand, after sufficient time the capacitor is fully charged and no current flows. Then, from Eq. 7-22,

$$E = 0 + v_C \tag{7-24}$$

Equations 7-23 and 7-24 state only the initial and final conditions for the simple charging circuit, and it becomes necessary to obtain expressions for v_C and i as

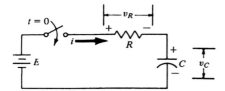

Figure 7-22   The charging circuit.

functions of time throughout the transient period. Returning to Eq. 7-22, we can substitute Eq. 7-21. for the current with the resulting equation

$$E = RC \frac{dv_C}{dt} + v_C \qquad (7\text{-}25)$$

This last equation is a differential equation relating the variable v_C to the known constant terms of the circuit E, R, and C. Solution of the differential equation yields the instantaneous voltage across the capacitor:

$$v_C = E(1 - e^{-t/RC}) \qquad (7\text{-}26)$$

The voltage across the resistor is then

$$v_R = E - v_C = E - E + Ee^{-t/RC} = Ee^{-t/RC} \qquad (7\text{-}27)$$

and

$$i = \frac{v_R}{R} = \frac{E}{R} e^{-t/RC} = I_0 e^{-t/RC} \qquad (7\text{-}28)$$

The expression for the current (Eq. 7-28) is also obtained by calculus from Eq. 7-21.

The factor of $e^{-t/RC}$ in all the transient equations clearly indicates that the current i_C and voltage v_R approach zero at the same exponential rate as v_C approaches E (see Fig. 7-23).

The product RC appearing in the exponential terms has the units of

$$RC = (\text{ohms})(\text{farads}) = \left(\frac{\text{volts}}{\text{ampere}} \right) \left(\frac{\text{coulombs}}{\text{volt}} \right)$$

$$= \left(\frac{\text{volts}}{\text{coulomb/second}} \right) \left(\frac{\text{coulombs}}{\text{volt}} \right) = \text{seconds}$$

thereby making the exponent of e a dimensionless quantity. This product is defined

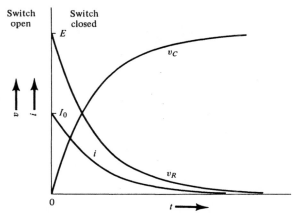

Figure 7-23   The charging transients.

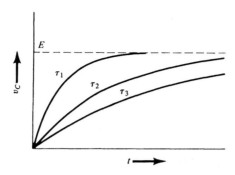

Figure 7-24 Effect of τ on the charging curve: $\tau_1 < \tau_2 < \tau_3$.

as the *time constant* of the charging circuit and has the symbol τ (tau):

$$\tau = RC \tag{7-29}$$

Since the time constant affects the exponential term, it determines how fast the curve changes. Figure 7-24 shows the relative effect of τ on the transient curve; for a larger τ the rate of change is less.

It is interesting to note the change that occurs in a time equal to one time constant. For $t = \tau = RC$, the exponential factor becomes $e^{-1} = 0.368$ so that

$$v_C = E(1 - 0.368) = 0.632E$$

and

$$i = I_0(0.368) = 0.368I_0$$

That is, *during one time constant, the voltage rises to 63.2% of its final value, while the current falls to 36.8% of its initial value.*

EXAMPLE 7-6

The switch in Fig. 7-22 is closed at $t = 0$. If $R = 50$ kΩ, $C = 40$ μF, and $E = 90$ V, what is the capacitor voltage 1 s after the switch is closed?

Solution

$$\tau = RC = (50 \times 10^3)(40 \times 10^{-6}) = 2\,\text{s}$$
$$v_C = E(1 - e^{-t/\tau}) = 90(1 - e^{-1/2})$$
$$= 90(1 - 0.606) = 35.4\,\text{V}$$

EXAMPLE 7-7

Using the information from Example 7-6, determine the time required for the voltage across C to reach 70 V.

Solution. $\tau = 2$ s. Substituting x temporarily for t/τ gives

$$v_C = 70\,\text{V} = E(1 - e^{-x}) = 90(1 - e^{-x})$$
$$\frac{70}{90} = 0.778 = 1 - e^{-x}$$

so that

$$e^{-x} = 1 - 0.778 = 0.222$$

Then

$$x = t/\tau = 1.505$$

and finally

$$t = 1.505\tau = 1.505(2) = 3.01 \text{ s}$$

Quite often the capacitor in an RC charging circuit has an initial charge on it and, therefore, has a voltage V_0 at $t = 0$. The polarity of the charge may be such that V_0 opposes the applied voltage or aids the applied voltage, as shown in Fig. 7-25. The current in each case is given in general by the expression

$$i = I_0 e^{-t/RC}$$

but the inital current depends on the initial voltage V_0. It should be apparent that for the opposing case

$$I_0 = \frac{E - V_0}{R}$$

and for the aiding case

$$I_0 = \frac{E + V_0}{R}$$

The instantaneous voltage across the capacitor must similarly be expressed in terms of the initial voltage V_0. The capacitor voltage may be obtained as the algebraic sum of the initial voltage plus a transient term expressing the additional voltage to which the capacitor may charge. The following example illustrates the case of a charged capacitor.

EXAMPLE 7-8

The switch in Fig. 7-25(a) is closed at $t = 0$, opened at $t = 1$ s, and held open for 1 s. It is alternately closed and opened at 1 s intervals. If $R = 40 \text{ k}\Omega$, $C = 40 \text{ }\mu\text{F}$, $E = 100$ V, and $V_0 = 0$ initially, plot a curve of i and v_C for the first 4 s.

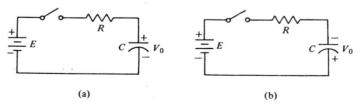

(a)   (b)

Figure 7-25 Initial charge on a capacitor: **(a)** opposing; **(b)** aiding.

Solution.   Sketches of the current and voltage vs. time are first obtained as follows.

$t = 0$ *to* $t = 1$ s *(switch closed)*
The capacitor is charging from 0 V to a voltage v_1 occurring at $t = 1$ s

$t = 1$ *to* $t = 2$ s *(switch open)*
The capacitor remains charged at the voltage level v_1. No current flows.

$t = 2$ *to* $t = 3$ s *(switch closed)*
The capacitor has an initial voltage v_1 and proceeds to charge through the additional difference in voltage $(E - v_1)$ to the voltage at the end of the interval v_3.

$t = 3$ *to* $t = 4$ s *(switch open)*
The capacitor remains charged at the voltage level v_3. No current flows. The curves are shown in Fig. 7-26. The mathematical expressions for the current and capacitor voltage and their values at the end of the 1 s intervals are then obtained.

$t = 0$ *to* $t = 1$ s

$$\tau = RC = (40 \times 10^3)(40 \times 10^{-6}) = 1.6 \text{ s}$$

$$i = I_0 e^{-t/RC} = \frac{E}{R} e^{-t/1.6} = \frac{100}{40 \times 10^3} e^{-t/1.6}$$

$$= 2.5 \, e^{-t/1.6} \text{ mA}$$

$$v_C = E(1 - e^{-t/RC}) = 100(1 - e^{-t/1.6}) \text{ V}$$

and at $t = 1$ s,

$$e^{-1/1.6} = e^{-0.626} = 0.535$$
$$i_1 = 2.5(0.535) \text{ mA} = 1.335 \, \text{mA}$$
$$v_1 = 100(1 - 0.535) \text{ V} = 46.5 \text{ V}$$

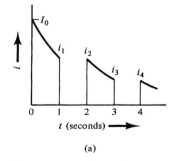

(a)

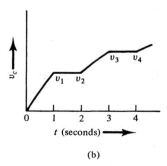

(b)

Figure 7-26   Curves for Example 7-8; **(a)** current; **(b)** capacitor voltage versus time.

$t = 1\ to\ t = 2$ s

No charging takes place. At $t = 2$ s,

$$i_2 = i_1 = 1.335\ \text{mA}$$
$$v_2 = v_1 = 46.5\ \text{V}$$

$t = 2\ to\ t = 3$ s

$$i = \frac{E - v_1}{R} e^{-t/1.6} = \frac{53.5}{40 \times 10^{-3}} e^{-t/1.6}$$
$$= 1.335 e^{-t/1.6}\ \text{mA}$$

$$v_C = v_2 + (100 - v_2)(1 - e^{-t/1.6})$$
$$= 46.5 + (53.5)(1 - e^{-t/1.6})$$

Notice that the last two expressions hold only for the interval $t = 2$ to $t = 3$ s, and the time variable actually begins at $t = 2$ s so that at $t = 3$ s, one second of time has elapsed since the start of the interval. Thus at $t = 3$ s,

$$i_3 = 1.335 e^{-0.626} = 1.355(0.535) = 0.714\ \text{mA}$$

$$v_3 = 46.5 + 53.5(1 - 0.535)$$
$$= 46.5 + 24.9 = 71.4\ \text{V}$$

$t = 3\ to\ t = 4$ s

No additional charging takes place.

Drill Problem 7-7 ■

A series combination of a 1 kΩ resistor and a 20 μF capacitor has what time constant?

Answer.   0.02 s.   ■ ■

Drill Problem 7-8 ■

A 5 μF capacitor, initially uncharged, is charged through a 27 kΩ resistor. **(a)** What is the time constant of the circuit? **(b)** How long after a voltage is connected to the circuit does it take for the capacitor to charge to 80% of the applied voltage?

Answer.   (a) 0.135 s. (b) 0.217 s.   ■ ■

Drill Problem 7-9 ■

A 0.02 μF capacitor initially uncharged, a 4 kΩ resistor, a switch, and a 5 V source are all connected in series. Determine the voltage across the capacitor and the current in the circuit 50 μs after the switch is closed.

Answer.   2.32 V; 670 μA.   ■ ■

Drill Problem 7-10 ■

Refer to Fig. 7-25(a) for which $E = 5$ V, $R = 4.7$ kΩ, and $C = 0.05$ μF. With the switch opened, the capacitor has an initial charge so that $V_0 = 2$ V. **(a)** What is the initial charge on the capacitor? **(b)** How long after the switch is closed does it take to increase the capacitor voltage to 4.6 V?

Answer. (a) 0.1 μC. (b) 474 μs. ■ ■

7.8 THE DISCHARGING TRANSIENT

Discharging is the process of removing charge from a previously charged capacitor with a subsequent decay in capacitor voltage. Consider Fig. 7-27.

If the switch is initially placed in position 1, the capacitor will charge toward the supply voltage E and after 5τ can be considered fully charged. If the switch is then placed in position 2, the capacitor is directly across the resistor so that the charge leaks through the resistor. The Kirchhoff voltage relationship for the discharge case is

$$iR + v_C = 0 \tag{7-30}$$

Substituting the expression for i into Eq. 7-30, we obtain

$$RC\frac{dv_C}{dt} + v_C = 0$$

This differential equation if solved by calculus yields the expression for v_C.

$$v_C = V_0 e^{-t/RC} \tag{7-31}$$

From Eq. 7-30 or by differentiation of Eq. 7-31 in accordance with the current equation (Eq. 7-21), the discharge current is

$$i = -\frac{V_0}{R} e^{-t/RC} \tag{7-32}$$

The minus sign indicates that the current is opposite to that shown in Fig. 7-27, that is, the discharge current has a direction opposite to that for the charge current. Of course, the V_0 appearing in the discharge equations is the initial voltage on the capacitor at the start of the discharge transient. The discharge equations are summarized by Fig. 7-28.

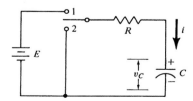

Figure 7-27 A charge–discharge circuit.

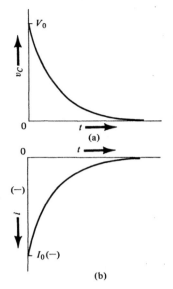

Figure 7-28 Discharge curves:
(a) voltage; **(b)** current.

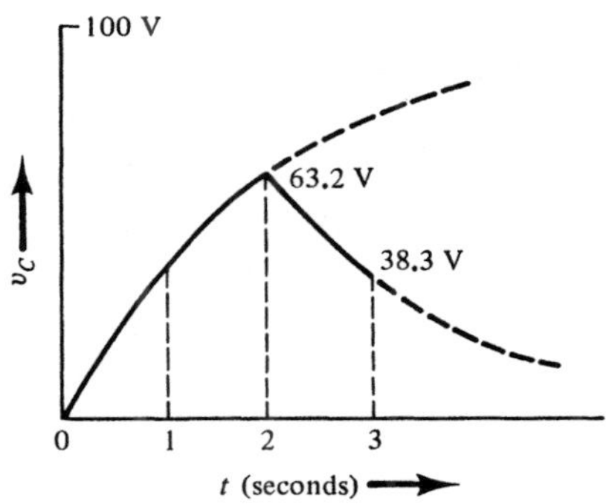

Figure 7-29 Voltage versus time for
Example 7-9

EXAMPLE 7-9

Refer to Fig. 7-27. The capacitor is initially uncharged, $E = 100$ V, $R = 2$ MΩ, and $C = 1$ μF. The switch is placed into position 1 for 2 s and is then placed into position 2. What is the capacitor voltage at the end of 1 s of discharge?

Solution. For both charging and discharging $\tau = RC = (2 \times 10^6)(1 \times 10^{-6}) = 2$ s. Thus 2 s after the switch is placed into position 1, $v_C = 63.2\%(E) = 63.2$ V. The switch is then thrown to position 2 and $V_0 = 63.2$ V, and 1 s after that,

$$v_C = V_0 e^{-t/2} = 63.2 e^{-1/2}$$
$$= 63.2(0.606) = 38.3 \text{ V}$$

The plot of the capacitor voltage is shown in Fig. 7-29.

EXAMPLE 7-10

A 10 μF capacitor is charged to 12 V and is then allowed to discharge through a 200 kΩ resistor. What current flows **(a)** after 3 s of discharge **(b)** after 5 s of discharge?

Solution. The time constant is $\tau = RC = (2 \times 10^5)(10 \times 10^{-6}) = 2$s. The currents are then calculated:

(a)

$$i = -\frac{V_0}{R} e^{-t/RC} = -\frac{12}{(200 \times 10^3)} e^{-3/2}$$
$$= -60 \text{ } \mu\text{A} (0.223) = -13.4 \text{ } \mu\text{A}$$

(b)
$$i = -60 \ \mu A \ e^{-5/2} = -60 \ \mu A \ (0.082)$$
$$= -4.93 \ \mu A$$

Drill Problem 7-11 ■

The capacitor in Fig. 7-27 is initially uncharged, and $E = 20$ V, $R = 2$ kΩ, and $C = 40 \ \mu F$. What is the capacitor voltage after 200 ms of charge followed by 100 ms of discharge?

Answer. 5.26 V. ■ ■

Drill Problem 7-12 ■

A 40 μF electrolytic capacitor is charged to 100 V and is then disconnected from the charging circuit. If the leakage resistance is 100 MΩ, how long does it take for the voltage to drop to 50 V?

Answer. 46.2 min. ■ ■

7.9 INITIAL AND FINAL VALUES

As was pointed out earlier, a transient RC state begins at a particular time $t = 0$. At that time the values of voltage and current in the RC circuit are referred to as the *initial values* or *initial conditions*. We denote the voltage across the capacitor at $t = 0$ as V_0 and the current flowing to the capacitor at $t = 0$ as I_0.

After the transition is complete (ideally, this occurs after infinite time, that is, at $t = \infty$, but practically it occurs after a few time constants), we have the *final values* or *final conditions* for the circuit. We refer to the final voltage and current as V_f and I_f.

We have studied the simple RC circuits in which the capacitor voltage increases toward the applied source voltage or decays toward 0 V. Now we shall study some cases for which a capacitor voltage increases to some other voltage than the applied source voltage or decays to a voltage other than zero. These situations are indicated in the graphs of Fig. 7-30.

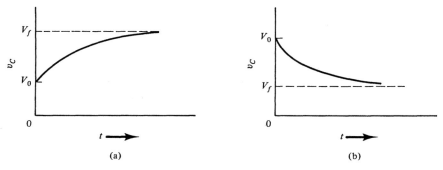

(a)   (b)

Figure 7-30 **(a)** Capacitor charge and **(b)** discharge curves in terms of initial and final values.

It should not be hard to recognize from Fig. 7-30 that at any time after $t = 0$ the equation that describes the capacitor voltage during charge is

$$v_C = V_0 + (V_f - V_0)(1 - e^{-t/\tau}) \qquad (7\text{-}33)$$

and that which describes the capacitor voltage during discharge is

$$v_C = V_f + (V_0 - V_f)\, e^{-t/\tau} \qquad (7\text{-}34)$$

Actually, with a little algebraic manipulation, Eq. 7-33 reduces to Eq. 7-34. However, do not let that confuse you! Since the V_f and V_0 values are at different levels, their effect in Eq. 7-34 is different for the charge and discharge cases. Thus, Eq. 7-34 can be used for both charge and discharge. However, whether the voltage increases or decreases exponentially, the capacitor current will decay as the capacitor charges or discharges. Thus, for either case

$$i_C = I_0 e^{-t/\tau} \qquad (7\text{-}35)$$

Once again, I_0 depends on the solution of Kirchhoff's equations at $t = 0$.

In evaluating initial and final conditions we must remember that an uncharged capacitor has an initial voltage $V_0 = 0$ and behaves as a short circuit. After charging, the capacitor voltage is V_f, no current flows, and the capacitor acts as an open circuit.

EXAMPLE 7-11

After being in position 1 for a long time, the switch represented in Fig. 7-31 is moved to position 2. Evaluate the initial and final values of voltage and current for the capacitor.

Solution.   While in position 1, the capacitor will have charged to $v_C = 12$ V and the current will have decayed to 0. When the switch is moved to position 2, $V_0 = 12$ V, and by inspection it is seen that $V_f = 15$ V.

We solve for I_0 by writing a KVL equation for $t = 0$:

$$-15 + I_0(100 \text{ k}\Omega) + V_0 = 0$$

$$I_0 = \frac{15 - V_0}{100 \text{ k}\Omega} = \frac{15 - 12}{100 \times 10^3} = 30 \ \mu\text{A}$$

Of course, $I_f = 0$.

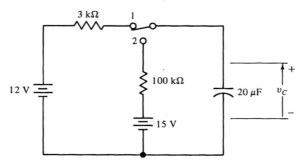

Figure 7-31   Circuit for Example 7-11.

EXAMPLE 7-12

Using the circuit and information from Example 7-11, determine the capacitor voltage and current 4 s after the switch is placed in position 2.

Solution.   First, we evaluate

$$\tau = RC = (100 \times 10^3)(20 \times 10^{-6}) = 2 \text{ s}$$

Then, with substitutions into Eq. 7-34, we have

$$v_C = v_f + (V_0 - V_f)e^{-t/\tau}$$
$$= 15 + (12 - 15)e^{-t/2} = 15 - 3e^{-4/2}$$
$$= 15 - 3(0.135) = 15 - 0.406 = 14.6 \text{ V}$$

From Eq. 7-35 we have

$$i_C = I_0 e^{-t/\tau} = (30 \ \mu\text{A})e^{-4/2} = 4.06 \ \mu\text{A}$$

Sometimes a transient problem actually involves different time constants for different parts of a multiple transient occurrence. For example, in Fig. 7-31 the capacitor charges through a 3 kΩ resistor when the switch is in position 1 and through a 100 kΩ resistor with the switch in position 2. Although in Example 7-12 we were only interested in the time constant when the switch was in position 2, we should recognize the circuit conditions for both switch conditions. Furthermore, the resistance through which a capacitor charges or discharges may not be as readily apparent as in our previous examples.

EXAMPLE 7-13

Determine the voltage equation v_C for $t \geq 0$ for the circuit of Fig. 7-32(a). The capacitor is initially uncharged.

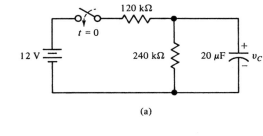

(a)

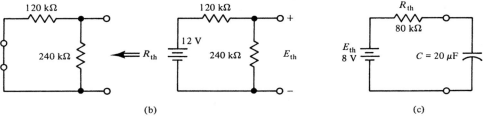

(b)   (c)

Figure 7-32   (a) Circuit for Example 7-13. **(b)** Determination of the Thevenin equivalent. **(c)** The equivalent charging circuit.

Solution.　For analysis at $t \geq 0$, we remove the capacitor and find a Thevenin equivalent for the remainder of the circuit. As noted for Fig. 7-32(b),

$$R_{th} = 120 \text{ k}\Omega \,\|\, 240 \text{ k}\Omega = \frac{(120)(240)}{360} = 80 \text{ k}\Omega$$

and by voltage division

$$E_{th} = 12 \text{ V}\left(\frac{240 \text{ k}\Omega}{360 \text{ k}\Omega}\right) = 8 \text{ V}$$

hence, we have the equivalent charging circuit of Fig. 7-32(c). Then

$$V_0 = 0, \qquad V_f = 8 \text{ V}$$

and 　　　　　　　　$\tau = R_{th}C = (80 \times 10^3)(20 \times 10^{-6}) = 1.6 \text{ s}$

Substituting into Eq. 7-34 we obtain

$$v_C = 8 + (0 - 8)e^{-t/1.6}$$
$$= 8 - 8e^{-t/1.6} \qquad \text{V}$$

Drill Problem 7-13 ■

The switch shown in Fig. 7-33 is placed in position 1 for 3 s and is then transferred to position 2. **(a)** What is the v_C as the switch is moved to position 2? **(b)** What is the final voltage across C? **(c)** What is v_C 2 s after the switch is placed in position 2? The capacitor is initially uncharged.

Answer.　(a) 45.1 V. (b) 80 V. (c) 67.2 V.　　　　　　　　　　■ ■

Drill Problem 7-14 ■

The switch shown in Fig. 7-33 is initially in the OFF position and the capacitor has a voltage of 60 V. The switch is placed in position 1 for 10 s and is then transferred to position 2. **(a)** What is the v_C as the switch is moved to position 2? **(b)** What initial current flows when the switch is moved to position 2? **(c)**What is v_C 2 s after the switch is placed in position 2?

Answer.　(a) 94.6 V. (b) 73 μA. (c) 85.4 V.　　　　　　　　■ ■

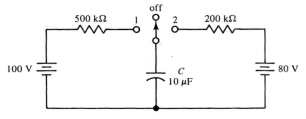

Figure 7-33　Drill Problems 7-13 and 7-14.

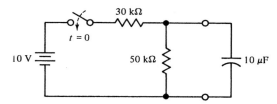

Figure 7-34 Drill Problem 7-15.

Drill Problem 7-15 ■

Determine the initial charging current and final voltage for the capacitor in Fig. 7-34. The capacitor is initially uncharged.

Answer. 0.333 mA, 6.25 V. ■ ■

7.10 ENERGY STORAGE

If a capacitor has no charge on it and hence no difference of potential between the plates, no work is needed to take a charge from one plate to the other. However, after a finite but small charge has been placed on the capacitor, work must be done on an infinitesimal charge dq in carrying it from one plate to another (the plates having a difference of potential V). The V, which is proportional to q, increases linearily as the charge increases (see Fig. 7-7), and hence the work changes for a given q.

In Chapter 2 the volt was defined as the work per unit charge. Alternatively, the small amount of work dw in moving the charge dq through a potential is

$$dw = V \, dq$$

This small amount of work is the area of the small rectangle of Fig. 7-35.

The total work done in charging a capacitor to a voltage V_1 by a transference of q_1 units of charge is given by the area under the curve,

$$W = \tfrac{1}{2} q_1 V_1$$

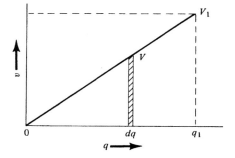

Figure 7-35 Voltage versus charge curve.

and since $q_1 = CV_1$

$$W = \tfrac{1}{2}CV_1^2 \qquad\qquad (7\text{-}36)$$

where W is in joules when C is in farads and V is in volts.

EXAMPLE 7-14

A 40 μF capacitor has an initial voltage of 20 V. If the voltage is increased to 40 V, what is the additional amount of energy stored?

Solution.   Using Eq. 7-36, we obtain

$$W_1 = \tfrac{1}{2}CV_1^2 = \tfrac{1}{2}(40 \times 10^{-6})(20)^2 = 8 \times 10^{-3} \text{ J}$$
$$W_2 = \tfrac{1}{2}CV_2^2 = \tfrac{1}{2}(40 \times 10^{-6})(40)^2 = 32 \times 10^{-3} \text{ J}$$
$$\Delta W = W_2 - W_1 = 24 \times 10^{-3} \text{ J}$$

7.11 SOME CAPACITOR APPLICATIONS

The energy storage feature makes the capacitor a useful device for the generation of a large amount of current for a very short time. For example, an electric spot welder may have a capacitor that charges through a high resistance so as not to draw a severely high current from a source. The capacitor then discharges by a short-duration pulse of very high current through the very low resistance of the welding materials. This is depicted in the circuit of Fig. 7-36, in which the capacitor serves as an energy source for the low resistance. The switch represents a more elaborate switching circuit.

The ability of a capacitor to oppose any change in voltage makes it very useful for spark or arc suppression. Normally, when a switch is opened there is a tendency for a spark or arc to form at the switch contacts. This arc tends to pit the switch contacts, thereby reducing the life of the switch. A capacitor connected across the contacts, as in Fig. 7-37, absorbs the energy that would otherwise create the arc. The resistor R may be needed to prevent welding at the switch contacts when the switch is closed and the capacitor is discharged.

Another application involving the opposition that a capacitor provides to a change in voltage is the *smoothing* of rectified *ac*. As you will study in electronics,

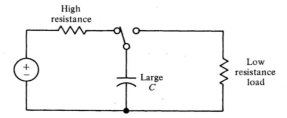

Figure 7-36   The capacitor used as an energy sink.

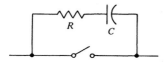

Figure 7-37 Spark suppression.

rectifiers convert alternating voltages to pulsating, unidirectional voltage. Thus, as depicted in Fig. 7-38(a), a pulsating voltage may appear across the rectifier load R. If we connect a capacitor across the output of the rectifier, as in Fig. 7-38(b), and properly choose the time constant for RC, we can allow a slow discharge of the capacitor during the time the pulse value is rapidly falling. Hence, we obtain a smoothing action of the load voltage; that is, the voltage becomes more uniform.

Finally, the series RC circuit can serve as a timing or time-delay circuit. Since a definite time must elapse for an initially uncharged capacitor to charge to a specified voltage, one may connect a sensing device across the capacitor so that when the voltage reaches the predetermined value the voltage operates the sensing device.

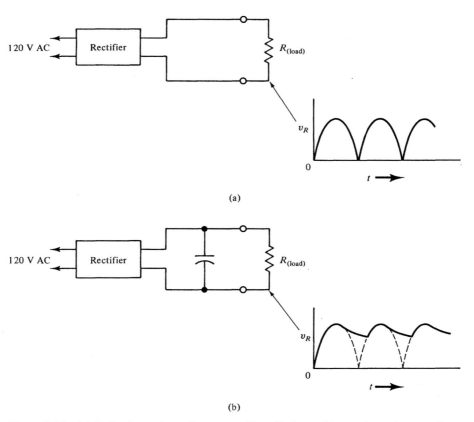

(a)

(b)

Figure 7-38 **(a)** Pulsating voltage from a rectifier. **(b)** Smoothing action of a capacitor.

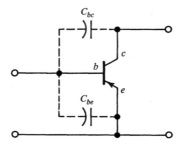

Figure 7-39   Stray capacitances for the bipolar transistor.

7.12 STRAY AND DISTRIBUTED CAPACITANCE

Recall that the effect of capacitance is present whenever there are charged surfaces separated by a dielectric. We should not be surprised that capacitive effects occur, even for the simple case of a single current-carrying wire resting against a metallic chassis. Capacitance that exists not by design but simply because two charged surfaces are close to each other is called *stray capacitance*. Although stray capacitance is usually negligible, exceptions occur when high-frequency ac or digital signals are present.

Many electronic devices have stray capacitances between their electrical regions. For example, the bipolar transistor has stray capacitances C_{be} and C_{bc} between the base, emitter, and collector regions ($b, e,$ and c) as depicted in Fig. 7-39. If a pulse is applied to the transistor, as indicated in Fig. 7-40, the stray capacitance C_{be} and the generator resistance R_g form an RC circuit. If the time constant of the circuit is

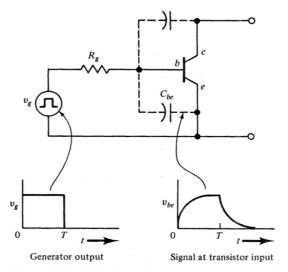

Generator output   Signal at transistor input

Figure 7-40   Pulse degradation due to stray capacitance in a transistor.

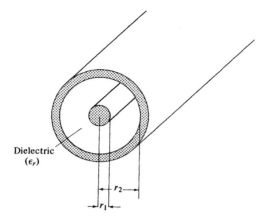

Dielectric
(ϵ_r)

Figure 7-41 Coaxial cable.

not low enough, the input voltage v_{be} varies considerably from the desired input pulse. The actual, degraded input signal possesses increasing and decreasing exponential sides as indicted in Fig. 7-40.

If stray capacitance is not localized but spread along lengths of wire or cable, we consider the capacitance to be *distributed*. Thus, communications personnel must consider the capacitance of transmission lines such as the coaxial cable of Fig. 7-41. For a coaxial cable with constants specified by Fig. 7-41 the capacitance is given by

$$C = \frac{0.0388\epsilon_r}{\log_{10}{(r_2/r_1)}} \qquad \left(\frac{\mu F}{mile}\right) \tag{7-37}$$

Similarly, power-line companies must consider the capacitance of parallel transmission wires, particularly for high-voltage situations.

QUESTIONS

1. What is capacitance?
2. What is an electric field?
3. What is a line of force?
4. State some characteristics of flux lines.
5. What is meant by electric field intensity and what are its units?
6. What is Gauss's law?
7. Flux density is defined in terms of what quantities?
8. What is dielectric polarization?
9. Define the terms "relative permittivity" and "absolute permittivity."
10. What is meant by dielectric constant?
11. What is dielectric strength?
12. What is the relationship between dielectric strength and dielectric constant?

13. When are small letters used for electrical quantities?
14. What is a farad?
15. What is the capacitance expression for a pair of parallel plates?
16. What is the effect of stacking plates?
17. What is fringing and what are its effects?
18. What is the difference between a polarized and nonpolarized electrolytic?
19. What is leakage current?
20. Describe the different types of capacitors.
21. How does the total capacitance of a series group of capacitors compare to the individual capacitance values?
22. What is the difference between a transient and a steady state?
23. What is meant by time constant?
24. How does the time constant affect the charge and discharge curves?
25. How does an initial charge affect the charging of a capacitor?
26. What are some applications for capacitors?

PROBLEMS

1. Two point charges of 4 μC each are separated in air by a distance of 1 m. What force does each exert on the other?

2. Sketch the lines of force for a system of four equal and like charges, each of which is at a corner of a square.

3. What is the electric field intensity at a point 1 m from a charge of 8 μC?

4. A positive charge of 5×10^{-8} C is suspended in air. Find (a) the flux emanating from the charge, (b) the electric field intensity 2×10^{-2} m from the charge, and (c) the flux density 2×10^{-2} m from the charge.

5. A voltage of 400 V is placed on a set of parallel plates separated by a distance of 1 cm. Find (a) the electric field intensity between the plates and (b) the force that a 1 μC charge would experience if it were in the field between the plates.

6. The electric field intensity at a point in space is 350 kV/m. (a) What is the force on an 8 μC charge placed at this point? (b) What is the flux density at this point?

7. The dielectric constant of a phenolic material is 6.3. What is the absolute permittivity of the phenolic?

8. What value of voltage must be connected to an 8 μF capacitor for it to obtain a charge of 1.2×10^{-3} C?

9. What charge is on a 24 μF capacitor when it is connected across a 250 V source?

10. What is the capacitance between two parallel plates, each 0.03 m on a side, if the dielectric separating them is 0.01 m thick and has a dielectric constant of 6?

11. A capacitor is constructed of two square plates, each 2×2 cm. If the dielectric separation is 0.01 m, compute the capacitance if the dielectric is (a) air and (b) mica.

12. A certain capacitor has a value of 25 μF. If the thickness of dielectric is doubled, what is the new capacitance?

13. A capacitor with air dielectric has a value of 0.02 μF. When the capacitor is inserted in transformer oil, the capacitance becomes 0.08 μF. What is the dielectric constant of the transformer oil?

14. What is the capacitance of a capacitor made of 25 plates of aluminum foil, each 1 × 2 cm, separated by layers of mica 2 mm thick?

15. Each plate of an air variable capacitor is equivalent to a semicircle 3.8 cm in diameter. The thickness of each plate is 0.038 cm. The overall outside length of the capacitor is 1.25 cm (see the top view of the capacitor in Fig. 7-42). If there are 11 plates, what is the capacitance when the plates are fully meshed?

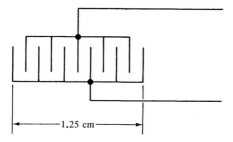

Figure 7-42 Problem 7-15.

16. What is the total capacitance of four 20 μF capacitors connected **(a)** in series and **(b)** in parallel?

17. What is the total capacitance of a 0.01 μF capacitor in series with a 0.02 μF capacitor? What is the voltage across each capacitor if 100 V is applied to the combination?

18. For the circuit of Fig. 7-43, what is **(a)** the total capacitance, **(b)** the voltage on each capacitor, and **(c)** the charge on each capacitor?

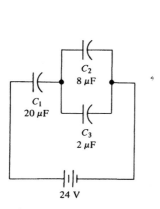

Figure 7-43 Problem 7-18.

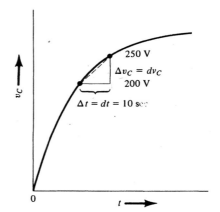

Figure 7-44 Problem 7-19.

19. The voltage across a 20 μF capacitor changes from 200 to 250 V in 10 s in accordance with with Fig. 7-44. If the charging curve is approximated by the dotted straight line shown, what average current flows during that 10 s, as determined by Eq. 7-21?

20. Referring to Fig. 7-22 and using $E = 90$ V and $C = 40$ μF, calculate the capacitor voltage 1 s after the switch is closed when R equals **(a)** 25 kΩ, **(b)** 100 kΩ.

21. The initial current for an RC charging circuit is given by E/R. Using Eq. 7-21, solve for the initial rate of change dv_C/dt. How does this initial slope or rate of change relate to the time constant?

22. In Fig. 7-22, $R = 22$ MΩ, $C = 16$ μF, and $E = 200$ V. If the capacitor is initially uncharged, determine the time required for the voltage across C to reach the values **(a)** 20 V, **(b)** 50 V, and **(c)** 100 V.

23. Using the data for Problem 20, calculate the current in the circuit 1 s after the switch is closed when R equals **(a)** 25 kΩ, **(b)** 100 kΩ.

24. Using the data for Problem 22, calculate the initial current. How much time must elapse for the current to decrease to one-half its initial value?

25. A capacitance of 20 μF and a resistance of 2 kΩ are connected in series with a 50 V source. Determine the time required for the current to drop to 10% of its initial value.

26. An 80 μF capacitor is charged to 632 V. A switch is closed, thereby discharging the capacitor through a 50 kΩ resistor. What will the capacitor voltage be after an elapse of **(a)** 1 s and **(b)** 4 s?

27. The switch of Fig. 7-45 is closed at $t = 0$ and opened at $t = 1$ s. The capacitor is initially uncharged. **(a)** What current flows just before the switch is opened? **(b)** What is the capacitor voltage after the switch is opened?

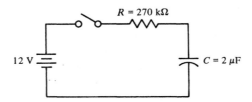

Figure 7-45   Problems 7-27 and 7-28.

28. The circuit of Fig. 7-45 is to be used to time a 5 s interval. The capacitor voltage is to trigger another circuit when it reaches 9.43 V. The capacitor is initially uncharged. What additional capacitance must be placed in parallel with the given value so that the trigger voltage is 9.43 V exactly 5 s after the switch is closed.

29. An uncharged capacitor is connected in series with a 12 V source and a resistor R. If the voltage is to increase to 8 V in 40 ms, what must be the time constant of the circuit?

30. A 20 μF capacitor is charged to 20 V and is then allowed to discharge through a 120 kΩ resistor. What current flows **(a)** after 3 s of discharge and **(b)** after 5 s of discharge?

31. Refer to Fig. 7-46 and consider the capacitor to be initially uncharged. Switch S_1 is closed at $t = 0$ and switch S_2 is closed 20 s later. **(a)** What is the capacitor voltage when switch S_2 is closed? **(b)** What is the capacitor current immediately before and immediately after S_2 is closed?

32. Suppose the capacitor shown in Fig. 7-47 is initially uncharged. **(a)** What are the initial and final values of currents i_1 and i_2? **(b)** What is the value of current i_2, 10 s after S_2 is closed?

33. Determine the initial current to and the final voltage across the capacitor of Fig. 7-48. The capacitor is initially uncharged.

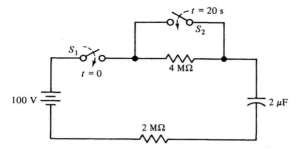

Figure 7-46 Problem 7-31.

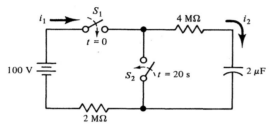

Figure 7-47 Problem 7-32.

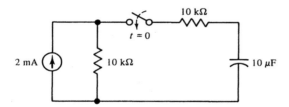

Figure 7-48 Problem 7-33.

34. The capacitor in Fig. 7-49 is initially uncharged. The switch is placed in position 1 for 2 s and then transferred to position 2. Consider the time at which the switch is transferred to position 2 as $t = 0$. **(a)** What are the initial values of voltage and current for the capacitor? **(b)** What are the final values of voltage and current for the capacitor?

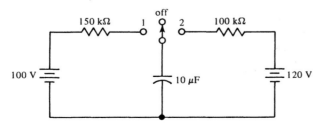

Figure 7-49 Problem 7-34.

35. Let the switch in Fig. 7-50 be closed at $t = 0$. What are the final voltages across the capacitors?

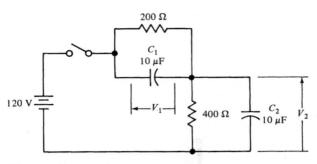

Figure 7-50   Problem 7-35.

36. To what final voltage does the capacitor in Fig. 7-51 charge? What is the time constant for the circuit?

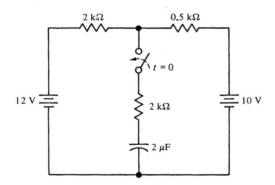

Figure 7-51   Problem 7-36.

37. Using Thevenin's theorem, find the maximum voltage to which the capacitor in Fig. 7-52 can charge. Through what charging resistance does it charge? What is the initial charging current?

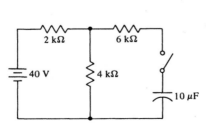

Figure 7-52   Problem 7-37.

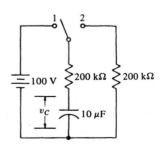

Figure 7-53   Problem 7-38.

38. The capacitor shown in Fig. 7-53 is initially uncharged. The switch is thrown first into position 1 for 1 s. Thereafter, it is thrown alternately into position 2 and back to position 1 at 1 s intervals. Plot the current and capacitor voltage for the first 4 s.

39. A coaxial cable has an inner conductor of AWG 14 wire. The outer conductor has negligible thickness and a diameter of $\frac{1}{4}$ in. If the dielectric has a relative constant of 2.0, what is the capacitance per mile for this cable?

40. An 80 μF capacitor and a 40 μF capacitor are connected in parallel across a 120 V source. Determine the energy stored in each.

41. A 3 μF capacitor is charged to 600 V. If the capacitor is discharged to 480 V, how much energy is removed from the capacitor?

42. Two capacitors, of 10 and 40 μF, are connected in series to a 100 V source. What energy is stored in each? What charge is stored on each?

Chapter 8

Magnetism and the Magnetic Circuit

8.1 MAGNETISM

The science of magnetism probably began with the early Greeks. Humans have long recognized the physical effects of magnetism, yet until this century no one understood why certain materials were magnetic. Even today many basic questions remain. It is now believed that all magnetic properties can be explained by quantum mechanics and electromagnetic theory as applied to systems of many atoms.

Magnetism is a property associated with materials that attract iron and iron alloys. Less familiar, but just as important, is the interrelationship between magnetism and electricity. A characteristic of this relationship, to be explained later, is that magnetism is associated with current-carrying conductors.

A body that possesses the property of magnetism is called a *magnet*. Magnets can be termed *natural* or *artificial*. A natural magnet is a material or a body that is magnetic in its original state; an artificial magnet is a body that possesses magnetism by induction. The lodestone (as it was called by early Greeks) is known today as the iron compound *magnetite* (Fe_3O_4) and is the principal natural magnet.

Certain materials (hardened-steel alloys) retain much of their magnetism long after the initial magnetizing force has been removed. These materials or bodies are called *permanent* magnets. Soft steel or iron retains only a small portion of the magnetism it receives by induction and is considered a *temporary* magnet. Because soft-steel alloys are so easy to magnetize and demagnetize, they are used as cores for *electromagnets*. The core of the electromagnet becomes a strong magnet only when electric current flows in the coil of wire surrounding the core.

Even though the first magnets were unattractive chunks of lodestone, early experimenters noticed some interesting phenomena. They noticed that freely suspended magnets orient themselves in a north–south direction. It was also found that pieces chipped from a magnet were magnetic, and magnet ends attracted and sometimes repelled other magnet ends. By experimentation it was found that certain substances, such as chromium, tungsten, and cobalt, when added in small amounts to steel, enhanced their magnetic properties. Similarly, heat treating was noticed to affect the magnetic property of a material.

Today, one has only to study the many electromagnetic devices, such as generators, motors, transformers, relays, and loudspeakers, to realize the many types and shapes of magnets.

8.2 THE MAGNETIC FIELD

In a classic experiment a sheet of paper is placed over a magnetic bar, iron filings are sprinkled onto the paper, and the paper is lightly tapped. The tapping allows the filings to arrange themselves in definite lines or paths (see Fig. 8-1).

As is evident in Fig. 8-1, a region of stress or force surrounds the magnet. This leads to the concept of a *magnetic field*, which is the region in space where a force acts upon a magnetic body. The filings and, thus, the force is concentrated at the ends; these regions are called *poles*.

If a bar magnet is suspended on a string, it orients itself so that one pole always points roughly toward geographic north and the other pole always points toward geographic south. The pole that seeks geographic north is called the north-seeking or simply *north pole* (N). Conversely, the pole seeking geographic south is called the south pole (S).

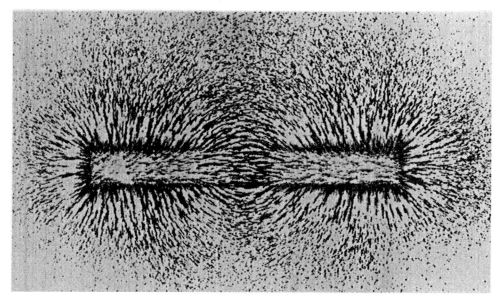

Figure 8-1   The field pattern around a bar magnet.

As with the electric field, it is convenient to use force or flux lines to describe a magnetic field. The *magnetic line of force* represents the path along which an isolated magnetic pole moves within a magnetic field. Faraday introduced the idea of flux lines only to simplify the analysis of force fields. One must remember that the field is continuous and exists between as well as along the lines of force.

Magnetic lines of force are considered by convention to have the following characteristics:

1. They form closed paths. (They travel from the north to the south pole outside the material and from the south to the north pole within a magnetized material.)
2. They repel each other. (They tend to separate.)
3. They tend to take the shortest path. (They act much like rubber bands exerting tension along their lengths.)

Using flux lines, we describe the magnetic field of the bar magnet as in Fig. 8-2. Because of the separation tendency, the lines do not cross one another.

The present theory of *ferromagnetism*, to be discussed later, substantiates the idea of continuous flux lines. Intuitively, it can be substantiated by breaking the bar magnet of Fig. 8-2 into two pieces (see Fig. 8-3). The reader may know from practical experience that two distinct magnets result. There is a force of attraction between the two pieces, indicating the tendency for the continuous flux lines to become as short as possible.

As noted in Fig. 8-3(a), there is a force of attraction between unlike poles. On the other hand, reversing one of the magnet pieces as in Fig. 8-3(b) causes a repelling action. Thus like poles repel.

It is not a simple matter to obtain "isolated" or "point source" magnetic poles. Charles Coulomb experimented with long, slender, magnetized needles separated

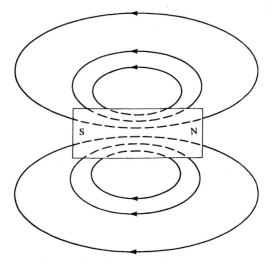

Figure 8-2   Flux lines around a bar magnet.

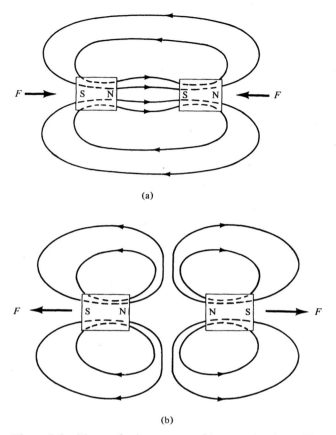

(a)

(b)

Figure 8-3 Pieces of a bar magnet: **(a)** attraction by unlike poles; **(b)** repulsion by like poles.

sufficiently to be considered point sources. He found that if the two "point source" magnetic poles had pole strengths of M_1 and M_2 and were separated by a distance r the force of attraction or repulsion was given by

$$F = k \frac{M_1 M_2}{r^2} = \left(\frac{1}{4\pi\mu}\right) \frac{M_1 M_2}{r^2} \tag{8-1}$$

The constant term k of Eq. 8-1 depends on the medium separating the poles. This dependence is also reflected by the constant μ, which is defined later as the *absolute permeability*.

Notice that the force equation for magnetic poles has the same form as the Coulomb equation for electric charges. In a parallel manner, we can define a *magnetic field intensity H* as the force per unit pole:

$$H = \frac{F}{M} \qquad \frac{\text{newtons}}{\text{webers}} \tag{8-2}$$

The *magnetic flux* or total number of lines of magnetic force is indicated by the Greek letter ϕ (phi). Often, the unit of flux is given as the *line* or *maxwell* (after James C. Maxwell, Scotland, 1831–1879). In SI units it is defined as the *weber* (after Wilhelm E. Weber, Germany, 1804–1891), where

$$1 \text{ weber} = 10^8 \text{ lines}$$

The number of flux lines passing perpendicularly through an area A is the *magnetic flux density B*. Mathematically, it is specified by

$$B = \frac{\phi}{A} \quad \frac{\text{webers}}{\text{square meters}} = \text{teslas} \tag{8-3}$$

It follows that in the SI units, flux density is in webers per square meters or teslas. However, the CGS unit, gauss, is still used frequently, and it should be noted that $1 \text{ T} = 10^4 \text{ G}$ (1 tesla = 10^4 gauss).

EXAMPLE 8-1

The flux density in a magnet is 0.5 T and the cross section of the magnet is 0.06 m^2. What is **(a)** the flux density in gauss (G) and **(b)** the flux in webers?

Solution

(a)
$$B = 0.5 \text{ T}\left(\frac{10^4 \text{ G}}{\text{T}}\right) = 5000 \text{ G}$$

(b)
$$\phi = BA = (0.5)(0.06) = 30 \times 10^{-3} \text{ Wb}$$

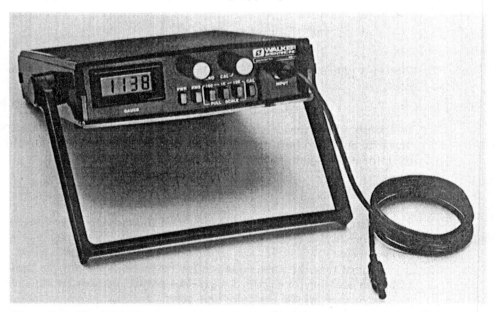

Figure 8-4   The MG-5D, a general-purpose portable laboratory gaussmeter. (Courtesy of Walker Scientific Inc.)

Magnetic flux density can be measured by an instrument called the gauss-meter. One such instrument designed for operation from either ac or internal "D" cells is the MG-5D portable laboratory gaussmeter of Fig. 8-4.

The MG-5D gaussmeter is a general-purpose portable gaussmeter designed to measure both constant (dc) and varying (ac) magnetic fields. The range of measurement is from ± 100 mG to ± 19.99 kG, but the range can be extended to 150 kG. Such an instrument is used to analyze magnetic fields and circuits like those we shall study in this chapter.

Drill Problem 8-1 ■

A flux density of 1.5 T exists in a 1×1 cm area. **(a)** What is the flux density in gauss? **(b)** What is the flux in webers?

Answer. (a) 15 kG. (b) 1.5×10^{-4} Wb. ■ ■

8.3 THE CURRENT-CARRYING CONDUCTOR

One of the most important events in the study of electricity was the discovery of the relationship between electricity and magnetism. In 1820, Hans Christian Oersted (Denmark, 1775–1851) discovered that a magnetic compass needle was deflected when placed near a current-carrying conductor.

Consider the simple experiment indicated by Fig. 8-5. In Fig. 8-5(a) a plane is passed perpendicular to a conductor and magnetic compasses placed on the plane all indicate the earth's magnetic field. However, when current flows in the conductor as in Fig. 8-5(b), the compass needles change direction and indicate that the paths of magnetic force are concentric circles around the conductor.

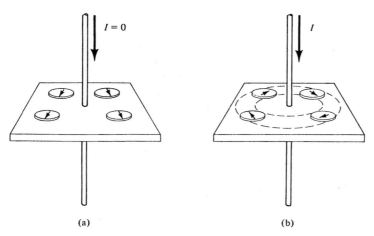

(a)   (b)

Figure 8-5 The conductor: **(a)** no current flow; **(b)** current flow and the resulting magnetic field.

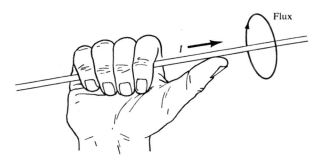

Figure 8-6   Right-hand rule.

The direction of the magnetic field around a current-carrying conductor is indicated by Fig. 8-5(b) and further specified by the following:

> **If a current-carrying conductor is grasped in the right hand with the thumb pointing in the direction of the conventional current, the fingers will then point in the direction of the magnetic lines of flux.**

This rule, called the *right-hand rule* is illustrated by Fig. 8-6.

Often, one sees an end view of the conductor so that the current is either "coming out of the page" or "going into the page." These two situations are indicated, respectively, by a dot (representing the point of the current arrow) or a cross (representing the tail of the current arrow). Thus, in Fig. 8-7 the left-hand conductor has current coming out of the page and the right-hand conductor has current going into the page.

Applying the right-hand rule to the conductors in Fig. 8-7, we find that a crowding of flux lines occurs in the area between the conductors. In turn, the flux lines tend to separate, with a resulting force on the conductors.

However, with the current in two parallel conductors traveling in the same direction as in Fig. 8-8, a force of attraction is exerted on the conductors. This is because of a contraction of flux lines between the conductors.

It was Ampere who discovered that the force existing between parallel conductors is directly proportional to their length l and the product of their currents $I_1 I_2$, and inversely proportional to the distance r between them. When the conductors are

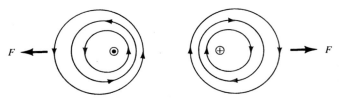

Figure 8-7   Magnetic field about two conductors, current directions opposing.

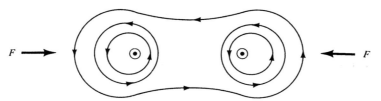

Figure 8-8 Magnetic field about two conductors, current directions similar.

in air or evacuated space, this force in newtons is given by

$$F = \frac{2 \times 10^{-7} I_1 I_2 l}{r} \tag{8-4}$$

In fact, Eq. 8-4 is used to define the ampere in SI units.

> **The ampere is the constant current that, if maintained in two straight parallel conductors that are of infinite length and negligible cross section and are separated from each other by a distance of 1 m in a vacuum, will produce between these conductors a force equal to 2×10^{-7} newton per meter of length.**

If the current-carrying conductor is formed into a single loop, as in Fig. 8-9, a magnetic field will be set up around the coil. Using the right-hand rule, we find that the magnetic field is concentrated within the center region of the coil.

The magnetic field of the conductor can be concentrated even more by winding the wire many times to form a helix, as in Fig. 8-10(a). If we apply the right-hand rule to the cross section of the helix, as in Fig. 8-10(b), we find that a portion of the flux produced by each turn will unite with a portion produced by the other turns. Thus the entire coil is linked so that the *electromagnet* or *solenoid*, as it is commonly called, exhibits the magnetic field of a bar magnet.

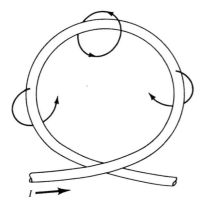

$I \longrightarrow$

Figure 8-9 Magnetic field of a coil.

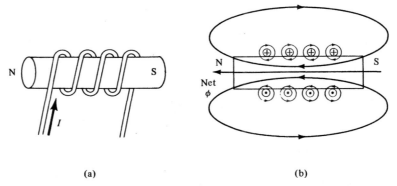

(a)   (b)

Figure 8-10   **(a)** Solenoid coil. **(b)** Its cross section showing the magnetic field.

Consistent with the magnetic field of Fig. 8-10, the direction of the lines of flux associated with a coil can be obtained by a *second right-hand rule*, which is as follows:

> **If a coil is grasped so that the fingers encircle the coil in the direction of the current in the coil, the thumb will then point in the direction of the magnetic flux in the center of the coil.**

It should be recognized that since flux lines flow externally from the north to the south pole, by the right-hand rule, the thumb direction indicates the north pole of the electromagnet formed by the coil.

8.4 MAGNETOMOTIVE FORCE

The value of flux ϕ that develops in a solenoid coil depends, as would be expected, on the current I and the number of turns N. The product of I and N is very important in magnetic considerations and is described by the term *magnetomotive force*, abbreviated mmf. Magnetomotive force is the force that causes flux to be established and is analogous to electromotive force for electric circuits. It is specified by the practical unit *ampere-turn* (At) and is given by the equation

$$\mathscr{F} = IN \qquad \text{At} \qquad (8\text{-}5)$$

The unit of mmf in SI units is the ampere, but it should be realized that this refers to a coil of one turn; that is, it is the mmf *per turn*. For clarity, the English unit of ampere-turn will be used in the text.

The solenoid coil is generally wound on a magnetic *core* that controls the path of the magnetic field. In turn, the path has a mean length l measured along its center line.

Analogous to electric field intensity, one can define *magnetic field intensity*. The magnetic field intensity or *magnetizing force H* is the mmf per unit length along the

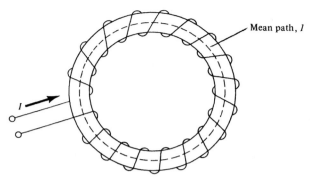

Mean path, *l*

I

Figure 8-11 Toroid for Example 8-2.

path of magnetic flux. The SI unit for the magnetizing force is ampere per meter, but for clarity the unit ampere-turn per meter will be used here. Thus,

$$H = \frac{\mathscr{F}}{l} \tag{8-6}$$

EXAMPLE 8-2

The steel toroid of Fig. 8-11 has a mean length of 0.09 m and a coil of 350 turns carrying a current of 1.2 A. What is the **(a)** mmf and **(b)** magnetic field intensity?

Solution

(a) $\mathscr{F} = IN = (1.2)(350) = 420$ At

(b) $H = \dfrac{\mathscr{F}}{l} = \dfrac{420}{9 \times 10^{-2}} = 4670$ At/m

It should be pointed out that in the CGS system the mmf is given in *gilberts* (Gb), where 1 Gb = 1.257 At. In the CGS system, the field intensity is in gilberts per centimeter, or oersted (Oe).

Although one may desire to use SI units exclusively, one must be able to recognize the other units that are often encountered. Table 8-1 compares the magnetic units, and Table 8-2 provides the necessary conversion factors.

For a particular magnetizing force H, a core will develop a particular flux ϕ and a flux density B. Flux density in a magnetic material plotted against magnetizing force forms a *magnetization curve*, also called a $B–H$ curve. Magnetization

TABLE 8-1 COMPARISON OF MAGNETIC UNITS

Quantity	SI unit	CGS unit	English unit
Flux (ϕ)	Weber	Maxwell	Line
Flux density (B)	Tesla (Wb/m^2)	Gauss (maxwell/cm^2)	Line per square inch
mmf ($\mathscr{F}$)	Ampere (per tesla)	Gilbert	Ampere-turn
Field intensity (H)	Amperes (per tesla) per meter	Oersted	Ampere-turn per inch

TABLE 8-2   CONVERSION FACTORS, MAGNETIC QUANTITIES

Quantity		Conversion factors	
ϕ	1 Wb	$= 10^8$ maxwells	$= 10^8$ lines
B	1 T	$= 10^4$ G	$= 6.45 \times 10^4$ lines/in.^{2}
$\mathscr{F}$	1 A (per tesla)	$= 0.796$ Gb	$= 1$ At
H	1 A (per tesla)/m	$= 1.26 \times 10^{-2}$ Oe	$= 2.54 \times 10^{-2}$ At/in.

curves are necessary when magnetic circuits are designed or analyzed because magnetic materials are nonlinear in their response to a magnetic force. Magnetization curves for several commercial metals are shown in Fig. 8-12.

Notice that the $B–H$ curves of Fig. 8-12 are semilogarithmic plots. Placing the field intensity on a logarithmic scale allows a greater accuracy at low values of H, where the curves would otherwise be almost vertical. Notice, too, that these particular curves have been specified in the CGS units. The $B–H$ curves for a few of the magnetic materials of Fig. 8-12 have been redrawn in Fig. 8-13 with a linear H scale and SI units. In particular, the transformer steel having 3.75% Si has been identified in the redrawn graph as *sheet steel*, the AS cast steel is identified simply as *cast steel*, and the AS cast iron is identified as *cast iron*.

Drill Problem 8-2 ■

A magnetomotive force of 810 At is created in a solenoid coil by a current of 1.8 A. What number of turns does the coil possess?

Answer.   450.   ■ ■

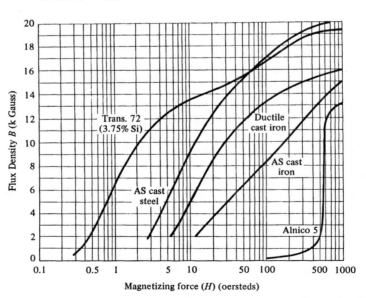

Figure 8-12   Flux density versus field intensity curves. (Courtesy of Hitachi Magnetics Corp., Edmore, Michigan.)

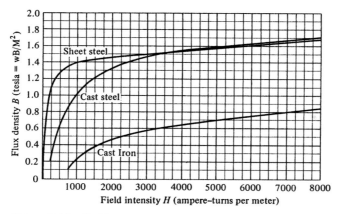

Figure 8-13 *B–H* curves.

Drill Problem 8-3 ■

A cast-iron toroid has a mean length of 10 cm. A 200-turn coil is wrapped on the toroid and develops 150 At. **(a)** What current must flow to the coil? **(b)** What is the magnetic field intensity? **(c)** Using a *B–H* curve, determine the approximate flux density.

Answer.   (a) 0.75 A. (b) 1500 At/m. (c) ≈ 0.35 T.   ■ ■

8.5 RELUCTANCE AND PERMEABILITY

The opposition to the establishment of magnetic flux is called *reluctance*. It is defined as the *ratio of the drop in mmf to the flux produced*. The symbol for reluctance is $\mathscr{R}$ so that

$$\mathscr{R} = \frac{\mathscr{F}}{\phi} \qquad \left(\frac{\text{ampere-turns}}{\text{weber}}\right)$$

Rearranging the preceding equation, we get the equation

$$\mathscr{F} = \phi\mathscr{R} \tag{8-7}$$

Equation 8-7 is analogous to Ohm's law if one considers that $\mathscr{F}$ is the magnetic voltage, ϕ the magnetic current, and $\mathscr{R}$ the magnetic resistance.

The reciprocal of reluctance, that is, the *magnetic conductance*, is called *permeance* $\mathscr{P}$:

$$\mathscr{P} = \frac{1}{\mathscr{R}} \tag{8-8}$$

In turn, the magnetic specific conductivity or permeance per unit length and cross section of a material defines the *absolute permeability*. Permeability is measured in *henrys per meter* and is given the symbol μ (mu).

Reluctance relates to the permeability, length l, and cross-sectional area A of the magnetic path by the following equation, which is completely analogous to the resistance formula of Chapter 3:

$$\mathcal{R} = \frac{l}{\mu A} \tag{8-9}$$

Although the unit of reluctance is the ampere-turn per weber, it is more properly defined by SI units as the ampere per weber, also called the reciprocal henry.

Equation 8-9 is suitable when analyzing air paths, but it is impractical for magnetic materials, since the permeability of these materials varies with the flux density. The *permeability of free space* μ_0 is $4\pi \times 10^{-7}$ H/m and is constant. The absolute permeability of another material can be expressed *relative* to the permeability of free space by

$$\mu = \mu_r \mu_0 \tag{8-10}$$

where μ_r is the dimensionless quantity called relative permeability.

The μ_r of nonmagnetic materials such as air, copper, wood, glass, and plastic is for all practical purposes equal to unity. On the other hand, the μ_r of magnetic materials such as cobalt, nickel, iron, steel, and their alloys is far greater than unity and, furthermore, is not constant.

Just as permittivity relates electric flux density to electric field intensity, permeability relates the magnetic flux density to the magnetic field intensity by the equation,

$$B = \mu H \tag{8-11}$$

Rearranging Eq. 8-11, we get $\mu = B/H$. The ratio of the flux density to field intensity at a particular point on the B–H curve is called the *normal permeability*. Over a limited range of magnetizing force ΔH, a corresponding change in flux density ΔB results. Then, an *incremental permeability* μ_Δ is defined over such a range as

$$\mu_\Delta = \Delta B / \Delta H \tag{8-12}$$

Obviously, the slope of any one of the B–H curves of Fig. 8-13 changes. The permeability or slope increases to a particular value, and then decreases at high field intensities and corresponding high flux densities. Plotting μ against B (or against H), we obtain a curve that clearly indicates the nonlinearity of μ (see Fig. 8-14).

EXAMPLE 8-3

Using the B–H curve of Fig. 8-13, compute and compare the relative normal permeability of cast steel at flux densities of (a) 0.8 T and (b) 1.4 T.

Solution.   Using Eqs. 8-10 and 8-11, we find

(a) $$\mu_r = \frac{\mu}{\mu_0} = \frac{B}{\mu_0 H} = \frac{0.8}{4\pi \times 10^{-7}\,(650)} = 978$$

(b) $$\mu_r = \frac{B}{\mu_0 H} = \frac{1.4}{4\pi \times 10^{-7}\,(2400)} = 465$$

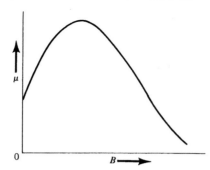

Figure 8-14 Variation of permeability for a magnetic material.

Drill Problem 8-4 ■

Using Fig. 8-13, determine the relative normal permeabilities of cast iron and cast steel at $B = 0.6$ T.

Answer. 955 and 140. ■ ■

Drill Problem 8-5 ■

Determine the approximate incremental permeability and its relative value for cast iron in the range 0.6–0.8 T.

Answer. 5.71×10^{-5} H/m; 45.4. ■ ■

8.6 PARA-, DIA-, AND FERROMAGNETIC MATERIALS

In order to classify materials as magnetic or nonmagnetic, it must be determined whether or not forces act on the material when the material is placed in a magnetic field. If a bar of any given material is suspended in a magnetic field, it will either turn at right angles to the field or align with the field. A material that turns at right angles to the field by producing a magnetic response opposite to the applied field is called *diamagnetic*. Those materials aligning themselves with the applied field are called *paramagnetic*. Within the paramagnetic class of materials is a special classification of materials called *ferromagnetic*. These materials are strongly attracted to magnets and exhibit paramagnetism to a phenomenal degree.

The effects of diamagnetism and paramagnetism are negligibly small, so that materials possessing these weak phenomena are said to be *nonmagnetic*. Diamagnetic materials such as silver, copper, and carbon have permeabilities slightly less than free space (for copper, $\mu_r = 0.999\ 998$). Paramagnetic materials such as aluminum and air have permeabilities slightly greater than that of free space (for air, $\mu_r = 1.000\ 000\ 4$). Of course, ferromagnetic materials such as iron, steel, cobalt and their alloys, with relative permeabilities extending into the hundreds and thousands, are said to be *magnetic*.

According to the now accepted view, the magnetic properties of matter are associated almost entirely with the spinning motion of electrons in the third shell of the atomic structure. An electron revolving in an orbit about the nucleus of an atom is equivalent to a tiny current loop that gives rise to a magnetic field. Also, a magnetic field is associated with the angular momentum of the electron's spin on its own axis.

In most atoms, there is a tendency for both the orbital and spin angular momentums to cancel each other by pair formation. For example, an electron spinning clockwise can pair with an electron spinning counterclockwise. Their total momentum and magnetism is then zero. Variations in this electron pairing account for the weak magnetism of the nonmagnetic materials. Diamagnetism results from an unbalance of the orbital pairing of electrons, whereas paramagnetism results from an unbalance of the spin pairing of electrons. Diamagnetic effects are masked quite frequently by paramagnetic effects if they are present.

Dia- and paramagnetic properties are attributed to individual atoms responding to a magnetic field, but ferromagnetic properties are unique in that they are attributed to the response of large groups of atoms. In ferromagnetic materials, certain internal forces make it possible for large groups of neighboring atoms to have their spin magnetic moments aligned parallel to each other. Containing approximately 10^{15} atoms and occupying approximately 10^{-8} cubic centimeters of volume, these groups are called *domains*.

In a bulk specimen of a ferromagnetic material, the magnetic moments of the many domains are randomly oriented so that the material appears unmagnetized. Where two domains come together, the change in direction of magnetization is not abrupt but, rather, a gradual transition over a region called a *domain boundary*. The concept of domain boundary is shown in Fig. 8-15. Arrows, like little compass needles, are used to show the moments of the two domains. Between the domains, the atomic magnets are progressively rotated, and in this unstable region they are easily affected by an external field.

Magnetization of a ferromagnetic material consists in changing the orientation of the domains so that they are no longer randomly oriented. Consider Fig. 8-16, which relates the magnetization process to the magnetization curve. An unmagnetized sample containing four domains is portrayed in (1). When a field H is applied, the domain boundaries are stretched with those domains whose magnetization is closely parallel to H increasing in size (2). As the field is continually increased, the favorable domains grow larger at the expense of the unfavorable ones (3). By a further increase in H, the domains are forced to align with the field (4). At this point, the magnetic material has made its maximum contribution in response to the

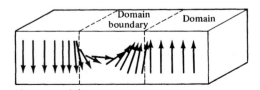

Figure 8-15 Two adjacent domains.

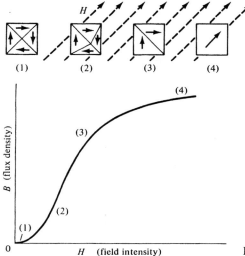

Figure 8-16   The magnetization process.

applied field so that a further increase in H results in a relatively small increase in flux density. Therefore, the material is said to be *magnetically saturated*.

When the external magnetizing force is removed from a magnetically saturated sample, some of the domains return to their original states; others do not. Depending on the remaining magnetism, one classifies the magnet material as temporary or permanent.

8.7 THE MAGNETIC CIRCUIT

The simple magnetic circuit consists of a source of mmf supplying flux through a continuous magnetic path. Depending on the type of magnetic path, magnetic circuits are either *linear* or *nonlinear*. When nonmagnetic materials are used for flux paths, μ is constant so that the reluctance can easily be calculated. In turn, B and H are linearly related. However, when ferromagnetic materials are used, μ is not constant and so the reluctance is not readily calculated. In turn, B and H have a nonlinear relationship and one must use a B–H curve in their determination.

In one type of problem, the flux is known and the mmf must be found. In another type of problem, the mmf is given and the flux established by that mmf must be found. For complex paths, this latter type of problem generally necessitates a trial and error solution. The following examples illustrate the two types of problems.

EXAMPLE 8-4

Find the mmf required to establish a flux of 1×10^{-4} Wb. in the plastic *toroidal* core of Fig. 8-17. The average circumference of the ring is 0.08 m, and the cross-sectional diameter of the doughnut-shaped core is 0.012 m.

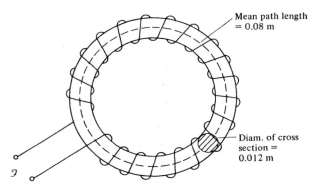

Mean path length = 0.08 m

Diam. of cross section = 0.012 m

Figure 8-17 Toroid for Examples 8-4 and 8-5.

Solution

$$A = \pi r^2 = \frac{\pi d^2}{4} = \frac{\pi (0.012)^2}{4} = 1.13 \times 10^{-4} \text{ m}^2$$

From Eqs. 8-7 and 8-9,

$$\mathscr{F} = \phi \mathscr{R} = \frac{\phi l}{\mu A} = \frac{(1 \times 10^{-4})(8 \times 10^{-2})}{(4\pi \times 10^{-7})(1.13 \times 10^{-4})}$$
$$= 5.64 \times 10^4 = 56\,400 \text{ At}$$

EXAMPLE 8-5

A sheet-steel core of the same dimensions is used instead of the plastic core of Example 8-4. Find the mmf needed to establish the same flux.

Solution. $A = 1.13 \times 10^{-4} \text{ m}^2$. Then

$$B = \frac{\phi}{A} = \frac{1 \times 10^{-4}}{1.13 \times 10^{-4}} = 0.886 \text{ T}$$

From the B–H curve, $H = 125$ At/m. Then from Eq. 8-6,

$$\mathscr{F} = Hl = (125)(0.08) = 10 \text{ At}$$

Notice from these examples that considerably less magnetic voltage is needed for the sheet-steel core because of its relative ease of passing flux. The second type of problem, finding the flux from a given mmf, is illustrated by Example 8-6.

EXAMPLE 8-6

A coil of 800 turns is wrapped on a cast-steel toroidal core. The core length is 0.5 m, and the cross-sectional area is $3.25 \times 10^{-4} \text{ m}^2$. What flux is established in the core when 2 A flows in the coil?

Solution. $\mathscr{F} = IN = 2(800) = 1600$ At. Then

$$H = \frac{\mathscr{F}}{l} = \frac{1600}{0.5} = 3200 \text{ At/m}$$

From the B–H curve, $B = 1.48$ T. Finally,

$$\phi = BA = (1.48)(3.25 \times 10^{-4}) = 4.81 \times 10^{-4} \text{ Wb}$$

When a coil surrounds a toroid, as in Fig. 8-17, some of the flux will leak from the coil into the air. This flux is called *leakage flux*. In the case of a non-magnetic core, it is important that the winding be spread over the entire core to insure that most of the flux remains in the core region. However, when the core material is largely ferromagnetic, the coil providing the mmf may be lumped or con-centrated in one region. In fact, in applications the coil is wound on a bobbin or spool. Figure 8-18(a) shows a group of bobbin-wound coils that are slipped onto the magnetic core as depicted in Fig. 8-18(b). Then, with a lumped mmf it becomes easier to consider the electrical analog of Fig. 8-18(c).

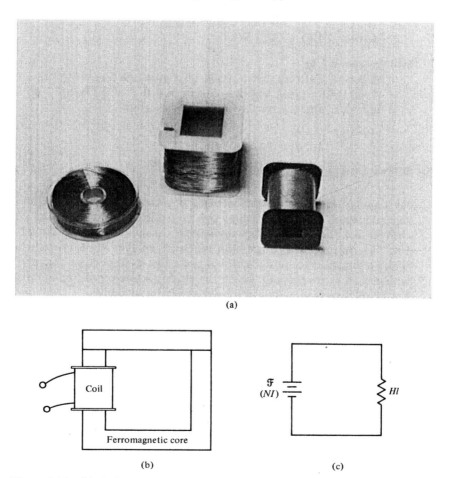

(a)

Coil

Ferromagnetic core

(b)

$\mathfrak{F}$
(*NI*)

Hl

(c)

Figure 8-18 **(a)** Coils of wire wound on spools **(b)** form the lumped mmf **(c)** that acts as a source in the electric analog.

The electrical analog of the magnetic circuit is based on Eqs. 8-5–8-7. Notice that even though Eq. 8-7 expresses the relationship analogous to Ohm's law, the form given by Eq. 8-6 is more useful.

$$\mathscr{F} = NI = Hl$$

Since the reluctance of the magnetic core is so much less than that of the surrounding air, to a fair approximation the flux can be considered the same throughout the core. Thus, flux leakage, even for magnetic circuits having air gaps, is often neglected.

Drill Problem 8-6 ■

A cast-iron core has a cross section of $\frac{1}{2}$ cm^2 and a mean length of 10 cm. If a coil placed on the core develops 100 At, determine the flux produced in the core. (Remember that differences in graph interpolation will lead to minor differences in an answer.)

Answer. 1.25×10^{-5} Wb.   ■ ■

Drill Problem 8-7 ■

A coil of 200 turns is wrapped on a sheet-steel core having a cross section of 2 cm^2 and a mean length of 20 cm. If a flux of 2.5×10^{-4} Wb is developed in the core, what current must flow in the coil?

Answer.   0.4 A.   ■ ■

8.8 SERIES AND PARALLEL MAGNETIC CIRCUITS

An analogy has been made between the simple magnetic circuit and the simple electric circuit. A further analogy can be made between series and parallel magnetic circuits and their electrical counterparts. Consider the series circuit of Fig. 8-19(a). The flux must pass through both materials so that the total reluctance is the sum of the reluctances around the closed path. Just as important, magnetic voltage is lost across each series element; hence the electrical analog of Fig. 8-19(b). Analogous to Kirchhoff's voltage equation for closed electrical paths,

$$\sum \mathscr{F} = 0 \qquad\qquad (8\text{-}13)$$

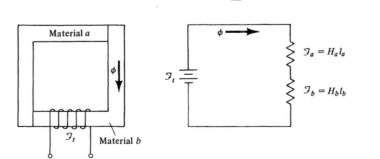

Figure 8-19   **(a)** A series magnetic circuit. **(b)** Its electrical analog.

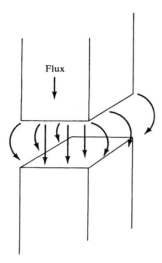

Figure 8-20 Fringing at an air gap.

For Fig. 8-19, then,

$$\mathscr{F}_t = H_a l_a + H_b l_b$$

One of the series elements of a magnetic circuit is often an air gap. In the ferrous portion of a magnetic circuit, the flux lines are confined for the most part by the very high permeability of the ferrous material. When an air gap exists as in Fig. 8-20, the flux lines passing through the gap are not confined to the projected area of the ferrous portion. Rather, the flux lines separate so that *fringing* results. Fringing increases the effective area, thereby decreasing the flux density.

For short air gaps, an empirical correction is to add the length of the gap to each cross-section dimension of the adjoining magnetic material. For example, if a core 5 × 4 cm has a 0.25 cm air gap, the effective area of the air gap A_g is $A_g = (5.25)(4.25) = 22.3$ cm². In the subsequent examples and problems, fringing is neglected unless otherwise noted.

EXAMPLE 8-7

A magnetic circuit composed of a cast-steel path and an air gap has the dimensions shown in Fig. 8-21. Compute the current necessary to support a flux of 2.5×10^{-4} Wb. Take fringing into account.

Solution. For the air gap,

$$A_g = (1.6 \times 10^{-2})^2 = 2.56 \times 10^{-4} \text{ m}^2$$

$$B = \frac{\phi}{A_g} = \frac{2.5 \times 10^{-4}}{2.56 \times 10^{-4}} = 0.98 \text{ T}$$

$$H = \frac{B}{\mu} = \frac{0.98}{4\pi \times 10^{-7}} = 7.8 \times 10^5 \text{ At/m}$$

$$(Hl)_g = (7.8 \times 10^5)(1 \times 10^{-3}) = 780 \text{ At}$$

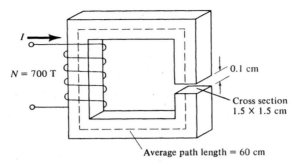

Figure 8-21 Example 8-7.

For the steel,

$$A = (1.5 \times 10^{-2})^2 = 2.25 \times 10^{-4} \text{ m}^2$$

$$B = \frac{\phi}{A} = \frac{2.5 \times 10^{-4}}{2.25 \times 10^{-4}} = 1.11 \text{ T}$$

From the graph of B versus H,

$$H = 1200 \text{ At/m}$$

$$(Hl)_s = (1200)(6 \times 10^{-1}) = 720 \text{ At}$$

Finally,

$$\mathscr{F} = (Hl)_g + (Hl)_s$$
$$= 780 + 720 = 1500 \text{ At}$$

and

$$I = \frac{\mathscr{F}}{N} = \frac{1500}{700} = 2.15 \text{ A}$$

A table is sometimes helpful in magnetic circuit problems because it enables one to keep track of the many quantities involved in a problem. A completed table for the preceding example is shown in Table 8-3.

In practice, circuits that appear to be parallel magnetic circuits are usually series–parallel circuits. For example, the magnetic circuit of Fig. 8-22(a) appears to be a parallel circuit, but the branch on which the coil is wound is in series with the parallel paths, hence the analog in Fig. 8-22(b).

The flux generated by the coil in Fig. 8-22 divides in order to pass through the two parallel branches. Analogous to Kirchhoff's current law, at a magnetic circuit

TABLE 8-3 DATA FOR EXAMPLE 8-7

Part	ϕ (Wb)	A (m²)	l (m)	B (T)	H (At/m)	Hl (At)	$\mathscr{F}$ (At)
Air gap	2.5×10^{-4}	2.55×10^{-4}	1×10^{-3}	0.98	7.8×10^5	780	1500
Steel	2.5×10^{-4}	2.25×10^{-4}	6×10^{-1}	1.11	1.2×10^3	720	

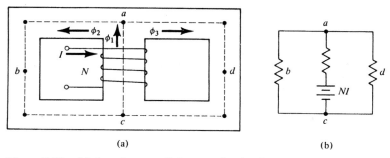

Figure 8-22 **(a)** A series–parallel magnetic circuit. **(b)** Its electrical analog.

junction,

$$\sum \phi = 0 \qquad (8\text{-}14)$$

and for Fig. 8-22,

$$\phi_1 = \phi_2 + \phi_3$$

EXAMPLE 8-8

The magnetic circuit of Fig. 8-22 has a coil of 1000 turns. Determine the current necessary to establish 1.5×10^{-4} Wb in the center branch. The core is a symmetrical sheet-steel core with the following cross sections and path lengths: $A_{ac} = 1.5 \times 10^{-4}$ m^2; $A_{abc} = A_{adc} = 0.6 \times 10^{-4}$ m^2; $l_{ac} = 8$ cm; $l_{abc} = l_{adc} = 18$ cm.

Solution. The circuit is symmetrical about the center line so that $\phi_2 = \phi_3 = \frac{1}{2}\phi_1$. The electrical analog suggests a magnetic voltage equation around only one loop; the left branch will be used.

For the center branch:

$$B = \frac{\phi}{A} = \frac{1.5 \times 10^{-4}}{1.5 \times 10^{-4}} = 1 \text{ T}$$

From the *B–H* curve,

$$H = 150 \text{ At/m}$$
$$(Hl)_{ac} = (150)(8 \times 10^{-2}) = 12 \text{ At}$$

For the left branch:

$$B = \frac{\phi}{A} = \frac{0.75 \times 10^{-4}}{0.6 \times 10^{-4}} = 1.25 \text{ T}$$

From the *B–H* curve,

$$H = 450 \text{ At/m}$$
$$(Hl)_{abc} = (450)(18 \times 10^{-2}) = 81 \text{ At}$$

Then

$$\mathscr{F} = (Hl)_{ac} + (Hl)_{abc} = 12 + 81 = 93 \text{ At}$$

Finally,

$$I = \frac{\mathscr{F}}{N} = \frac{93}{1000} = 93 \text{ mA}$$

Drill Problem 8-8 ■

A magnetic circuit consists of a sheet-steel core and an air gap and has the dimensions shown in Fig. 8-21. Compute the current necessary to support a flux of 2.5×10^{-4} Wb. Take fringing into account. Comment on this answer in comparison to that for Example 8-7.

Answer.   1.33 A; less current is required since the sheet steel magnetizes more readily than the cast steel of Example 8-7.   ■ ■

8.9 MAGNETIC CORE LOSSES

Consider a demagnetized ferromagnetic core. With the application and subsequent increase of a magnetizing force, the core becomes saturated along the *normal magnetization curve OA* of Fig. 8-23. When the magnetization force is reduced to zero, the flux density will not decrease along the same curve *AO*, but along a different curve *AC*. This difference is due to the tendency of the material to retain some of its magnetism. At point *C* the magnetizing force is essentially removed, yet the core has some *residual flux density* or *residual induction* B_r. In order to reduce this residual magnetism to zero, one must apply a field intensity in the opposite direction. This demagnetizing force H_c is known as the *coercive force*. Materials with a low coercive force are indicative of soft magnetic materials; materials with a high coercive force are indicative of permanent magnets.

Increasing the demagnetizing force results in magnetic saturation in the opposite direction (point *D*). Starting from saturation in the negative direction, subsequent

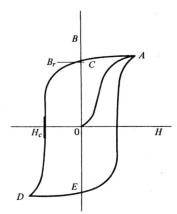

Figure 8-23   The hysteresis loop.

magnetization in the positive direction completes the portion of the curve labeled *DEA*. The complete curve of the magnetization cycle, *ACDEA* is called the *hysteresis loop*.

The area within the hysteresis loop is a product of *B* and *H* and, in terms of units, is

$$\left(\frac{\text{webers}}{\text{square meter}}\right)\left(\frac{\text{newtons}}{\text{weber}}\right) = \left(\frac{\text{newtons}}{\text{square meter}}\right)\left(\frac{\text{meters}}{\text{meter}}\right) = \frac{\text{joules}}{\text{cubic meter}}$$

That is, the area of the hysteresis loop represents the energy per unit volume that must be used per magnetization cycle to move the domains. With appropriate constants, the hysteresis loss can be given in watts per unit volume. An empirical relationship developed by Charles P. Steinmetz (American, German-born, 1865–1923) gives the hysteresis loss as

$$P_h = k_h f B_m^n \tag{8-15}$$

where P_h is in watts per unit volume, k_h is a constant term, f is the number of magnetization cycles per second, B_m is the maximum flux density, and n is the Steinmetz constant, often taken as 1.6.

It follows that the greater the energy required to magnetize a sample, the greater the energy needed to demagnetize it. Large hysteresis loops are therefore desirable for permanent magnets, because the large hysteresis loop represents a large storage of energy.

As will be pointed out in the next chapter, a changing magnetic field induces an emf in any conducting material in that field. Such emfs create within a magnetic core circulating or *eddy currents*. The eddy currents encounter the electrical resistance of the core producing a power loss proportional to $i^2 R$. Although the eddy current values cannot be determined directly, the power loss has been found empirically to be given by

$$P_e = k_e f^2 B_m^2 \tag{8-16}$$

where P_e is the eddy current loss in watts per unit volume and k_e a constant; f and B_m are as previously defined.

In order to reduce the magnitude of eddy currents and hence reduce the power lost in a core, magnetic cores are constructed by stacking thin *laminations* (see Fig. 8-24). The laminations are insulated from each other by a thin coat of varnish

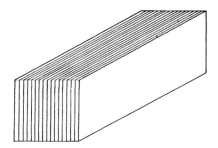

Figure 8-24   A laminated core.

or lacquer. Notice in Fig. 8-24 that a cross section of the laminated core produces a composite area of magnetic material and insulation. Then the effective area for flux flow is less than the composite cross section. In conclusion, the combined hysteresis and eddy current loss is known as the *core loss*.

Drill Problem 8-9 ■

The hysteresis and eddy current losses in a transformer are 6 and 36 W, respectively, when the transformer is used at 60 magnetization cycles/s. What is the total hysteresis and eddy current loss when the transformer is used at 50 magnetization cycles/s?

Answer.   30 W.   ■ ■

8.10 PERMANENT MAGNET CIRCUITS

Even though permanent magnets are not directly associated with electric circuits, they are indirectly associated with many electromagnetic devices. This section provides a brief introduction to the use of permanent magnets.

In the design of permanent magnet circuits, the second quadrant of the hysteresis loop, called the *demagnetization curve*, is used. Some demagnetization curves are shown in Fig. 8-25. The curves shown are for three members of the Alnico family, so called because the members are steel alloys containing aluminum, nickel, and cobalt. Although the Alnico magnetic is familiar to many of us, it is not the only type of permanent magnet available. Other permanent magnets are ceramic in nature and contain various oxides of barium, strontium, and iron.

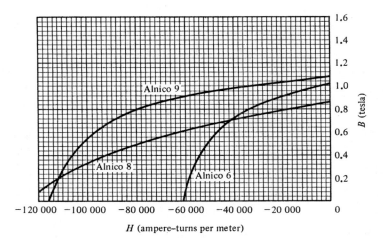

Figure 8-25   Demagnetization curves. (Courtesy of Hitachi Magnetics Corp., Edmore, Michigan.)

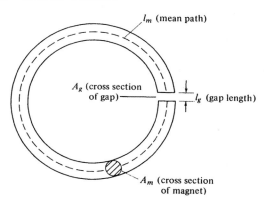

Figure 8-26 Permanent magnet circuit.

Consider the permanent magnet circuit and the corresponding dimensions in Fig. 8-26. Since the total mmf around the closed magnetic path is zero,

$$H_m l_m + H_g l_g = 0 \qquad (8\text{-}17)$$

where H_m and H_g are the magnetization forces in the magnet and gap, respectively. Solving Eq. 8-17 for H_m, we find that

$$H_m = -H_g \left(\frac{l_g}{l_m}\right)$$

or

$$H_m = \frac{-B_g}{\mu_0} \left(\frac{l_g}{l_m}\right) \qquad (8\text{-}18)$$

Neglecting leakage, all the flux of the magnet passes through the air gap so that

$$\phi = B_m A_m = B_g A_g$$

or

$$B_g = B_m \left(\frac{A_m}{A_g}\right) \qquad (8\text{-}19)$$

By substitution of Eq. 8-19 into Eq. 8-18 we obtain the ratio of flux density to field intensity in the magnet,

$$\frac{B_m}{H_m} = -\frac{\mu_0 A_g l_m}{A_m l_g} \qquad (8\text{-}20)$$

The relationship in Eq. 8-20 is a straight line whose intersection with the demagnetization curve determines the actual B and H of the magnet. The following example illustrates its use.

EXAMPLE 8-9

The magnetic circuit of Fig. 8-26 has the dimensions $l_m = 5.26$ cm, $A_m = 1.05$ cm^2, $l_g = 0.5$ cm, and $A_g = 2$ cm^2. If the magnet has the demagnetization curve of Fig. 8-27, calculate the flux in the air gap.

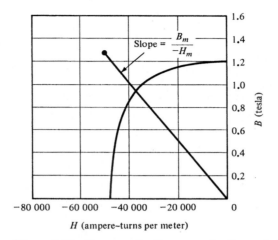

Figure 8-27   Demagnetization curve for Example 8-9.

Solution.   From Eq. 8-20,

$$\frac{B_m}{H_m} = -\frac{(4\pi \times 10^{-7})(2)(5.26)}{(1.05)(0.5)} = \frac{-2.51}{10^5} = \frac{-1.26}{50\,000}$$

A line is drawn on the demagnetization curve through the origin and the point where $B = 1.26$ T and $H = -50\,000$. The flux density of the magnet is given at the intersection of the curves as $B_m = 0.95$ T. Then

$$\phi = B_m A_m = (0.95)(1.05 \times 10^{-4}) \approx 1 \times 10^{-4} \text{ Wb}$$

The line from the origin that has the slope determined by Eq. 8-20 is called a *load line.* The actual numeric value for the slope is called the *permeance coefficient.* From Eq. 8-20, it is seen that the permeance coefficient depends not on the magnetic material but only on the magnetic circuit geometry.

8.11 MAGNETIC CIRCUIT APPLICATIONS

The operation of many electromechanical devices depends on magnetic fields produced by electromagnets or permanent magnets. Three of the many types of electromechanical devices are shown in Fig. 8-28.

The meter mechanism shown in Fig. 8-28(a) is known as a dc permanent-magnet-type meter. As shown in the simplified sketch of Fig. 8-28(b), a permanent magnet supplies magnetic flux lines that interact with a coil that itself generates a sensitive electromagnetic field. This mechanism will be discussed more thoroughly in Chapter 17, on measurement.

A second example of a practical magnetic circuit is the relay. Two small relays suitable for circuit board mounting are shown in Fig. 8-28(c) with their covers re-

(a)

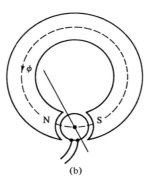

(b)

Figure 8-28 Three types of electromechanical devices relying on magnetic flux: **(a)** The dc permanent magnet meter mechanism. **(b)** The flux in the meter mechanism. **(c)** Relays. **(d)** Flux in a relay circuit. **(e)** The dc permanent magnet motor. **(f)** Flux in a dc motor.

moved. From a magnetic circuit aspect, a relay depends on a coil generating a flux that closes an air gap against the action of a spring-loaded armature. This is indicated in Fig. 8-28(d).

A third application of magnetic circuits is the electric motor. A small dc motor is shown in Fig. 8-28(e). Clearly identified in that figure is a set of permanent magnet poles and an armature that contains a set of coils. As depicted in Fig. 8-28(f), the permanent magnets produce a magnetic field. When the armature coils are energized, the two fields interact. The interaction results in a rotation of the armature. Other

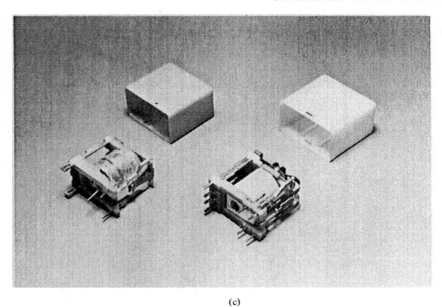

(c)

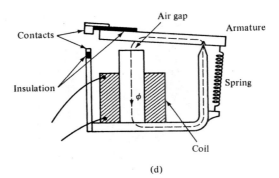

(d)

Figure 8-28   (continued)

applications of magnetic circuits include the permanent magnet speaker and the transformer.

One of the characteristics of flux lines is that they exert a tension along their length. When the electromagnet of the relay in Fig. 8-28 is energized, the flux lines in the air gap exert a force tending to close that gap. This force is given by the equation

$$F = \frac{B^2 A}{2\mu_0} \tag{8-21}$$

The force F, is in newtons when the flux density B is in tesla; the common area on both sides of the air gap A is in square meters.

(e)

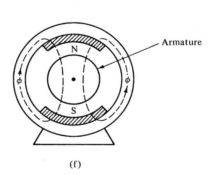

(f)

Figure 8-28 (continued)

EXAMPLE 8-10

The relay of Fig. 8-28 has an air gap of 10^{-3} m². A force of 2.8 lb is needed to close the gap. What flux density is needed?

Solution. $F = 2.8$ lb $= 12.5$ N.

$$B^2 = \frac{2\mu_0 F}{A} = \frac{2(4\pi \times 10^{-7})(12.5)}{10^{-3}} = 3.14 \times 10^{-2}$$

Then

$$B = \sqrt{3.14 \times 10^{-2}} = 0.177 \text{ T}$$

Drill Problem 8-10 ■

A relay needs a flux of 0.4×10^{-4} Wb to close its 2 mm air gap. The relay has a core and air gap area of 1 cm². What force is developed across the gap?

Answer.   6.37 N.   ■ ■

QUESTIONS

1. What is magnetism?
2. What is the difference between a permanent and a temporary magnet?
3. What is meant by (a) the field of a magnet and (b) the poles of a magnet?
4. What are magnetic lines of force? What are their characteristics?
5. Define magnetic field intensity. What are its units?
6. Define magnetic flux density. What are its units?
7. How is the direction of the magnetic flux around a current-carrying conductor found?
8. How is the ampere defined?
9. How are the poles of an electromagnet determined?
10. What is meant by the term "mmf"?
11. What is the magnetizing force H?
12. Describe the B–H curve.
13. What is (a) reluctance and (b) permeance?
14. What is (a) absolute permeability and (b) relative permeability?
15. How is incremental permeability found?
16. What are the differences between para-, dia-, and ferromagnetic materials?
17. What is the domain theory of magnetism?
18. What is meant by magnetic saturation?
19. What is leakage flux?
20. Describe the magnetic circuit. How is it analogous to the electric circuit?
21. What are the magnetic equivalents of Kirchhoff's voltage and current laws?
22. What is fringing? How is fringing accounted for in air-gap calculations?
23. What is meant by (a) residual induction and (b) coercive force?
24. Describe the hysteresis loop.
25. What is meant by (a) hysteresis, (b) eddy current, and (c) core losses?
26. What is a demagnetization curve and how is it used?

PROBLEMS

1. The basic unit of pole strength is the weber. Calculate the force of repulsion between the two north poles of Fig. 8-3(b) if the two poles are 10 cm apart and each have a pole strength of 4×10^{-4} Wb. For air, the constant k in SI units is 6.33×10^4 (Eq. 8-1).

2. Two bar magnets are placed in a straight line with their north poles 5 cm apart. **(a)** What is the force of repulsion if $M_1 = 25.1 \times 10^{-5}$ Wb and $M_2 = 6.28 \times 10^{-5}$ Wb? **(b)** What is the field intensity at the M_2 north pole as a result of the M_1 north pole?

3. A magnetic steel bar has a flux of 1.44×10^{-3} Wb and a cross-sectional area of 1.6×10^{-3} m^2. What is the flux density in teslas and gauss?

4. A magnetic flux of 150 000 maxwells is uniformly distributed over a 5×10 cm area. What is the flux density in teslas and gauss?

5. Convert the following magnetic quantities to English units: **(a)** 2000 A (per tesla), **(b)** 2×10^{-6} Wb, **(c)** 8000 G.

6. Convert the following quantities to CGS units: **(a)** 400 At, **(b)** 1.28 T, **(c)** 2000 At/in.

7. Determine the direction of current flow in the coils of Fig. 8-29.

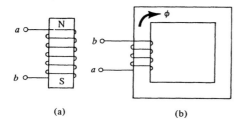

(a)   (b)   **Figure 8-29**   Problem 8-7.

8. A force of 2.88 N exists between two conductors that run parallel to each other for a distance of 24 m. The conductors are in air and have a spacing between centers of 0.15 m. Calculate the current flowing in the conductors.

9. Two bus bars having a center-to-center spacing of 0.1 m have a force of separation of 1.4 N when carrying 200 A to and from a load. With all other factors unchanged, what is the force of separation when the current becomes 300 A?

10. A coil of 240 turns develops an mmf of 1000 At. What current flows in the coil?

11. What field intensity exists in a toroid having a mean length of 40 cm, a cross-sectional area of 10 cm^2, and a coil of 300 turns of wire carrying a current of 1.2 A?

12. What magnetic voltage must be applied to a core of mean length 0.15 m if the field intensity of the core is 8000 At/m?

13. A cast-iron toroid has a mean length of 12 cm and a coil of 500 turns carrying a current of 0.75 A. What is the **(a)** mmf and **(b)** field intensity?

14. The reluctance of an electromagnet is 2×10^4 At/Wb. If the electromagnet has a coil of 600 turns carrying a current of 80 mA, what flux is in the electromagnet?

15. An air gap exists in a magnetic path. The air gap has a cross section of 24×10^{-4} m^2 and a length of 0.8 cm. What is the reluctance of the air gap?

16. What is the permeance of the air gap in Problem 15?

17. A cast-iron toroid of 4 cm^2 cross-sectional area and 0.2 m mean length has a normal permeability of 1200. Compute the flux generated when a coil of 500 turns is wrapped on the toroid and energized with 0.5 A.

18. The flux density of an iron core is 0.6 T when the field intensity is 3250 At/m. What are the absolute and relative permeabilities of the iron?

19. Using Fig. 8-13, find the normal relative permeabilities of cast iron, cast steel, and sheet steel at a field intensity of 1000 At/m.

20. Find and compare the incremental relative permeabilities of cast steel in the ranges 500–750 and 6500–6750 At/m.

21. What mmf is required to produce a flux of 2×10^{-4} Wb in a cast-steel ring of mean circumference 100 cm and cross section 5 cm²?

22. A cast-iron toroid of 4×10^{-4} m² cross section and a mean diameter of 10 cm is wound with 5 turns of wire/cm. What current is required to produce a flux of 2.5×10^{-4} Wb?

23. Find the flux in the magnetic core of Fig. 8-30.

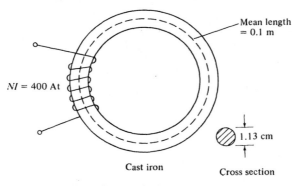

Mean length = 0.1 m

$NI = 400$ At

1.13 cm

Cast iron   Cross section

Figure 8-30   Problem 8-23.

24. Find the current necessary to produce a flux of 1.2×10^{-4} Wb in the core of Fig. 8-31.

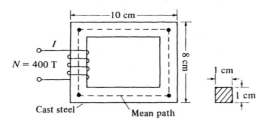

10 cm

8 cm

I

$N = 400$ T

1 cm

1 cm

Cast steel   Mean path

Figure 8-31   Problem 8-24.

25. Neglecting fringing, what current I is necessary to establish a flux of 6×10^{-4} Wb in the air gap of Fig. 8-32.

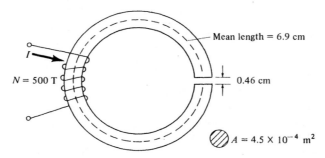

Mean length = 6.9 cm

I

$N = 500$ T

0.46 cm

$A = 4.5 \times 10^{-4}$ m²

Figure 8-32   Problem 8-25.

26. Determine the flux in the air gap of Fig. 8-33. This involves a trial and error solution. For the first trial, solve for the flux assuming that all the magnetic voltage is lost across the air gap. Using this value of flux, find the total mmf and compare to the actual mmf. If not within 10% of the actual mmf, assume other values of flux. Do not neglect fringing.

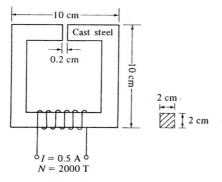

Figure 8-33 Problem 8-26.

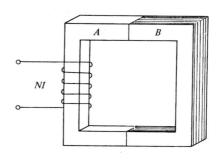

Figure 8-34 Problem 8-27.

27. In Fig. 8-34, material A is cast steel having an area of 0.00125 m² and a length of 0.25 m. Material B is laminated sheet steel having an effective area of 0.00112 m² and a length of 0.30 m. Determine the mmf required to produce a magnetic flux of 1×10^{-3} Wb.

28. Refer to Example 8-8. If an air gap of 0.2 cm is cut in the center branch, what current is now necessary for a flux of 1.5×10^{-4} Wb in that branch? Neglect fringing.

29. Find the mmf needed to establish a flux of 1.8×10^{-4} Wb in the center branch of the symmetrical cast-steel core of Fig. 8-35.

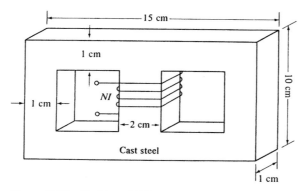

Figure 8-35 Problem 8-29.

30. A certain magnetic core has a core loss of 8.4 W, of which one-third is eddy current loss, when the magnetizing field is changed 60 times a second. Determine the core loss if the magnetizing field is changed only 50 times a second.

31. A core operated at 60 magnetization cycles/s has a hysteresis loss of 1.5 W and an eddy current loss of 0.9 W. What would the total core loss be at 25 magnetization cycles/s?

32. A permanent magnet circuit (see Fig. 8-26) used Alnico 9 material. Neglecting fringing, the circuit has the following dimensions $l_m = 30$ cm, $l_g = 0.8$ cm, and $A_g = A_m = 1.25 \times 10^{-4}$ m². Calculate the flux in the air gap.

33. A ceramic permanent-magnet material contains iron oxide and has the demagnetization curve of Fig. 8-36(a). A magnetic circuit with dimensions as in Fig. 8-36(b) is formed. **(a)** What is the permeance coefficient? **(b)** What is the flux in the air gap?

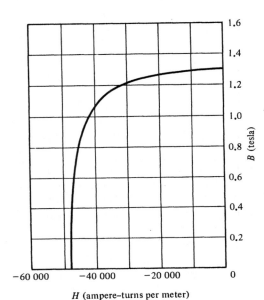

(a)

(b)

Figure 8-36 Problem 8-33: **(a)** Demagnetization curve; **(b)** magnetic circuit.

34. What force of attraction exists between the poles of Fig. 8-37 if the flux in the air gap is 3×10^{-4} Wb?

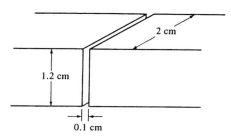

Figure 8-37 Problem 8-34.

35. A relay has a 2.5 mm air gap that requires a force of 11.2 N to close. If the area of the core and gap are each taken as 2 cm^2, what flux density is needed for the relay?

36. What flux must the relay coil of Fig. 8-28(d) develop in order for the force on the armature to be 5 N? The core that the coil is wound on has an area of 0.5×10^{-4} m^2.

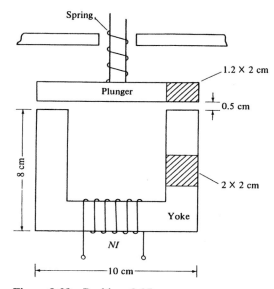

Figure 8-38 Problem 8-37.

37. Shown in Fig. 8-38 is a magnetic circuit that operates as a solenoid. A total force of 17.9 lb must be exerted on the plunger before it moves. What mmf is required for operation if the yoke and plunger are cast steel? Neglect fringing.

Chapter 9

Inductance

9.1 INDUCTION

In the preceding chapter it was shown that a magnetic field surrounds a current-carrying conductor. In this chapter *electromagnetic induction* and the related property of *inductance* are discussed. The phenomenon of electromagnetic induction is characterized by a generated emf whenever a conductor cuts through flux lines. This phenomenon can easily be observed if one performs the experiment shown in Fig. 9-1, where a conductor is connected to a sensitive electrical meter, the galvanometer. If the conductor is passed down through the magnetic field as indicated, an emf is generated with the polarity noted. Simultaneously, a current flows through the galvanometer in the direction shown. If the motion of the conductor ceases, the galvanometer pointer returns to the zero position, indicating no induced emf. However, when the conductor is moved upward through the flux, the emf and resultant

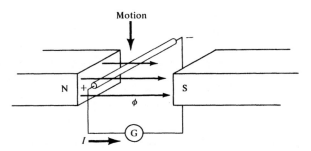

Figure 9-1 Electromagnetic induction.

galvanometer deflection are opposite to that created by the downward motion of the conductor.

Further demonstrations indicate that the magnitude of generated emf depends directly on the speed or rate at which the flux lines are cut. Then, too, the induced emf is proportional to the number of conductors or turns of a coil connected in series.

The same results are noticed if one holds the conductor stationary and moves the magnetic field. A study of practical generators reveals, for instance, that some generators utilize conductors moving in the field of stationary magnets, whereas other generators utilize magnets moving past stationary conductors.

Such observations led Michael Faraday (England, 1791–1867) to develop the following statement:

When the magnetic flux linking a coil changes, a voltage proportional to the rate of flux change is induced in the coil.

Known as *Faraday's law*, the above statement is given mathematically by Eq. 9-1, where e is the induced emf, in volts, N the number of turns, and $d\phi/dt$ is the rate of change in flux (in webers per second):

$$e = -N\frac{d\phi}{dt} \tag{9-1}$$

When the flux is increasing, $d\phi$ is positive and the average voltage determined by Eq. 9-1 is negative. For decreasing flux, $d\phi$ is negative and the average voltage is positive.

EXAMPLE 9-1

The flux linking a coil changes from 0 to 1.8×10^{-4} Wb in $\frac{1}{10}$ s. If -3.6 V is developed during the flux change, how many turns are there in the coil?

Solution

$$\frac{d\phi}{dt} = \frac{(1.8 \times 10^{-4} - 0)}{0.1} = 1.8 \times 10^{-3} \text{ Wb/s}$$

$$N = \frac{-(-3.6)}{1.8 \times 10^{-3}} = 2000 \ T$$

The negative sign in Eq. 9-1 represents the mathematical significance of *Lenz's law*. Heinrich Lenz (Russia, 1804–1865) discovered that *the direction of an induced emf is always such that the resulting magnetic field of the induced current will oppose the change in flux, giving rise to the emf.* This law is based on the conservation of energy. If the flux from the induced voltage and the subsequent current were in the same direction as the inducing flux, each would aid the other, so that an infinite emf would be developed.

Figure 9-2 illustrates Lenz's law. As the bar magnet is brought closer to the coil, the flux in the coil increases in the direction shown. By Lenz's law, the coil tends to oppose the change in flux by producing opposing flux; hence the establishment of

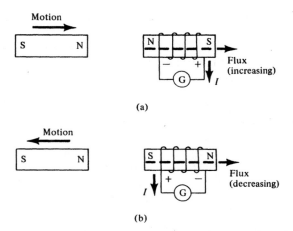

Figure 9-2   Generation of emf: **(a)** increasing flux; **(b)** decreasing flux.

the north and south poles. A galvanometer completes the electric circuit and allows a current to flow. The induced current, in turn, has a direction specified by the right-hand solenoid rule.

When the bar magnet is withdrawn as shown in Fig. 9-2(b), the flux decreases and an emf is established in accordance with Faraday's law. But the emf and its resultant current are induced so as to keep the flux from decreasing. Thus a south pole is established adjacent to the north pole of the bar magnet, and the force of attraction between these poles opposes the motion of the bar magnet.

With regard to Fig. 9-1, if the length of the conductor, the velocity of the conductor, and the magnetic field are mutually perpendicular, Faraday's law may be represented by

$$e = vBl \tag{9-2}$$

The voltage e is in volts when the mutually perpendicular quantities of v, B, and l are in meters per second, teslas, and meters, respectively. The polarity of the generated emf can be obtained by the *right-hand* or *generator rule* which is stated as follows (see Fig. 9-3):

> **When the first three fingers of the right hand are extended at right angles to each other so that the thumb points in the direction of motion of the conductor and the forefinger points in the direction of magnetic flux, then the center finger points to the positive terminal of the emf or in the direction of current flow when the emf is connected to an external circuit.**

Drill Problem 9-1 ■

The flux in a generator coil of 200 turns changes by 5×10^{-4} Wb in 4 ms. What voltage is developed in the coil?

Answer.   25 V.   ■ ■

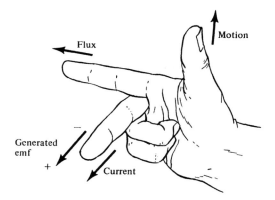

Figure 9-3   The generator rule.

9.2 SELF-INDUCTANCE

Consider the coil of Fig. 9-4. When the switch is closed, the current will attempt to increase, thereby causing an increase in flux. In turn, a *counter emf* is induced that opposes the change from zero current. Notice that in Fig. 9-4 the current is represented as an instantaneous current. From the previous chapter, we know that the instantaneous flux depends on the instantaneous current,

$$\phi = \frac{Ni}{\mathscr{R}} \tag{9-3}$$

Equation 9-3 indicates that for a particular coil, an infinitesimal change in flux $d\phi$ is the result of an infinitesimal change in current di. Then

$$d\phi = \frac{N}{\mathscr{R}}\, di \tag{9-4}$$

Introducing Eq. 9-4 into Eq. 9-1, we obtain an expression that shows that the emf generated by a coil occurs because of an opposition to current change,

$$e = -\frac{N^2}{\mathscr{R}}\frac{di}{dt} \tag{9-5}$$

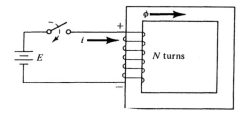

Figure 9-4   Counter emf.

The property of a coil or an electric circuit to oppose a change in current through that coil or circuit is called *inductance*. The SI unit for inductance is the *henry*, named in honor of Joseph Henry (United States, 1797–1878). A coil or circuit has an inductance of one henry (1 H) when a current change of 1 A/s induces a counter emf of 1 V. Thus,

$$e = -L \frac{di}{dt} \qquad (9\text{-}6)$$

where L is the symbol for inductance in henrys. The minus sign indicates only that the induced voltage is a counter emf and is in opposition to the applied voltage, as indicated by Fig. 9-4.

Hereafter, whenever we use either Eq. 9-1 or Eq. 9-6 we will drop the minus sign. However, we must not forget that the voltage given by either equation represents a counter emf and has a polarity so as to oppose a change in current. As evident from Fig. 9-4, *when the current increases the counter emf behaves as a voltage drop in the direction of the increasing current.* Conversely, a decreasing current produces a counter emf opposite to that shown in Fig. 9-4. In the latter case, the counter emf behaves as a source in that it attempts to maintain the flow of current. Consistent with our convention of using the letter symbol v for a voltage drop, we refer to the coil voltage drop as v_L.

Comparison of Eqs. 9-5 and 9-6 reveals that the inductance of a coil depends on the number of turns and the reluctance of the core of the coil.

$$L = \frac{N^2}{\mathscr{R}} \qquad (9\text{-}7)$$

Substitution for the reluctance relationship then produces

$$L = \frac{\mu A N^2}{l} \qquad (9\text{-}8)$$

where L is in henrys when μ is in henrys per meter, A is in square meters, and l is in meters.

Notice from Eq. 9-8 that inductance depends only on the physical relationships defining the coil, that is, on the core material, cross section of core, number of turns, and coil length. However, the permeability μ is nonlinear for ferrites. Although it is not suited for all inductance calculations, Eq. 9-8 can be used for toroid coils and single-layer coils where the length exceeds the diameter by a factor greater than 10. For other coil configurations you should consult an engineering handbook as most have empirical coil design formulas. Other formulas usually contain terms comparable to those found in Eq. 9-8.

EXAMPLE 9-2

A coil of 200 turns of wire is wound on a steel core having a mean length of 0.1 m and a cross section of 4×10^{-4} m^2. The relative permeability at the rated current of the coil is 1000. Determine the inductance of the coil.

Solution

$$L = \frac{\mu A N^2}{l} = \frac{(10^3)(4\pi \times 10^{-7})(4 \times 10^{-4})(2 \times 10^2)^2}{1 \times 10^{-1}} = 0.201 \text{ H}$$

EXAMPLE 9-3

If the current in a coil of 0.1 H changes as shown in Fig. 9-5(a), determine the voltage across the coil.

Solution. The coil response depends on the *change in current*.

    t = 0 to t = 2 ms
        There is no change in current,

$$\frac{di}{dt} = 0 \qquad v_L = 0$$

    t = 2 to t = 4 ms

$$\frac{di}{dt} = \frac{(0.5 - 0) \text{ A}}{(4 - 2) \text{ ms}} = \frac{0.5 \text{ A}}{2 \text{ ms}} = 0.25 \times 10^3 \text{ A/s}$$

$$v_L = L\frac{di}{dt} = (0.1)(0.25 \times 10^3) = 25 \text{ V}$$

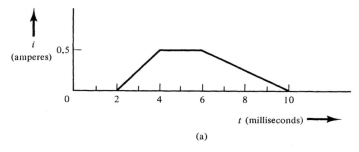

(a)

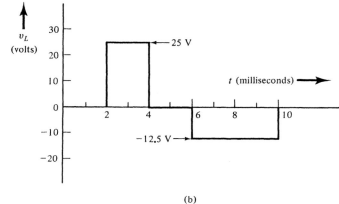

(b)

Figure 9-5 Example 9-3: **(a)** The coil current; **(b)** the coil voltage.

t = 4 to t = 6 ms

$$\frac{di}{dt} = 0 \qquad v_L = 0$$

t = 6 to t = 10 ms

$$\frac{di}{dt} = \frac{(0 - 0.5)\ \text{A}}{(10 - 6)\ \text{ms}} = \frac{-0.5}{4}\ \text{A/ms} = -0.125 \times 10^3\ \text{A/s}$$

$$v_L = L\frac{di}{dt} = (0.1)(-0.125 \times 10^3) = -12.5\ \text{V}$$

after 10 ms

$$\frac{di}{dt} = 0 \qquad v_L = 0$$

hence, we have the voltage diagram of Fig. 9-5(b).

In the preceding presentation, only one coil and its own opposition to current flow is considered. The resulting inductance is called self-inductance or simply inductance. Often, two coils are linked magnetically so that a change in current in one coil produces an emf in the second coil. The term *mutual inductance* is then used to describe the occurrence.

A physical device specifically designed to have inductance is called an *inductor*, a *choke*, or a *coil*. Commercially available inductors can be classified according to the type of core and whether the inductor is fixed or variable.

In general, air or ceramic core coils are used in radio and other electronic applications. The inductances of these inductors are generally in the microhenry to millihenry range. On the other hand, iron cores are used in chokes for power supplies and audio applications. High inductances can be obtained, in the range of henrys, but core losses become significant.

To obtain high inductances for radio applications, powdered iron cores are frequently used. The advantage of higher permeability is obtained, yet the losses are reduced. The powdered iron core is usually in the form of a cylinder called a slug,

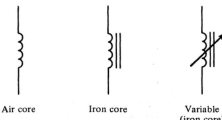

Air core   Iron core   Variable
(iron core)   **Figure 9-6**   Inductor symbols (ideal coils).

whose position can be varied within the actual coil. By varying the position, one obtains a variation in inductance.

The symbols for inductors are shown in Fig. 9-6 and some typical examples are shown in Figs. 9-7(a) and 9-7(b). The inductors shown in Fig. 9-7(a) have inductances in the microhenry range and are used for radio applications. In particular, the inductor in the center has a ferrite core for variable tuning. A typical toroid inductor appears in Fig. 9-7(b). It has an inductance of approximately 1 H and is used for audio applications.

It should be pointed out that any inductor is made of wire that itself possesses resistance. This resistance is distributed throughout the coil and is essentially in series with the *pure* or *ideal inductance* of the coil. Additionally, the coil has stray capacitance between its windings. Hence, a general circuit representing the practical inductor contains coil resistance and stray capacitance, as noted in Fig. 9-8.

The stray capacitance of a coil becomes important only in radio, TV, and microwave applications, so we may neglect it in our work. Furthermore, when the

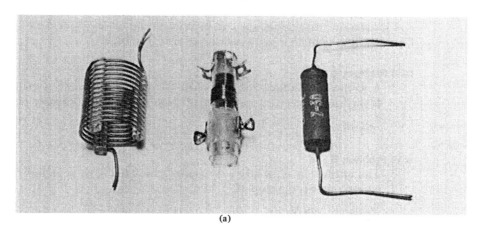

(a)

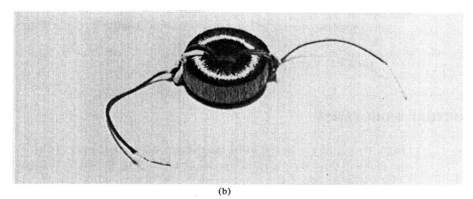

(b)

Figure 9-7 Typical inductors. **(a)** From left to right: air core, variable slug core, molded composition core. **(b)** Toroid with a ferromagnetic core.

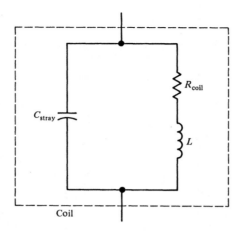

Figure 9-8 Coil resistance and stray capacitance may be significant in the practical inductor.

coil resistance is negligible, the coil behaves as a pure inductor. In subsequent work, if no mention is made of coil resistance, it is neglected.

Drill Problem 9-2 ■

A ferrite core coil of 100 turns has a length of 5 cm and a diameter of 2 mm. If the inductance is 0.5 mH, what is the relative permeability of the core?

Answer. 633.   ■ ■

Drill Problem 9-3 ■

The current in a 20 mH coil changes by 0.6 mA in a period of 1.5 μs. What counter emf is developed?

Answer. −8 V.   ■ ■

Drill Problem 9-4 ■

If a current change of 0.4 A/s produces a counter emf of 2 V across a coil, what is the inductance?

Answer. 5 H.   ■ ■

9.3 MUTUAL INDUCTANCE

Often, two coils are in the same neighborhood so that part of the flux produced by one coil will pass through the second coil. (See Fig. 9-9.) The two coils are then said to be mutually coupled.

Each of the coils in Fig. 9-9 has a self-inductance, since each is capable of producing an induced voltage in its own winding as a result of a change in its own

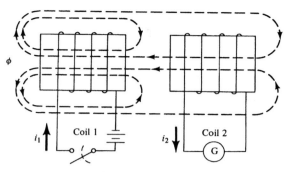

Figure 9-9   Mutually coupled coils.

flux. Because of the mutual coupling, a voltage is also induced in the second coil as a result of a flux change in the first coil.

When the switch in Fig. 9-9 is closed, the current in coil 1 increases, this increase being limited only by the coil's resistance. As the current increases, the flux linking coil 2 to coil 1 also increases. Then, by the laws of Faraday and Lenz, a voltage is induced in coil 2, given by

$$e_2 = -M_{12}\frac{di_1}{dt} \tag{9-9}$$

where e_2 is the induced voltage in coil 2, M_{12} the *mutual inductance* from coil 1 to coil 2, and di_1/dt the rate of change of current in coil 1.

In accordance with Eq. 9-9, two coils have a mutual inductance of 1 H if current changing at a rate of 1 A/s in one coil induces a voltage of 1 V in a second coil.

When the current in the first coil decreases, the magnetic fields in both coils collapse. Then the induced voltage in coil 2 is opposite to that developed when the switch is first closed.

If the dc source and switch were interchanged with the galvanometer, a voltage e_1 would be induced in coil 1 as a result of a changing current in coil 2. Then

$$e_1 = -M_{21}\frac{di_2}{dt} \tag{9-10}$$

In Eq. 9-10, the term M_{21} is the mutual inductance from coil 2 to coil 1. Ordinarily, $M_{12} = M_{21}$ so that the subscripts can be dropped;

$$M = M_{12} = M_{21} \tag{9-11}$$

In a coupled circuit, the coil or windings to which electrical energy is directly applied is the *primary winding* and the coil in which an emf is induced by mutual coupling is the *secondary winding*. The secondary winding usually has a load connected to it. Not all the primary flux may link the secondary coil; therefore, it is proper to define a *coefficient of coupling* k:

$$k = \frac{\phi_m}{\phi_p} \qquad (9\text{-}12)$$

where ϕ_m is the mutual flux linking the primary to the secondary and ϕ_p is the total flux in the primary. Obviously, the coefficient of coupling is never greater than 1. Depending on whether a little or much of the flux of one coil links another, the coils are said to be loosely coupled or tightly coupled, respectively. The coefficient of coupling is increased by bringing the coupled coils closer together and by aligning their axes. Introducing a magnetic core also increases the coupling, since the leakage flux is minimized.

Mutual inductance can be obtained in terms of the self-inductances. First, by equating Eqs. 9-1 and 9-6, we obtain an alternative defining equation for inductance,

$$e = -N\frac{d\phi}{dt} = -L\frac{di}{dt}$$

so that

$$L = N\frac{d\phi}{di} \qquad (9\text{-}13)$$

It follows that the mutual inductances are by Eq. 9-13,

$$M_{12} = N_2 k\left(\frac{d\phi_1}{di_1}\right)$$

and

$$M_{21} = N_1 k\left(\frac{d\phi_2}{di_2}\right)$$

Then the product $M_{12}M_{21}$ is

$$M_{12}M_{21} = M^2 = N_1 N_2 k^2\left(\frac{d\phi_1}{di_1}\right)\left(\frac{d\phi_2}{di_2}\right) \qquad (9\text{-}14)$$

Since the self-inductances are

$$L_1 = N_1\frac{d\phi_1}{di_1} \qquad L_2 = N_2\frac{d\phi_2}{di_2}$$

Eq. 9-14 becomes

$$M^2 = k^2 L_1 L_2$$

or

$$k = \frac{M}{\sqrt{L_1 L_2}} \qquad (9\text{-}15)$$

EXAMPLE 9-4

Coils 1 and 2 are placed near each other and have 200 and 800 turns, respectively. A change of current of 2 A in coil 1 produces flux changes of 2.5×10^{-4} Wb in coil 1 and 1.8×10^{-4} Wb in coil 2. Determine **(a)** the self-inductance of coil 1, **(b)** the coefficient of coupling, and **(c)** the mutual inductance.

Solution

(a) $\qquad L_1 = N_1\left(\dfrac{d\phi_1}{di_1}\right) = 200\left(\dfrac{2.5 \times 10^{-4}}{2}\right) = 2.5 \times 10^{-2} = 25 \text{ mH}$

(b) $\qquad k = \dfrac{\phi_2}{\phi_1} = \dfrac{1.8 \times 10^{-4}}{2.5 \times 10^{-4}} = 0.72 = 72\%$

(c) $\qquad M = N_2 k\left(\dfrac{d\phi_1}{di_1}\right) = 800(0.72)\left(\dfrac{2.5 \times 10^{-4}}{2}\right) = 72 \text{ mH}$

There are times when two coils must be near each other but coupling is not desired. The coils are then shielded from each other by magnetic enclosures called magnetic shields. Conversely, there are times when two coils are deliberately coupled. The network of these two magnetically coupled coils is a *transformer*. We shall study the transformer in more detail in Chapter 19.

Drill Problem 9-5 ■

Two coils have inductances of 2 and 1 H, respectively. If they are coupled with a coefficient of coupling of 0.95, what is the mutual inductance?

Answer.   1.34 H.   ■ ■

9.4 SERIES AND PARALLEL INDUCTORS

Two mutually coupled coils can be connected in series with their mmf's aiding or opposing. Two series-connected coils with aiding flux are shown in Fig. 9-10. With the connection of Fig. 9-10, the total inductance of coil 1, L_{1t}, equals the self-inductance L_1 plus the mutual inductance M caused by the flux from coil 2 linking coil 1:

$$L_{1t} = L_1 + M$$

Similarly, the total inductance of coil 2, L_{2t}, equals the self-inductance plus the mutual inductance linking coil 1 to coil 2:

$$L_{2t} = L_2 + M$$

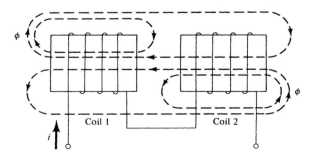

Figure 9-10   Series-connected coils aiding flux.

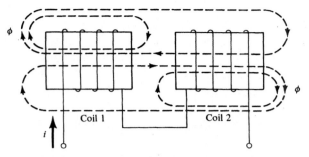

Figure 9-11   Series-connected coils opposing flux.

The total inductance for two coils connected in series with additive flux is thus

$$L_t = L_{1t} + L_{2t}$$

or $$L_t = L_1 + L_2 + 2M \qquad (9\text{-}16)$$

However, if the mutually coupled coils are connected so that the mutual flux opposes the flux of self-inductance (see Fig. 9-11), inductances L_{1t} and L_{2t} are

$$L_{1t} = L_1 - M$$

and $$L_{2t} = L_2 - M$$

Then the total inductance for two series-opposing coils is

$$L_t = L_1 + L_2 - 2M \qquad (9\text{-}17)$$

Of course, if no mutual coupling exists between the coils, the total inductance is the sum of the self-inductances only. Thus, with no mutual coupling, the total inductance L_t is given, in general, by

$$L_t = L_1 + L_2 + L_3 + \cdots + L_n \qquad (9\text{-}18)$$

When the winding directions of coupled coils are shown as in Figs. 9-10 and 9-11, it is easy to determine whether the connections are additive or subtractive. For schematic diagrams, a *dot notation* is used to indicate the winding directions relative to each other. Two additively coupled coils are shown schematically in Fig. 9-12(a); two subtractively coupled coils are shown in Fig. 9-12(b). Notice that when current flows into both (or out of both) dotted terminals, the connection is understood to be additive. Conversely, when current flows into one dotted terminal and out of the other dotted terminal, the windings are connected subtractively.

        (a)   (b)

Figure 9-12   Dot notation for coupled coils: **(a)** additive connection; **(b)** subtractive connection.

EXAMPLE 9-5

A 10 H inductor and a 9 H inductor are connected as in Fig. 9-12(a). The mutual inductance is 8 H. What is the total inductance and the coefficient of coupling?

Solution

$$L_t = L_1 + L_2 + 2M = 10 + 9 + 16 = 35 \text{ H}$$

$$k = \frac{M}{\sqrt{L_1 L_2}} = \frac{8}{\sqrt{90}} = 0.843$$

When two mutually coupled coils are connected in parallel, finding the total inductance is more difficult. The total inductance for this situation is given by the following equation:

$$L_t = \frac{L_1 L_2 - M^2}{L_1 + L_2 \pm 2M} \tag{9-19}$$

The negative sign in the denominator is used when the coils are aiding, and the positive sign is used when the coils are opposing. When the coupling is zero, $M = 0$, and Eq. 9-19 reduces to the product over the sum relationship,

$$L_t = \frac{L_1 L_2}{L_1 + L_2} \tag{9-20}$$

It follows that the total inductance for parallel inductances is given by the following equation (similar to that for parallel resistances):

$$\frac{1}{L_t} = \frac{1}{L_1} + \frac{1}{L_2} + \frac{1}{L_3} + \cdots + \frac{1}{L_n} \tag{9-21}$$

Drill Problem 9-6 ■

Determine **(a)** the total inductance for the coils connected as in Fig. 9-13, and **(b)** their coefficient of coupling.

Answer. (a) 27 mH. (b) 0.082.   ■ ■

Drill Problem 9-7 ■

Determine the total inductance for the coil groups of Fig. 9-14. There is no mutual coupling.

Answer. (a) 6.5 mH. (b) 3.08 mH.   ■ ■

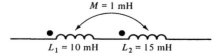

$M = 1 \text{ mH}$

$L_1 = 10 \text{ mH}$ $L_2 = 15 \text{ mH}$ **Figure 9-13**   Drill Problem 9-6.

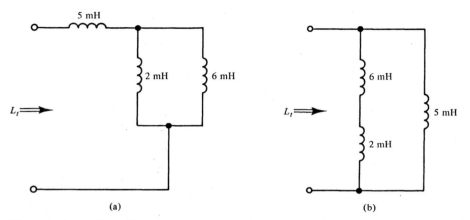

(a)   (b)

Figure 9-14   Drill Problem 9-7.

9.5 THE INCREASING *RL* TRANSIENT CURRENT

By generating a counter electromotive force, the inductor opposes a change in current; thus the current in a coil must change gradually. Then a change from a state of zero coil current to a finite and constant coil current implies a transient state, during which time the current varies with time. Remembering that a coil has a finite, if perhaps small, resistance and that an *RC* circuit produces an exponential transient, one might expect the *RL* circuit also to produce an exponential transient. Indeed, this is the case.

Consider the *RL* circuit of Fig. 9-15. At any time after the switch is closed, Kirchhoff's voltage relationship is given by

$$E = v_R + v_L \tag{9-22}$$

where v_R and v_L are the instantaneous voltage drops across the resistor and coil, respectively.

The transient begins when the switch is closed, at a time called $t = 0$ s. At that instant, the current is zero because the coil immediately opposes a change in current

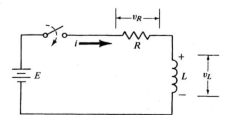

Figure 9-15   The *RL* circuit.

by producing the counter voltage

$$v_L = L\frac{di}{dt} \tag{9-23}$$

Since the polarity of v_L is established in Fig. 9-15, the minus sign of Eq. 9-6 is not used. Also, the inductor voltage is considered a drop in voltage, hence, the symbol v_L. At time $t = 0$, $i = 0$, so that $v_R = iR = 0$. Then by Eq. 9-22

$$E = 0 + L\frac{di}{dt} \tag{9-24}$$

From Eq. 9-24, the rate at which the current begins to change is given by

$$\frac{di}{dt}(t = 0^+) = \frac{E}{L} \tag{9-25}$$

where $t = 0^+$ signifies the time immediately after the switch is closed.

If there was no resistance in the circuit, the current would continue to increase with the slope E/L. But the resistance in Fig. 9-15, which represents the coil resistance, the external resistance, or both, limits the current. This maximum current I_m occurs when $v_L = 0$. For the coil voltage to be zero, a theoretically infinite time must pass so that there is no further current change. Then by the voltage equation,

$$E = iR + 0$$

and

$$I_m = E/R \tag{9-26}$$

The instantaneous currents at times other than zero and infinity are found by the solution of the voltage equation written in differential form,

$$E = iR + \frac{di}{dt} \tag{9-27}$$

The solution of Eq. 9-27 is

$$i = \frac{E}{R}(1 - e^{-Rt/L}) \tag{9-28}$$

One recognizes Eq. 9-28 as an increasing exponential curve, where the maximum value of current is E/R (Fig. 9-16).

Notice that the inductance appears only in the exponential term, thereby determining how fast the curve changes. The quantity L/R appearing as a reciprocal in Eq. 9-28 has the units of

$$\frac{L}{R} = \frac{\text{henrys}}{\text{ohm}} = \left(\frac{\text{volt}\cdot\text{seconds}}{\text{ampere}}\right)\left(\frac{1}{\text{ohm}}\right) = \text{seconds}$$

making the exponent of e dimensionless. This fraction is defined as the time constant of the *RL* circuit,

$$\tau = L/R \tag{9-29}$$

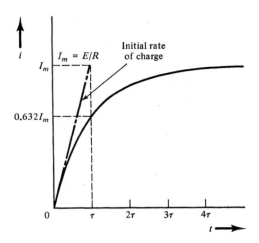

Figure 9-16 The RL increasing transient current.

The reader should recall from RC circuits that after one time constant, the increasing exponential curve is 63.2% of its final value. One time constant is also the time that would be required for the current to reach its maximum value were it to increase at the original rate of change (see Fig. 9-16). After 5τ, the transient for all practical purposes is complete.

Since $v_R = iR$, the voltage across the resistor also increases. Multiplying Eq. 9-28 by R, we obtain

$$v_R = E(1 - e^{-Rt/L}) \tag{9-30}$$

Then by the voltage equation,

$$v_L = E - v_R = E - E + Ee^{-Rt/L}$$
$$= Ee^{-Rt/L} \tag{9-31}$$

These two voltage equations are illustrated in Fig. 9-17.

The voltage v_L may be found more rigorously by differentiation of the current equation according to Eq. 9-23. Thus,

$$v_L = L\frac{di}{dt} = L\frac{d}{dt}\left[\frac{E}{R} - \frac{E}{R}e^{-Rt/L}\right]$$
$$= L\left[0 - \frac{E}{R}\left(-\frac{R}{L}e^{-Rt/L}\right)\right] = Ee^{-Rt/L}$$

Reference is made to the universal exponential curves presented in Fig. 7-21. The exponent x in the equations for the referenced curves equals t/τ.

EXAMPLE 9-6

A coil having a resistance of 15 Ω and an inductance of 0.6 H is connected to a 120 V line. **(a)** What is the rate at which the current is increasing at the instant

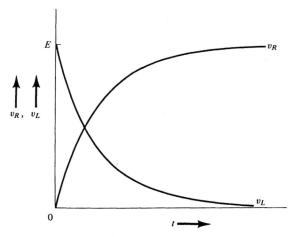

Figure 9-17 v_L and v_R versus time.

the coil is connected to the line? **(b)** What is the maximum current that can flow? **(c)** What is the current 0.1 s after the coil is connected to the line?

Solution

(a)
$$\frac{di}{dt} = \frac{E}{L} = \frac{120}{0.6} = 200 \text{ A/s}$$

(b)
$$I_m = \frac{E}{R} = \frac{120}{15} = 8 \text{ A}$$

(c)
$$\tau = \frac{L}{R} = \frac{0.6}{15} = 0.04 \text{ s}$$

$$i = I_m(1 - e^{-t/\tau}) = 8(1 - e^{-0.1/0.04})$$
$$= 8(1 - e^{-2.5}) = 8(1 - 0.083)$$
$$= 8(0.917) = 7.33 \text{ A}$$

Drill Problem 9-8 ■

A 6 V relay coil has a resistance of 40 Ω. What is the maximum current that the relay allows?

Answer. 0.15 A. ■■

Drill Problem 9-9 ■

A 12 V relay has an inductance of 1.5 H and a resistance of 60 Ω. **(a)** What maximum coil current flows? **(b)** How much time elapses between the time the voltage is applied until the time when the current reaches 0.15 A?

Answer. (a) 0.2 A. (b) 34.6 ms. ■■

A 12 V relay with an inductance of 1.5 H has a resistance of 10 Ω. What resistance must be added in series to change the time constant of the circuit to 100 ms?

Answer.   5 Ω.   ■ ■

9.6 THE DECAYING *RL* TRANSIENT CURRENT

Just as the current in an inductor cannot rise instantly, it cannot decrease instantly; it must decay gradually in an exponential manner. The basic *RL* decay circuit is shown in Fig. 9-18. Prior to the decay transient, switch S_1 is closed and S_2 is open. If the switches are in these positions for enough time, a steady state current $I_m = E/R$ flows in the coil. At time $t = 0$, switch S_1 is opened as S_2 is simultaneously closed. Accordingly, the coil tries to maintain the current I_m by releasing the energy stored in the magnetic field. The resistor, in turn, dissipates the energy released by the coil so that the field of the coil *collapses* gradually, as does the circuit current.

The equation for the decaying current is found by solution of the voltage equation

$$iR + L\frac{di}{dt} = 0 \tag{9-32}$$

Thus

$$i = I_m e^{-Rt/L} \tag{9-33}$$

where I_m is the current flowing in the coil just prior to the start of the decay transient.

Again the time constant τ is $\tau = L/R$, and since the current is a decaying exponential, the time constant is the time required for the current to decay to 36.8% of its original value.

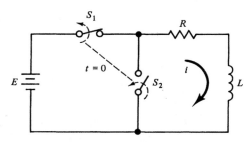

S_1 opened, S_2 closed at $t = 0$ **Figure 9-18**   The decaying *RL* circuit.

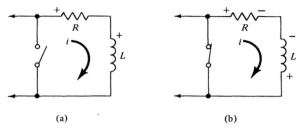

Figure 9-19   Change in coil voltage: **(a)** increasing field; **(b)** decreasing field.

The voltage across the resistor is a decaying exponential curve, since $v_R = iR$; that is,

$$v_R = Ee^{-Rt/L} \qquad (9\text{-}34)$$

As shown in Fig. 9-19, when the coil develops a counter emf to prevent the collapse of the magnetic field, the emf is opposite in polarity to that established during the build up of the field. But the coil voltage is the same magnitude as the resistor voltage; hence, we see the voltage curves of Fig. 9-20.

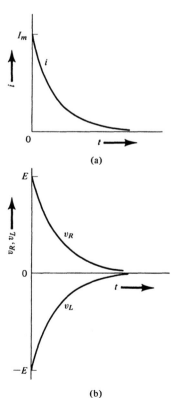

Figure 9-20   **(a)** Current; **(b)** voltage curves for the decaying field.

EXAMPLE 9-7

A relay has a resistance of 30 Ω and an inductance of 0.25 H. A current of 4 A flows in the relay when it is suddenly shorted. What current flows in the relay 0.02 s after it is shorted?

Solution

$$\tau = \frac{L}{R} = \frac{0.25}{30} = 8.34 \times 10^{-3} \text{ s}$$

$$i = I_m e^{-t/\tau} = 4e^{-20 \times 10^{-3}/8.34 \times 10^{-3}}$$

$$= 4e^{-2.4} = 4(0.091) = 0.36 \text{ A}$$

Drill Problem 9-11 ■

A motor field coil carrying 3 A is shorted by a 20 Ω resistor. The coil has a resistance of 10 Ω and an inductance of 10 H. One second after the short occurs, what is **(a)** the current and **(b)** the voltage across the 20 Ω resistor?

Answer. (a) 0.149 A. (b) 2.99 V.   ■ ■

9.7 INITIAL AND FINAL VALUES

In the two preceding sections we studied the simple RL transient circuit for which the inductor current either increased from zero to a finite maximum value I_m or decayed from a finite maximum value I_m to zero. But there is no reason to think that all RL transient currents must involve a change to or from a zero value of current. Now let us expand our ideas of the initial and final values to the general RL transient circuit.

Since a coil opposes any change in current we state that *a coil has a current flow immediately after the start of a transient period that exactly equals that immediately before the start of the transient period.* We denote the voltage across the coil at $t = 0$ as V_0 and the current through the coil at $t = 0$ as I_0. We call these values at $t = 0$ the *initial values.* In turn, when the RL transient is complete, we have values of voltage and current that we call the *final values.* We label the final values of voltage and current V_f and I_f, respectively.

When evaluating initial and final conditions we must remember that the inductor voltage is maximum at the start of the transient period. However, the initial current depends on the circuit and may or may not be zero. Furthermore, after an inductive field is established, the inductor current is I_f, no inductor voltage is present, and the coil acts as a short circuit.

The determination of the initial voltage V_0 is more involved because the voltage can jump instantaneously, and even change polarity, as the coil attempts to maintain a given current flow. To evaluate V_0 we write a KVL at $t = 0$ and use the initial current as one of the known values.

EXAMPLE 9-8

The switch of Fig. 9-21(a) has been in position 1 for a long time. At a time called $t = 0$ it is placed into position 2. Determine the **(a)** initial coil current, **(b)** initial coil voltage, and **(c)** final coil current.

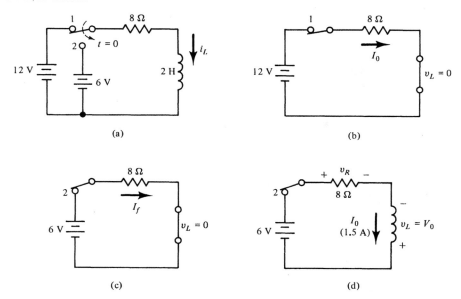

Figure 9-21 Example 9-8: **(a)** the original circuit; circuits for determining **(b)** the initial current, **(c)** the final current, and **(d)** the initial inductor voltage.

Solution. (a) We determine the initial current by observing the final value of current for the switch in position 1. As shown in Fig. 9-21(b), the coil has no voltage drop because the previous transient is considered finished. Hence, the coil behaves as a short. Then

$$I_0 = \frac{12 \text{ V}}{8 \text{ }\Omega} = 1.5 \text{ A}$$

(b) Immediately after the switch is placed into position 2, the current I_0 is maintained. [See Fig. 9-21(c).] Writing a KVL equation we obtain

$$v_R - V_0 - 6 = 0$$

or
$$V_0 = v_R - 6 = I_0(8 \text{ }\Omega) - 6$$
$$= 12 - 6 = 6 \text{ V}$$

(c) After the transient the coil again appears as a short as depicted in Fig. 9-21(d). Then

$$I_f = \frac{6 \text{ V}}{8 \text{ }\Omega} = 0.75 \text{ A}$$

Note that in part (b) of Example 9-8 the polarity of V_0 was arbitrarily selected to aid the 6 V source in maintaining the current I_0. Of course, the solution of the KVL equation indicates the polarity choice to be correct as evidenced by the positive value for V_0.

For the general RL transient circuit, the inductor current will vary exponentially, as indicated in Fig. 9-22. Note that the inductor current versus time curves have

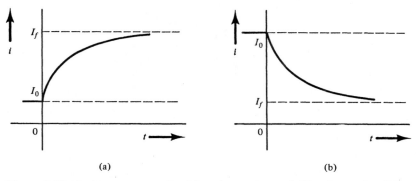

(a)

(b)

Figure 9-22 Inductor current for **(a)** an increasing and **(b)** a decreasing RL transient circuit.

the same form as the capacitor voltage versus time curves studied in Chapter 7. Hence we can describe the curves of Fig. 9-22 by an equation similar in form to Eq. 7-34. (The voltage terms are replaced by current terms.) Thus, the inductor current for *either* the increasing or collapsing field is given by the equation,

$$i_L = I_f + (I_0 - I_f)e^{-t/\tau} \tag{9-35}$$

Furthermore, whether the current increases or decreases, the coil voltage will always decay as the current reaches its final value. Thus,

$$v_L = V_0 e^{-t/\tau} \tag{9-36}$$

As was pointed out in Example 9-8, the value V_0 depends on the solution of a KVL equation at $t = 0$.

EXAMPLE 9-9

The switch S_1 in Fig. 9-23 is closed for a very long time. At a time called $t = 0$, switch S_2 is closed. What current flows in the coil 8 ms after S_2 is closed?

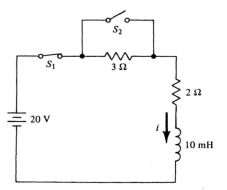

Figure 9-23 Example 9-9: circuit.

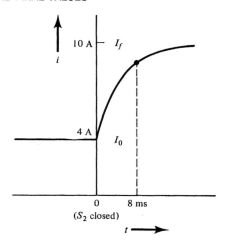

Figure 9-24 Example 9-9: i versus t.

Solution. The transient period begins when S_2 is closed.

Before S_2 is closed the transient due to S_1 being closed is completed; the current $i = I_0 = 20/5 = 4$ A.

After S_2 is closed the 3 Ω resistor is shorted so that the current can increase to a maximum value of $I_f = 20/2 = 10$ A. (See Fig. 9-24.) The current is then given by Eq. 9-35, where

$$\tau = \frac{L}{R} = \frac{10 \times 10^{-3}}{2} = 5 \text{ ms}$$

Then 8 ms after S_2 is closed,

$$i_L = 10 + (4 - 10)e^{-8/5} = 10 - 6e^{-1.6}$$
$$= 10 - 6(0.202) = 8.79 \text{ A}$$

Drill Problem 9-12 ■

Determine the initial and final values of **(a)** the coil current and **(b)** the coil voltage for the circuit of Fig. 9-25. The switch is placed in position 2 at $t = 0$ after being in position 1 for a long time.

Answer. (a) 1 and 2 A. (b) 6 and 0 V. ■ ■

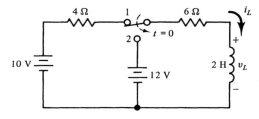

Figure 9-25 Drill Problems 9-12 and 9-13.

Drill Problem 9-13 ■

Determine the equations of coil current and coil voltage for the circuit and conditions described in Drill Problem 9-12.

Answer. $i_L = 2 - e^{-3t}$ A; $v_L = 6e^{-3t}$ V. ■ ■

9.8 ENERGY STORAGE

During the period that a magnetic field is being established by a coil, a portion of the energy supplied by the source is stored in the field. Of course, part of the energy supplied by the source is dissipated by the circuit resistance.

It was shown that the energy stored by a capacitor is $\frac{1}{2}CV^2$. It is not surprising then that the energy stored in a magnetic field is, by analogy,

$$W = \tfrac{1}{2}LI^2 \tag{9-37}$$

where W is the energy in joules, L the inductance in henrys, and I the current in amperes.

The energy equation can also be obtained by integration. The instantaneous power is found by multiplying the instantaneous voltage and current (see Fig. 9-26). Thus $p = vi$. The infinitesimal energy being stored is in turn the power × an infinitesimal period of time,

$$dw = p\,dt$$

that is, the energy is the area under the p versus t curve (the shaded part of Fig. 9-26). In calculus, the operation of integration allows one to sum the infinitesimal energy from $t = 0$ to $t = \infty$ (as the current changes from 0 to I A).

$$W = \int_0^\infty dw = \int_0^\infty p\,dt = \int_0^\infty \left(L\frac{di}{dt} \right)(i)\,dt$$

$$= \int_0^I Li\,di = \tfrac{1}{2}Li^2\big|_0^I = \tfrac{1}{2}LI^2$$

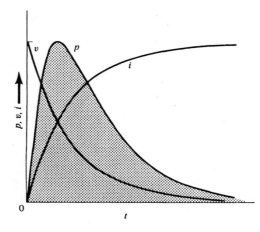

Figure 9-26 Instantaneous power in an inductor.

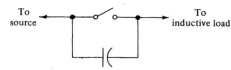

Figure 9-27   Capacitor used for arc suppression.

EXAMPLE 9-10

A coil with an inductance of 50 mH and a resistance of 2 Ω is connected to a 12 V source. What energy is stored in the field?

Solution

$$I_f = \frac{E}{R} = \frac{12}{2} = 6 \text{ A}$$

$$W = \tfrac{1}{2}LI^2 = \tfrac{1}{2}(50 \times 10^{-3})(36) = 0.9 \text{ J}$$

The energy released by the interruption of an inductive current is sometimes destructive. A rapidly collapsing field will generate a high counter emf across switch contacts, causing arcing and burning of the contacts. To eliminate this effect, one can connect a capacitor across the switch contacts to oppose any change in voltage at the contacts (see Fig. 9-27). A *discharge resistor* is often added in parallel with inductive circuits so that when the supply voltage is disconnected the inductive decay is not too rapid.

Drill Problem 9-14 ■

The current in the 16 H field coil of a motor changes from 8 to 7.5 A. What is the change in stored energy in the field coil?

Answer.   −62 J.   ■ ■

Drill Problem 9-15 ■

The switch in Fig. 9-28 is closed at $t = 0$. **(a)** What voltage finally appears across the coil? **(b)** What energy is finally stored in the coil?

Answer.   (a) 14.1 V. (b) 1.76 J.   ■ ■

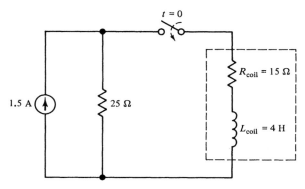

Figure 9-28   Drill Problem 9-15.

QUESTIONS

1. What characterizes electromagnetic induction?
2. What is Faraday's law?
3. What is Lenz's law?
4. What three mutually perpendicular quantities are specified by the right-hand rule?
5. What is self-inductance?
6. Define the unit of inductance.
7. How does inductance vary with the length and number of turns of a coil?
8. What is mutual inductance?
9. When are coils mutually coupled?
10. What is the coefficient of coupling and how does it relate to mutual inductance?
11. What is meant by primary and secondary windings?
12. What are the expressions for total inductance when two inductors are connected **(a)** series aiding and **(b)** series opposing?
13. What is the significance of dot notation for inductors?
14. Describe some types of inductors.
15. Discuss the rise of current in an inductive circuit.
16. What is the inductive time constant?
17. How does the resistance affect the final current in a rising *RL* circuit?
18. Discuss the fall of current in an inductive circuit.
19. How are the *RL* curves affected by the time constant?
20. How does one find the amount of energy stored in an inductor?

PROBLEMS

1. Determine the induced emf for a 500-turn coil when the flux changes by 30×10^{-5} Wb in 0.02 s.
2. What is the change in flux if a voltage of 2.4 V is induced in a 600-turn coil in 0.01 s?
3. A voltage of 9.6 V is generated by a flux change of 5×10^{-3} Wb in 1/60 s. How many turns must the coil have in this case?
4. Determine the direction of current flow in the resistances of Fig. 9-29.
5. The conductor in Fig. 9-1 is moving downward at a speed of 0.5 m/s. What emf is induced in the conductor if the conductor is 30 cm long and the flux density is 0.8 T?
6. An emf of 0.2 V is induced in a 10-cm-long conductor that is moving perpendicular to a magnetic field at a rate of 0.5 m/s. What is the strength of the magnetic field?
7. The current in a coil of 120 turns and 0.48 H changes by 0.5 A in 25 m/s. What emf is induced?
8. A 300-turn coil is wound on a plastic toroid having a cross section of 1×10^{-3} m^2 and a mean length of 40 cm. What is the inductance of the coil?

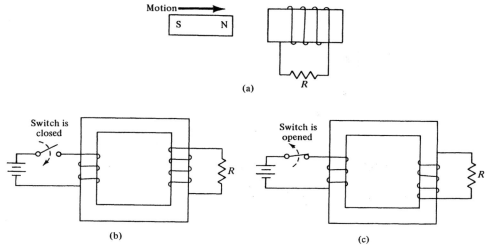

(a)

(b)

(c)

Figure 9-29   Problem 9-4.

9. An air-wound coil of 560 turns has an inductance of 360 mH. If 120 turns are removed from the coil, everything else remaining constant, what is its new inductance?

10. What is the inductance of a radio coil having 300 turns wound on a fiber core 1 cm in diameter and 4 cm long?

11. If a powdered iron core with a relative permeability of 300 is placed inside the coil of Problem 10, what is the inductance?

12. Many empirical formulas for calculating inductance can be found in handbooks. One such formula gives the approximate inductance of the single-layer, air core coil of Fig. 9-30 as

$$L = \frac{r^2 N^2}{9r + 10l} \qquad \mu H$$

where L is the inductance, N is the number of turns, and r and l are the radius and length, respectively, in inches. How many turns must be spaced on a form 1.35 in. long with a radius of 0.5 in. if an inductance of 12.5 μH is desired?

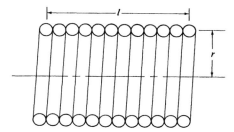

Figure 9-30   Cross section of an air-wound core for Problem 9-12.

13. Two coils, 1 and 2, are placed near each other and have 300 and 600 turns, respectively. A change in current of 1.5 A in coil 2 produces flux changes of 1.2×10^{-4} Wb in coil 2 and 0.9×10^{-4} Wb in coil 1. Determine **(a)** the self-inductance of coil 2, **(b)** the coefficient of coupling, and **(c)** the mutual inductance.

14. Find the total inductance and the coefficient of coupling for the coils of Fig. 9-31.

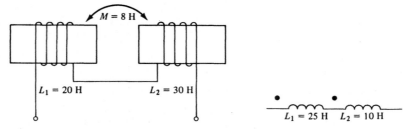

$L_1 = 20$ H $L_2 = 30$ H

Figure 9-31   Problem 9-14.

$L_1 = 25$ H $L_2 = 10$ H

Figure 9-32   Problem 9-15.

15. Two coils are connected in series, as shown in Fig. 9-32. Determine the total series inductance if **(a)** the coefficient of coupling is 80% and **(b)** the coefficient of coupling is zero.

16. Two coils are connected in series-aiding. The total inductance is 800 mH. The coils are then connected in series opposing, and the inductance is found to be 500 mH. The inductance of the first coil is 350 mH. **(a)** What is the inductance of the second coil? **(b)** What is the mutual inductance?

17. Find the total inductance for the series-connected coils of Fig. 9-33.

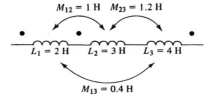

$M_{12} = 1$ H $M_{23} = 1.2$ H

$L_1 = 2$ H $L_2 = 3$ H $L_3 = 4$ H

$M_{13} = 0.4$ H

Figure 9-33   Problem 9-17.

18. Determine the total inductances for the circuits of Fig. 9-34. No mutual inductance is present.

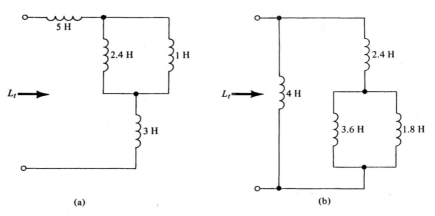

5 H

2.4 H   1 H

$L_t \longrightarrow$

3 H

(a)

2.4 H

$L_t \longrightarrow$ 4 H

3.6 H   1.8 H

(b)

Figure 9-34   Problem 9-18.

19. A coil having a resistance of 15 Ω and an inductance of 0.6 H is connected to a 120 V line. What is the current through the coil 0.05 s after the coil is connected to the line?

20. A coil of 12 Ω resistance and 0.36 H inductance is connected to a voltage source. What time is required for the current to build up to 63.2% of its final value?

21. A relay coil has 450 Ω resistance and 6.8 H inductance. How soon after 22 V is applied to the coil does the relay operate if it takes 40 mA to operate the relay?

22. Given the circuit of Fig. 9-35, if the switch is closed at $t = 0$, determine the time required for the current i to reach 95% of its final value.

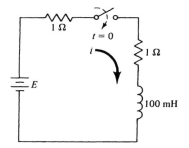

Figure 9-35 Problem 9-22.

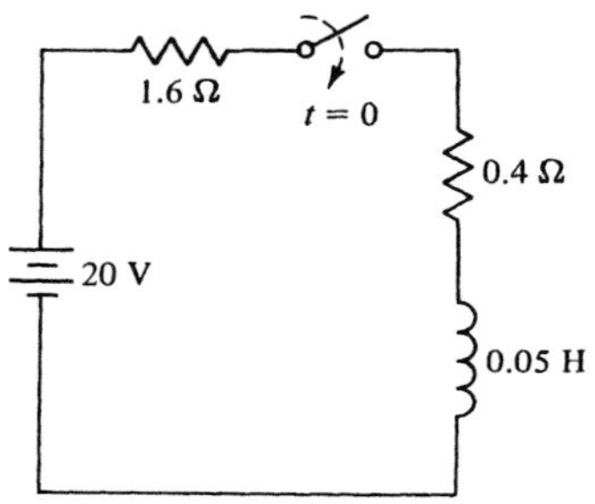

Figure 9-36 Problems 9-23 and 9-30.

23. The switch in Fig. 9-36 is closed at $t = 0$. Determine (a) the coil current and (b) the coil voltage 50 ms later.

24. A coil of 18 mH inductance and 9 Ω resistance is shorted with 400 mA flowing through the coil. What is the coil current 2.5 ms after the short occurs?

25. Switches S_1 and S_2 of Fig. 9-23 have been closed so that all previous transients have passed. At a time $t = 0$, switch S_2 is opened. (a) Find the expression for the current as a function of time. (b) Find the value of current 4 ms after S_2 is opened.

26. Switches S_1 and S_2 of Fig. 9-23 have been closed so that all previous transients have passed. Beginning with a time $t = 0$, switch S_2 is opened and closed at 4 ms intervals. Plot the coil current versus time for the first 20 ms.

27. Determine the initial and final values of (a) coil current and (b) coil voltage for the coil in Fig. 9-37. The switch is closed at $t = 0$.

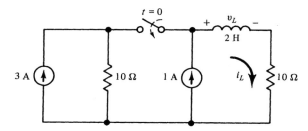

Figure 9-37 Problem 9-27.

28. Determine the initial and final values of (a) coil current and (b) coil voltage for the circuit of Fig. 9-38. The switch is in position 1 for a long time and is placed in position 2 at $t = 0$.

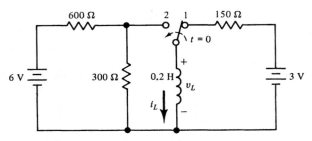

Figure 9-38   Problems 9-28 and 9-29.

29. Determine the coil current and coil voltage for the circuit and conditions described in Problem 9-28 at a time $t = 3$ ms.

30. What current flows in the circuit of Fig. 9-36 five time constants after the switch is thrown? What energy is stored in the coil at that time?

31. Determine the energy stored in the field of a motor when a current of 2.4 A flows in the field. The inductance of the field is 28 H.

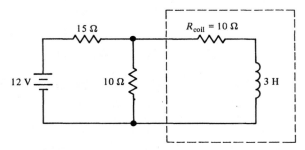

Figure 9-39   Problem 9-32.

32. Given the circuit of Fig. 9-39 with the practical coil: **(a)** What is the voltage measured across the coil? **(b)** How much energy is stored in the field of the coil?

Chapter 10

Alternating Current Relationships

10.1 THE GENERATION OF ALTERNATING CURRENT

In an earlier chapter, alternating voltages and currents were described as voltages and currents that periodically change direction. Compared to direct current, alternating current provides lower transmission losses for high power, larger capacity generators, and easier transformation of voltage and current levels.

A large portion of electrical theory deals with the particular alternating voltages and currents described as *sinusoidal*. A sinusoidal quantity or simple harmonic quantity is the product of a constant and either the sine or cosine of an angle. Sinusoidal quantities are highly desirable in engineering work, since they are easy to manipulate mathematically. One may add, subtract, multiply, divide, integrate, and differentiate sinusoidal quantities; the result is always another sinusoidal quantity. Thus electrical engineers often strive for devices that produce sinusoidal voltages and currents.

It is instructive to analyze the generation of alternating current by the simple, single-turn generator of Fig. 10-1, in which a single turn of wire rotates in the magnetic field set up by the permanent magnets. In turn, as the coil cuts the flux of the magnetic field, an emf is induced in the coil. A combination of slip rings and brushes provides the connections for current flow to the external resistance. The polarity of the generated emf is determined by the right-hand rule relating the mutually perpendicular quantities of motion, flux, and current or generated emf.

One can obtain an idea of the magnitude and polarity of the generated emf as the coil is rotated by analyzing the motion and flux at selected positions. In Fig. 10-2(a), the coil is in the vertical position with conductor *b* at the top and *a* at the bottom (position 1). The instantaneous motion of each conductor is with the field; therefore, no flux lines are cut and the instantaneous emf is zero. After 45° of rotation, the coil is in position 2, as seen in Fig. 10-2(b). A component of the motion now cuts

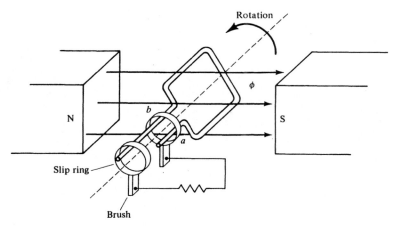

Figure 10-1   The simple ac generator.

the flux so that voltages are generated at ends *a* and *b* with the designated polarity. Further rotation brings the conductors to position 3, as in Fig. 10-2(c). At this time, all of the motion is perpendicular to the flux so that a maximum cutting of flux lines results. Hence a maximum voltage is developed, still with conductor *a* negative and conductor *b* positive. In Fig. 10-2(d), position 4, the movement of conductor *b* is still

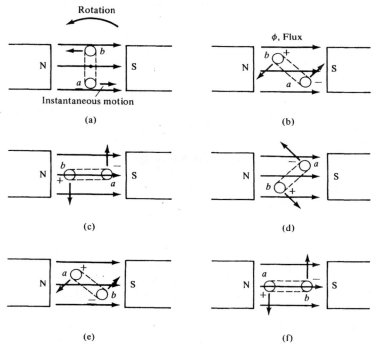

Figure 10-2   The coil of a generator at positions **(a)** 1; **(b)** 2; **(c)** 3; **(d)** 4; **(e)** 5; **(f)** 6.

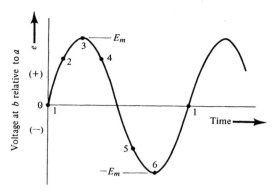

Figure 10-3   Plot of emf versus time.

downward and that of conductor a is upward. The instantaneous voltages still have the same polarity, but the magnitudes have decreased. Finally, after 180° of rotation, the coil is vertical so that the instantaneous voltages are zero, but conductor a is now at the top and b is at the bottom.

Further rotation from 180 to 360° (relative to position 1) produces downward motion of conductor a and upward motion of conductor b. It follows, as illustrated by Figs. 10-2(e) and 10-2 (f), that conductor a is now positive and b is negative. In position 5, the generated emf is not maximum, but in position 6 it is maximum. Further coil rotation brings the coil back to position 1.

Since the coil position is a function of time, the instantaneous voltage at coil end b referenced to a can be plotted against time, as in Fig. 10-3. All of the instantaneous positions and the corresponding voltages make a continuous curve, with the six positions of Fig. 10-2 providing only six points on the curve.

In physics, a wave is defined as a disturbance that is propagated in a medium in such a manner that at any point in the medium, the displacement is a function of time. In electric circuits, a *wave* is a variation of current or voltage in the circuit. A *waveform* is a graph or plot of voltage or current versus time (or a time-dependent variable).

The particular waveform of Fig. 10-3 describes a *sine wave*. It is not particularly difficult to show why the generator of Fig. 10-1 generates a sine wave. From Eq. 9-2, the voltage developed in a conductor of length l moving in a magnetic field of density B is

$$e = Blv_n \tag{10-1}$$

A subscript n is added to the velocity symbol to indicate that *normal* or *perpendicular* velocity is used. In Fig. 10-4, the coil of a simple generator is shown displaced by an angle α from its original vertical position. At that instant, either of the conductors is moving at a velocity v called the *tangential velocity*, since it is tangent to the circle formed by the rotating conductors. Only a part of this velocity v_n results in actual motion of the conductor perpendicular to the magnetic field. By Fig. 10-4, v_n is seen to relate to velocity v by the trigonometric sine function so that

$$v_n = v \sin \alpha$$

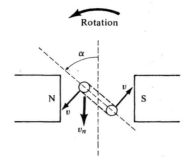

Figure 10-4 Coil velocity of the simple generator.

Substitution of this latter equation into Eq. 10-1 results in the emf equation for one conductor,

$$e = Blv \sin \alpha \qquad (10\text{-}2)$$

and for two sides of the coil in series,

$$e = 2Blv \sin \alpha \qquad (10\text{-}3)$$

Since B, l, and v are constant terms, the coefficient of the sine term can be replaced by a single constant E_m,

$$e = E_m \sin \alpha \qquad (10\text{-}4)$$

A sinusoidal quantity is stated mathematically in the same form as Eq. 10-4. The *amplitude* of a sinusoidal quantity is the *peak* or *maximum value* that the quantity attains. Since the sine function has a maximum value of 1, E_m is the maximum value or amplitude of the emf wave described by Fig. 10-3 and Eq. 10-4. Notice that a capital letter with a subscript m is used to denote the amplitude.

In Fig. 10-4, the angle α increases from 0° to 360° for one revolution of the coil; that is, circular motion can be divided into 360 uniform sectors. An alternative unit of angular displacement is the *radian* (1 rad = 57.3°). Using this relationship, we see that 2π rad = 360°.

Since most rotational motion is specified in radian measure, the *rotational* or *angular velocity* is stated in radians per second. The letter symbol for angular velocity is the lowercase Greek omega, ω. It follows that the angular displacement α at any instant equals the angular velocity × the time:

$$\alpha = \omega t \qquad (10\text{-}5)$$

Thus, the general equations for a sinusoidal voltage and a sinusoidal current are, respectively,

$$e = E_m \sin \omega t$$
$$i = I_m \sin \omega t$$

where e and i are the instantaneous values of voltage and current and E_m and I_m are the *maximum* or *peak* values.

EXAMPLE 10-1

A certain voltage is described by the equation $e = 120 \sin 377t$. What is the instantaneous emf when $t = 0.01$ s?

Solution

$$\omega t = 377(0.01) = 3.77 \text{ rad}$$

$$3.77 \text{ rad} = 3.77 \text{ rad}\left(\frac{360°}{2\pi \text{ rad}}\right) = 216°$$

Then

$$e = 120 \sin 216° = 120(-0.588) = -70.5 \text{ V}$$

EXAMPLE 10-2

A sinusoidal current has an instantaneous value of 20 mA at a time $t = 0.0011$s. If $\omega = 6283$ rad/s, what is the peak value of current?

Solution

$$\omega t = 6283(0.0011) = 6.91 \text{ rad}$$

$$= 6.91 \text{ rad}\left(\frac{360°}{2\pi \text{ rad}}\right) = 396°$$

$$i = 20 \text{ mA} = I_m \sin 396° = I_m \sin 36° = I_m(0.588)$$

$$I_m = \frac{20 \text{ mA}}{0.588} = 34 \text{ mA}$$

Since the sine wave repeats itself every 360° or every 2π rad, any angle greater than 360° or 2π rad can be thought of as an integral number of repetitions plus a fraction of the 360°. Thus, in Example 10-2 the angle corresponding to ωt is described as either 396° or 36°. This is illustrated in Fig. 10-5. Many scientific calculators allow angular measure to be entered in either degree or radian form. However, many computers allow angular measure to be entered only in radian measure.

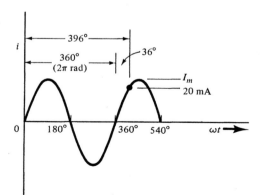

Figure 10-5 The sinusoidal current described in Example 10-2.

Drill Problem 10-1 ■

A sinusoidal current is described as $i = 200 \sin 377t$ mA. What is the instantaneous current at **(a)** t = 0.01 s and **(b)** $t = 0.12$ s?

Answer. (a) −118 mA. (b) 190 mA.   ■ ■

10.2 WAVEFORMS AND FREQUENCY

A *periodic waveform* is a repetitive waveform, that is, one that repeats itself after given time intervals. The waveform need not be sinusoidal to be repetitive; for example, the triangular and square waves of Fig. 10-6 repeat themselves after the same time intervals as the sine wave of that figure. The smallest possible nonrepetitive portion of a periodic waveform is a *cycle*. The time interval between successive repetitions or cycles is called the *period T*. It should be noted that a sinusoidal quantity passes through one complete sequence of values for every 2π rad of change of ωt.

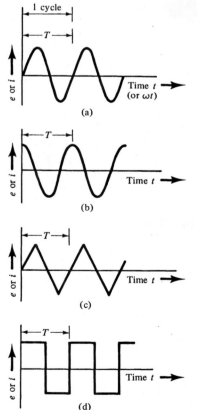

Figure 10-6 Various repetitive waveforms; **(a)** sine wave; **(b)** cosine wave; **(c)** triangular wave; **(d)** square wave.

The *frequency f* of a time-varying, periodic quantity is the number of cycles occurring in unit of time. The SI unit for frequency is the *hertz* (Hz), in honor of Heinrich R. Hertz (Germany, 1857–1894). One hertz equals one cycle per second. Thus, the frequency is the reciprocal of the period *T*:

$$f = 1/T \qquad (10\text{-}6)$$

In producing one cycle of sinusoidal voltage, a generating conductor passes through 360 electrical time degrees in a period of time $T = 1/f$. Then the angular frequency ω is

$$\omega = \frac{2\pi}{T} = \frac{2\pi}{1/f} = 2\pi f \qquad (10\text{-}7)$$

where ω is in radians per second when *f* is in hertz.

EXAMPLE 10-3

What are the frequency and period of the voltage described by the equation, $e = 120 \sin 377t$?

Solution

$$f = \frac{\omega}{2\pi} = \frac{377}{2\pi} = 60 \text{ Hz}$$

$$T = \frac{1}{f} = \frac{1}{60} = 0.0167 \text{ s}$$

Not all sinusoidal waves have a value of zero at $t = 0$. The *initial phase angle* is the phase angle of the periodic quantity at the instant measurement of the quantity begins ($t = 0$). Thus the cosine wave of Fig. 10-6 might be described as a sine wave with a 90° or $\pi/2$ rad "head start." Then the cosine wave can be expressed as a displaced sine wave; that is, $\cos \omega t = \sin(\omega t + 90°)$. Similarly, the sine wave can be described as a cosine wave that has been "delayed" by 90° or $\pi/2$ rad; that is, $\sin \omega t = \cos(\omega t - 90°)$. If the general sine expression for voltage (or current) is modified to take the initial phase angle into account, the result is

$$e = E_m \sin(\omega t \pm \phi) \qquad (10\text{-}8)$$

where ϕ is the initial phase angle, and the plus sign is used for an advanced sine wave and the minus sign for a delayed sine wave. It should be noted that both the ωt and ϕ terms in Eq. 10-8 must be in the same units (radians or degrees) to be correctly added.

Frequently, two sinusoidal quantities with the same period are compared. The *phase difference* is the fractional part of a period by which the corresponding values of the two quantities are separated. In addition, the terms "leading" or "lagging" are used to describe waves that are advanced or delayed. In Fig. 10-6, for example, the

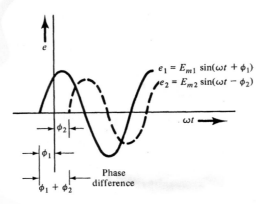

$$e_1 = E_{m1} \sin(\omega t + \phi_1)$$
$$e_2 = E_{m2} \sin(\omega t - \phi_2)$$

Figure 10-7   Two out-of-phase sinusoids.

sine wave lags the cosine wave by a phase difference of 90°. Alternatively, the cosine wave leads the sine wave by 90°.

As another example of phase difference between sinusoids, consider two voltages, e_1 and e_2, defined by the graphs of Fig. 10-7. First, notice that e_1 represents a sine wave advanced by the angle ϕ_1, and e_2 represents a sine wave delayed by the angle ϕ_2. Then, the phase difference is the total angular displacement between the two waves, that is, in this case, $\phi_1 + \phi_2$. Notice that we can define the phase difference between two sinusoids, even though they have different magnitudes. Phase difference can also be defined between two dissimilar quantities such as voltage and current. However, we must limit the definition of phase difference to waveforms or quantities of the same frequency. Referring to Fig. 10-7, we can say either that e_1 *leads* e_2 or that e_2 *lags* e_1 by the angle $\phi_1 + \phi_2$.

A *pulse* is defined as an abrupt step in voltage or current followed by an equal and opposite abrupt step in voltage or current. Ideally, the pulse is rectangularly shaped as indicated in Fig. 10-8. Thus, at a time t_0, the voltage or current rises to a given value along the *leading pulse edge*. At a later time, t_1 the voltage or current falls back to zero along the *trailing pulse edge*.

The time during which the amplitude is not zero, that is $t_1 - t_0$, is called the *pulse duration* or *pulse width* t_w. The pulse depicted in Fig. 10-8 is positive, but we can also have negative pulses.

Ideally, a pulse can be generated by a simple switch closure followed by a switch opening. However, if the switching is done electronically, it is possible to get repetitive pulses, as indicated by the waveform in Fig. 10-9. As with any repetitive wave-

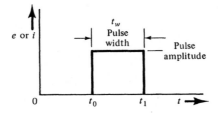

Figure 10-8   A rectangular pulse.

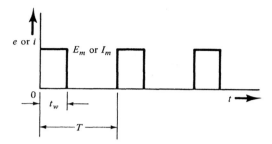

Figure 10-9 Repetitive pulses.

form, we can define a period T and a frequency f, both related by Eq. 10-6. In the case of repetitive pulses, the frequency is also called the *pulse repetition rate*.

Another term used to describe repetitive pulse waveforms is the *duty cycle*. Duty cycle is the ratio of "on time" to the total period. That is, it is the ratio of pulse width to period:

$$\text{duty cycle} = t_w/T \tag{10-9}$$

Although, because it is a ratio, the duty cycle is a unitless decimal quantity, it is usually expressed as a percent. In addition to its use in describing digital signals, the term "duty cycle" is also used to describe the operation of motors subject to repetitive "start and stop" operations.

EXAMPLE 10-4

A digital timing circuit generates the signal shown in Fig. 10-10. Determine the pulse repetition frequency and the duty cycle.

Solution. $T = 0.5 \ \mu s$. Then,

$$f = \frac{1}{T} = \frac{1}{0.5 \times 10^{-6}} = 2 \times 10^6 = 2 \ \text{MHz}$$

and the duty cycle is

$$\frac{t_w}{T} = \frac{0.04 \times 10^{-6}}{0.5 \times 10^{-6}} = 0.08 = 8\%$$

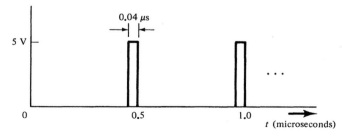

Figure 10-10 Example 10-4.

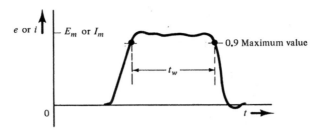

Figure 10-11   The practical pulse.

Although we ideally indicate that the sides of a pulse or a square wave are vertical, they may not be in a practical situation. The actual waveform of a positive pulse might appear as in Fig. 10-11. Definition of the pulse width is now vague. One way of specifying the width or duration is simply to specify the time that the pulse is above the 90% level.

For large power generation of alternating voltages and currents, the electro-mechanical generator depicted in Fig. 10-1 is used. However, for electronic circuits, electronic signal or function generators are used. Two such electronic signal generators are shown in Fig. 10-12 and Fig. 10-13.

The signal generator of Fig. 10-12 is called a *function generator* because it electronically generates square and triangular voltages in addition to sinusoidal voltages. It has a frequency range of 1–100 kHz and generates voltage levels up to 10 V peak-to-peak.

The signal generator of Fig. 10-13 is a *pulse generator* because it electronically generates either a single pulse or repetitive pulses. It can generate repetitive pulses in a frequency range from 0.5 Hz to 5 MHz. The output voltage level is variable to 10 V.

The symbols we use for alternating voltage sources appear in Fig. 10-14.

The frequencies encountered in electrical work are virtually limitless. The lowest

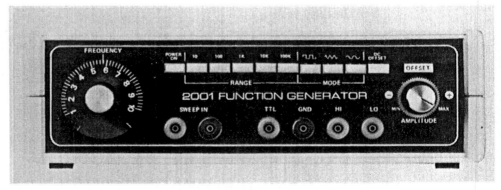

Figure 10-12   The model 2001 Function Generator. (Courtesy of Global Specialties Corp.)

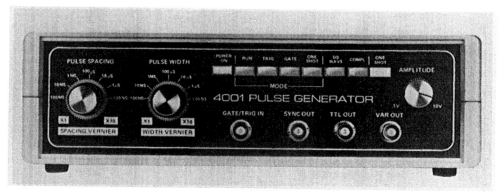

Figure 10-13 The model 4001 Pulse Generator. (Courtesy of Global Specialties Corp.)

frequency is zero frequency ac (direct current). Next are the *power frequencies*, the most prevalent being 60 Hz. The frequencies ranging from approximately 20 Hz to 20 kHz are called the *audio frequencies*, since they are audible to the human ear when transformed into sound by loudspeakers. Frequencies from 20 kHz to those in the gigahertz range are called *radio frequencies* and are associated with *electromagnetic radiation*. The total and apparently limitless range of frequencies is known as the *frequency spectrum*, part of which is shown in Fig. 10-15.

Drill Problem 10-2 ■

Determine the frequency and period of the voltage and current described by the following equations: **(a)** $e = 10 \sin 6283t$ V; **(b)** $i = 14.14 \sin(31\,416t)$ mA.

Answer. (a) 1 kHz; 1 ms. (b) 5 kHz; 200 μs. ■ ■

Drill Problem 10-3 ■

Determine the phase difference between the two voltages and the two currents.

(a)
$$e_1 = 10 \sin(377t + 30°) \text{ V}$$
and
$$e_2 = 20 \sin(377t - 10°) \text{ V}$$

(b)
$$i_1 = 2 \cos(6283t) \text{ mA}$$
and
$$i_2 = 1.4 \sin(6283t + 28°) \text{ mA}$$

Answer. (a) 40°. (b) 62°. ■ ■

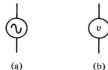

(a) (b)

Figure 10-14 Schematic symbols: **(a)** sinusoidal voltage source; **(b)** nonsinusoidal alternating voltage source.

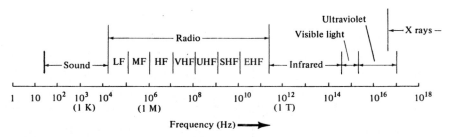

Figure 10-15   Frequency spectrum.

Drill Problem 10-4 ■

Determine the frequency, period, and duty cycle for the waveforms of Fig. 10-16.

Answer.   (a) 50 Hz; 20 ms; 25%. (b) 50 kHz; 20 μs; 80%.   ■ ■

10.3 THE AVERAGE VALUE

Since the instantaneous value of any particular alternating voltage or current is constantly changing, it is often desirable to know the average value of the voltage or current. The average value for any variable is obtained from a plot of the variable versus time by dividing the area under the curve by the length of the curve. In particular, the *average voltage* or *average current*, E_{av} or I_{av}, is usually obtained over a complete cycle.

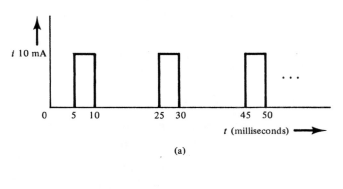

(a)

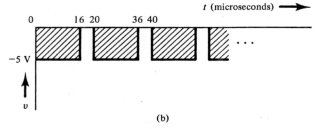

(b)

Figure 10-16   Drill Problems 10-4 and 10-5.

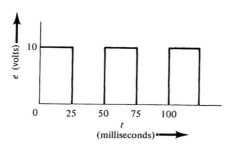

Figure 10-17 Example 10-5.

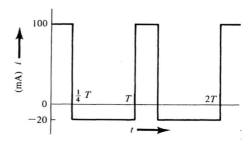

Figure 10-18 Example 10-6.

$$E_{av}(\text{or } I_{av}) = \frac{\text{area under curve for one cycle}}{T} \qquad (10\text{-}10)$$

where T is the period.

EXAMPLE 10-5

What is the period and average value for the waveform of Fig. 10-17?

Solution. The wave repeats itself every 50 ms; therefore, $T = 50$ ms.

$$E_{av} = \frac{\text{area}}{T} = \frac{(10)(25)}{50} = 5 \text{ V}$$

EXAMPLE 10-6

Calculate the average value for the current waveform of Fig. 10-18.

Solution

$$I_{av} = \frac{\text{area}}{T} = \frac{(100)(\frac{1}{4}T) + (-20)(\frac{3}{4}T)}{T}$$

$$= \frac{25T - 15T}{T} = 10 \text{ mA}$$

The average value of a sinusoid taken over a complete cycle is obviously zero, since the area under the positive half of the cycle equals the area under the negative half of the cycle.

However, sine pulses equivalent to a half-cycle of a sine wave are frequently encountered, particularly in rectifier circuits. With this in mind, the "average value" of a sine wave is defined as the average value of one complete loop ($\frac{1}{2}$ cycle). If the area formula of a sine pulse (or another unusual shape) is not available, one may either find an approximate area by using simple geometric shapes or obtain the actual area by calculus.

EXAMPLE 10-7

Using the rectangular and triangular shapes superimposed on the sine pulse of Fig. 10-19, obtain an approximate area for the sine pulse.

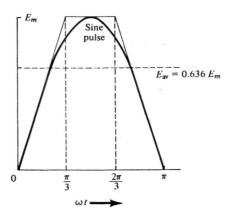

Figure 10-19   A sine pulse and one method of approximation.

Solution

$$\text{Area} = \tfrac{1}{2}(E_m)\left(\frac{\pi}{3}\right) + E_m\left(\frac{\pi}{3}\right) + \tfrac{1}{2}(E_m)\left(\frac{\pi}{3}\right)$$

$$= \left(\frac{\pi}{3} + \frac{\pi}{3}\right)E_m = \frac{2\pi}{3}\,E_m \approx 2.1E_m$$

By using calculus, one finds the area under the sine pulse actually to be $2E_m$. Using this exact value, we find the average value to be

$$E_{av} = \frac{\text{area}}{\text{length}} = \frac{2E_m}{\pi} = 0.636E_m \qquad (10\text{-}11)$$

Thus the "average value" of a sine wave is taken as 0.636 × the maximum value and refers only to a half-cycle, as shown in Fig. 10-19.

Those familiar with integral calculus may desire to find the average value of a voltage (or current) by

$$E_{av} = \frac{1}{T}\int_0^T e\,dt \qquad (10\text{-}12)$$

Drill Problem 10-5 ■

Determine the average values of the waveforms of Fig. 10-16.

Answer.   (a) 2.5 mA. (b) −4 V.   ■ ■

Drill Problem 10-6 ■

A current depicted in Fig. 10-20 flows through a 20 Ω resistor. Determine **(a)** the average current flow, and **(b)** the average voltage drop across the resistor.

Answer.   (a) 0.8 A. (b) 16 V.   ■ ■

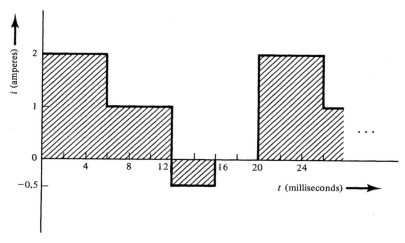

Figure 10-20 Drill Problems 10-6 and 10-7.

10.4 THE EFFECTIVE OR ROOT-MEAN-SQUARE VALUE

Since an alternating current varies periodically in strength from zero to some maximum value, one may wonder about the ampere value used to specify a particular current. The maximum value has a particular meaning for sinusoidal waves but does not give any indication of the manner in which the current varies in attaining that value for a nonsinusoidal wave. Similarly, the average value is not a good way of specifying ac values, because the average for a sinusoid over a whole cycle is zero.

Recall that the heating effect in a resistance is proportional to the square of the current and is, therefore, independent of the direction of current flow. It is appropriate that the heating effect of alternating current is taken as the basis for the definition of an ac ampere.

> **An alternating current is said to have an effective value of 1 A when it develops the same amount of heat in a given resistance as would be produced by a direct current of 1 A in the same resistance over the same time.**

The effective value of a sinusoidal current is considered first. If the effective alternating current is designated I_{eff}, then, from the definition, the power developed by this current equals the power developed by a constant direct current I:

$$P = I_{eff}^2 R = I^2 R \qquad (10\text{-}13)$$

Since the current is constantly changing, the power is constantly changing, and the instantaneous power p at any time is given by

$$p = i^2 R = (I_m \sin \omega t)^2 R = I_m^2 R \sin^2 \omega t \qquad (10\text{-}14)$$

From trigonometry,

$$\sin^2 \omega t = \tfrac{1}{2}(1 - \cos 2\omega t) \tag{10-15}$$

Substitution of Eq. 10-15 into Eq. 10-14 results in

$$p = \frac{I_m^2 R}{2}(1 - \cos 2\omega t) \tag{10-16}$$

The instantaneous power from Eq. 10-16, has as its plot a negative cosine wave displaced so that at no time is the instantaneous power negative (see Fig. 10-21). The actual power dissipated equals the average of the power curve which, in turn, equals $I_m^2 R/2$. Equating the average power to the dc power of Eq. 10-13, one obtains

$$I_{\text{eff}}^2 R = \frac{I_m^2 R}{2}$$

or

$$I_{\text{eff}}^2 = \frac{I_m^2}{2} \tag{10-17}$$

Taking the square root of both sides, we obtain

$$I_{\text{eff}} = \frac{I_m}{\sqrt{2}} = 0.707 I_m \tag{10-18}$$

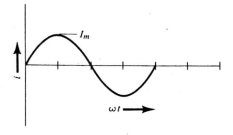

(a)

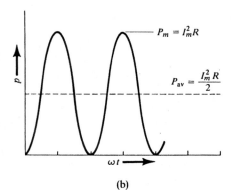

(b)

Figure 10-21   Resistor power: **(a)** instantaneous current; **(b)** instantaneous power.

In an analogous way, the effective value of voltage V_{eff} for a sine wave is

$$V_{eff} = \frac{V_m}{\sqrt{2}} = 0.707 V_m$$

The *effective value* of the voltage or current is often called the *root-mean-square* (rms) *value,* because the letters taken in reverse signify the method we use in finding the effective value. We square the initial waveform point by point, obtain the mean or average of the squared waveform, and then obtain the square root of this mean value.

Mathematically, the effective value of current (similarly for voltage) is

$$I_{eff} = \sqrt{\frac{1}{T} \int_0^T i^2 \, dt} \tag{10-19}$$

However, as long as the squared wave shapes are simple geometric shapes, it is not necessary to use calculus to find the effective value.

EXAMPLE 10-8

Find the effective value for the current waveform of Example 10-6.

Solution. The waveform is squared point by point, hence giving the i^2 curve of Fig. 10-22:

$$\text{average value of } i^2 \text{ curve} = \frac{(10^4)(\frac{1}{4}T) + (400)(\frac{3}{4}T)}{T}$$

$$= \frac{2500T + 300T}{T} = 2800 \text{ mA}^2$$

$$I_{eff} = \sqrt{2800 \text{ mA}^2} = 53 \text{ mA}$$

Since an effective value has an equivalent dc value, the subscripts rms and eff are usually dropped. The effective values of alternating voltages and currents are then represented by V and I, respectively. Alternating current devices are rated in effective values; however, one must not forget that, even though these devices are rated in effective values, they must be able to withstand the maximum or peak value of the

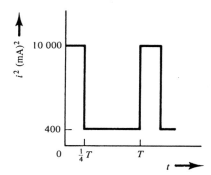

Figure 10-22 Example 10-8.

voltage or current. For sinusoidal quantities, this peak value equals $\sqrt{2}$ (or 1.414) × the effective value.

It should also be pointed out that ammeters and voltmeters used in ac circuits are generally calibrated in effective values. In fact, some meter movements actually respond to the square of the current, whereas other movements respond to the average value of a current.

It should also be pointed out that it is always possible to express any of the preceding nonsinusoidal waves as an infinite series of sine waves of different frequencies and amplitudes. (This will be discussed in Chapter 16.)

EXAMPLE 10-9

A lamp is connected across a 60 Hz household circuit. If the voltage across the lamp is measured as 120 V, **(a)** what is the peak voltage value, and **(b)** what is a mathematical expression for the voltage?

Solution.   The effective voltage is $V = 120$ V. Then

(a) $V_m = V\sqrt{2} = 120\sqrt{2} = 170$ V

(b) $\omega = 2\pi f = 2\pi(60) = 377$

$$v = V_m \sin \omega t = 170 \sin 377t$$

Drill Problem 10-7 ∎

Determine the effective current value for the current waveform of Fig. 10-20. Compare the effective value to the average value of 0.8 A.

Answer.   1.24 A. (The effective value is greater.)   ∎ ∎

Drill Problem 10-8 ∎

Determine the effective values of the following sinusoids:
(a) $10 \sin(377t + 30°)$V;
(b) $2 \cos(6283t)$ mA;
(c) $14.14 \sin(1000t)$ mV.

Answer.   (a) 7.07 V. (b) 1.414 mA. (c) 10 mV.   ∎ ∎

10.5 RESISTANCE AND ALTERNATING CURRENT

Even though alternating current is being considered, Ohm's law is still valid. However, one must generalize the law so that the voltage and current are instantaneous values, that is,

$$v = iR \tag{10-20}$$

Consider that a sine wave source of voltage is connected to a pure resistance, as in Fig. 10-23(a). Consistent with previous notation, the ac voltage at the source is designated by an e; the voltage at the load is designated by a v.

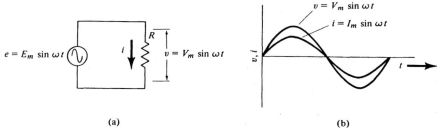

Figure 10-23 **(a)** Resistive circuit; **(b)** voltage and current relationship.

As indicated by Fig. 10-23(b), when the voltage is zero, the current is zero; when the voltage is maximum, the current is maximum. In other words,

the voltage and current are in phase in a resistive circuit.

By Ohm's law, if the current is

$$i = I_m \sin \omega t$$

then the instantaneous voltage drop across the resistor is

$$v = iR = I_m R \sin \omega t$$

The maximum voltage V_m equals $I_m R$,

$$V_m = I_m R \tag{10-21}$$

Furthermore, from the effective values, $I = I_m/\sqrt{2}$ and $V = V_m/\sqrt{2}$, it follows that

$$V = IR \tag{10-22}$$

Since the voltage and current are in phase, the instantaneous power is at all times positive, regardless of direction of current flow. In the case of a pure resistance, the average power, or simply power, is the product of the effective values of voltage and current.

$$P = VI \tag{10-23}$$

Substitution of Eq. 10-22 yields the alternative forms

$$P = I^2 R = V^2/R \tag{10-24}$$

A formal derivation of the power in a sinusoidal ac circuit with any phase displacement angle ϕ between the voltage and current appears in Chapter 12, on single-phase circuits.

EXAMPLE 10-10

A toaster-oven is rated 1600 W at 120 V ac. Except for a negligible timing and control current, all the current is due to the resistive heating element. Determine **(a)** the current for rated conditions and **(b)** the resistance.

Solution.   For a pure resistive load, (a) $P = VI = 1600$ W, or

$$I = \frac{P}{V} = \frac{1600 \text{ W}}{120 \text{ V}} = 13.33 \text{ A}$$

(b) Then,

$$R = \frac{V}{I} = \frac{120 \text{ V}}{13.33 \text{ A}} = 9 \ \Omega$$

Drill Problem 10-9 ■

A 120 V ac appliance is rated at 1200 W. The appliance consists of a heating element considered a pure resistance. **(a)** Determine the resistance and **(b)** the current.

Answer.   (a) 12 Ω. (b) 10 A.   ■ ■

10.6 CAPACITANCE AND ALTERNATING CURRENT; REACTANCE

Consider a pure capacitance connected to a sinusoidal ac source, as in Fig. 10-24(a). The voltage across the capacitor is

$$v = V_m \sin \omega t$$

From Chapter 7, the instantaneous current flowing to a capacitor is

$$i = C \frac{dv}{dt}$$

This expression says that the current is proportional to the slope of the voltage curve (dv/dt). In this case, the current is proportional to the slope of a sine wave.

As shown in Fig. 10-24(b), when the sine wave of voltage passes the zero point, its slope is a maximum and the magnitude of current is a maximum. On the other hand, when the voltage wave is at its peak value, its slope is zero and the current is zero. In fact, *the slope of a sine wave is a cosine wave*, as indicated by Fig. 10-24(b). Hence,

in the pure capacitive circuit, the current leads the voltage by a phase angle of 90°.

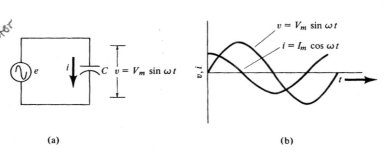

I associated with capacitor
V associated with inductor

(a)   (b)

Figure 10-24 (a) Capacitive circuit; **(b)** voltage and current relationship.

A formal derivation by calculus yields the phase relationship very readily. Substituting the sine wave expression for voltage into the current equation, we obtain

$$i = C \frac{dv}{dt} = C \frac{d}{dt} (V_m \sin \omega t) = \omega C V_m \cos \omega t \qquad (10\text{-}25)$$

With an alternating voltage of effective value V connected to the capacitance, an effective value of current I flows. The current does not flow through the dielectric, of course, but flows externally from plate to plate as the capacitance alternately charges and discharges. However, a certain relationship does exist between the effective values of voltage and current.

From Eq. 10-25, the maximum current is

$$I_m = \omega C V_m \qquad (10\text{-}26)$$

or

$$\frac{V_m}{I_m} = \frac{1}{\omega C} \qquad (10\text{-}27)$$

The quantity $1/\omega C$ is called the *capacitive reactance* and is a measure of the opposition to alternating current. Capacitive reactance has the units volts/ampere or *ohms* and is represented symbolically by X_C.

$$X_C = \frac{1}{\omega C} = \frac{1}{2\pi f C} \qquad \Omega \qquad (10\text{-}28)$$

Remembering that the effective values are related to the maximum values by the same ratio, we can write from Eq. 10-27,

$$V_C = I_C X_C \qquad (10\text{-}29)$$

The subscript C is added to the symbols for effective voltage and current in Eq. 10-29 as a reminder that the equation is for the capacitive case.

Since the pure capacitance cannot dissipate power, the product of capacitive voltage and current is not power. On the other hand, the capacitance alternately stores and releases power as it alternately charges and discharges. This power is called *capacitive reactive power* and equals the product, $V_C I_C$. The reactive power is symbolized by the letter Q and has units of *reactive volt-amperes (var)*. *Thus*

$$Q = V_C I_C = I_C^2 X_C = V_C^2 / X_C \qquad \text{var} \qquad (10\text{-}30)$$

EXAMPLE 10-11

In Fig. 10-24(a), $C = 2\ \mu F$ and the source supplies 1 kHz at an effective value of 10 V. **(a)** What current flows? **(b)** What is the reactive power?

Solution

(a)
$$X_C = \frac{1}{\omega C} = \frac{1}{2\pi (10^3)(2 \times 10^{-6})} = 79.5\ \Omega$$

$$I_C = \frac{V_C}{X_C} = \frac{10}{79.5} = 0.126\ A$$

(b)
$$Q = V_C I_C = (10)(0.126) = 1.26\ \text{var}$$

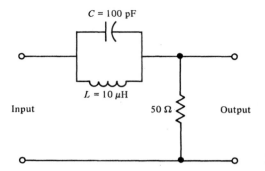

Input   50 Ω   Output

Figure 10-25   Example 10-12.

EXAMPLE 10-12

The network shown in Fig. 10-25 is called a band-stop filter. (We shall study its purpose in Chapter 13.) For the value of capacitance given, determine the capacitive reactance when an input signal of the following frequency is applied to the network: **(a)** 5 kHz, **(b)** 5 MHz, and **(c)** 50 MHz.

Solution.   A portion of the input voltage will appear across C.
(a) At 5 kHz

$$X_C = \frac{1}{\omega C} = \frac{1}{2\pi(5 \times 10^3)(100 \times 10^{-12})} = 318 \text{ k}\Omega$$

(b) At 5 MHz

$$X_C = \frac{1}{\omega C} = \frac{1}{2\pi(5 \times 10^6)(100 \times 10^{-12})} = 318 \ \Omega$$

(c) At 50 MHz

$$X_C = \frac{1}{\omega C} = \frac{1}{2\pi(50 \times 10^6)(100 \times 10^{-12})} = 31.8 \ \Omega$$

Notice that at low frequencies a capacitor has a relatively high reactance and thus impedes current flow more than at high frequencies, where it has a low value of reactance. If we plot the magnitude of reactance versus frequency for a given value of capacitance, we obtain the graph of Fig. 10-26. As the frequency is decreased toward the limit of $f = 0$, which corresponds to direct current, the reactance increases toward the limit of $|X_C| = \infty$. Conversely, as the frequency is increased toward the limit of $f = \infty$, the reactance decreases toward the limit of $|X_C| = 0$. Thus, we say that a capacitor "blocks" direct current. As we saw in Chapter 7, it actually charges to a given voltage, after which no further current flows. On the other hand, at a theoretically infinite frequency the capacitor appears as a short.

Drill Problem 10-10 ■

A capacitor-start ac motor uses a capacitor of 190 μF. What is the capacitive reactance at **(a)** 50 Hz and **(b)** 60 Hz?

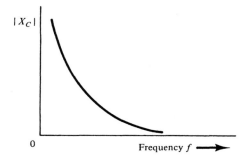

Figure 10-26 Magnitude of capacitive reactance versus frequency for fixed capacitance.

Answer. (a) 16.7 Ω. (b) 13.9 Ω. ∎ ∎

Drill Problem 10-11 ∎

At what frequency will a 100 pF capacitor have a reactance of 1 kΩ?

Answer. 1.59 MHz. ∎ ∎

Drill Problem 10-12 ∎

A capacitor of 50 μF is connected across a 60 Hz, 120 V ac line. **(a)** What current flows to the capacitor? **(b)** What is the reactive power for the capacitor?

Answer. (a) 2.26 A. (b) 271 var. ∎ ∎

10.7 INDUCTANCE AND ALTERNATING CURRENT; REACTANCE

The study of a pure inductance connected to a sinuosoidal ac source follows that for the capacitance with the roles of voltage and current interchanged.

Consider a sinusoidal current to be flowing in the pure inductance of Fig. 10-27(a); that is,

$$i = I_m \sin \omega t$$

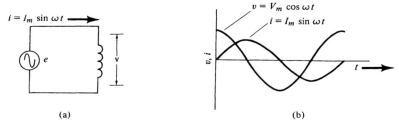

Figure 10-27 **(a)** Inductive circuit; **(b)** voltage and current relationship.

Since the sine wave of current is constantly changing, the coil constantly produces a counter emf given by

$$v = L\frac{di}{dt}$$

The voltage across the inductor is proportional to the slope of the sine wave of current and is therefore a cosine wave, as shown in Fig. 10-27(b). Thus,

in the pure inductive circuit, the current lags the voltage by a phase angle of 90°.

A formal derivation using calculus yields the phase relationship very readily. Substituting the equation for current into the equation for voltage, we find that

$$v = L\frac{di}{dt} = L\frac{d}{dt}(I_m \sin \omega t) = I_m\omega L \cos \omega t \tag{10-31}$$

The quantity $\omega L I_m$ is the maximum value of voltage across the inductor (it occurs at $t = 0$),

$$V_m = I_m\omega L \tag{10-32}$$

The quantity ωL is called the *inductive reactance* and is a measure of the opposition to alternating current. Inductive reactance, like capacitive reactance, is measured in ohms. The symbol X_L is used to denote inductive reactance.

$$X_L = \omega L = 2\pi f L \tag{10-33}$$

Since the maximum values of Eq. 10-32 are related to the effective values by the same ratio, we can write

$$V_L = I_L X_L \tag{10-34}$$

The subscript L is added to the voltage and current symbols for the inductive case.

EXAMPLE 10-13

The voltage across a 1 H inductor is $e = 10 \sin 200\,t$. What is the expression for instantaneous current?

Solution

$$X_L = \omega L = (200)1 = 200\ \Omega$$

$$I_m = \frac{V_m}{X_L} = \frac{10}{200} = 0.05\ \text{A}$$

In the inductance i lags e by 90°, a phase angle of $-90°$ denotes the delay so that

$$i = I_m \sin(\omega t - 90°) = 0.05 \sin(200t - 90°)$$

As with pure capacitance, the pure inductance cannot dissipate power. Rather, the inductance alternately stores and releases power as its magnetic field alternately builds up and collapses. This *inductive reactive power* equals the product $V_L I_L$. As before, the letter symbol Q and units of reactive volt-amperes are used for the reactive power. Thus,

$$Q = V_L I_L = I_L^2 X_L = V_L^2 / X_L \qquad \text{var} \qquad (10\text{-}35)$$

EXAMPLE 10-14

Determine the stored power for the inductor of Example 10-13.

Solution. From Example 10-13, $V_m = 10$ V and $I_m = 0.05$ A. The effective values of voltage and current are

$$V = 0.707 V_m = 0.707(10) = 7.07 \text{ V}$$

$$I = 0.707 I_m = 0.707(0.05) = 0.0354 \text{ A}$$

Then $Q = V_L I_L = 7.07(0.0354) = 0.25$ var. Alternatively,

$$Q = \frac{V_L^2}{X_L} = \frac{(7.07)^2}{200} = 0.25 \text{ var}$$

EXAMPLE 10-15

Refer to the filter network of Fig. 10-25. For the given value of inductance, determine the inductive reactance when an input signal of the following frequency is applied to the network: **(a)** 5 kHz; **(b)** 5 MHz; **(c)** 50 MHz.

Solution. As a portion of the input voltage appears across L, a reactance develops.

(a) At 5 kHz

$$X_L = \omega L = 2\pi(5 \times 10^3)(10 \times 10^{-6}) = 0.314 \ \Omega$$

(b) At 5 MHz

$$X_L = \omega L = 2\pi(5 \times 10^6)(10 \times 10^{-6}) = 314 \ \Omega$$

(c) At 50 MHz

$$X_L = \omega L = 2\pi(50 \times 10^6)(10 \times 10^{-6}) = 3.14 \text{ k}\Omega$$

Notice that at low frequencies an inductor has a relatively low reactance, whereas at high frequencies it has a relatively high reactance. Inductive reactance behaves just the opposite of capacitive reactance. If we plot the magnitude of inductive reactance as a function of frequency for a given value of inductance, we obtain the graph in Fig. 10-28. As the frequency increases toward $f = \infty$, the inductive reactance increases toward the limit $|X_L| = \infty$. Conversely, as the frequency decreases to $f = 0$, which corresponds to direct current, the reactance decreases toward the limit $|X_L| = 0$. Thus, an inductor acts like a short for direct current; but, as in Chapter 9, only after the dc transient period is complete. Finally, when we compare the two graphs of reactance versus frequency, we should note that the X_L relationship to frequency is linear and the X_C to frequency relationship is nonlinear and inverse.

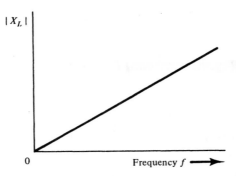

$|X_L|$

0   Frequency f ➝

Figure 10-28 Magnitude of inductive reactance versus frequency for a fixed inductance.

Drill Problem 10-13 ■

A coil having resistance of 100 Ω and an inductance of 0.8 H is connected to a 60 Hz source. A current of 0.3 A flows through the coil. **(a)** What power is dissipated in the resistive portion of the coil? **(b)** What power is stored in the purely inductive part of the coil?

Answer. (a) 9 W. (b) 27.2 var.   ■ ■

Drill Problem 10-14 ■

What is the reactance of a 50 mH coil **(a)** at 1 kHz and **(b)** at 2 MHz?

Answer. (a) 314 Ω. (b) 628 kΩ.   ■ ■

QUESTIONS

1. What is an alternating voltage or an alternating current?
2. What is meant by sinusoidal alternating current?
3. Explain the operation of a single-loop ac generator.
4. What is a wave and a waveform?
5. What is meant by the peak value of a sine wave?
6. What is the significance of a negative value of voltage?
7. State the general equations for sinusoidal voltage and current.
8. What is a periodic waveform?
9. What is frequency and what are its units?
10. How does the period relate to the frequency?
11. What is meant by phase difference?
12. What is the significance of the terms "leading" and "lagging"?
13. Describe the frequency spectrum.
14. What is meant by the average value of a waveform and how is it obtained?
15. What is meant by the average value of a sine wave?
16. What is meant by the effective value of a wave?

17. What is the effective value of a sine wave?

18. Is the instantaneous power delivered to a resistor in an ac circuit constant?

19. What is capacitive reactance?

20. What is the phase relationship between voltage and current in a capacitor?

21. What is capacitive reactive power and how is calculated?

22. What is inductive reactance?

23. What is the phase relationship between voltage and current in an inductor?

24. In what units is inductive reactive power measured?

PROBLEMS

1. A convenient way to construct a sine wave is to draw a circle with a radius equal to the amplitude or maximum value of the sine wave as shown in Fig. 10-29. As the radius is rotated through a complete revolution, the vertical projection of the radius is considered the instantaneous value and is plotted against a linear scale of degrees. Using this method and 30° angular displacements, construct a sine wave.

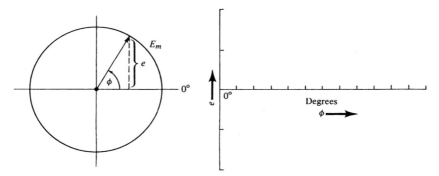

Figure 10-29   Problem 10-1.

2. A sine wave has a maximum value of 325 V. Determine the instantaneous voltage when ωt equals **(a)** 30° and **(b)** 0.5 rad.

3. A sine wave has a maximum value of 5 A. Find the instantaneous current when ωt equals **(a)** 75° and **(b)** 240°.

4. An alternating voltage is given by the expression $e = 120 \sin(3141t + \pi/2)$. What is the instantaneous voltage when $t = 5$ ms?

5. The equation for a sinusoidal current is $i = 75 \sin(6280t + 90°)$ mA. What is the instantaneous current at $t = 0.002$ s?

6. Using a horizontal scale of ms, sketch a 25 Hz and 60 Hz sine wave on the same plot. Use the same amplitude for both waves.

7. Find the frequency and the period for the sine waves of Problems 4 and 5.

8. How many seconds are required for a 60 Hz sine wave to pass from its zero value to one-half its maximum value?

9. The instantaneous value of a sinusoidal current is 36 mA at $\omega t = 135°$. What is the maximum value of the current?

10. Write the mathematical expressions for the waveforms of Fig. 10-30.

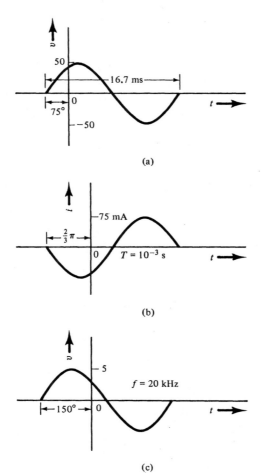

(a)

(b)

(c)   **Figure 10-30** Problem 10-10.

11. What are the period and frequency for each of the waveforms of Fig. 10-31?

12. A 10 μs pulse is repeated every 100 μs. What is the duty cycle?

13. Determine the period, frequency, and duty cycle of the waveform of Fig. 10-32.

14. Determine the average value of the waveform of Fig. 10-32.

15. Given the waveform of Fig. 10-33, notice that a dc pulse value is added to a sinusoid.
(a) What is the average value of the waveform? **(b)** What is the frequency of the sinusoid?
(c) What is the mathematical expression for the sinusoid?

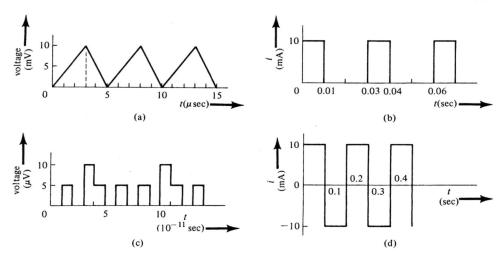

(a)

(b)

(c)

(d)

Figure 10-31 Problem 10-11.

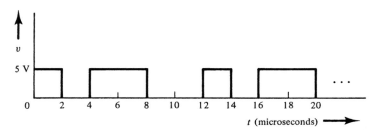

Figure 10-32 Problems 10-13 and 10-14.

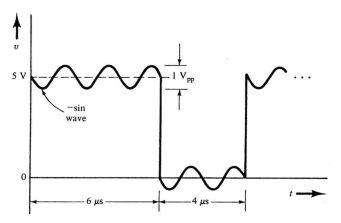

Figure 10-33 Problem 10-15.

16. Find the average values for each of the waveforms of Fig. 10-31.

17. Using familiar geometric shapes, approximate the average value of the exponential waveform of Fig. 10-34.

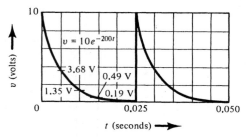

Figure 10-34 Problem 10-17.

18. What are the effective values of the following sinusoids?

(a) $v = 120 \sin 377t$ V.   (b) $i = 0.002 \cos(6280t + 45°)$ A.

(c) $v = 70.7 \sin(377t + 10°)$ V.

19. Find the average and effective values for the waveforms of Fig. 10-35.

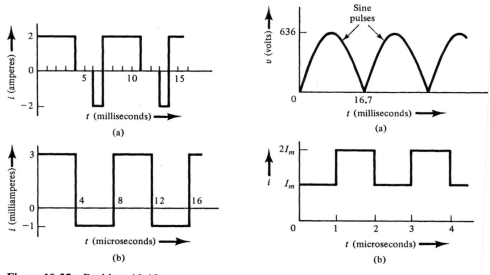

Figure 10-35 Problem 10-19.   **Figure 10-36** Problem 10-20.

20. Find the effective values for the waveforms of Fig. 10-36.

21. A certain hoist motor has the following duty cycle: the motor takes 40 A for 20 s, after which the power is off for 40 s. It then takes 30 A for 10 s, after which the power is off for 30 s. If the duty cycle is repeated continuously, what effective current rating must the motor have?

22. What effective current flows through a 12 Ω resistor connected to a voltage of $e = 120 \sin 377t$?

23. When a certain alternating voltage is applied to a 10 Ω resistor, the resistor dissipates 360 W. What are the effective values of the voltage and current?

24. A 150 W soldering iron has a resistance of 60 Ω. What effective current does the iron require?

25. A resistance of 100 Ω is connected to an ac source of 200 V. What are the effective and maximum values of the current?

26. An alternating current of 3.5 A flows through a 500 Ω resistance. What is the voltage and power?

27. Determine the reactance of a 200 μF capacitor at 60 Hz and 1 kHz.

28. A capacitor of 50μF has a 500 V, 60 Hz voltage applied. What current flows?

29. At what frequency does a 4 μF capacitor have a reactance of 80 Ω?

30. The voltage across a capacitor is $v = 141 \sin(3140t + 15°)$. If the capacitance is 0.01 μF, what is the expression for instantaneous current?

31. A capacitance of 30 μF is connected to a 120 V, 60 Hz source. What is the reactive power?

32. A current of $i = 3 \sin(377t + 15°)$ flows in a capacitor having a reactance of 500 Ω. **(a)** What is the capacitance? **(b)** What is the voltage equation?

33. Determine the reactance of a 1.4 H coil at 60 Hz and 1 kHz.

34. At what frequency will an 80 μH coil have a reactance of 18 500 Ω?

35. A coil with negligible resistance is connected to a 10 V, 1 kHz source and draws 1 mA. What are **(a)** the reactance, **(b)** the inductance, and **(c)** the power dissipated?

36. An inductor of 16 H is connected to a 110 V, 60 Hz source. **(a)** What current flows? **(b)** What is the reactive power?

37. An inductor has a reactive power of 1 kvar when connected to a 60 Hz source. If the current is 10 A, what are **(a)** the voltage and **(b)** the inductance?

38. What is the instantaneous current in an inductor of 0.02 H when the voltage is $v = 80 \sin(1000t + 105°)$ V?

39. At what frequency will a 0.001 H coil and a 6 pF capacitor have the same magnitude of reactance?

40. At what frequency will a 40 mH inductance and a 600 pF capacitance have the same magnitude of reactance?

Chapter 11

Phasors and Phasor Algebra

11.1 PHASOR REPRESENTATION OF ALTERNATING CURRENT

In the preceding chapter, voltage and current relationships were developed for sinusoidal emfs applied to pure resistance, capacitance, and inductance. Even in the simple cases involving only individual and pure circuit parameters, the use of instantaneous sinusoidal expressions may prove to be cumbersome. Certainly with combinations of resistance, capacitance, and inductance, the voltage and current expressions become even more cumbersome. In this chapter a technique that facilitates the mathematics involving sinusoidal quantities is presented.

Early on in physics, one learns to classify a physical quantity as either a scalar or a vector. A *scalar* is a quantity that is completely specified by a magnitude, that is, describable by a number. On the other hand, a *vector* is a quantity that is specified by both a magnitude and a direction. Examples of scalar quantities are gallons, temperature, density, and ohms, each of which is added to or subtracted from similar quantities by ordinary algebraic methods. Force and velocity are common examples of vector quantities, since they require a consideration of direction of action in addition to magnitude. In contrast to scalars, vectors are not necessarily added or subtracted algebraically but must be combined in such a way as to take into account both magnitude and direction. The associated mathematics is referred to as *vector algebra*.

In the study of sinusoidal quantities, one is interested in the *frequency*, the *amplitude*, and the *phase angle*, usually in relation to other sinusoidal quantities. One can try to categorize the sinusoidal quantity as a vector quantity if one allows the usual vector properties of length and direction to represent the amplitude and phase angle, respectively. However, the amplitude of a sinusoidal function constantly changes and the phase angle represents a time displacement, not a space displacement. Yet with some modifications the vector model can be used.

If a line having a magnitude A is rotated counterclockwise at a constant angular frequency ω, as in Fig. 11-1(a), the projection that this line makes on the vertical or y axis is a sine wave, as in Fig. 11-1(b). Notice that one revolution of the line segment or radius corresponds to one cycle of the projected wave. The positive x axis, that part extending to the right of the intersection of the horizontal and vertical axes, is called the *reference axis*. Phase angles are referred to this axis, and by convention are *positive when taken in the counterclockwise direction* and *negative when taken in the clockwise direction*.

Since, at any instant of time, the line segment of Fig. 11-1 has a magnitude A and a phase angle θ, one can think of the associated sinusoidal wave as the projection of a *rotating vector*. These rotating vectors, which are employed to represent the time variation of sinusoidal quantities, are called *phasors*.

The two-dimensional plot of Fig. 11-1(a), which shows the magnitude and phase angle for the associated sinusoidal wave, is known as a *phasor diagram*. A phasor diagram is particularly useful for showing the phase difference between two or more sinusoidal quantities, so most phasor diagrams will have two or more phasors. When various voltage and current phasors are represented on the same diagram, all voltage phasors are drawn to the same scale and all current phasors are drawn to another scale.

The phasor in Fig. 11-1 is stopped or "frozen" in one of its many possible positions. By referring to the sinusoidal development in the figure, it is seen that this position corresponds to the initial phase angle of the sinusoidal quantity. In general, however, the phasor angle may not indicate an initial phase angle, since in many cases a phasor diagram is drawn with one particular phasor lying along the reference axis, regardless of the initial phase angle of the associated wave. A phasor lying on the reference axis is a *reference phasor*; the choice of a reference phasor is explained later.

It is evident that the phasor of Fig. 11-1 has a magnitude equal to the maximum value of the associated wave. Although phasor diagrams can be drawn to represent maximum values, they are customarily drawn in terms of effective values.

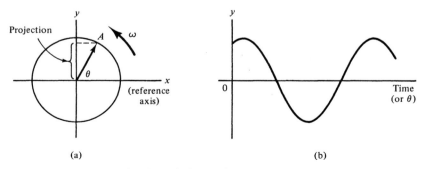

(a)   (b)

Figure 11-1   **(a)** Rotating line; **(b)** its vertical projection.

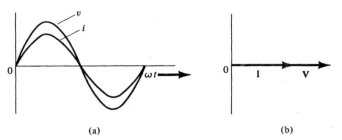

(a)   (b)

Figure 11-2 **(a)** Voltage and current relationship for a resistance; **(b)** related phasor diagram.

It is important to remember that when the waves of two sinusoidal quantities of the same frequency are plotted against time, any phase difference remains constant throughout time. Inasmuch as the associated phasors rotate at the same angular frequency, the phase difference between the phasors is also constant. With phasor quantities of different frequencies, the phase angles change with time so that a phasor diagram is meaningless.

In the preceding chapter it was explained that when a sinusoidal voltage is applied to a resistance, the resulting current is a sinusoid that is in phase with the voltage. This information is indicated by a phasor diagram as in Fig. 11-2.

On the other hand, for a capacitance the sinusoidal current leads the voltage by an angle of 90°, as shown in the time plot and phasor diagram of Fig. 11-3. For an inductance, the voltage leads the current by an angle of 90°, hence, the phasor relationships of Fig. 11-4.

Although the current phasor is used as the reference phasor for the capacitive and inductive cases in Figs. 11-3 and 11-4, the voltage phasor can also be used as a reference. This corresponds to a rotation of each phasor diagram until the phasor voltage of each diagram is along the reference axis. The choice of reference generally depends on the circuit configuration for which a phasor diagram is desired. Since the current is common to all parts of a series circuit, *it is customary to use the current as a reference phasor for a series circuit.* The voltage is common to all

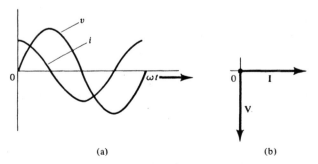

(a)   (b)

Figure 11-3 **(a)** Voltage and current relationship for a capacitance; **(b)** related phasor diagram

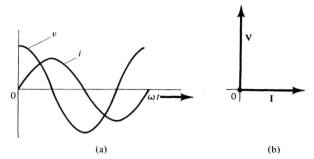

Figure 11-4 **(a)** Voltage and current relationship for an inductance; **(b)** related phasor diagram.

parts of a parallel circuit so that *for a parallel circuit it is customary to use the voltage as a reference.*

There is only a minor difference between a phasor diagram and the vector diagram the reader may be acquainted with from physics. The phasor diagram represents the phasors at one instant of time; the vector diagram represents the vectors without regard to time. Otherwise, the mathematics of vector algebra is applicable to phasors. However, to remind one that ac quantities and their phasor representations are time relations and not space relations, the term "phasor algebra" is subsequently used rather than the term "vector algebra."

In printed matter, vector or phasor quantities are sometimes printed in italics or in boldface type. The phasor quantities in this text are in boldface type. Thus one describes the sinusoidal voltage

$$e = 141.4 \sin(\omega t + 67°) \quad \text{V} \tag{11-1}$$

by the phasor quantity

$$\mathbf{E} = 100 \angle 67° \text{ V} \tag{11-2}$$

As is customary, the phasor magnitude in Eq. 11-2 is the effective value of the associated sinusoidal wave, and the angle is the phase angle at time $t = 0$.

Drill Problem 11-1 ■

Write the phasor expressions for the following waves at $t = 0$.
(a) $i = 15 \sin(377t + 75°)$ A.
(b) $e = 2 \cos(377t - 25°)$ V.
(c) $i = 20.2 \sin(1000t - 52°)$ mA.
(d) $e = 100 \sin(377t - 120°)$ V.

Answer

(a) $10.6 \angle 75°$ A.

(b) $1.41 \angle 65°$ V.

(c) $14.3 \angle -52°$ mA.

(d) $70.7 \angle -120°$ V. ■ ■

Drill Problem 11-2 ■

Write the sine wave expressions for the following phasors if the frequency is as stated.

(a) $f = 60$ Hz, $\mathbf{I} = 3 \angle 50°$ A.
(b) $f = 1$ kHz, $\mathbf{V} = 2 \angle -45°$ V.
(c) $f = 50$ Hz, $\mathbf{I} = 2 \angle -\pi/2$ rad A.
(d) $f = 10$ kHz, $\mathbf{V} = 4 \angle 60°$ V.

Answer

(a) $i = 4.24 \sin(377t + 50°)$ A.

(b) $v = 2.83 \sin(6283t - 45°)$ V.

(c) $i = 2.83 \sin(314t - 90°)$ A.

(d) $v = 5.66 \sin(6.283 \times 10^4 t + 60°)$ V.   ■ ■

11.2 POLAR AND RECTANGULAR FORMS OF PHASORS

In Eq. 11-2, the phasor quantity lying in the coordinate plane is defined by its magnitude and the angle it makes with the reference axis. Phasors defined in this manner are said to be in *polar form*. The magnitude and angle of a phasor quantity are also known as the *modulus* and the *argument*, respectively.

As previously noted, angles are considered positive when measured counterclockwise from the reference axis. Then if one wishes to rotate a phasor counterclockwise, one adds the displacement angle to the phase angle of the original phasor; the modulus remains the same. For example, in Fig. 11-5, the phasor **A** is rotated through 90°. It follows that when a phasor is rotated through ±180° the original phasor is *reversed*, as indicated in Fig. 11-6. Mathematically, the reversal of a phasor is the *negative of a phasor* and is identified by a minus sign preceding the phasor:

$$-\mathbf{A} = -(A \angle \theta) = A \angle \theta \pm 180° \tag{11-3}$$

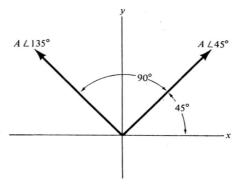

Figure 11-5 Rotation of a phasor (90° counterclockwise).

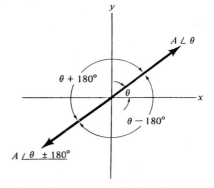

Figure 11-6 Reversal of a phasor.

In Eq. 11-3 notice that italic symbols are used to represent only the magnitude of a phasor.

It should be pointed out that the polar form presented here is a "shorthand" version of what is known as the exponential form of a phasor, a form that is discussed in Section 11.7.

A phasor lying in a coordinate plane may also be specified in magnitude and direction by the projections it makes on the horizontal and vertical axes. These projections are known as the *rectangular coordinates*, and a phasor defined in terms of its horizontal and vertical components is said to be in *rectangular form*.

In order to express a phasor in rectangular form and vice versa, one must be familiar with trigonometry because the horizontal component, vertical component, and the phasor itself form a right triangle. For example, if the phasor **A** of Fig. 11-7 has horizontal and vertical components of b and c, respectively, the magnitude of the phasor **A** is given by the *Pythagorean theorem* as

$$A = \sqrt{b^2 + c^2} \qquad (11\text{-}4)$$

In trigonometry, the sine (sin), cosine (cos), and tangent (tan) are defined in terms of the two shorter sides and the larger side (called the *hypotenuse*) as

$$\sin \theta = \frac{\text{opposite side}}{\text{hypotenuse}} = \frac{c}{A} \qquad (11\text{-}5)$$

$$\cos \theta = \frac{\text{adjacent side}}{\text{hypotenuse}} = \frac{b}{A} \qquad (11\text{-}6)$$

$$\tan \theta = \frac{\text{opposite side}}{\text{adjacent side}} = \frac{c}{b} \qquad (11\text{-}7)$$

Knowing the phasor components, we may use an inverse trigonometric operation to obtain the phase angle θ:

$$\theta = \tan^{-1} \frac{c}{b} \qquad (11\text{-}8)$$

where $\tan^{-1}$ means *the angle whose tangent is.*

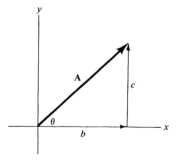

Figure 11-7 A phasor and its rectangular components.

Reference to Fig. 11-7 and the trigonometric functions reveals that the horizontal and vertical components of a phasor may be expressed, respectively, as

$$b = A \cos \theta \qquad (11-9)$$

and
$$c = A \sin \theta \qquad (11-10)$$

In order to express a phasor as a sum of two rectangular components, one must distinguish the horizontal and vertical components. In physics, the components of vector forces are often distinguished by subscripts. In electrical engineering, the rectangular components of a phasor are distinguished by the use of the symbol j in front of the vertical component. Those components lacking the j are understood to be horizontal, whereas those having a j are understood to be vertical. Thus the phasor of Fig. 11-7 may be described in rectangular form as

$$\mathbf{A} = b + jc$$

The symbol j is more than a "tag" to distinguish the vertical from the horizontal component. It is a *mathematical operator*, because it indicates that the quantity to which it is attached is subjected to a particular mathematical procedure or operation. The j *operator* indicates that the phasor quantity, to which it is applied as a multiplying factor, has been rotated through a 90° counterclockwise rotation.

The significance of the j operator is best shown by its successive application to a phasor originally lying along the reference axis. In Fig. 11-8, if the phasor **A**, originally along the reference axis, is rotated by application of the j operator, it forms the phasor $j\mathbf{A}$ positioned 90° counterclockwise from **A**. In this position, $j\mathbf{A}$ is a new phasor that if operated on by j produces the new vector $j^2\mathbf{A}$. Since this phasor lies to the left of the origin on the negative horizontal axis,

$$j^2\mathbf{A} = -\mathbf{A} \qquad (11-11)$$

Thus,
$$j^2 = -1 \qquad (11-12)$$

and
$$j = \sqrt{-1} \qquad (11-13)$$

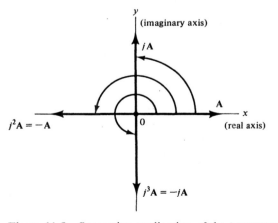

Figure 11-8   Successive application of the j operator.

Application of the j operator to $j^2\mathbf{A}$ then leads to the phasor $j^3\mathbf{A}$, which equals $-j\mathbf{A}$. Notice from this last statement that just as a $+j$ produces a 90° counterclockwise rotation, a $-j$ is equivalent to a 90° clockwise rotation.

In mathematics, the square root of -1 is called the imaginary number i. From Eq. 11-13, we see that j is identical to the imaginary number i. However, in electrical engineering the j is used to avoid misinterpretation with the symbol for instantaneous current. Because of these mathematical implications, the horizontal axis is frequently called the real axis; the vertical axis, the j axis or imaginary axis. Correspondingly, the horizontal component of a phasor is often called the real component; the vertical component, the j, imaginary, or quadrature component.

A *complex number* is defined in mathematics as a number of the form $b + jc$, where j is specified by Eq. 11-13. A phasor in polar form can be changed to the rectangular or complex form by the relation

$$A \angle \theta = A \cos \theta + jA \sin \theta \tag{11-14}$$

We now have at our disposal Eqs. 11-4 and 11-8, which allow us to convert a phasor in rectangular form to polar form. Conversely, Eq. 11-14 allows us to convert a phasor in polar form to rectangular form. Both these operations are easily performed with a hand calculator or a personal computer. Although calculators and computers might eliminate many of the intermediate steps in the conversion process, those steps are included in the text examples so that the conversion process can be closely followed. The Appendix has a collection of computer programs in the BASIC language. One of the programs is for polar to rectangular conversion of phasors; another is for rectangular to polar conversion of phasor quantities.

EXAMPLE 11-1

Convert the phasors of Fig. 11-9 from rectangular form to polar form:
(a) $\mathbf{A} = 5 + j5$, **(b)** $\mathbf{A} = -20 - j40$, **(c)** $\mathbf{A} = 4 - j3$.

Solution

(a) $A = \sqrt{(5)^2 + (5)^2} = \sqrt{50} = 7.07.$
 $\theta = \tan^{-1} \frac{5}{5} = 45°.$
 $\mathbf{A} = 7.07 \angle 45°.$

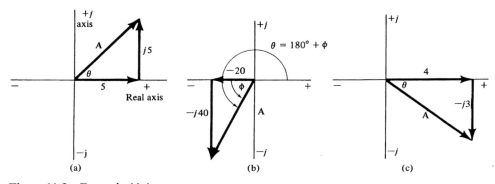

(a)   (b)   (c)

Figure 11-9 Example 11-1.

(b) $A = \sqrt{(20)^2 + (40)^2} = \sqrt{2000} = 44.8.$

$\phi = \tan^{-1}\frac{40}{20} = \tan^{-1} 2 = 63.4°.$

$\theta = \phi + 180° = 63.4° + 180° = 243.4°.$

$\mathbf{A} = 44.8 \angle 243.4°.$

(c) $A = \sqrt{(4)^2 + (3)^2} = \sqrt{25} = 5.$

$\theta = \tan^{-1}\frac{3}{4} = 36.8°.$

$\mathbf{A} = 5 \angle -36.8°.$

EXAMPLE 11-2

Convert the phasors of Fig. 11-10 from polar form to rectangular form:
(a) $\mathbf{A} = 13 \angle 112.6°$, (b) $\mathbf{A} = 9.5 \angle 73°$, (c) $\mathbf{A} = 1.2 \angle -152°$.

Solution

(a) $\qquad \mathbf{A} = 13 \angle 112.6° = 13 \cos 112.6° + j13 \sin 112.6°$

$\qquad\qquad = -13 \cos 67.4° + j13 \sin 67.4° = -5 + j12$

(b) $\qquad \mathbf{A} = 9.5 \angle 73° = 9.5 \cos 73° + j9.5 \sin 73°$

$\qquad\qquad = 2.77 + j9.08$

(c) $\qquad \mathbf{A} = 1.2 \angle -152° = 1.2 \cos(-152°) + j1.2 \sin(-152°)$

$\qquad\qquad = -1.2 \cos 28° - j1.2 \sin 28° = -1.06 - j0.564$

EXAMPLE 11-3

Reverse the phasor $\mathbf{A} = 10 \angle -30°$ and express in rectangular form.

Solution

$$-\mathbf{A} = -10 \angle -30° = 10 \angle 150° = 10 \cos 150° + j10 \sin 150°$$
$$= -10 \cos 30° + j10 \sin 30° = -8.66 + j5$$

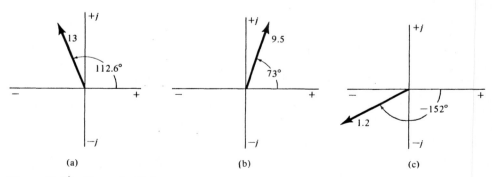

(a)   (b)   (c)

Figure 11-10   Example 11-2.

Alternately,

$$\mathbf{A} = 10\cos(-30°) + j10\sin(-30°)$$
$$= 8.66 - j5$$

Then

$$-\mathbf{A} = -8.66 + j5$$

Drill Problem 11-3 ■

Convert the following phasor representations from rectangular to polar form: **(a)** $6 + j11.3$. **(b)** $-2.5 - j3.6$. **(c)** $-3.2 + j4.5$. **(d)** $-28 - j22$.

Answer. (a) $12.8 \angle 62°$. (b) $4.38 \angle -124.8°$. (c) $5.52 \angle 125.4°$. (d) $35.6 \angle -141.8°$.
■ ■

Drill Problem 11-4 ■

Convert the following phasor representations from polar to rectangular form: **(a)** $22 \angle 45°$. **(b)** $0.06 \angle \pi/2$ rad. **(c)** $9.2 \angle -120°$. **(d)** $220 \angle +120°$.

Answer. (a) $15.6 + j15.6$. (b) $j0.06$. (c) $-4.6 - j7.97$. (d) $-110 + j190$.
■ ■

Drill Problem 11-5 ■

Reverse the phasor expressions given in Drill Problem 11-4 and express them in polar and rectangular form.

Answer

(a) $22 \angle -135°$, $-15.6 - j15.6$.

(b) $0.06 \angle -90°$, $-j0.06$.

(c) $9.2 \angle 60°$, $4.6 + j7.97$.

(d) $220 \angle -60°$, $110 - j190$.
■ ■

11.3 PHASOR ADDITION

Similar or like phasor quantities, that is, only voltages or only currents can be added either graphically or by the use of a complex algebra that uses phasor components.

In the preceding phasor diagrams, a directed line segment or arrow is used to graphically describe a phasor quantity. The end of the line segment with the arrow head is called the head; the opposite end is called the tail. *Phasors specified in polar form cannot be added directly unless one uses a complete graphical solution.* (An exception occurs when the phasors being added have the same phase angle.) One graphical technique, the *polygon method*, consists of first drawing each phasor to scale. If we begin with any one of the phasors, in any order of succession, we find that the tail of each successive phasor is attached to the head of the preceding one. The line drawn to complete the triangle or polygon (from the tail of the first phasor

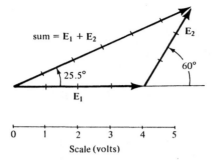

Figure 11-11   Example 11-4.

to the head of the last phasor) equals the sum, sometimes called the *resultant*. With a scale and protractor, we can then measure the magnitude and phase angle of the resultant.

EXAMPLE 11-4

Graphically solve for the sum of the phasor voltages $E_1 = 4 \angle 0°$ V and $E_2 = 3 \angle 60°$ V.

Solution. E_1 is chosen as the first phasor and E_2 is connected to it, as shown in Fig. 11-11. The resultant or sum is then measured to be $6.1 \angle 25.5°$ V.

EXAMPLE 11-5

Graphically solve for the sum $I_t = I_1 + I_2 + I_3 + I_4$, where $I_1 = 8 \angle 0°$ A, $I_2 = 10 \angle 45°$ A, $I_3 = 11 \angle 150°$ A, and $I_4 = 16 \angle 200°$ A.

Solution. Starting with I_1, we add the phasors successively, as shown in Fig. 11-12. The sum I_t is measured: $I_t = 11.9 \angle 143°$ A.

Although a graphical solution is always possible, the accuracy of the solution is obviously limited. A more accurate solution is obtained by representing the phasors

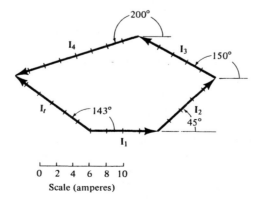

Figure 11-12   Example 11-5.

in complex number or rectangular form, the resultant or sum of which is obtained by a simple addition of the complex quantities. In complex algebra, the horizontal components and the vertical (j) components are separately added and the net rectangular sum can then be converted to the polar form if desired. Thus, if

$$\mathbf{A} = a + jb \quad \text{and} \quad \mathbf{B} = c + jd$$

their sum, $\mathbf{C}$, in rectangular form is given by the equation,

$$\mathbf{C} = \mathbf{A} + \mathbf{B} = (a + jb) + (c + jd) \tag{11-15}$$

where the components a, b, c, and d are the rectangular components, including appropriate minus or plus signs. Further, by combining the real and imaginary terms, we have

$$\mathbf{C} = (a + c) + j(b + d) \tag{11-16}$$

It is recommended that a sketch be made of the phasor quantities, even if the addition is carried out by complex algebra. The sketch will provide an approximate solution as a check.

EXAMPLE 11-6

Using complex algebra, add the two phasors of Example 11-4.

Solution

$$\mathbf{E}_1 = 4 \angle 0° = 4 + j0 \text{ V}$$

$$\mathbf{E}_2 = 3 \angle 60° = 1.5 + j2.6 \text{ V}$$

$$\mathbf{E}_t = 5.5 + j2.6$$

$$= \sqrt{(5.5)^2 + (2.6)^2} \angle \tan^{-1} \frac{2.6}{5.5}$$

$$= 6.1 \angle 25.3° \text{ V}$$

EXAMPLE 11-7

Find $\mathbf{I}_t$ in Example 11-5 by complex algebra.

Solution

$$\mathbf{I}_1 = 8 \angle 0° = 8 + j0$$

$$\mathbf{I}_2 = 10 \angle 45° = 7.07 + j7.07$$

$$\mathbf{I}_3 = 11 \angle 150° = -9.53 + j5.5$$

$$\mathbf{I}_4 = 16 \angle 200° = -15 - j5.48$$

$$\mathbf{I}_t = -9.46 + j7.09$$

$$= \sqrt{(9.46)^2 + (7.09)^2} \angle 180 - \tan^{-1} \frac{7.09}{9.46}$$

$$= 11.8 \angle 143.2° \text{ A}$$

Drill Problem 11-6 ■

Graphically add the two phasor voltages $E_1 = 120 \angle 0°$ V and $E_2 = 120 \angle 60°$ V.

Answer. $E_t \approx 208 \angle 30°$ V.   ■ ■

Drill Problem 11-7 ■

Using complex algebra, add the two phasors of Drill Problem 11-6 and express the answer in polar form.

Answer. $E_t = 208 \angle 30°$ V.   ■ ■

Drill Problem 11-8 ■

Using either graphical construction or complex algebra, determine the sum of the following phasors and express in polar form: $I_1 = 10 \angle 60°$ A, $I_2 = 5 \angle 30°$ A, $I_3 = 8 \angle -90°$ A, and $I_4 = 6 \angle 0°$ A.

Answer. $15.65 \angle 11.6°$ A.   ■ ■

11.4 PHASOR SUBTRACTION

As defined by Eq. 11-3, the negative of a phasor is another phasor whose direction is reversed, that is, changed by 180°. In phasor algebra, the process of subtraction is accomplished by changing the sign of the quantity to be subtracted and then adding. Symbolically,

$$\mathbf{A} - \mathbf{B} = \mathbf{A} + (-\mathbf{B}) \qquad (11\text{-}17)$$

It follows that since the subtraction process is accomplished by an addition in accordance with Eq. 11-17, phasors cannot be subtracted in polar form unless they are subtracted by a complete graphical solution. The alternative is to convert the phasors to rectangular form before subtracting. Then, if

$$\mathbf{A} = a + jb \quad \text{and} \quad \mathbf{B} = c + jd$$

their difference, $\mathbf{C}$, in rectangular form is given by the equation

$$\mathbf{C} = \mathbf{A} - \mathbf{B} = (a + jb) - (c + jd) \qquad (11\text{-}18)$$

where the components a, b, c, and d are the rectangular components, including the appropriate minus or plus signs.

EXAMPLE 11-8

Graphically find $\mathbf{V}_t = \mathbf{V}_1 - \mathbf{V}_2$ if $\mathbf{V}_1 = 35.4 \angle 75°$ V and $\mathbf{V}_2 = 25 \angle 120°$ V.

Solution.   The phasor $-\mathbf{V}_2$ is constructed and added to $\mathbf{V}_1$ as in Fig. 11-13. Then $\mathbf{V}_t$ is measured: $\mathbf{V}_t = 25 \angle 30°$ V.

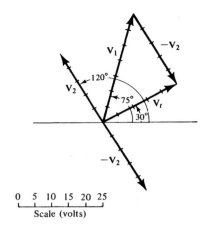

Figure 11-13 Example 11-8.

EXAMPLE 11-9

Perform the subtraction of Example 11-8 by complex algebra.

Solution

$$\mathbf{V}_1 = 35.4 \angle 75° = 9.15 + j34.2 \text{ V}$$

$$-\mathbf{V}_2 = 25 \angle -60° = +12.5 - j21.7 \text{ V}$$

$$\mathbf{V}_t = 21.65 + j12.5 \text{ V}$$
$$= 25 \angle 30° \text{ V}$$

Drill Problem 11-9 ■

Determine the phasor difference $\mathbf{V}_t = \mathbf{V}_1 - \mathbf{V}_2$ if $\mathbf{V}_1 = 69.3 \angle 90°$ and $\mathbf{V}_2 = 69.3 \angle 210°$.

Answer. $120 \angle 60°$. ■ ■

Drill Problem 11-10 ■

Determine $\mathbf{I}_4$ if $\mathbf{I}_1 = \mathbf{I}_2 + \mathbf{I}_3 + \mathbf{I}_4$ and $\mathbf{I}_1 = j2.1$, $\mathbf{I}_2 = 1 - j1.8$, and $\mathbf{I}_3 = 1.2 \angle 45°$.

Answer. $3.57 \angle 121.2°$. ■ ■

11.5 PHASOR MULTIPLICATION

Unlike the addition and subtraction of phasors, the multiplication of phasors is carried out in either the polar or the rectangular mode. The product of two phasors in polar form is equal to the product of their magnitudes and the algebraic sum of their arguments. That is, if

$$\mathbf{A} = A \angle \alpha \quad \text{and} \quad \mathbf{B} = B \angle \beta$$

then
$$\mathbf{AB} = AB \angle (\alpha + \beta) \tag{11-19}$$

EXAMPLE 11-10

Multiply $\mathbf{A} = 8 \angle -50°$ and $\mathbf{B} = 3 \angle 30°$.

Solution
$$\mathbf{AB} = (8 \angle -50°)(3 \angle 30°) = 24 \angle -50° + 30° = 24 \angle -20°$$

Phasors expressed in rectangular form are treated as ordinary binomials for mathematical operations. Therefore, the product of two phasors in rectangular form equals the sum of the possible products of the components. One must remember that when the term j^2 appears, it can be replaced by -1. In a similar fashion, any higher power of j should be changed to a simpler form.

It follows that for rectangular expressions multiplication is defined by
$$\mathbf{AB} = (a + jb)(c + jd) \tag{11-20}$$
In general, the multiplication indicated by Eq. 11-20 proceeds term by term so that
$$\begin{aligned} \mathbf{AB} &= ac + jbc + jad + j^2 bd \\ &= (ac - bd) + j(bc + ad) \end{aligned} \tag{11-21}$$

EXAMPLE 11-11

Multiply $\mathbf{A} = 5.14 - j6.13$ and $\mathbf{B} = 2.6 + j1.5$; convert the results to polar form.

Solution
$$\begin{aligned} \mathbf{AB} &= (5.14 - j6.13)(2.6 + j1.5) \\ &= 13.4 - j15.9 + j7.7 - j^2 9.2 \\ &= 13.4 - j8.2 + 9.2 = 22.6 - j8.2 \\ &= 24 \angle -20° \end{aligned}$$

EXAMPLE 11-12

Multiply $\mathbf{A} = 2 + j0$, $\mathbf{B} = 3 - j4$, and $\mathbf{C} = 2 + j3$.

Solution
$$\begin{aligned} \mathbf{ABC} &= (2 + j0)(3 - j4)(2 + j3) = (6 - j8)(2 + j3) \\ &= 12 - j16 + j18 - j^2 24 = 12 + j2 + 24 \\ &= 36 + j2 \end{aligned}$$

Drill Problem 11-11 ∎

Given the phasors $\mathbf{A} = 10 \angle 30°$, $\mathbf{B} = 2 \angle 45°$, and $\mathbf{C} = 2.5 \angle 60°$, perform the following multiplications and express the answers in polar form: **(a)** $\mathbf{AB}$. **(b)** $(\mathbf{B} + \mathbf{C})\mathbf{A}$. **(c)** $\mathbf{ABC}$. **(d)** $\mathbf{B}(\mathbf{A} + \mathbf{C})$.

Answer.   (a) $20 \angle 75°$. (b) $44.5 \angle 83.3°$. (c) $50 \angle 135°$. (d) $24.4 \angle 80.8°$ ∎ ∎

Drill Problem 11-12 ■

Multiply the following and leave the results in rectangular form: **(a) A** $= 2 + j4$ and **B** $= 4 + j2$. **(b) A** $= (5 + j2)$, **B** $= 1 + j4$, and **C** $= -j2$.

Answer.   (a) $j20$. (b) $44 + j6$.   ■ ■

11.6 PHASOR DIVISION

As with multiplication, the division of phasor quantities is most easily carried out in polar form, but can also be accomplished in rectangular form.

The quotient of two phasors in polar form equals the quotient of their magnitudes and the difference of their arguments (the angle in the denominator is subtracted from that in the numerator). In mathematical form, if we have

$$\mathbf{A} = A \angle \alpha \quad \text{and} \quad \mathbf{B} = B \angle \beta$$

then

$$\frac{\mathbf{A}}{\mathbf{B}} = \frac{A}{B} \angle (\alpha - \beta) \tag{11-22}$$

EXAMPLE 11-13

Divide **A** $= 1 \angle 0°$ by **B** $= 8 \angle 45°$.

Solution

$$\frac{\mathbf{A}}{\mathbf{B}} = \frac{1}{8} \frac{\angle 0°}{\angle 45°} = 0.125 \angle -45°$$

Two complex quantities that differ only in the signs of their j components are called *complex conjugate quantities*, and either is called the conjugate of the other. An asterisk is often used to symbolize the conjugate; that is, **A*** is the conjugate of **A**. For example, if **A** $= 3 + j4$, then **A*** $= 3 - j4$. As indicated by Fig. 11-14, the conjugate **A*** of a phasor **A** is the image of **A** with respect to the horizontal or real axis. It follows that in polar form the conjugate of **A** $= A \angle \theta$ is the phasor **A*** $= A \angle -\theta$.

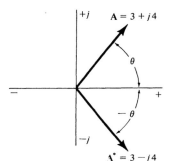

Figure 11-14   A phasor and its conjugate.

When finding the quotient of two complex quantities, we must first *rationalize* the denominator. Rationalization of the denominator is the process of eliminating the symbol j from the denominator by multiplying both the numerator and the denominator by the complex conjugate of the denominator. This process leaves a *rational number* (one without the symbol j) in the denominator, as illustrated in the following example.

EXAMPLE 11-14

Rationalize the denominator of the following expression:

$$\frac{1}{\mathbf{A}} = \frac{1}{3 + j4}$$

Solution.   The conjugate of the denominator is $3 - j4$ so that

$$\frac{1}{\mathbf{A}} = \frac{1}{3 + j4}\left(\frac{3 - j4}{3 - j4}\right) = \frac{3 - j4}{9 + j12 - j12 - j^2 16}$$

$$= \frac{3 - j4}{9 + 16} = \frac{3 - j4}{25} = 0.12 - j0.16$$

Notice that the answer in Example 11-14 is the rectangular form of the reciprocal of **A**. With only a little effort the expression can be converted to the polar form and compared to what is obtained if the phasor **A** is first converted to the polar form and then divided into $1 \angle 0°$ in accordance with Eq. 11-22.

The general equation expressing the quotient of two phasors, each in rectangular form, is

$$\frac{\mathbf{A}}{\mathbf{B}} = \frac{(a + jb)}{(c + jd)} = \frac{\mathbf{AB^*}}{\mathbf{BB^*}} = \frac{(a + jb)(c - jd)}{(c + jd)(c - jd)}$$

$$= \frac{(a + jb)(c - jd)}{c^2 + d^2} = \frac{(a + jb)(c - jd)}{B^2} \tag{11-23}$$

Note that when a complex number is multiplied by its conjugate as in Example 11-14 and Eq. 11-23, the result is a real number equal to the square of the magnitude of the complex number. That is,

$$\mathbf{BB^*} = |B|^2 \tag{11-24}$$

EXAMPLE 11-15

Divide $\mathbf{A} = 1 - j2$ by $\mathbf{B} = 3 + j4$.

Solution

$$\frac{\mathbf{A}}{\mathbf{B}} = \frac{1 - j2}{3 + j4} = \frac{1 - j2}{3 + j4}\left(\frac{3 - j4}{3 - j4}\right)$$

$$= \frac{3 - j6 - j4 + j^2 8}{9 + j12 - j12 - j^2 16} = \frac{3 - j10 - 8}{9 + 16}$$

$$= \frac{-5 - j10}{25} = -0.2 - j0.4$$

Drill Problem 11-13 ■

Simplify the following phasor expressions:

(a) $\dfrac{25 \angle 30°}{40 \angle -52°}$.   (b) $\dfrac{(3.2 \angle 15°)(10.4 \angle 32°)}{5.8 \angle -12°}$.

Answer. (a) 0.625 $\angle$ 82°. (b) 5.74 $\angle$ 59°.   ■ ■

Drill Problem 11-14 ■

Determine the complex conjugate for each of the following:
(a) 25 $\angle$ 40°. (b) $0.2 - j0.5$. (c) $-0.2 + j0.5$. (d) $-0.5 - j0.2$.

Answer. (a) 25 $\angle -40°$. (b) $0.2 + j0.5$. (c) $-0.2 - j0.5$. (d) $-0.5 + j0.2$.   ■ ■

Drill Problem 11-15 ■

Given the phasors **A** $= 10 \angle 30°$, **B** $= 2 \angle 45°$, and **C** $= 2.5 \angle 60°$, simplify the following expressions and express the answers in polar form:

(a) $\dfrac{\mathbf{A}^*}{\mathbf{ABC}}$.   (b) $\dfrac{\mathbf{B} + \mathbf{C}}{\mathbf{A}}$.   (c) $\dfrac{\mathbf{B}}{\mathbf{A} + \mathbf{C}}$.

Answer. (a) 0.2 $\angle -165°$. (b) 0.446 $\angle$ 23.3°. (c) 0.164 $\angle$ 9.2°.   ■ ■

Drill Problem 11-16 ■

Simplify the following complex expressions by using complex algebra and express the answers in rectangular form:

(a) $\dfrac{2 + j3}{1 - j2}$.   (b) $\dfrac{(3 + j4)(5 - j2)}{(3 + j4) + (5 - j2)}$.

Answer. (a) $-0.8 + j1.4$. (b) $3.11 + j0.97$.   ■ ■

11.7 THE PHASOR EXPONENTIAL FORM

The polar form of a phasor introduced in Eq. 11-2 is actually a special case of a more general mathematical form called the *exponential form*. The exponential form is based on

$$e^{\pm j\theta} = \cos \theta \pm j \sin \theta \tag{11-25}$$

where e is the base of the natural logarithm. Equation 11-25 is known as the Euler formula, in honor of Leonard Euler, the Swiss mathematician who in the 1700s developed the relation. The Euler equation is related to the rectangular and polar form of phasors by the following equivalence:

$$\mathbf{A} = a + jb = A \angle \alpha = A \cos \alpha + jA \sin \alpha$$
$$= Ae^{j\alpha} \tag{11-26}$$

Notice in Eq. 11-26 that the same information contained in the polar form (magnitude and angle) is contained in the exponential form. The exponential form is a bit more clumsy, but through its use we see why the angles add when we multiply two phasor quantities.

EXAMPLE 11-16

Given that $\mathbf{A} = A \angle \alpha$ and $\mathbf{B} = B \angle \beta$, determine their product using the exponential form.

Solution

$$\mathbf{A} = A \angle \alpha = Ae^{j\alpha} \qquad \mathbf{B} = B \angle \beta = Be^{j\beta}$$

Then $\mathbf{AB} = Ae^{j\alpha}Be^{j\beta}$ and since exponents add for multiplication,

$$\mathbf{AB} = ABe^{j(\alpha + \beta)}$$

Also, $\mathbf{AB} = AB \angle (\alpha + \beta)$. Note that we substitute the exponent symbol e for the angle symbol $\angle$.

EXAMPLE 11-17

Express the following phasors in exponential form: **(a)** $3 + j3$. **(b)** $25 \angle -30°$.

Solution

(a) $3 + j3 = 3\sqrt{2} \angle 45° = 4.24e^{j45°}$.
(b) $25 \angle -30° = 25e^{-j30°}$.

It was stated in Section 11.1 that it is customary to use effective values for phasor representations. Certainly that practice is recommended whenever we have rms instrument readings of voltage and current. However, there are times when we obtain voltage and current measurements from an *oscilloscope*, an electronic instrument that can display the waveform of a varying sinusoidal voltage or current. Then it is relatively easy to obtain the maximum or peak value. It follows that unless we simultaneously have an instrument with rms readings, it is easier to use maximum values for our phasor representations. There is nothing wrong with using *either* effective values or maximum values for phasors as long as we indicate which form we are using. In this text a subscript m is added to indicate phasors of peak or maximum values. Phasors without the subscript m are effective value phasors.

EXAMPLE 11-18

Determine the voltage expression for the sum $v_t = v_1 + v_2$, where $v_1 = 100 \sin(\omega t + 30°)$ V and $v_2 = 100 \sin(\omega t + 60°)$ V.

Solution. First we obtain phasor expressions for the sinusoids,

$$\mathbf{V}_{m1} = 100 \angle 30° \text{ V} \qquad \mathbf{V}_{m2} = 100 \angle 60° \text{ V}$$

As shown in Fig. 11-15,

$$\begin{aligned}
\mathbf{V}_{mt} &= 100 \angle 30° + 100 \angle 60° \\
&= 100 \cos 30° + j100 \sin 30° + 100 \cos 60° + j100 \sin 60° \\
&= 86.6 + j50 + 50 + j86.6 \\
&= 136.6 + j136.6 = 193 \angle 45° \text{ V}
\end{aligned}$$

Thus, $v_t = 193 \sin(\omega t + 45°)$ V

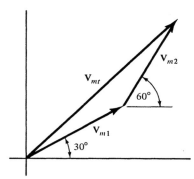

Figure 11-15 Phasor diagram of maximum values for Example 11-18.

The phasor mathematics remains the same whether we are working with effective or maximum values.

QUESTIONS

1. What is the difference between a scalar and a vector quantity?
2. What is the reference axis?
3. How are phase angles measured with respect to the reference axis?
4. What is a phasor quantity?
5. What is a phasor diagram?
6. What is meant by a reference phasor and how is it generally chosen?
7. What are two ways of representing phasors?
8. What is the significance of a negative phasor quantity?
9. What is the significance of the j operator?
10. How are phasors changed from polar to rectangular form?
11. How are phasors changed from rectangular to polar form?
12. How are phasor quantities added?
13. How are phasor quantities subtracted?
14. How are phasor quantities multiplied **(a)** in polar form and **(b)** in rectangular form?
15. What is a complex conjugate quantity?
16. How are phasor quantities divided **(a)** in polar form and **(b)** in rectangular form?

PROBLEMS

1. Write the phasor expressions that represent the following waves at $t = 0$:
   (a) $e = 120 \sin(\omega t + 30°)$ V. **(b)** $e = 50 \cos(\omega t + 10°)$ V.
   (c) $i = 2 \sin(1000t - 78°)$ mA. **(d)** $i = 10 \sin(377t + 25°)$ A.

2. Write the sine wave expressions for the following phasors if the frequency is 60 Hz:
   (a) $\mathbf{I} = 25 \angle 42°$ A.
   (b) $\mathbf{I} = 80 \angle -20°$ mA.
   (c) $\mathbf{V} = 70.7 \angle 90°$ V.
   (d) $\mathbf{V} = 35.4 \angle +120°$ V.

3. Write the negative of each of the phasors in Problem 11-2 and sketch a phasor diagram showing each original phasor and its negative.

4. Convert the following from rectangular to polar form:
   (a) $9 + j4$.
   (b) $50 + j86.6$.
   (c) $-50 + j86.6$.
   (d) $-1 + j2$.
   (e) $-3 - j4$.
   (f) $4 - j8$.
   (g) $0.024 - j0.016$.
   (h) $-2 - j2$.

5. Convert the following from polar to rectangular form:
   (a) $32 \angle -33°$.
   (b) $1.2 \angle 45°$.
   (c) $22 \angle 120°$.
   (d) $15.6 \angle -190°$.
   (e) $28.3 \angle -135°$.
   (f) $0.5 \angle 70°$.
   (g) $80 \angle -150°$.
   (h) $100 \angle 5°$.

6. Graphically determine the sum for the following:
   (a) $A = 80 \angle 0° + 100 \angle 60°$.
   (b) $B = 1.2 \angle 0° + 1.2 \angle 120°$.
   (c) $C = 8 \angle -30° + 6 \angle 90° + 4 \angle 180°$.

7. Perform the indicated addition and express the answers in polar form for the following:
   (a) $(-2 - j3) + (1 + j2)$.
   (b) $(0.3 - j0.5) + (0.4 + j0.3)$.
   (c) $(8 - j2) + (10 \angle 90°)$.
   (d) $(1 + j0.6) + (-0.7 + j0.2) + (0.6 - j0.4)$.
   (e) $(6 + j2) + (-5 + j3) + (-4 - j7)$.
   (f) $80 \angle 0° + 100 \angle 60°$.
   (g) $1.2 \angle 0° + 1.2 \angle 120°$.

8. Graphically find $\mathbf{V}_t = \mathbf{V}_1 - \mathbf{V}_2$ for the following:
   (a) $\mathbf{V}_1 = 80 \angle 90°$ V. $\mathbf{V}_2 = 60 \angle 0°$ V.
   (a) $\mathbf{V}_1 = 3 \angle 60°$ V. $\mathbf{V}_2 = 2.1 \angle -160°$ V.

9. Perform the subtractions of Problem 11-8 by rectangular phasors. Then convert the answers to polar form and compare them to the graphical solutions.

10. Perform the following subtractions and leave the answers in rectangular form:
    (a) $(3 - j5) - (4 + j3)$.
    (b) $(-1 + j4) - (4 + j3)$.
    (c) $2 \angle -30° - 1.2 \angle 45°$.
    (d) $130 \angle 45° - 90 \angle 30°$.

11. Find the following products and leave the answers in the same forms:
    (a) $(30 \angle 20°)(2 \angle -45°)$.
    (b) $(7 \angle 15°)(5 \angle -30°)$.
    (c) $(110 \angle 120°)(2 \angle 45°)$.
    (d) $(1 \angle 80°)(2.5 \angle -45°)(2 \angle -15°)$.
    (e) $(-1 - j1)(1 + j1)$.
    (f) $(-2 + j5)(-1 - j2)$.
    (g) $(3 - j4)(4 + j3)$.
    (h) $(1 - j4)(0.3 - j0.2)$.

12. Perform the following divisions and leave the answers in the same forms:

    (a) $\dfrac{50 \angle -53°}{3.6 \angle -123°}$.

    (b) $\dfrac{4 \angle 105°}{6.2 \angle 60°}$.

    (c) $\dfrac{3 - j4}{2 + j3}$.

    (d) $\dfrac{2 - j8}{1 + j4}$.

    (e) $\dfrac{1 + j0}{5 - j10}$.

    (f) $\dfrac{-j45}{0.636 - j0.636}$.

13. If $A = 1 + j4$, $B = 2 - j8$, and $C = 5 \angle 53.1°$, find $(A + B)C$.

14. If $Z_1 = 50 \angle 45°$ and $Z_2 = 100 \angle -70°$, find $Z_t = Z_1 Z_2/(Z_1 + Z_2)$.

15. Rationalize the following denominators:

   (a) $\dfrac{1}{5 + j8}$.

   (b) $\dfrac{1}{(3 + j4)(2 - j1)}$.

16. Evaluate the following determinant expressions and express the answers in polar form:

   (a) $\begin{vmatrix} 2 + j1 & j6 \\ 2 - j1 & 4 \end{vmatrix}$.

   (b) $\begin{vmatrix} 10 \angle 60° & 2 \angle 0° \\ 8 \angle 30° & 4 \angle -90° \end{vmatrix}$.

17. Evaluate the following quantities and express the results in polar form:

   (a) $3e^{j90°}(15 - j15)$.

   (b) $5e^{j30°}(2 - j4)/(3 + j3)(2e^{-j15°})$.

18. Determine the voltage or current expression as a displaced sine wave for each of the following:

   (a) $v_1 = 15 \sin 300t + 10 \cos 300t$.

   (b) $i_2 = 2 \sin(366t - \pi/4) + 1 \sin 366t$.

   (c) $i_t = 3 \sin \omega t + 4 \sin(\omega t - 90°)$.

Chapter 12

Single-Phase Alternating Current Circuits

12.1 THE SINGLE-PHASE CIRCUIT

An electric circuit that is energized by a single ac source is called a *single-phase circuit*. As was pointed out earlier, if an ac circuit consists of a voltage source and a resistance only, the resulting current is in phase with the voltage and the ratio of voltage to current equals the resistance. On the other hand, if a single-phase circuit contains only capacitance or inductance, the resulting current *leads* or *lags* the voltage by exactly 90° and the ratio of voltage to current equals the reactance.

Clearly, in a single-phase circuit containing both resistance and reactance, the ratio of voltage to current does not equal either the resistance or the reactance. Rather, it represents the total opposition of the circuit to the flow of alternating current. This total opposition, due to a combination of resistance and reactance, is called *impedance* and is specified by the symbol **Z**.

Since the impedance is a ratio of volts to amperes, it is appropriately described in ohms. Thus,

$$\mathbf{Z} = \frac{\mathbf{V}}{\mathbf{I}} \tag{12-1}$$

where **V** and **I** are the phasor values of voltage and current.

In Eq. 12-1, the impedance function mathematically relates two phasor quantities; to do this completely, the impedance must be specified by both a magnitude and an angle. Impedance, though, is not a sinusoidally varying quantity; it is therefore not a phasor quantity. However, as is shown later, impedance does result from a vector combination of resistance and reactance; hence one uses the vector notation in Eq. 12-1. The impedance magnitude is the ratio of voltage magnitude to current magnitude, and the impedance angle is the difference in phase angle between the voltage and current.

A single-phase circuit is shown by the schematic of Fig. 12-1(a), where the load is represented by the impedance **Z** and its general symbol. In turn, the general time plot of voltage and current and the phasor diagram for the circuit appear in Figs. 12-1(b) and 12-1(c), respectively.

Although the voltage of the source and, for that matter, any other voltage or current in the sinusoidally driven circuit are constantly changing in polarity, it becomes necessary to "freeze" the source voltage while we analyze the circuit. We do this by placing polarity marks on the source terminals, as shown in Fig. 12-1(a). The polarity signs indicate that, at that particular time, the instantaneous voltage is going positive as traced from the minus to plus terminal. On the other hand, at the load, we consider that an instantaneous voltage drop occurs in the direction of the current. Both these notations are needed if we are to write meaningful voltage equations.

EXAMPLE 12-1

The voltage and current across an ac load are $v = 2 \sin \omega t$ V and $i = 20 \sin(\omega t - 50°)$ mA. What is the vector expression for the load impedance?

Solution. We use the maximum values for the phasors: $\mathbf{V}_m = 2 \angle 0°$ V and $\mathbf{I}_m = 20 \angle -50°$ mA. Then

$$|\mathbf{Z}| = \frac{\mathbf{V}_m}{\mathbf{I}_m} = \frac{2 \angle 0°}{(20 \angle -50°)(10^{-3})} = 100 \angle 50° \ \Omega$$

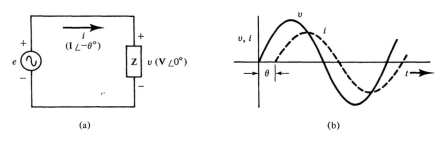

(a)　　　　　　　　　　　　　　(b)

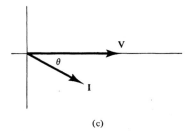

(c)

Figure 12-1 (a) Single-phase circuit; (b) general time plot of voltage and current; (c) phasor diagram of voltage and current.

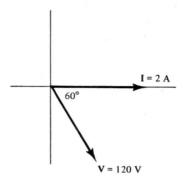

Figure 12-2   Example 12-2.

EXAMPLE 12-2

The voltage and current in an ac circuit are specified by the phasor diagram shown in Fig. 12-2. **(a)** Determine the circuit impedance. **(b)** Does the phasor diagram indicate capacitive or inductive behavior?

Solution.   Here $\mathbf{I} = 3 \angle 0°$ and $\mathbf{V} = 120 \angle -60°$ so that

(a)
$$\mathbf{Z} = \frac{\mathbf{V}}{\mathbf{I}} = \frac{120 \angle -60°}{3 \angle 0°} = 40 \angle -60° \ \Omega$$

(b) Since **I** leads **V**, the circuit exhibits capacitive behavior.

As discussed in the last chapter and illustrated by Examples 12-1 and 12-2, we may have either effective (rms) or maximum values as the given or measured quantities. The phasor mathematics is then performed in either form.

Drill Problem 12-1 ■

Determine the impedance and comment on whether the associated circuit exhibits capacitive or inductive properties for the following:

(a) $\mathbf{V} = 120 \angle 30° \ \text{V}, \mathbf{I} = 10 \angle -15° \ \text{A}.$

(b) $v = 10 \sin 1000t \ \text{V}, i = 2 \sin(1000t + 90°) \ \text{mA}.$

(c) $\mathbf{V} = 80 \angle -90° \ \text{mV}, \mathbf{I} = 20 \angle -10° \ \mu\text{A}.$

Answer.   (a) $12 \angle 45° \ \Omega$, inductive.   (b) $5 \angle -90° \ \text{k}\Omega$, purely capacitive. (c) $4 \angle -80° \ \text{k}\Omega$, capacitive.   ■ ■

12.2 KIRCHHOFF'S LAWS

In our study of ac circuits, we shall consider combinations of resistance, capacitance, and inductance connected in various series, parallel, and series–parallel configurations. It is appropriate to ask whether the various laws, theorems, and concepts we developed in the study of dc circuits (Chapters 5 and 6) are applicable to alternating current. The answer is that with some modification, those laws, theorems, and concepts are indeed applicable to ac circuits. An example is Eq. 12-1 which relates voltage to current and is similar to Ohm's law for direct current. Because of its similarity to Ohm's law, Eq. 12-1 is sometimes called Ohm's law for alternating current.

Two of the most important laws in the study of dc circuits, Kirchhoff's laws, are also applicable to ac circuits. Because the laws are satisfied at any instant of time, we can write them in terms of instantaneous quantities. Thus, KVL is expressed

$$\sum v = 0$$

and KCL is expressed

$$\sum i = 0$$

We must remember that these are algebraic summations around a closed loop and at a single node, respectively. Furthermore, with the use of mathematics, it can be shown that KVL and KCL are satisfied at any instant of time if one takes into account not only the magnitude of the voltages and the currents, respectively, but also their phase relationships. Thus, for sinusoidal voltages in phasor form we can state KVL as follows:

The phasor sum of all voltages around a closed loop is equal to zero.

For sinusoidal currents in phasor form we can state KCL as follows:

The phasor sum of all currents at a node is equal to zero.

Let us now consider the use of KVL and KCL with a few examples.

EXAMPLE 12-3

An ac circuit with corresponding voltages is shown in Fig. 12-3. Determine the unknown voltage.

Solution. Writing a KVL equation, we obtain

$$10 \angle 0° = \mathbf{V}_1 + \mathbf{V}_2 = 5 \angle 30° + \mathbf{V}_2$$

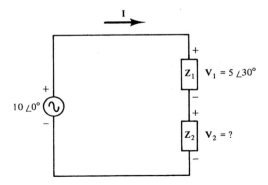

Figure 12-3 Example 12-3.

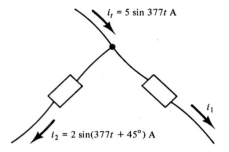

Figure 12-4   Example 12-4.

Solving for $\mathbf{V}_2$, we have

$$\mathbf{V}_2 = 10\angle 0° - 5\angle 30° = 10 + j0 - (4.33 + j2.5)$$
$$= 5.67 - j2.5 = 6.2\angle -23.8° \text{ V}$$

EXAMPLE 12-4

Determine the current i_1 as a function of time for the network of Fig. 12-4.

Solution.   We first convert the currents i_t and i_2 to phasor form utilizing maximum values.

$$\mathbf{I}_{mt} = 5\angle 0° \text{ A} \qquad \mathbf{I}_{m2} = 2\angle 45° \text{ A}$$

Then by KCL,

$$\mathbf{I}_{m1} = \mathbf{I}_{mt} - \mathbf{I}_{m2} = 5\angle 0° - 2\angle 45°$$
$$= 5 + j0 - 1.414 - j1.414$$
$$= 3.58 - j1.41 = 3.85\angle -21.5° \text{ A}$$

Thus $i_1 = 3.85 \sin(377t - 21.5°)$ A.

Drill Problem 12-2 ■

Determine the phasor voltage $\mathbf{V}_t$ for the circuit of Fig. 12-5.

Answer.   105.8 $\angle 20.9°$ V.   ■ ■

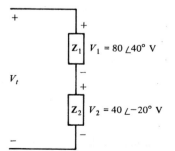

Figure 12-5   Drill Problem 12-2.

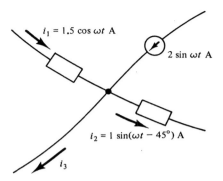

Figure 12-6   Drill Problem 12-3.

Drill Problem 12-3 ■

Determine the current i_3 for the circuit of Fig. 12-6.

Answer.   2.55 $\sin(\omega t + 59.6°)$ A.   ■ ■

12.3 THE SERIES *RL* CIRCUIT

In Fig. 12-7(a), a sinusoidal voltage $v_t = V_m \sin \omega t$ is applied to a series *RL* circuit. The Kirchhoff equation relating the voltages around the circuit is the differential equation

$$v_t = V_m \sin \omega t = iR + L \frac{di}{dt} \tag{12-2}$$

The current expression obtained as a solution to the preceding equation is, not surprisingly, a displaced sine wave

$$i = I_m \sin(\omega t + \theta) \tag{12-3}$$

where θ is a negative angle (*i* lags *v*). It follows that the resistor voltage v_R and the inductor voltage v_L are also sinusoidal. As shown in Fig. 12-7(b), the resistor voltage

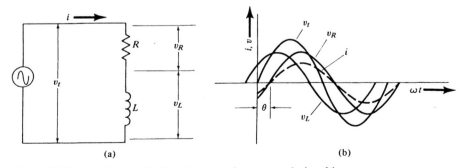

Figure 12-7   **(a)** *RL* circuit; **(b)** voltage and current relationship.

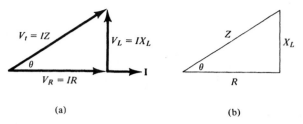

Figure 12-8   *RL* circuit: **(a)** Phasor diagram; **(b)** impedance diagram.

is in phase with the current, the inductor voltage leads the current by 90°, and their instantaneous sum equals the total applied voltage.

With the knowledge that a sinusoidal current flows in the circuit of Fig. 12.7, we can more conveniently express the voltage relationship in phasor terms. With the current as a reference, $V_R = IR$ is in phase with **I** and $V_L = IX_L$ leads **I** by 90°; hence

$$\mathbf{V}_t = V_R + jV_L \tag{12-4}$$

and we have the phasor diagram of Fig. 12-8(a).

It is evident from the phasor diagram that the total voltage equals the hypotenuse of the triangle formed by the resistor and inductor voltages, so

$$V_t = \sqrt{V_R^2 + V_L^2} \tag{12-5}$$

and

$$\theta = \tan^{-1}\frac{V_L}{V_R} \tag{12-6}$$

By Eq. 12-1, the total voltage equals the current × the impedance of the *RL* circuit; that is, $\mathbf{V}_t = \mathbf{IZ}$. Dividing the sides of the triangle formed by the phasor voltages by the current *I*, we obtain a similar triangle relating *R*, X_L, and *Z*. This triangle, shown in Fig. 12-8(b), is called an *impedance diagram*. Because *R*, X_L, and *Z* are not phasor quantities, no arrowheads are shown on the impedance diagram. However, the impedance function can be thought of as a *vector* sum of resistance and reactance so that it may algebraically relate the phasor voltage and current. Then

$$\mathbf{Z} = R + jX_L \tag{12-7}$$

or

$$\mathbf{Z} = \sqrt{R^2 + X_L^2} \angle \tan^{-1}\frac{X_L}{R} \tag{12-8}$$

EXAMPLE 12-5

A resistance of 40 Ω is connected in series with an inductance of 0.12 H. What current flows when the combination is connected to a 60 Hz voltage of 120 V?

Solution

$$X_L = \omega L = 377(0.12) = 45.2 \ \Omega$$

$$\mathbf{Z} = R + jX_L = 40 + j45.2$$

$$= \sqrt{40^2 + 45.2^2} \angle \tan^{-1}\frac{45.2}{40} = 60.4 \angle 48.5° \ \Omega$$

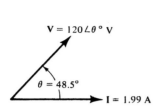

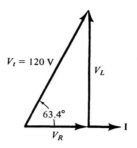

Figure 12-9 Phasor diagram: Example 12-5.

Figure 12-10 Phasor diagram: Example 12-6.

Considering the voltage phasor to be at an angle of 0°, we find that

$$\mathbf{I} = \frac{\mathbf{V}}{\mathbf{Z}} = \frac{120 \angle 0°}{60.4 \angle 48.5°} = 1.99 \angle -48.5° \text{ A}$$

Therefore, $\mathbf{I} = 1.99$ A and lags $\mathbf{V}$ by 48.5°.

Notice in Example 12-5 that the angle of the phasor voltage is not initially given but is conveniently chosen as 0°. As expected for an inductive circuit, the current is found to lag the voltage. However, when one sketches the phasor diagram for Example 12-5, one uses current as the reference phasor; hence we use the phasor diagram of Fig. 12-9.

EXAMPLE 12-6

A series *RL* circuit with $R = 5 \ \Omega$ and $X_L = 10 \ \Omega$ is connected to a 120 V source. What are the voltages across the resistor and inductor?

Solution

$$\mathbf{Z} = 5 + j10 = \sqrt{5^2 + 10^2} \ \angle \tan^{-1} \frac{10}{5} = 11.2 \angle 63.4° \ \Omega$$

The impedance angle of 63.4° is the same angle by which the voltage leads the current (see Fig. 12-10). Then

$$V_R = 120 \cos 63.4° = 53.7 \text{ V}$$
$$V_L = 120 \sin 63.4° = 107.3 \text{ V}$$

As was pointed out in Chapter 9, a practical coil has inductance and resistance, both of which are distributed throughout the coil winding. Because of the distributed nature of the inductance and resistance, we cannot measure the separate inductive and resistive voltage drops within the coil. However, we can measure the total coil voltage that will lead the coil current by the impedance angle of the coil.

EXAMPLE 12-7

The coil in Fig. 12-11 is known to have an impedance of $64.1 \angle 51.4 \ \Omega$. **(a)** Determine the coil and resistor voltages. **(b)** Sketch the phasor diagram for these voltages.

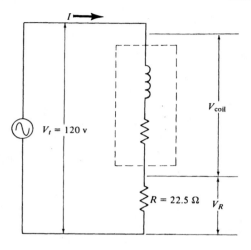

Figure 12-11   Circuit for Example 12-7.

Solution

$$\mathbf{Z}_t = \mathbf{Z}_{coil} + 22.5 \ \Omega = 64.1 \ \angle 51.4° + 22.5$$
$$= 40 + j50 + 22.5 = 62.5 + j50$$
$$= 80 \ \angle 38.7° \ \Omega$$

$$\mathbf{I} = \frac{\mathbf{V}_t}{\mathbf{Z}_t} = \frac{120 \ \angle 0°}{80 \ \angle 38.7°} = 1.5 \ \angle -38.7° \ \text{A}$$

(a) $V_{coil} = IZ_{coil} = (1.5)(64.1) = 96.2$ V (leads **I** by 51.4°). $V_R = IR = (1.5)(22.5) = 33.7$ V (in phase with **I**).

(b) With the current as reference, $\mathbf{V}_t$ leads **I** by the total impedance angle of 38.7°; hence the phasor diagram is that of Fig. 12-12.

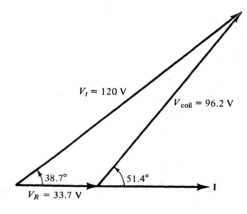

Figure 12-12   Phasor diagram: Example 12-7.

EXAMPLE 12-8

Compare the impedance that the series combination of a 10 Ω resistor and a 1 mH coil has at 100 Hz, 1 kHz, and 10 kHz.

Solution

At 100 Hz

$$X_L = \omega L = 2\pi(100)(1 \times 10^{-3}) = 0.628 \ \Omega$$

$$\mathbf{Z} = \sqrt{10^2 + 0.628^2} \ \angle \tan^{-1} \frac{0.628}{10}$$

$$\approx 10 \ \angle 3.6° \ \Omega$$

At 1 kHz

$$X_L = \omega L = 2\pi(10^3)(1 \times 10^{-3}) = 6.28 \ \Omega$$

$$\mathbf{Z} = \sqrt{10^2 + 6.28^2} \ \angle \tan^{-1} \frac{6.28}{10}$$

$$= 11.8 \ \angle 32° \ \Omega$$

At 10 kHz

$$X_L = \omega L = 2\pi(10^4)(1 \times 10^{-3}) = 62.8 \ \Omega$$

$$\mathbf{Z} = \sqrt{10^2 + 62.8^2} \ \angle \tan^{-1} \frac{62.8}{10}$$

$$= 63.6 \ \angle 80.9° \ \Omega$$

In Chapter 10 we noted that reactance is frequency-dependent. Example 12-8 points out the frequency dependence of the impedance which depends on the reactance. Notice that at a relatively low frequency the resistance of the circuit dominates the impedance expression and the phase angle is close to 0°. Hence, the current is nearly in phase with the voltage. At higher frequencies the inductive reactance term becomes dominant, the impedance angle approaches 90°, and the current lags the voltage considerably. In the next chapter, we shall study these frequency effects in more depth.

Drill Problem 12-4 ■

What resistance is connected in series with a 0.5 H coil if the combination at 60 Hz presents an impedance of 300 Ω?

Answer.   233 Ω.   ■ ■

Drill Problem 12-5 ■

At what frequency will a coil with a resistance of 2 Ω and an inductance of 100 μH have an impedance of 5 Ω?

Answer.   7.29 kHz.   ■ ■

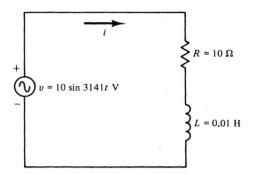

Figure 12-13   Drill Problem 12-6.

Drill Problem 12-6 ■

Determine **(a)** the mathematical expression of the current i for the circuit depicted in Fig. 12-13 and **(b)** the power dissipated in the resistor.

Answer.   (a) $i = 0.304 \sin(3141t - 72.3°)$ A. (b) 0.46 W.   ■ ■

12.4 THE SERIES *RC* CIRCUIT

The study of a series *RC* circuit when a sinusoidal voltage $V_t = V_m \sin \omega t$ is applied follows the same development as the study of the *RL* circuit. In particular, the voltage equation for the series *RC* circuit in Fig. 12-14(a) is

$$v_t = V_m \sin \omega t = iR + \frac{1}{C} \int i \, dt \qquad (12\text{-}9)$$

The current equation, which satisfies Eq. 12-9, is

$$i = I_m \sin(\omega t + \theta) \qquad (12\text{-}10)$$

where, θ is a positive angle (i leads v). Since the current is sinusoidal, the resistor voltage v_R and the capacitor voltage v_C are also sinusoidal. These two voltages and their relationship to the current and total voltage are shown in Fig. 12-14(b).

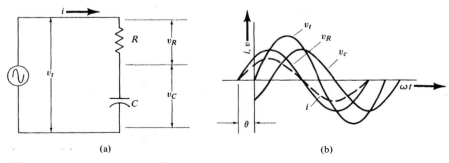

Figure 12-14   **(a)** *RC* circuit; **(b)** voltage and current relationship.

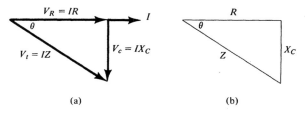

(a)   (b)

Figure 12-15   *RC* circuit: **(a)** Phasor diagram; **(b)** impedance diagram.

In complex form, the total voltage is specified by

$$\mathbf{V}_t = V_R - jV_C \tag{12-11}$$

in accordance with the phasor diagram of Fig. 12-15(a). From the trigonometric relationships,

$$V_t = \sqrt{V_R^2 + V_C^2} \tag{12-12}$$

and

$$\theta = -\tan^{-1}\frac{V_C}{V_R} \tag{12-13}$$

As in the *RL* case, the current *I* is a common factor in each of the sides of the triangle formed by the phasor voltages of Fig. 12-15(a). A similar triangle is obtained by dividing the common factor *I* from the voltages; hence the impedance diagram is that of Fig. 12-15(b).

Notice that in the series *RC* case that the impedance diagram is below the horizontal side so that

$$\mathbf{Z} = R - jX_C \tag{12-14}$$

and

$$\mathbf{Z} = \sqrt{R^2 + X_C^2}\angle - \tan^{-1}\frac{X_C}{R} \tag{12-15}$$

EXAMPLE 12-9

A series *RC* combination is connected to a 110 V, 60 Hz voltage. Measurements indicate a current of 3 A and a power loss of 66 W in the resistance. Find **(a)** the impedance in polar form and **(b)** the value of the capacitance.

Solution.   The resistance and impedance magnitude are first found:

$$Z = \frac{V}{I} = \frac{110}{3} = 36.7\ \Omega$$

$$R = \frac{P}{I^2} = \frac{66}{9} = 7.33\ \Omega$$

Referring to Fig. 12-15(b), we can write

$$\theta = -\cos^{-1}\frac{R}{Z} = -\cos^{-1}\frac{7.33}{36.7} = -78.5°$$

and $X_C = Z \sin 78.5° = 36.7 \sin 78.5° = 35.9$ Ω.

(a) $\qquad\qquad\qquad \mathbf{Z} = 36.7 \angle -78.5°$ Ω

(b) $\qquad\qquad\qquad C = \dfrac{1}{\omega X_C} = \dfrac{1}{377(35.9)} = 0.074 \times 10^{-3} = 74\ \mu F$

EXAMPLE 12-10

A 1 kΩ resistor is in series with a 0.1 μ F capacitor. Determine the rectangular and polar expressions for the impedance at 100 Hz, 1 kHz, and 10 kHz.

Solution

At 100 Hz

$$X_C = \frac{1}{\omega C} = \frac{1.}{2\pi(100)(0.1 \times 10^{-6})} = \frac{1}{2\pi(0.1 \times 10^{-4})}$$
$$= 15.9\ k\Omega$$

Then, in rectangular form

$$\mathbf{Z} = R - jX_C = 1 - j15.9\ k\Omega$$

In polar form,

$$\mathbf{Z} = \sqrt{1^2 + 15.9^2} \angle -\tan^{-1} 15.9$$
$$\approx 15.9 \angle -86°\ k\Omega$$

At 1 kHz

$$X_C = \frac{1}{\omega C} = \frac{1}{2\pi(10^3)(0.1 \times 10^{-6})} = 1.59\ k\Omega$$

Then, in rectangular form

$$\mathbf{Z} = R - jX_C = 1 - j1.59\ k\Omega$$

In polar form,

$$\mathbf{Z} = \sqrt{1^2 + 1.59^2} \angle -\tan^{-1} 1.59 = 1.88 \angle -57.8°\ k\Omega$$

At 10 kHz

$$X_C = \frac{1}{\omega C} = \frac{1}{2\pi(10^4)(0.1 \times 10^{-6})} = 159\ \Omega$$

Then, in rectangular form,

$$\mathbf{Z} = R - jX_C = 1 - j0.159\ k\Omega$$

In polar form,

$$\mathbf{Z} = \sqrt{1^2 + (0.159)^2} \angle -\tan^{-1} 0.159$$
$$= 1.01 \angle -9.03°\ k\Omega$$

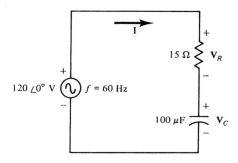

Figure 12-16 Drill Problems 12-7 and 12-8.

The results of Example 12-10 emphasize that at low frequencies a series *RC* circuit has an impedance that is dominantly capacitive in nature. Conversely, at high frequencies the capacitive reactance becomes negligible and the impedance of the combination appears to be resistive.

Drill Problem 12-7 ■

What phasor current flows in the circuit of Fig. 12-16?

Answer.   3.93 ∠ 60.5° A.   ■ ■

Drill Problem 12-8 ■

Determine for the circuit of Fig. 12-16 **(a)** the phasor voltages $\mathbf{V}_R$ and $\mathbf{V}_C$ and **(b)** the reactive power for the capacitor.

Answer.   (a) 58.9 ∠ 60.5° V; 104 ∠ − 29.5° V. (b) 410 vars.   ■ ■

Drill Problem 12-9 ■

What value of capacitance must be connected in parallel with the capacitor of Fig. 12-16 in order to have the current increase to a 5 A magnitude? What is the impedance angle for this situation?

Answer.   41.5 μF: − 51.3°.   ■ ■

12.5 THE SERIES *RLC* CIRCUIT

When a sinusoidal voltage is applied to a series *RLC* circuit, the resulting current is also sinusoidal. In terms of the phasor quantities of Fig. 12-17(a), the voltage across the resistor is the product IR and is in phase with the current. The voltage across the inductor equals the product IX_L and leads the current by 90°; whereas the voltage across the capacitor equals the product IX_C and lags the current by 90°. Thus the inductive and capacitive voltages are out of phase with each other by 180° and subtract, as shown in the phasor diagram of Fig. 12-17(b),

$$\mathbf{V}_t = V_R + j(V_L - V_C) \tag{12-16}$$

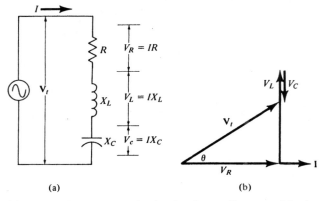

(a)   (b)

Figure 12-17 (a) *RLC* circuit; (b) phasor diagram with phasors added tip to tail.

In terms of an impedance diagram as in Fig. 12-18, the total impedance is given by

$$\mathbf{Z}_t = R + j(X_L - X_C)$$
$$= R + jX_t \tag{12-17}$$

or
$$\mathbf{Z}_t = \sqrt{R^2 + X_t^2} \angle \tan^{-1} \frac{X_t}{R} \tag{12-18}$$

where $X_t = X_L - X_C = $ *total or net reactance.*

It follows that if $X_L > X_C$, *the circuit is inductive* and appears electrically equivalent to a single resistor in series with an inductor. The voltage leads the current as in the case shown in the phasor diagram of Fig. 12-17. On the other hand, if $X_C > X_L$, *the circuit is capacitive* and appears electrically equivalent to a single resistor in series with a capacitor so that the voltage lags the current.

Of special note is the case in which $X_L = X_C$. The total impedance reduces to a value of $R \angle 0°$, and the *RLC* circuit responds as a purely resistive circuit. With this condition, the voltage and current are in phase and the circuit said to be in *series resonance*. Resonance is more fully explained in the next chapter.

EXAMPLE 12-11

In Fig. 12-17(a), $R = 5 \, \Omega$, $X_L = 8 \, \Omega$, and $X_C = 14 \, \Omega$. Find the current and the voltage across each element if the applied voltage is 15 V.

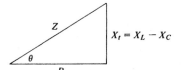

Figure 12-18 Impedance diagram for a series *RLC* circuit.

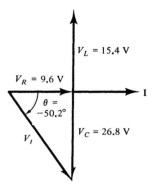

Figure 12-19   Phasor diagram:
Example 12-11.

Solution

$$\mathbf{Z} = R + jX_t = 5 + j(8 - 14) = 5 - j6$$
$$= 7.82 \angle -50.2° \; \Omega$$

$$\mathbf{I} = \frac{\mathbf{V}}{\mathbf{Z}} = \frac{15 \angle 0°}{7.82 \angle -50.2°} = 1.92 \angle +50.2° \; \text{A}$$

$$V_R = IR = 1.92(5) = 9.6 \; \text{V}$$

$$V_L = IX_L = 1.92(8) = 15.4 \; \text{V}$$

$$V_C = IX_C = 1.92(14) = 26.8 \; \text{V}$$

Notice that it is possible in an ac circuit for the magnitude of voltage across a circuit element to exceed the applied voltage.

In Example 12-11, the V_L and V_C both exceed the applied voltage, yet V_L cancels part of V_C, as indicated by the phasor diagram of Fig. 12-19.

Drill Problem 12-10 ■

Refer to Fig. 12-17 and consider that $R = 10 \; \Omega$, $X_L = 20 \; \Omega$, $X_C = 6 \; \Omega$, and $\mathbf{V}_t = 20 \angle 0°$ V. **(a)** Is the overall circuit reactance inductive or capacitive? **(b)** What is the phasor current? **(c)** What are the phasor voltages $\mathbf{V}_R$, $\mathbf{V}_L$, and $\mathbf{V}_C$?

Answer.   (a) Inductive. (b) $1.16 \angle -54.5°$ A. (c) $11.6 \angle -54.5°$ V; $23.2 \angle 35.5°$ V; $6.96 \angle -144.5°$ V.   ■ ■

12.6 SERIES AND PARALLEL IMPEDANCES

It should be apparent from the examples thus far that when an ac series circuit contains more than one resistance or reactance that the total resistance equals the sum of the resistances and the total reactance equals the algebraic sum of the reactances

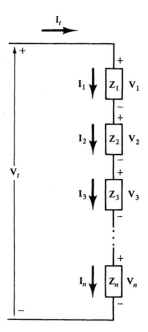

Figure 12-20   Series impedances.

(X_L, plus; X_C, minus). Consider the series connection of impedances in Fig. 12-20. From KVL we have

$$\mathbf{V}_t = \mathbf{V}_1 + \mathbf{V}_2 + \mathbf{V}_3 + \cdots + \mathbf{V}_n$$

Each separate impedance has a voltage equal to the current × the impedance so that

$$\mathbf{V}_t = \mathbf{I}_t\mathbf{Z}_t = \mathbf{I}_1\mathbf{Z}_1 + \mathbf{I}_2\mathbf{Z}_2 + \mathbf{I}_3\mathbf{Z}_3 + \cdots + \mathbf{I}_n\mathbf{Z}_n$$

Furthermore, all of the currents in the series connection are the same. Thus, we can state that the total impedance of any number of impedances in series is the complex sum of the individual impedances:

$$\mathbf{Z}_t = \mathbf{Z}_1 + \mathbf{Z}_2 + \mathbf{Z}_3 + \cdots + \mathbf{Z}_n \tag{12-19}$$

Notice that the series *RL*, *RC*, and *RLC* cases are just three special cases of impedance connections. They were chosen to introduce the concept of impedance because they represent simple combinations of series resistance and reactance.

EXAMPLE 12-12

Determine for the circuit of Fig. 12-21, **(a)** the total impedance, **(b)** the current **I**, and **(c)** the phasor voltages $\mathbf{V}_1$ and $\mathbf{V}_2$.

Solution

(a)
$$\mathbf{Z}_t = 20 \angle 30° + 15 \angle -10°$$
$$= 17.32 + j10 + 14.77 - j2.6$$
$$= 32.1 + j7.4 = 32.9 \angle 13° \ \Omega$$

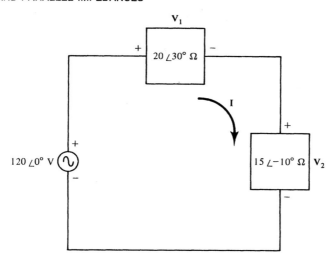

Figure 12-21 Example 12-12.

(b)

$$I = \frac{V}{Z_t} = \frac{120 \angle 0°}{32.9 \angle 13°} = 3.65 \angle -13° \text{ A}$$

(c) Then

$$V_1 = IZ_1 = (3.65 \angle -13°)(20 \angle 30°) = 73 \angle 17° \text{ V}$$

and $\quad V_2 = IZ_2 = (3.65 \angle -13°)(15 \angle -10°) = 54.8 \angle -23° \text{ V}$

In the ac parallel circuit, we may have branches that contain only resistance, inductance, capacitance, or some combination of these. In considering the ac parallel circuit we may use generalized branch impedances, as in Fig. 12-22.

With reference to Fig. 12-22, Kirchhoff's current law requires that

$$I_t = I_1 + I_2 + I_3$$

Each branch current equals the applied voltage V_t divided by the branch impedance so that

$$I_t = \frac{V_t}{Z_1} + \frac{V_t}{Z_2} + \frac{V_t}{Z_3}$$

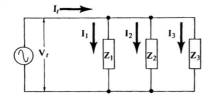

Figure 12-22 Parallel circuit.

In turn, the circuit offers a total impedance, $\mathbf{Z}_t$, relating the total current $\mathbf{I}_t$ to the total voltage.

$$\mathbf{I}_t = \frac{\mathbf{V}_t}{\mathbf{Z}_t} = \frac{\mathbf{V}_t}{\mathbf{Z}_1} + \frac{\mathbf{V}_t}{\mathbf{Z}_2} + \frac{\mathbf{V}_t}{\mathbf{Z}_3}$$

Thus the total or equivalent impedance of any number of impedances in parallel is given by

$$\frac{1}{\mathbf{Z}_t} = \frac{1}{\mathbf{Z}_1} + \frac{1}{\mathbf{Z}_2} + \frac{1}{\mathbf{Z}_3} + \cdots \tag{12-20}$$

The case of two impedances in parallel occurs frequently enough to deserve special attention. The impedance relationship from Eq. 12-20 is

$$\frac{1}{\mathbf{Z}_t} = \frac{1}{\mathbf{Z}_1} + \frac{1}{\mathbf{Z}_2}$$

and if we solve for $\mathbf{Z}_t$ in terms of $\mathbf{Z}_1$ and $\mathbf{Z}_2$, we obtain a familiar product over sum equation:

$$\mathbf{Z}_t = \frac{\mathbf{Z}_1 \mathbf{Z}_2}{\mathbf{Z}_1 + \mathbf{Z}_2} \tag{12-21}$$

EXAMPLE 12-13

Given the single-node pair of Fig. 12-23, **(a)** what phasor voltage is developed and **(b)** what power is dissipated in the resistor?

Solution. (a) The total impedance is calculated by Eq. 12-21

$$\mathbf{Z}_t = 15\ \Omega \,\|\, (j20\ \Omega) = \frac{(15)(j20)}{15 + j20} = \frac{300\ \angle\, 90°}{25\ \angle\, 53.1°}$$

$$= 12\ \angle\, 36.9°\ \Omega$$

Then

$$\mathbf{V} = \mathbf{I}\mathbf{Z}_t = (0.1\ \angle\, 0°)(12\ \angle\, 36.9°) = 1.2\ \angle\, 36.9°\ \text{V}$$

(b) $P = V^2/R = (1.2)^2/15 = 0.096$ W.

EXAMPLE 12-14

Refer to Fig. 12-24. **(a)** Determine the total impedance. **(b)** Determine $\mathbf{I}_t$, $\mathbf{I}_1$, $\mathbf{I}_2$, and $\mathbf{I}_3$. **(c)** Sketch a phasor diagram of the currents.

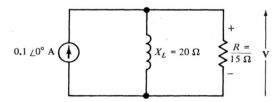

Figure 12-23   Example 12-13.

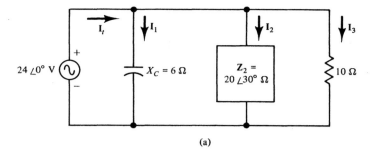

(a)

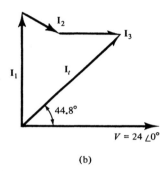

(b)

Figure 12-24   Example 12-14: **(a)** Circuit; **(b)** phasor diagram.

Solution

(a)
$$\frac{1}{\mathbf{Z}_t} = \frac{1}{6 \angle -90°} + \frac{1}{20 \angle 30°} + \frac{1}{10}$$
$$= 0.167 \angle 90° + 0.05 \angle -30° + 0.1$$
$$= j0.167 + 0.043 - j0.025 + 0.1$$
$$= 0.143 + j0.142 = 0.202 \angle 44.8° \text{ S}$$

$$\mathbf{Z}_t = \frac{1}{0.202 \angle 44.8°} = 4.95 \angle -44.8° \ \Omega$$

(b)
$$\mathbf{I}_t = \frac{24 \angle 0°}{4.95 \angle -44.8°} = 4.85 \angle 44.8° \text{ A}$$

$$\mathbf{I}_1 = \frac{24 \angle 0°}{6 \angle -90°} = 4 \angle 90° \text{ A}$$

$$\mathbf{I}_2 = \frac{24 \angle 0°}{20 \angle 30°} = 1.2 \angle -30° \text{ A}$$

$$\mathbf{I}_3 = \frac{24 \angle 0°}{10} = 2.4 \angle 0° \text{ A}$$

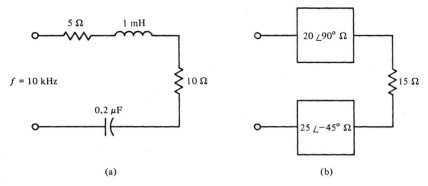

(a)   (b)

Figure 12-25   Drill Problem 12-11.

(c) The phasor diagram appears in Fig. 12-24(b). Note that the voltage is used as the reference since it is common to all elements. Alternatively, in this example we calculate the total current by first adding the branch currents. Then the impedance can be determined by dividing the total current into the voltage. Thus, you will find that, as with dc resistive circuits, ac impedance circuits can often be solved in a variety of ways.

Drill Problem 12-11 ■

Determine the total impedance for each of the networks of Fig. 12-25.

Answer.   (a) 22.5 ∠−48.2° Ω. (b) 32.8 ∠4.2° Ω.   ■ ■

Drill Problem 12-12 ■

Determine the **(a)** total impedance, **(b)** phasor current, and **(c)** phasor coil voltage for the circuit of Fig. 12-26.

Answer.   (a) 16.2 ∠21.8° Ω. (b) 0.617 ∠−21.8° A. (c) 5.82 ∠36.2° V.   ■ ■

Drill Problem 12-13 ■

Determine the total impedance for the networks of Fig. 12-27.

Answer.   (a) 141 ∠−45° Ω. (b) 3.86 ∠−15° kΩ.   ■ ■

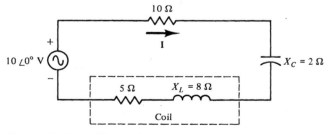

Figure 12-26   Drill Problem 12-12.

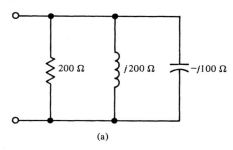

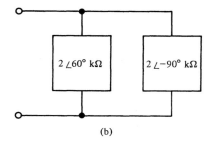

(a)

(b)

Figure 12-27 Drill Problem 12-13.

12.7 VOLTAGE AND CURRENT DIVISION

Circuits containing both series and parallel impedance combinations can be reduced to one equivalent impedance in the same manner that comparable resistance combinations are reduced. It is often desirable when considering these impedance combinations to find the voltage across or the current through a particular impedance. This is easily done by applying the voltage and current division techniques developed in Chapter 5. Now, however, one must use impedances, phasor voltages, and currents.

If, as in Fig. 12-28, a group of impedances are in series, *the voltage across a particular impedance* $\mathbf{Z}_n$ *is the total voltage* $\mathbf{V}_t \times$ *the ratio* $\mathbf{Z}_n/\mathbf{Z}_t$.

$$\mathbf{V}_n = \mathbf{V}_t\left(\frac{\mathbf{Z}_n}{\mathbf{Z}_t}\right) \tag{12-22}$$

For two impedances in parallel, *the branch current in either impedance equals the total current* $\mathbf{I}_t \times$ *the ratio of the opposite impedance to the sum of the two*

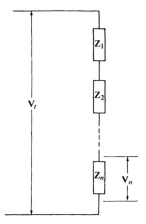

Figure 12-28 Voltage division.

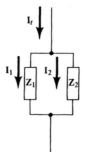

Figure 12-29   Current division.

impedances. For Fig. 12-29 then

$$\mathbf{I}_1 = \mathbf{I}_t\left(\frac{\mathbf{Z}_2}{\mathbf{Z}_1 + \mathbf{Z}_2}\right) \tag{12-23}$$

and

$$\mathbf{I}_2 = \mathbf{I}_t\left(\frac{\mathbf{Z}_1}{\mathbf{Z}_1 + \mathbf{Z}_2}\right) \tag{12-24}$$

EXAMPLE 12-15

Refer to Fig. 12-30. Using current division, determine **(a)** the current in the resistor and **(b)** the power dissipated in the resistor.

Solution

(a)
$$\mathbf{I}_1 = \mathbf{I}_t\frac{\mathbf{Z}_2}{\mathbf{Z}_1 + \mathbf{Z}_2} = 0.1 \angle 0°\left(\frac{j20}{15 + j20}\right)$$

$$= 0.1 \angle 0°\left(\frac{20 \angle 90°}{25 \angle 53.1°}\right) = 0.08 \angle 36.9° \text{ A}$$

(b) Then $P = I_1^2 R = (0.08)^2(15) = 0.096$ W.

Drill Problem 12-14 ∎

Using voltage or current division, determine the unknown quantities specified in Fig. 12-31.

Answer.   (a) $4.47 \angle 63.4°$ V. (b) $2.24 \angle -75.4°$ A.   ∎ ∎

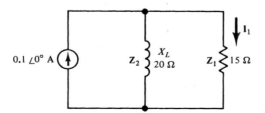

Figure 12-30   Example 12-15.

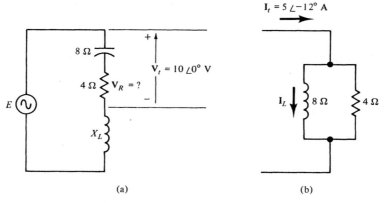

(a)   (b)

Figure 12-31 Drill Problem 12-14.

12.8 SERIES–PARALLEL CIRCUITS

Circuits that contain both series and parallel impedance combinations are called *series–parallel circuits.* As for the resistive series–parallel circuits of Chapter 5, analysis involving series–parallel impedances begins with a recognition of which parts are in series and which parts are in parallel. Then any desired unknown quantities are determined by using the same reduction techniques suggested earlier for resistive circuits. Some of these techniques are illustrated in the examples that follow.

EXAMPLE 12-16

Determine the total impedance for the circuit in Fig. 12-32(a).

Solution. With a reduction as in Fig. 12-32(b),

$$\mathbf{Z}_1 = \frac{(R_2)(jX_L)}{R_2 + jX_L} = \frac{(30\text{ k})(j40\text{ k})}{(30\text{ k}) + (j40\text{ k})} = \frac{(1200\text{ k}^2)\angle 90°}{(50\text{ k})\angle 53.1°}$$

$$= 24\angle 36.9°\text{ k}\Omega = (19.2 + j14.4)\text{ k}\Omega$$

$$\mathbf{Z}_t = 10\text{ k} + \mathbf{Z}_1 = 10\text{ k} + 19.2\text{ k} + j14.4\text{ k}$$

$$= 29.2 + j14.4\text{ k} = 32.6\angle 26.2°\text{ k}\Omega$$

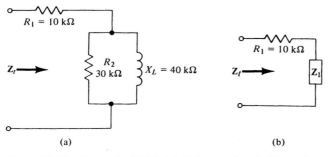

(a)   (b)

Figure 12-32 Example 12-16: **(a)** Original circuit; **(b)** equivalent circuit.

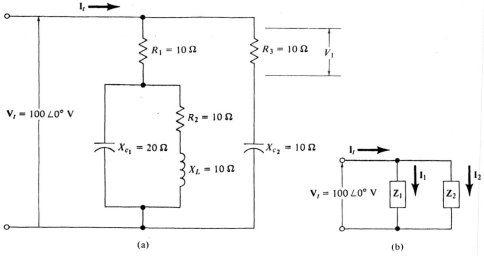

Figure 12-33   Example 12-17: **(a)** Original circuit; **(b)** Equivalent circuit.

EXAMPLE 12-17

Determine **(a)** the total current I_t and **(b)** the voltage V_1 for the circuit of Fig. 12-33(a).

Solution.   With reference to the equivalent circuit of Fig. 12-33(b),

(a)
$$Z_1 = 10 + \frac{(-j20)(10 + j10)}{(10 + j10 - j20)} = 10 + \frac{(200 - j200)}{(10 - j10)}$$

$$= 10 + 20 = 30 \ \Omega$$

$$Z_2 = 10 - j10 = 14.14 \ \angle -45° \ \Omega$$

$$I_1 = \frac{V_t}{Z_1} = \frac{100 \ \angle 0°}{30 \ \angle 0°} = 3.33 \ \angle 0° \ A$$

$$I_2 = \frac{V_t}{Z_2} = \frac{100 \ \angle 0°}{14.14 \ \angle -45°} = 7.07 \ \angle +45° = 5 + j5 \ A$$

and
$$I_t = I_1 + I_2 = 3.33 + 5 + j5 = 8.33 + j5$$

$$= 9.72 \ \angle 31° \ A$$

(b) By voltage division,

$$V_1 = V_t \left(\frac{10}{10 - j10} \right) = 100 \ \angle 0° \left(\frac{10 \ \angle 0°}{14.14 \ \angle -45°} \right) = 70.7 \ \angle 45° \ V$$

Drill Problem 12-15 ■

Determine **(a)** the total impedance and **(b)** the power dissipation in the $10 \ \Omega$ resistor for the circuit of Fig. 12-34.

Answer.   (a) $7.07 \ \angle -45° \ \Omega$. (b) 40 W.   ■ ■

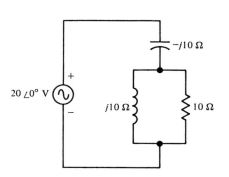

Figure 12-34 Drill Problem 12-15.

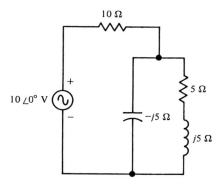

Figure 12-35 Drill Problem 12-16.

Drill Problem 12-16 ■

For the circuit of Fig. 12-35, determine **(a)** the total impedance and **(b)** the inductor current.

Answer.   (a) 15.8 $\angle -18.4°$ Ω. (b) 0.63 $\angle -71.6°$ A.   ■ ■

12.9 ADMITTANCE

The reciprocal of impedance is called *admittance*. Admittance is measured in siemens and is specified by the letter symbol **Y**:

$$\mathbf{Y} = \frac{1}{\mathbf{Z}} \tag{12-25}$$

From Eq. 12-20, it is apparent that for parallel branches the total or equivalent admittance $\mathbf{Y}_t$ equals

$$\mathbf{Y}_t = \mathbf{Y}_1 + \mathbf{Y}_2 + \mathbf{Y}_3 + \cdots \tag{12-26}$$

Admittance, like impedance, is a complex quantity; therefore, Eq. 12-26 represents an addition of complex numbers.

Since the admittance is complex in form, it generally has both a real and an imaginary part. The real part of the admittance term is called *conductance* and is specified by the letter G; the imaginary part is called *susceptance* and is specified by the letter B. Both G and B are measured in siemens. Thus

$$\mathbf{Y} = G \pm jB \tag{12-27}$$

or

$$\mathbf{Y} = \sqrt{G^2 + B^2} \angle \pm \tan^{-1} \frac{B}{G} \tag{12-28}$$

The susceptance term arises from the presence of inductive or capacitive elements; as shown in the following example, a positive sign in Eq. 12-27 indicates capacitive susceptance B_C and a negative sign indicates inductive susceptance B_L.

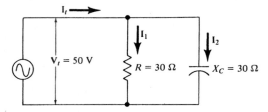

Figure 12-36   Circuit for Example 12-18.

EXAMPLE 12-18

Find the admittance and total current for the circuit of Fig. 12-36 when the applied voltage is 50 V.

Solution

$$\text{For branch 1:} \quad \mathbf{Z}_1 = 30 \ \Omega$$
$$\text{For branch 2:} \quad \mathbf{Z}_2 = -j30 \ \Omega$$

Then

$$\mathbf{Y}_t = \mathbf{Y}_1 + \mathbf{Y}_2 = \frac{1}{30} + \frac{1}{-j30}$$

$$= 0.0333 + j0.0333 = 0.0471 \ \angle +45° \ \text{S}$$

$$\mathbf{I}_t = \frac{\mathbf{V}_t}{\mathbf{Z}_t} = \mathbf{V}_t\mathbf{Y}_t = (50 \ \angle 0°)(0.0471 \ \angle +45°)$$

$$= 2.35 \ \angle 45° \ \text{A}$$

Notice in the preceding example that the current leads the voltage as expected for a capacitive circuit. Also, as indicated before, a capacitive element results in a positive susceptance term in the complex admittance expression. An *admittance diagram* relating conductance and capacitive susceptance then extends above the horizontal side, as shown in Fig. 12-37(a). However, for a circuit having inductive susceptance, the diagram extends below the horizontal side as shown in Fig. 12-37(b).

If a branch has only resistance, as in branch 1 of Example 12-18, the admittance consists of only a conductance given by $1/R$. On the other hand, if a branch has only reactance, as in branch 2 of Example 12-18, the admittance consists of only a suscep-

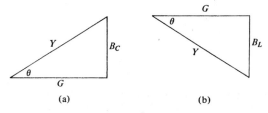

Figure 12-37   Admittance diagram; **(a)** Capacitive susceptance; **(b)** inductive susceptance.

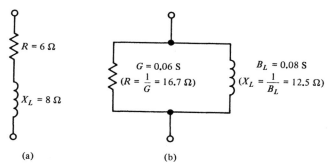

Figure 12-38 **(a)** A series circuit; **(b)** its parallel equivalent.

tance given by $B = 1/X$. If, however, a branch contains a combination of resistance and reactance, the conductance and susceptance cannot be found by simply obtaining the reciprocals of resistance and reactance. Rather, the admittance is found in a manner illustrated by the following example.

EXAMPLE 12-19

A certain branch of a parallel circuit consists of a $6\ \Omega$ resistance in series with an $8\ \Omega$ inductive reactance. Find the admittance of the branch.

Solution

$$\mathbf{Z} = R + jX_L = 6 + j8$$
$$\mathbf{Y} = \frac{1}{\mathbf{Z}} = \frac{1}{6+j8}\left(\frac{6-j8}{6-j8}\right) = \frac{6-j8}{36+64}$$
$$\mathbf{Y} = 0.06 - j0.08 = 0.10\ \angle -53.1°\ \text{S}$$

After we obtain the admittance expression for a particular branch, as in Example 12-19, we can *synthesize* or "build" the admittance by the parallel connection of a single conductance of value G and a single susceptance of value B. This circuit is the *parallel equivalent circuit*. The circuit of Fig. 12-38(b) is the parallel equivalent circuit of the series branch of Example 12-19. In turn, the series circuit of Fig. 12-38(a) is the *series equivalent circuit* of the parallel circuit, since it represents the simplest series combination of a resistance and reactance equivalent to the parallel circuit.

Drill Problem 12-17 ■

Determine the admittance, conductance, and susceptance of the circuits shown in Fig. 12-39.

Answer. (a) $0.125\ \angle 40°$, 0.096, $j0.080$ S. (b) $0.083\ \angle -53.1°$, 0.05, $-j0.0667$ S.

■ ■

Drill Problem 12-18 ■

Determine the **(a)** admittance and **(b)** total current of the circuit of Fig. 12-40.

Answer. (a) $0.106\ \angle -45°$ S. (b) $1.06\ \angle -45°$ A.

■ ■

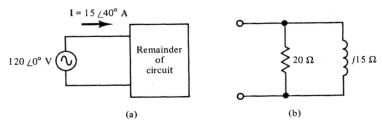

Figure 12-39 Drill Problem 12-17.

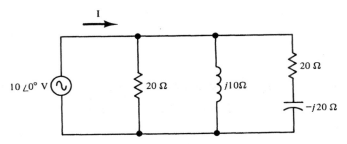

Figure 12-40 Drill Problem 12-18.

12.10 APPARENT, REAL, AND REACTIVE POWER

When a sinusoidal voltage is applied to a pure resistance, the current is sinusoidal and in phase with the voltage. Hence the actual dissipated power, as shown in Chapter 10, equals the product of the effective values of voltage and current. In the case of a pure reactive element, the voltage and current are out of phase by exactly 90°, and the product of the effective values of voltage and current equals the reactive power that is alternately stored and released. One might then expect that if a circuit has both resistance and reactance, both dissipated and reactive power result.

Consider the general single-phase circuit with a sinusoidal voltage $v = V_m \sin \omega t$ applied to it. A current $i = I_m \sin(\omega t + \theta)$ results and is *leading*, that is, θ is positive for a capacitive-type circuit and is *lagging*, that is, θ is negative for an inductive-type circuit. The instantaneous power is

$$p = vi = V_m I_m \sin \omega t \sin(\omega t + \theta)$$

Using the trigonometric identity,

$$\sin \alpha \sin \beta = \tfrac{1}{2}[\cos(\alpha - \beta) - \cos(\alpha + \beta)]$$

in the power expression, one finds

$$p = \frac{V_m I_m}{2} [\cos(\omega t - \omega t - \theta) - \cos(\omega t + \omega t + \theta)]$$

$$= \frac{V_m I_m}{\sqrt{2}} [\cos(-\theta) - \cos(2\omega t + \theta)]$$

$NO \; \sqrt{}$

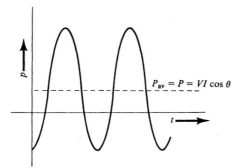

Figure 12-41   Instantaneous power curve.

Because $V = V_m/\sqrt{2}$, $I = I_m/\sqrt{2}$, and $\cos \theta = \cos(-\theta)$,

$$p = VI \cos \theta - VI \cos(2\omega t + \theta) \tag{12-29}$$

The second term of Eq. 12-29 represents a negative cosine wave of twice the frequency of the applied voltage; since the average value of a cosine wave is zero, this term contributes nothing to the average power. However, the first term is of particular importance because the terms V, I, and $\cos \theta$ are all constant and do not change with time. In fact, it is noted in the plot of Eq. 12-29 (see Fig. 12-41) that the constant term $VI \cos \theta$ is the average value of the instantaneous power. Thus the average value of power P is given by

$$P = VI \cos \theta \tag{12-30}$$

where V and I are the effective values of voltage and current and θ is the phase angle between the voltage and current. Since the phase angle for a single-phase circuit is always between $\pm 90°$, $\cos \theta \geq 0$ and $P \geq 0$. In the special cases of only resistance or only reactance, Eq. 12-30 reduces to $P = VI$ and $P = 0$, respectively.

The term $\cos \theta$ is called the *power factor* of the circuit and the angle θ is sometimes referred to as the *power factor angle*. In the inductive circuit, where the current lags the voltage, the power factor is described as a *lagging power factor*. In the capacitive circuit, where the current leads the voltage, the power factor is said to be a *leading power factor*.

Because the product VI in Eq. 12-30 does not represent either average power in watts or reactive power in vars, it is defined by a new term, *apparent power*. The product VI, called apparent power, has the unit of *volt-ampere* (VA) and is indicated by the letter S. Thus,

$$P = S \cos \theta \tag{12-31}$$

EXAMPLE 12-20

An impedance $\mathbf{Z} = 3 + j4 \ \Omega$ is connected to a 100 V ac source. Determine the apparent and actual power.

Solution

$$Z = 3 + j4 = 5 \angle 53.1° \ \Omega$$

$$I = \frac{V}{Z} = \frac{100 \angle 0°}{5 \angle 53.1°} = 20 \angle -53.1° \ A$$

$$S = VI = (100)(20) = 2000 \ VA$$

$$P = S \cos \theta = 2000 \cos 53.1°$$
$$= 1200 \ W$$

Alternately, the power equals I^2R.

$$P = I^2R = (20)^2(3) = 400(3) = 1200 \ W$$

Equation 12-31 suggests a Pythagorean relationship between the actual power and the apparent power. Recall that a series *RL* circuit has a voltage phasor diagram that forms a right triangle, as shown in Fig. 12-42(a). Multiplying each side by a magnitude of current *I*, we recognize a similar triangle, where the real power is along the horizontal axis, the reactive power is along the vertical axis, and the apparent power is the hypotenuse. Such a triangle, as shown in Fig. 12-42(b), is called a *power triangle*. Then

$$Q = S \sin \theta \tag{12-32}$$

and

$$S = \sqrt{P^2 + Q^2} \tag{12-33}$$

EXAMPLE 12-21

Using the data from Example 12-20, determine the reactive power.

Solution

$$Q = S \sin \theta = 2000 \sin 53.1° = 1600 \ var$$

Alternatively,

$$Q = I^2X_L = (20)^2(4) = 400(4) = 1600 \ var$$

Perhaps you are wondering whether the apparent power is equal to the phasor product of voltage and current. It is not! Power calculations depend on the angular

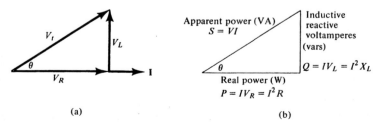

(a)   (b)

Figure 12-42   **(a)** Voltage diagram: *RL* series circuit; **(b)** the power triangle.

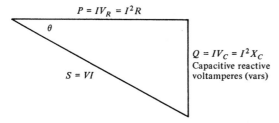

$$P = IV_R = I^2R$$

θ

$S = VI$

$Q = IV_C = I^2X_C$
Capacitive reactive
voltamperes (vars)

Figure 12-43 The power triangle for the capacitive circuit.

difference between the phasor voltage and current. In order to obtain the difference angle between voltage and current, we must use the conjugate of the voltage or current. Thus, we define the phasor product of the voltage and the *conjugate* of the current as *complex power*. That is,

$$\mathbf{S} = \mathbf{VI}^* = P + jQ \tag{12-34}$$

where $\mathbf{S}$ = complex power in volt-amperes,
        $\mathbf{V}$ = phasor voltage, and
        $\mathbf{I}^*$ = conjugate of the phasor current.

In general, Eq. 12-34 provides in polar form the apparent power magnitude $S = |\mathbf{S}|$ and the angle of the power triangle. In rectangular form it provides the real term P and the reactive power term Q.

EXAMPLE 12-22
Using the data $\mathbf{V} = 100 \angle 0°$ V and $\mathbf{I} = 20 \angle -53.1°$ from the preceding examples, determine the complex power and compare to the preceding examples.

Solution

$$\mathbf{S} = \mathbf{VI}^* = (100 \angle 0°)(20 \angle 53.1°)$$
$$= 2000 \angle 53.1° \text{ VA} = 1200 \text{ W} + j1600 \text{ var}$$

The magnitudes S, P, and Q are the same as in the preceding examples.

If we have a series capacitive circuit, the current leads the voltage and the power triangle is below the axis as shown in Fig. 12-43. If a circuit contains both capacitance and inductance, the net or total reactive power Q_t is the *difference between the capacitive reactive power and the inductive reactive power*. In such a case, the capacitance returns energy *to* the circuit, while the inductance simultaneously takes energy *from* the circuit. A certain amount of reactive power is thus exchanged or traded back and forth between the capacitance and inductance.

Obviously, when two or more resistances are in an ac circuit, the total average power P_t equals the sum of the individual powers.

EXAMPLE 12-23
Given the series circuit of Fig. 12-44, determine **(a)** the power factor, **(b)** the current, and **(c)** the power triangle.

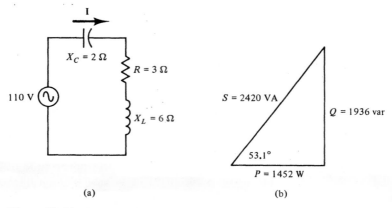

(a)   (b)

Figure 12-44   Example 12-23: **(a)** Circuit; **(b)** power triangle.

Solution

(a)
$$\mathbf{Z} = R + j(X_L - X_C) = 3 + j(6 - 2)$$
$$= 3 + j4 = 5 \angle 53.1° \; \Omega$$

Therefore $\theta = 53.1°$ and $\cos \theta = 0.6$.

(b)
$$\mathbf{I} = \frac{\mathbf{V}}{\mathbf{Z}} = \frac{110 \angle 0°}{5 \angle 53.1°} = 22 \angle -53.1° \; \text{A}$$

(c)
$$\mathbf{S} = \mathbf{VI}^* = (110 \angle 0°)(22 \angle 53.1°)$$
$$= 2420 \angle 53.1° \; \text{VA} = 1452 \; \text{W} + j1936 \; \text{var}$$

Alternatively,

$$P = I^2 R = (22)^2(3) = 1452 \; \text{W}$$
$$Q = Q_t = I^2 X_L - I^2 X_C$$
$$= (22)^2(6 - 2) = 1936 \; \text{var}$$

The power triangle is shown as Fig. 12-44(b).

EXAMPLE 12-24

For the circuit of Fig. 12-45, determine the total **(a)** real power, **(b)** reactive power, **(c)** apparent power, and **(d)** power factor.

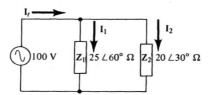

Figure 12-45   Example 12-24.

Solution

$$\mathbf{I}_1 = \frac{\mathbf{V}}{\mathbf{Z}_1} = \frac{100 \angle 0°}{25 \angle 60°} = 4 \angle -60° \text{ A}$$

$$\mathbf{I}_2 = \frac{\mathbf{V}}{\mathbf{Z}_2} = \frac{100 \angle 0°}{20 \angle 30°} = 5 \angle -30° \text{ A}$$

$$\mathbf{S}_1 = \mathbf{VI}_1^* = (100 \angle 0°)(4 \angle 60°) = 400 \angle 60°$$
$$= 200 + j346$$

$$\mathbf{S}_2 = \mathbf{VI}_2^* = (100 \angle 0°)(5 \angle 30°) = 500 \angle 30°$$
$$= 433 + j250$$

(a) $P_t = P_1 + P_2 = 200 + 433 = 633$ W.

(b) $Q_t = Q_1 + Q_2 = 346 + 250 = 596$ var.

(c) $S_t = \sqrt{P_t^2 + Q_t^2} = \sqrt{633^2 + 596^2} = 869$ VA.

(d) $\cos \theta = P_t/S_t = 633/869 = 0.728$.

Notice that even though $P_t = \sum P$ and $Q_t = \sum Q$, $S_t \neq \sum S$. We must first obtain P_t and Q_t in order to find S_t.

Finally it should be noted that the power triangle for the inductive circuit is in the first quadrant, just as it is for the impedance triangle. Conversely, the power triangle for the capacitive circuit is in the fourth quadrant, just as it is for the impedance triangle in the capacitive case.

It would be equally correct to specify a power triangle for the inductive circuit in the fourth quadrant and one for the capacitive circuit in the first quadrant. Of course, these latter triangles would be similar to the associated admittance diagrams. As many of our "power" circuits are parallel circuits, this latter set of power triangles is sometimes preferable. The two formats should not be confused, however.

Drill Problem 12-19 ■

Determine the complex power and the power factor for a circuit with the following phasor quantities: $\mathbf{V} = 120 \angle 30°$ V, $\mathbf{I} = 10 \angle -10°$ A.

Answer. 919 W $+ j771$ var; 0.766.   ■ ■

Drill Problem 12-20 ■

A single-phase motor is supplied with 1400 W from a 220 V, 60 Hz line. The motor operates at a power factor of 0.88. **(a)** What current flows to the motor? **(b)** What is the apparent power?

Answer. (a) 7.23 A. (b) 1591 VA.   ■ ■

Drill Problem 12-21 ■

Determine the **(a)** total power, **(b)** reactive power, **(c)** apparent power, and **(d)** power factor angle for the circuit depicted in Fig. 12-46.

Answer. (a) 2230 W. (b) 650 var. (c) 2323 VA. (d) 16.3°.   ■ ■

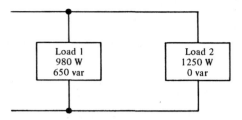

Figure 12-46   Drill Problem 12-21.

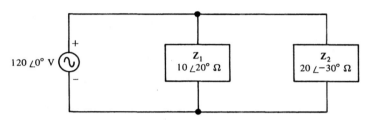

Figure 12-47   Drill Problem 12-22.

Drill Problem 12-22 ■

Determine the total complex power, impedance angle, and power factor for the circuit of Fig. 12-47.

Answer.   1977 W $+ j$132 var; 3.82°; 0.998.   ■ ■

12.11 POWER FACTOR CORRECTION

Most industrial and many residential electrical loads are inductive, that is, they operate at a lagging power factor. The consequence of a lagging power factor is that a larger apparent power is required compared to that required for a similar load operating at unity power factor. This is easily seen by referring to the power triangle of Fig. 12-42.

In Fig. 12-42, an apparent power S is required to produce a power P for a load operating at a power factor angle of θ. However, if the same power P is maintained while the power factor angle of θ is brought closer to 0°, the hypotenuse of the power triangle becomes smaller until, in the limiting case, it equals the power side of the triangle. The apparent power needed to supply a given true power is then a minimum. A lower apparent power is desirable, especially for electric utility companies, since for a given voltage a lower line current and a greater efficiency result.

Although it is not possible to change the inductive nature of a load itself, it is possible to connect a capacitive load in parallel with the inductive load in order to cancel some inductive reactive power, thereby decreasing the apparent power. This addition of a capacitor in parallel with an inductive load so as to reduce the apparent power while not altering the voltage or current to the original load is known as *power factor correction*.

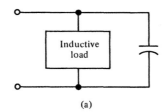

(a)

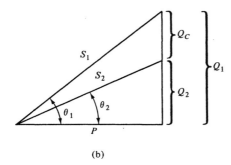

(b)

Figure 12-48　Power factor correction:
(a) Circuit; **(b)** power triangle.

A power factor correction circuit and its associated power triangle are shown in Fig. 12-48. In reference to the power triangle of Fig. 12-48, the parallel capacitor supplies a reactive power Q_C that cancels some of the original reactive power Q_1 leaving a net inductive reactive power Q_2. Accordingly, the apparent power is decreased from S_1 to S_2. Notice that if an overall power factor of unity is desired, the capacitor is chosen so that its reactive power equals the reactive power of the inductive load.

EXAMPLE 12-25

A load operating at a lagging power factor of 0.7 dissipates 2 kW when connected to a 220 V, 60 Hz power line. What value of capacitance is needed to correct the power factor to 0.9?

Solution.　Referring to the data given and Fig. 12-48,

$$\theta_1 = \cos^{-1} 0.7 = 45.6°$$
$$\theta_2 = \cos^{-1} 0.9 = 26°$$

Then

$$Q_1 = P \tan \theta_1 = 2000 \tan 45.6° = 2040 \text{ var}$$
$$Q_2 = P \tan \theta_2 = 2000 \tan 26° = 975 \text{ var}$$

and
$$Q_C = Q_1 - Q_2 = 2040 - 975 = 1065 \text{ var}$$

From $Q_C = V^2/X_C$,

$$X_C = \frac{V^2}{Q_C} = \frac{(220)^2}{1065} = \frac{4.84 \times 10^4}{1.065 \times 10^3} = 45.4 \ \Omega$$

and
$$C = \frac{1}{\omega X_C} = \frac{1}{377(45.4)} = 58.3 \times 10^{-6} = 58.3 \ \mu F$$

Drill Problem 12-23 ■

Determine the value of capacitance needed to correct the power factor of the motor load of Drill Problem 12-20 to a 0.95 power factor.

Answer. 16.2 μF.   ■ ■

Drill Problem 12-24 ■

What parallel capacitance is needed to correct a load of 250 kVA at 0.95 lagging power factor to unity power factor?

Answer. 78.1 kvar.   ■ ■

12.12 EFFECTIVE RESISTANCE

The ohmic or dc resistance of a circuit R_{dc} is the resistance described in terms of conductor geometry, that is $R = \rho(l/A)$. Although the ohmic resistance correctly defines resistance in a dc circuit, it does not adequately define resistance in an ac circuit. To understand why this is so, consider a more general definition for resistance.

Resistance is the property of an electric circuit that determines for a given current the rate at which electric energy is converted into heat or radiant energy.

A resistance value is such that the product of it and the square of the current gives the rate of energy conversion. Certain energy conversion effects, which depend on frequency, exist in the ac circuit, thereby making the resistance appear higher than that for the zero frequency or the dc case. The equivalent resistance component, due to these special energy conversion effects plus the dc resistance, equals the *ac* or *effective resistance* R_e. In terms of the total power in a circuit, R_e is defined as

$$R_e = \frac{P}{I^2} \tag{12-35}$$

Two of the effects that make the effective resistance greater than the dc resistance are *hysteresis* and *eddy current losses*, both of which are discussed in Chapter 8. As was pointed out in that chapter, if any magnetic materials are located within a magnetic field, such as that created by a wire-carrying alternating current, some of the electric energy is dissipated as heat in the magnetic material. Also, any conducting materials located within the magnetic field have eddy currents induced in them so that some additional electric energy is lost. Both hysteresis and eddy current effects are important, even at power frequencies.

Another effect due to a conductor carrying alternating current is characterized by a reduction in the effective cross section of the conductor itself. The magnetic field produced by a conductor carrying ac extends within the conductor and is greatest at the center of the conductor. As a result, the counter emf produced at the center of the conductor is greater than that near the outer surface. Thus, the current is not uniform throughout the cross section, but crowds to the outer surface, thereby reducing the effective cross section and increasing the resistance. This effect, the *skin effect*, is not appreciable at power frequencies, except in conductors with a large cross-sectional area ($>250\,000$ cm). However, at higher frequencies, the skin effect is more pronounced. In fact, at frequencies in the gigahertz range, hollow conductors are used since the current travels essentially on the surface.

Another energy loss that adds to the effective resistance is the *electromagnetic radiation* that may result from the circuit functioning as an antenna. This effect depends on the physical dimensions of the circuit and the frequency of ac and is not appreciable below radio frequencies.

Lastly, any dielectric within an ac electrostatic field exhibits internal losses. The mechanism by which energy is lost in a dielectric is called *dielectric hysteresis*. This effect is important mostly at high frequencies.

EXAMPLE 12-26

A coil with an iron core has a dc resistance of 15 Ω. When connected to a 60 Hz source, the following measurements are noted: $V = 120$ V, $I = 1.2$ A, and $P = 35$ W. Find the effective resistance and the coil impedance.

Solution

$$R_e = \frac{P}{I^2} = \frac{35}{(1.2)^2} = 24.3 \ \Omega$$

$$\mathbf{Z} = \frac{V}{I} \angle \cos^{-1} \frac{P}{S}$$

$$= \frac{120}{1.2} \angle \cos^{-1} \frac{35}{(120)(1.2)}$$

$$= 100 \angle \cos^{-1} 0.243 = 100 \angle 75.9° \ \Omega$$

Unless one wants specifically to consider the difference between the ohmic and effective resistances, as in Example 12-26, it is not necessary to use subscripts with the resistance symbol R. Rather, it is assumed that if a dc circuit is being considered, any specified resistance value is an ohmic value; if an ac circuit is being considered, any specified resistance value is an effective value.

Drill Problem 12-25 ■

A toroid coil used in an audio filter has a dc resistance of 22 Ω. At a certain frequency the coil dissipates 4 mW of power when 12 mA of current flows through it. By what percent does the resistance increase from its dc value to the given ac value?

Answer. 26.4%. ■ ■

QUESTIONS

1. What is a single-phase circuit?
2. What is impedance and what are its symbol and its unit?
3. Why is impedance not a phasor quantity?
4. State Kirchhoff's laws for ac circuits. How do they differ from those for simple dc circuits?
5. Comment on the relationship for R, X_L, and Z in the series ac circuit. How do they relate to an impedance diagram?
6. When sketching a phasor diagram for a series circuit, why do you choose the current as the reference phasor?
7. Describe the R, X_C, Z impedance diagram.
8. When is an RLC circuit inductive? When is it capacitive?
9. What condition exists when a circuit is in series resonance?
10. How does one calculate the total impedance for a number of impedances in series?
11. How does one calculate the total impedance for a number of impedances in parallel?
12. What is admittance, its unit, and its symbol?
13. How do admittances combine when in parallel?
14. What are conductance and susceptance? What are their symbols?
15. Describe the admittance diagram for both parallel RL and RC circuits.
16. What is meant by a parallel equivalent circuit?
17. What is meant by a series equivalent circuit?
18. When two impedances are in parallel, what familiar mathematical format is used for the total impedance?
19. How are voltage and current division techniques applied to ac circuits?
20. The instantaneous power expression for an ac circuit consists of two terms; describe these terms and their significance.
21. Define power factor and power factor angle.
22. What is meant by (a) a lagging power factor and (b) a leading power factor?
23. Define apparent power and give its unit and its symbol.
24. What is a power triangle? How is it similar to the impedance diagram?
25. Why is a lagging power factor undesirable?
26. What is power factor correction?
27. What is the difference between dc and ac resistance?
28. What are the effects that are accounted for in the effective resistance value?

PROBLEMS

1. Determine the unknown quantities for the circuits of Fig. 12-49.
2. Using Kirchhoff's laws, solve for the phasor quantities specified in Fig. 12-50.

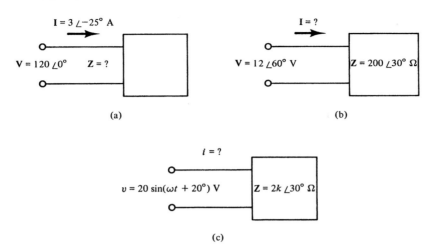

(a)

(b)

(c)

Figure 12-49 Problem 12-1.

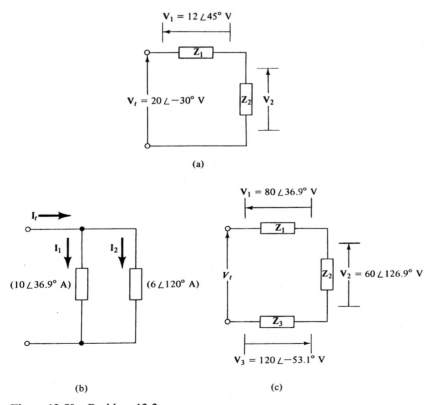

(a)

(b)

(c)

Figure 12-50 Problem 12-2.

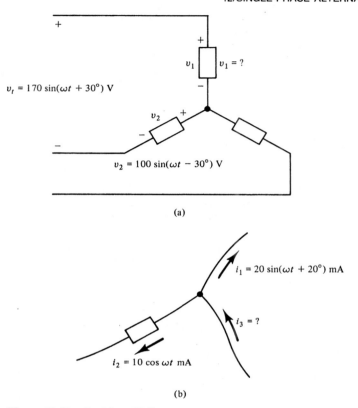

(a)

(b)

Figure 12-51   Problem 12-3.

3. Determine the unknown voltage or current expressions for the networks of Fig. 12-51.

4. A coil having an inductance of 0.14 H and a resistance of 12 Ω is connected across a 110 V, 25 Hz line. Find **(a)** the impedance, **(b)** the current, and **(c)** the angle between the current and voltage.

5. A resistance of 12 Ω is connected in series with an inductance of 0.1 H. What current flows when the combination is connected to a 110 V, 60 Hz voltage?

6. A pure inductance of 5 H is connected in series with a 2 Ω resistor. The combination is connected to a 110 V, 60 Hz source. What current flows in the circuit?

7. A coil having a reactance of 12 Ω and a resistance of 5 Ω is connected in series with a resistor of 11 Ω. What are the voltages across the coil and the resistor when a 110 V, 60 Hz voltage is applied?

8. A certain impedance has a current of 2 A when a voltage of 26 V, 60 Hz is dropped across it. If the current lags the voltage by 15°, find the series resistance and inductance that will satisfy these conditions.

9. A series combination of a resistance and a capacitance produces a 2 A current that leads the applied voltage by 80°. If the magnitude of applied voltage is 110 V (60 Hz), what are the resistance and capacitance?

10. A certain resistance and capacitance, when connected in series, have an impedance of $72 \angle -30°$ Ω at 60 Hz. What will the impedance be at 25 Hz?

11. Determine **(a)** the phasor voltages $\mathbf{V}_R$ and $\mathbf{V}_C$ and **(b)** the power dissipated in the circuit of Fig. 12-52.

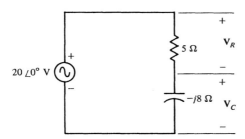

Figure 12-52   Problem 12-11.

12. The filter circuit in Fig. 12-53 is designed so that $\mathbf{V}_{out}$ has a phase angle of $\angle -60°$. What capacitance is needed?

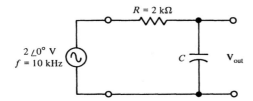

Figure 12-53   Problem 12-12.

13. Determine the voltage across each of the components in the circuit of Fig. 12-54.

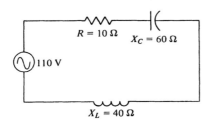

Figure 12-54   Problem 12-13.

14. Given the circuit of Fig. 12-55, determine the voltages $\mathbf{V}_1$ and $\mathbf{V}_2$.

15. A series circuit consists of a 2 kΩ resistor, a 1 μH inductor, and a 6 pF capacitor. At what frequency will the total impedance be purely resistive?

16. Determine the total impedance for each of the circuits of Fig. 12-56.

17. Calculate $\mathbf{Z}_t$ for the parallel circuits shown in Fig. 12-57.

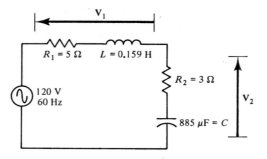

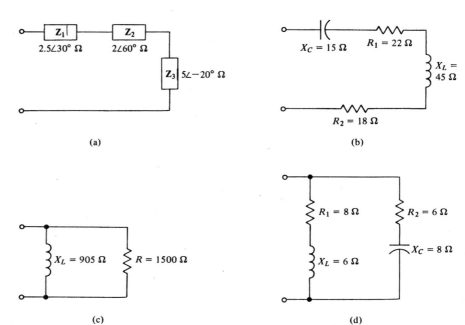

Figure 12-55   Problem 12-14.

(a)

(b)

(c)

(d)

Figure 12-56   Problem 12-16.

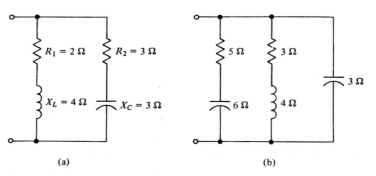

(a)

(b)

Figure 12-57   Problem 12-17.

18. Find the total current I_t in polar form for the circuit of Fig. 12-58.

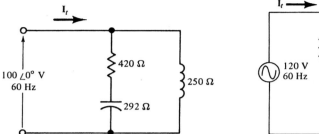

Figure 12-58   Problem 12-18.

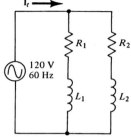

Figure 12-59   Problem 12-19.

19. In the circuit of Fig. 12-59, branch 1 has a current of 5 A and a power loss of 100 W, while branch 2 has a current of 10 A and a power loss of 500 W. Find the total line current.

20. Calculate I_1, I_2, and I_t for the circuit of Fig. 12-60.

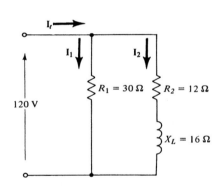

Figure 12-60   Problem 12-20.

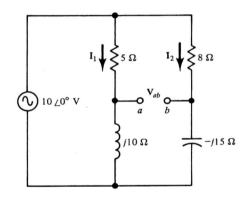

Figure 12-61   Problem 12-21.

21. Determine the currents I_1 and I_2 and the voltage V_{ab} for the circuit of Fig. 12-61.

22. If the current I_2 in Fig. 12-62 is $4.63 \angle 68.2°$ A, find the current I_1.

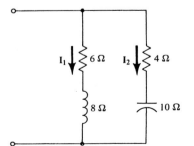

Figure 12-62   Problem 12-22.

23. Using voltage division, solve for the voltages across Z_1 of Fig. 12-56(a) and R_1 of Fig. 12-56(b) if 100 $\angle 0°$ V is applied to each of these circuits.

24. Using current division, solve for the branch currents for the circuits of Fig. 12-63.

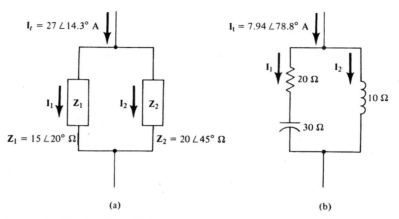

(a)   (b)

Figure 12-63   Problem 12-24.

25. Determine the total impedance for the circuit of Fig. 12-64 if $Z_1 = 13.4 \angle 65.3°$ Ω, $Z_2 = 19.2 \angle -51.3°$ Ω, and $Z_3 = 17.86 \angle 26.6°$ Ω.

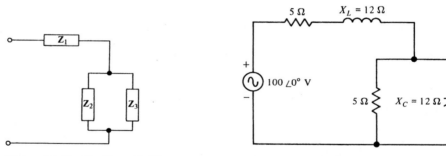

Figure 12-64   Problem 12-25.   **Figure 12-65**   Problem 12-26.

26. Determine the **(a)** total impedance and **(b)** the voltage V_1 for the series–parallel circuit of Fig. 12-65.

27. Determine the phasor current I_1 and voltage V_1 for the circuit of Fig. 12-66.

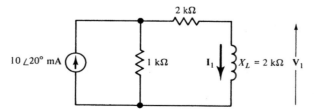

Figure 12-66   Problem 12-27.

28. Determine the expression for total current i_t for the circuit of Fig. 12-67.

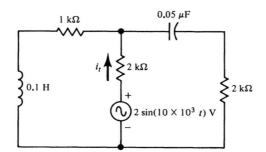

Figure 12-67   Problem 12-28.

29. Determine the phasor quantities I_t and V_1 for the circuit of Fig. 12-68.

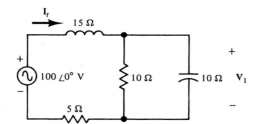

Figure 12-68   Problem 12-29.

30. Using admittance values for the parallel branches of Fig. 12-56, determine Y_t for the parallel circuits.

31. Determine the unknown quantities for each of the circuits of Fig. 12-69.

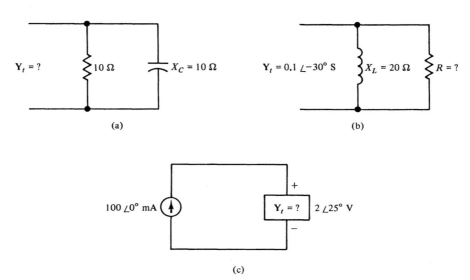

Figure 12-69   Problem 12-31.

32. What value of capacitance is needed in the circuit of Fig. 12-70 so that the total admittance is pure conductance (i.e., so that the phase angle is 0°)?

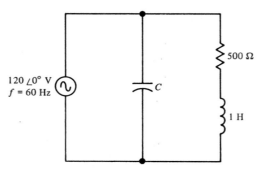

Figure 12-70   Problem 12-32.

33. A coil is placed in series with a resistance of 38 Ω. When the combination is connected to a 220 V, 60 Hz source, the current is 4.2 A and the power is 670 W. What is the power factor?

34. Two inductors having resistances of $R_1 = 3$ Ω and $R_2 = 2$ Ω, and inductances of $L_1 = 0.01$ H and $L_2 = 0.02$ H are connected in series with a 100 V, 60 Hz source. What are **(a)** the apparent power and **(b)** the true power?

35. Calculate the power dissipated by a load connected to a 2300 V ac line if the load is 200 kVA operating at a lagging power factor of 0.9.

36. A series circuit connected to a 200 V 60 Hz line consists of a capacitor with 30 Ω reactance, a resistor of 44 Ω, and a coil with 90 Ω reactance and 36 Ω resistance. Determine **(a)** the power factor and **(b)** the real power.

37. With 5 A in the circuit of Fig. 12-71, what is the **(a)** true power, **(b)** reactive power, **(c)** apparent power, and **(d)** power factor?

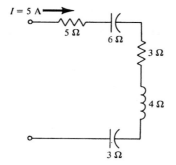

Figure 12-71   Problem 12-37.

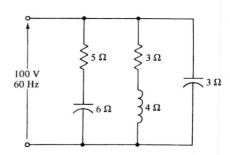

Figure 12-72   Problem 12-38.

38. A voltage of 100 V is applied to the parallel impedances of Fig. 12-72. What is the **(a)** true power, **(b)** reactive power, **(c)** apparent power, and **(d)** power factor?

39. Find the **(a)** total power, **(b)** reactive power, **(c)** apparent power, and **(d)** power factor for the system portrayed by the block diagram of Fig. 12-73.

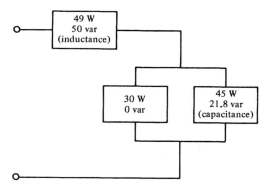

Figure 12-73   Problem 12-39.

40. A circuit with an admittance of $0.02 \angle 50°$ S has a voltage of $120 \angle 0°$ V across it. What are **(a)** the complex power, **(b)** the real power, and **(c)** the reactive power?

41. Determine the total **(a)** complex power, **(b)** real power, and **(c)** reactive power for the circuit of Fig. 12-74.

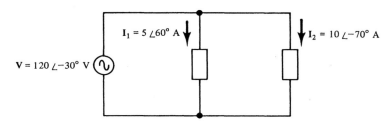

Figure 12-74   Problem 12-41.

42. A certain 120 V, 60 Hz motor has a power input of 575 W at 80% lagging power factor, when operating at maximum efficiency. At a certain location, only 240 V at 60 Hz is available. A resistor of suitable current-carrying capacity is placed in series so that the motor receives only 120 V at its terminals. Determine **(a)** the value of series resistance needed and **(b)** the total amount of power supplied to the motor and resistor.

43. An 800 kVA load operates at a 0.6 lagging power factor. It is desired to correct the power factor to unity. What (in kilovars) parallel capacitance is necessary?

44. A load draws 6 kW of power when an apparent power of 8 kVA is supplied from a 200 V, 60 Hz source. What value of parallel capacitance is needed to obtain a unity power factor?

45. It is desired to correct the power factor of an industrial plant from 2400 kVA at 0.67 lagging power factor to a 0.95 lagging power factor. Determine **(a)** the capacitance in kilovars needed and **(b)** the kilo-volt-amperes of the plant after correction.

46. Laboratory measurements show that a particular coil allows 1.1 A to flow through it when a 98 V, 60 Hz voltage is applied. If a power measurement indicates 2.18 W, what are **(a)** the effective resistance and **(b)** the impedance of the coil?

47. With 5 V dc applied to a certain circuit, a current of 0.1 A flows. If the dc voltage is replaced by an ac supply of 60 Hz, it is found that a voltage of 10 V is needed to cause a current of 0.1 A to flow in the same circuit. If the power taken by the circuit with ac supplied is 0.6 W, **(a)** what value of effective resistance does the circuit possess and **(b)** what power loss is attributed to hysteresis and eddy current loss if the increased resistance is caused by these factors?

48. A certain coil has a total core loss of 30 W at 60 Hz; of this, 4 W is eddy current loss. Using the hysteresis and eddy current relationships developed in Chapter 8, determine the core loss if the coil is used at 25 Hz.

49. If the current in the coil of Problem 48 is $i = 1.8 \sin(157t + 58°)$ A, what contribution does the core loss make to the value of effective resistance at that particular frequency?

50. A coil with an iron core has a current of 1.5 A when connected to a 120 V, 60 Hz source. When a pure resistance of 40 Ω is connected in series with the coil, the current drops to 1.2 A. What are the effective resistance and inductance of the coil?

Chapter 13

Frequency Selective Circuits and Resonance

13.1 FREQUENCY SELECTIVE CIRCUITS

It is apparent that the impedance, and hence the response, of any circuit containing reactive elements depend upon frequency. Thus any circuit containing reactive elements can be called a frequency selective circuit, since it provides a certain response to certain frequencies. Even though the term *frequency selective circuit* applies to any circuit containing reactive elements, it is often used to denote only a circuit specifically designed to separate different frequencies. A circuit possessing this property of frequency discrimination is alternately called an electric wave filter or, simply a *filter*.

When only a broad discrimination is desired, a single reactive element is sometimes placed in the path of the current. For example, a capacitor appears as an open circuit to direct current (zero-frequency alternating current) but as a very low reactance or possible short to high-frequency alternating current. On the other hand, an inductor passes direct current but limits the flow of current for high-frequency alternating current.

Before considering the frequency selective nature of the *RLC* circuit, it is instructive to consider the selective nature of the simpler *RC* circuit. In Fig. 13-1, an

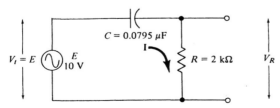

Figure 13-1 *RC* circuit.

TABLE 13-1   DATA FOR THE *RC* CIRCUIT OF FIG. 13-1

f (Hz)	R (kΩ)	X_C (kΩ)	$\mathbf{Z}$ (kΩ)	$\mathbf{I}$ (mA)	V_R (V)
10	2	200	200 $\angle -89.4°$	0.05 $\angle +89.4°$	0.1
100	2	20	20.1 $\angle -84.3°$	0.498 $\angle +84.3°$	1.0
200	2	10	10.2 $\angle -78.7°$	0.980 $\angle +78.7°$	1.96
500	2	4	4.47 $\angle -63.4°$	2.23 $\angle +63.4°$	4.46
1 k	2	2	2.83 $\angle -45°$	3.54 $\angle +45°$	7.07
2 k	2	1	2.24 $\angle -26.6°$	4.47 $\angle +26.6°$	8.94
5 k	2	0.4	2.04 $\angle -11.3°$	4.90 $\angle +11.3°$	9.80
10 k	2	0.2	2.01 $\angle -5.7°$	4.96 $\angle +5.7°$	9.92
100 k	2	0.02	2.0 $\angle -0.6°$	5.0 $\angle +0.6°$	10.0

RC circuit is connected to an ac variable-frequency source that is capable of supplying a constant voltage of 10 V at any desired frequency.

As the frequency of the source is varied, the capacitive reactance varies. It follows that the impedance and current also vary and are given by

$$\mathbf{Z} = R - jX_C \tag{13-1}$$

and

$$\mathbf{I} = \frac{\mathbf{E}}{R - jX_C} = I \angle \theta \tag{13-2}$$

Computations of the capacitive reactance, impedance, and current for Fig. 13-1 at several values of frequency enable one to prepare a table such as Table 13-1. Table 13-1 also includes the value of the resistor voltage $V_R = IR$ at each frequency. This voltage is of interest because it represents the useful voltage received by the resistor when it is considered a load.

The magnitudes of current and resistor voltage are plotted against frequency in Fig. 13-2. A plot of voltage or current vs. frequency is a *frequency response curve*.

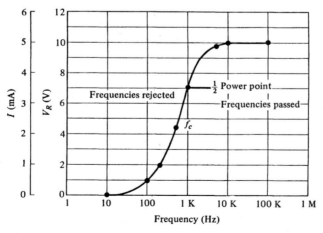

Figure 13-2   Frequency response curve for the circuit of Fig. 13-1.

As in Fig. 13-2, the frequency scale of a response curve is generally logarithmic so that many decades of frequency can be observed.

A measure of the usefulness of a frequency selective circuit is the current or voltage delivered by the circuit at a particular frequency or range of frequencies. The response curve of Fig. 13-2 shows that at low frequencies a relatively small current flows; hence the voltage developed across the resistor is small. At relatively high frequencies, the capacitive reactance is negligible and the current, limited only by the resistance, is a maximum:

$$I_m = \frac{E}{R} \qquad (13\text{-}3)$$

(I_m in Eq. 13-3 is the maximum effective value of current that flows in the circuit and is not to be confused with the I_m previously defined as the maximum value of a sinusoidal wave.) Also, the maximum power P_m is delivered to the resistor only at relatively high frequencies (above 20 kHz in Fig. 13-2) and is

$$P_m = I_m^2 R = \frac{E^2}{R} \qquad (13\text{-}4)$$

If we start at a frequency where the maximum power is delivered to the resistor and vary the frequency away from the range of maximum power, we reach a frequency where the power delivered to the resistor is one-half the maximum power. This frequency is the *half-power* or *cutoff frequency* f_c, where the power is

$$P = I^2 R \qquad (13\text{-}5)$$

Also,

$$P = \tfrac{1}{2}P_m = \tfrac{1}{2}I_m^2 R \qquad (13\text{-}6)$$

Equating Eqs. 13-5 and 13-6, we obtain

$$I^2 R = \tfrac{1}{2}I_m^2 R$$

so that

$$I = \frac{1}{\sqrt{2}} I_m = 0.707 I_m \qquad (13\text{-}7)$$

where I_m is defined as in Eq. 13-3.

The significance of Eq. 13-7 is that the cutoff frequency is the frequency at which the current is 0.707 × the maximum current. A similar relationship exists for the voltage.

The frequencies at which the currents or voltages are greater than 0.707 × I_m or V_m are said to be *selected* or *passed* by the circuit. The frequencies at which the currents or voltages are less than 0.707 × I_m or V_m are said to be *rejected*.

The cutoff frequency in Fig. 13-2 occurs at 1 kHz so that frequencies above 1 kHz are passed and those below it are rejected. Because of this, the *RC* configuration of Fig. 13-1 acts as a *high-pass filter*.

By referring to Eq. 13-2 or Table 13-1, we find that the capacitive reactance equals the resistance for $I = 0.707 I_m$. With this in mind, we can design the *RC* circuit for a particular f_c.

EXAMPLE 13-1

The capacitance of Fig. 13-1 is changed so that an $f_c = 10$ kHz results. Calculate the needed C.

Solution. At f_c,

$$X_C = R = 2 \text{ k}\Omega$$

Then

$$\frac{1}{2\pi f_c C} = 2 \times 10^3$$

or $$C = \frac{1}{2\pi(10 \times 10^3)(2 \times 10^3)} = 0.0795 \times 10^{-7} = 0.00\ 795\ \mu\text{F}$$

If one desires a frequency response characteristic for which the high frequencies are rejected and the low frequencies are passed, that is, a *low-pass filter*, the *RL* configuration of Fig. 13-3(a) can be used. The frequency response curve of Fig. 13-3(b) follows from an analysis of this *RL* circuit.

We can define a cutoff frequency f_c for the low-pass filter as well as a high-pass filter. Above the cutoff frequency, the signals are considered rejected. The cutoff frequency corresponds to that frequency at which the power delivered to the resistor is one-half the maximum power. With the circuit of Fig. 13-3, this occurs when the inductive reactance equals the resistance.

A number of variations of high and low-pass filters are used in practical frequency selection circuits. One variation is obtained by interchanging the positions of the resistor and capacitor in the circuit of Fig. 13-1. This results in a low-pass filter. In fact, this latter low-pass filter would be preferable to the *RL* low-pass filter because the cost and size of a capacitor are usually less than for a comparable coil in the *RL* circuit. You will undoubtedly study the simple *RC* and *RL* filters to a greater extent in your electronics courses.

Drill Problem 13-1 ■

An *RC* circuit such as that in Fig. 13-1 consists of a 10 kΩ resistor and a 0.1 μF capacitor. What is the cutoff frequency for the circuit?

Answer. 159 Hz.   ■ ■

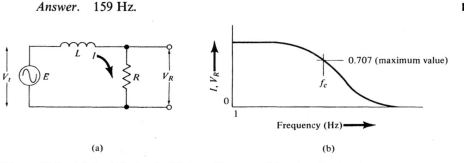

               (a)   (b)

Figure 13-3   **(a)** An *RL* circuit; **(b)** *I* or V_R versus f for the circuit.

Drill Problem 13-2 ■

An *RL* circuit such as that in Fig. 13-3 is used as a frequency selection network with a cutoff frequency of 360 Hz. If the resistance value is 300 Ω, what is the inductance value?

Answer.   0.133 H.   ■ ■

13.2 THE DECIBEL

A unit often used to indicate the loss of power in a filter network or a communication system is the *bel*. Named after Alexander Graham Bell, the bel is the *common logarithm of the ratio of two powers*. The *decibel*, abbreviated dB, equals one-tenth of a bel and is given by

$$dB = 10 \log_{10} \frac{P_o}{P_i} \qquad (13\text{-}8)$$

where P_i is the input power and P_o is the output power. With $P_o = P_i$, the power ratio is one and the decibel level as obtained from the preceding equation is 0 dB. However, with $P_o < P_i$, the power ratio is less than 1 and the decibel value is negative. Thus, we have a loss in the network with respect to the 0 dB case. We refer to this reduction in power level of a signal as *attenuation*.

At a half-power frequency, the power delivered to the resistor is one-half the maximum available power. Therefore, if the maximum power is taken as 0 dB, the decibel loss at a half-power frequency is

$$dB = 10 \log_{10} \frac{\frac{1}{2}P_m}{P_m} = 10 \log_{10} \tfrac{1}{2}$$
$$= (10)(-0.3) = -3 \text{ dB}$$

Accordingly, the half-power frequency is sometimes defined as the *3 dB frequency* or the frequency at which the power is 3 dB down (at a loss) from the maximum delivered power.

Consider the filter or communications network depicted by the four-terminal network of Fig. 13-4. The input voltage and current are identified as V_i and I_i, respectively, and the output voltage and current are identified as V_o and I_o, respectively. The input resistance offered at the input is denoted as R_i. Then

$$P_i = \frac{V_i^2}{R_i} \quad \text{and} \quad P_o = \frac{V_o^2}{R_o}$$

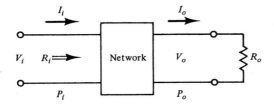

Figure 13-4   Voltage and current relationships for a general four-terminal network.

If both of these expressions are substituted into Eq. 13-8 and if $R_i = R_o$, we obtain

$$dB = 10 \log_{10}\left(\frac{V_o}{V_i}\right)^2$$

or
$$dB = 20 \log_{10}\frac{V_o}{V_i} \qquad (13\text{-}9)$$

Instead, if the relationships

$$P_i = I_i^2 R_i \quad \text{and} \quad P_o = I_o^2 R_o$$

are substituted into Eq. 13-8 and if $R_i = R_o$, we obtain the equation

$$dB = 10 \log_{10}\left(\frac{I_o}{I_i}\right)^2$$

or
$$dB = 20 \log_{10}\frac{I_o}{I_i} \qquad (13\text{-}10)$$

We noted in Chapter 4 that all physical systems have less output power than input power. This is also true of electronic amplifiers, in which some desired signal is amplified at the expense of the dc power delivered by a power supply. However, we can refer to *only* the desired signal and define an increase of signal power, voltage, or current level as *gain*. Then, when we use the decibel equations, we obtain a positive decibel level indicating the amount of signal gain.

EXAMPLE 13-2

Refer to the voltage data of Table 13-1 for the *RC* high-pass filter and determine the decibel loss at 500 Hz.

Solution. Here, $V_i = 10$ V. At 500 Hz, $V_o = V_R = 4.46$ V. Then

$$dB = 20 \log_{10}\frac{V_o}{V_i} = 20 \log_{10}\frac{4.46}{10}$$
$$= (20)(-0.3507) = -7.01 \qquad \text{(loss)}$$

EXAMPLE 13-3

The input power to a communications cable is 1 W and the output is $\frac{1}{10}$ W. Compute the cable loss in decibels.

Solution

$$dB = 10 \log_{10}\frac{P_o}{P_i} = 10 \log_{10}\frac{0.1}{1}$$
$$= 10 \log_{10} 0.1 = -10 \qquad \text{(loss)}$$

Drill Problem 13-3 ■

At a particular frequency the input voltage to a filter is 12 V and the output is 5 V. Express the voltage ratio in decibels.

Answer. -7.6 dB.   ■ ■

Drill Problem 13-4 ■

A power amplifier increases a signal level by 9 dB. If the input signal level is 4 W, what is the output level?

Answer.   31.8 W.   ■ ■

Drill Problem 13-5 ■

The resistance and capacitance of an *RC* high-pass filter are 300 Ω and 200 pF. What is the voltage loss in decibels at 1 MHz?

Answer.   −9.04 dB.   ■ ■

13.3 SERIES RESONANCE

A series circuit containing *R*, *L*, and *C* is in *resonance* when the current in the circuit is in phase with the total voltage across the circuit. Depending on the particular values of *R*, *L*, and *C*, resonance occurs at one distinct frequency. Because of its distinct frequency characteristics, the series resonant circuit is one of the most important frequency selective circuits.

Consider a series *RLC* circuit connected to a variable-frequency source, as in Fig. 13-5(a). The impedance of the circuit is

$$\mathbf{Z} = R + j(X_L - X_C)$$

If the circuit voltage and current are in phase, that is, the circuit is in resonance, the reactive (j) term of the impedance expression equals zero. A graph of the reactances X_L, X_C, and $X_L - X_C$ versus frequency appears in Fig. 13-5(b).

Since X_L is directly proportional to the frequency, its graph appears in Fig. 13-5(b) as a straight line passing through the origin. On the other hand, X_C is inversely related to frequency, so its graph is a hyperbola. It follows that near $f = 0$, the total reactance is capacitive and the current leads the voltage. As the frequency is increased, the capacitive nature diminishes until at $f = f_r$ the total

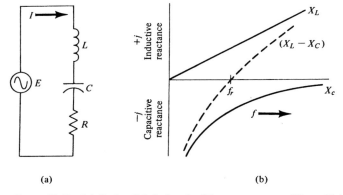

(a)   (b)

Figure 13-5 **(a)** Series *RLC* circuit; **(b)** net reactance ($X_L - X_C$) versus f.

reactance is zero. At frequencies greater than f_r, the total reactance is inductive and the current lags the voltage.

Since at resonance $X_L = X_C$, the resonant frequency f_r is determined using the values of L and C:

$$2\pi f_r L = \frac{1}{2\pi f_r C}$$

$$f_r^2 = \frac{1}{(2\pi)^2 LC}$$

or
$$f_r = \frac{1}{2\pi\sqrt{LC}} \qquad\qquad (13\text{-}11)$$

Although frequency is discussed as the variable, it is apparent from Eq. 13-11 that any of the three quantities f, L, or C can be varied to produce resonance.

Since the reactive term is zero at resonance, the impedance is a minimum and equal to the resistance R. The current given by

$$I = \frac{1}{\sqrt{R^2 + (X_L - X_C)^2}}$$

is a maximum. The following examples provide an indication of the response of the series RLC circuit for a variation of frequency.

EXAMPLE 13-4

In Fig. 13-5(a), $E = 10$ V, $R = 2$ kΩ, $C = 0.0795$ μF, and $L = 0.318$ H. Find the frequency at which the circuit resonates.

Solution

$$f_r = \frac{1}{2\pi\sqrt{LC}} = \frac{1}{2\pi\sqrt{(0.318)(7.95 \times 10^{-8})}}$$

$$= \frac{1}{2\pi\sqrt{2.53 \times 10^{-8}}} = 10^3 = 1\,\text{kHz}$$

EXAMPLE 13-5

Using the values of E, R, L, and C given in Example 13-4, calculate the magnitudes of impedance and current at various frequencies. Plot the impedance and current response curves.

Solution. At selected frequencies, the reactances for the capacitor and inductor are calculated. The impedance magnitude is found using the equation

$$\mathbf{Z} = \sqrt{R^2 + (X_L - X_C)^2}$$

Then, the current is obtained at each frequency and is listed in a resonance data table, such as Table 13-2. When the impedance and current magnitudes in Table 13-2 are plotted on semilogarithmic scales, we obtain the graphs of Fig. 13-6.

TABLE 13-2   RESONANCE DATA FOR EXAMPLE 13-5

f (Hz)	R (kΩ)	X_L (kΩ)	X_C (kΩ)	$X_L - X_C$ (kΩ)	Z (kΩ)	I (mA)
10	2	0.02	200.0	-200.0	200.0	0.05
100	2	0.2	20.0	-19.8	19.9	0.50
200	2	0.4	10.0	-9.6	9.8	1.02
500	2	1.0	4.0	-3.0	3.6	2.78
800	2	1.6	2.5	-0.9	2.2	4.55
1 k	2	2.0	2.0	0.0	2.0	5.0
1.25 k	2	2.5	1.6	$+0.9$	2.2	4.55
2 k	2	4.0	1.0	$+3.0$	3.6	2.78
5 k	2	10.0	0.4	$+9.6$	9.8	1.02
10 k	2	20.0	0.2	$+19.8$	19.9	0.50
100 k	2	100.0	0.02	$+200.0$	200.0	0.05

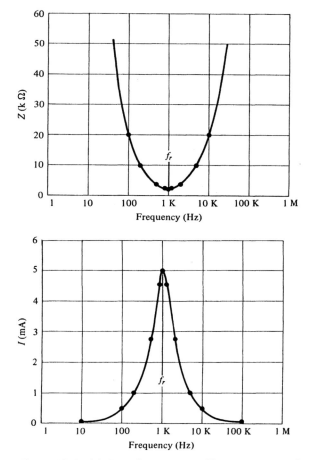

Figure 13-6 **(a)** Impedance curve; **(b)** current curve for a series resonant circuit (data from Example 13-5).

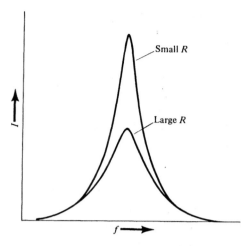

Figure 13-7 Effect of resistance on the resonant curve.

Notice from the curves of Example 13-5 that the impedance is a minimum and the current is a maximum at resonance. Therefore, the circuit is said to be *tuned* to the frequency at which resonance occurs.

Because the current at resonance is limited only by the resistance, a change in resistance is accompanied by a change in the peak value for the current response curve. The effect of a resistance change on the current response curve is shown in Fig. 13-7, where it is indicated that a smaller circuit resistance results in a sharper response.

There are an infinite number of *LC* combinations that result in the same product, and hence, from Eq. 13-11, the same resonant frequency. By changing *L* and *C* while maintaining a constant resonant frequency, we change the *L/C* ratio. An increase in the *L/C* ratio also sharpens the current response curve, as indicated in Fig. 13-8. Notice that even though the sides or "skirts" of the resonance curve

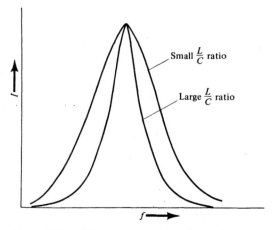

Figure 13-8 Effect of the *L/C* ratio on the resonant curve.

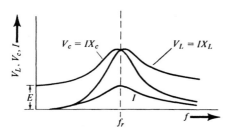

Figure 13-9 I, V_C, and V_L versus f.

are changed, the maximum current remains unchanged since the circuit resistance has not been changed.

It is interesting to note how the voltages across the inductance and capacitance vary with frequency. Obviously, at $f = 0$, $X_L = 0$, $X_C = \infty$, and all the voltage is across the capacitance. At $f = \infty$, the converse is true. Near resonance, the inductive and capacitive voltages are approximately equal and peak, as shown in Fig. 13-9. The capacitive voltage peaks slightly before resonance, because X_C decreases at a rate greater than the rate of current increase. On the other hand, the inductive voltage peaks slightly after resonance, because X_L increases at a rate greater than the rate of current decrease.

As indicated by Fig. 13-9 and the example that follows, the capacitive and inductive voltages near resonance may be larger in magnitude than the source voltage. Yet it should be remembered that the reactive voltages are subtracted from each other, since they are out of phase with each other by 180°.

EXAMPLE 13-6

A series RLC circuit resonates at 600 Hz. If $E = 25$ V, $R = 50\ \Omega$, and $X_L = X_C = 400\ \Omega$ at resonance, what are the voltage magnitudes V_R, V_C, and V_L?

Solution. At f_r,

$$I = \frac{E}{R} = \frac{25}{50} = 0.5 \text{ A}$$
$$V_R = E = IR = 25 \text{ V}$$
$$V_C = V_L = IX_C = (0.5)(400) = 200 \text{ V}$$

As was pointed out earlier, if the frequency is not varied, resonance may be obtained by varying either the capacitance or inductance. Current and voltage curves for these cases are shown in Fig. 13-10.

Drill Problem 13-6 ■

A coil with an inductance of 200 μH and a resistance of 5 Ω is connected in series with a capacitor to form a series circuit resonant at 500 kHz. What is the capacitor value?

Answer. 507 pF.　　　　　　　　　　　　　　　　　　　　　　　■ ■

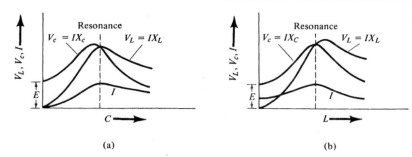

Figure 13-10 I, V_C, and V_L versus (a) capacitance and (b) inductance.

Drill Problem 13-7 ■

The series resonant circuit of Drill Problem 13-6 has an applied voltage of 2 mV at 500 kHz. **(a)** What current flows at resonance? **(b)** What is the voltage across the capacitor?

Answer.   (a) 0.4 mA. (b) 251 mV.   ■ ■

Drill Problem 13-8 ■

Determine the resonant frequency of a series circuit with the following elements: $R = 12\ \Omega$, $L = 0.1$ H, $C = 0.47\ \mu$F.

Answer.   734 Hz.   ■ ■

13.4 Q, SELECTIVITY, AND BANDWIDTH

At resonance, the reactive energy is alternately transferred, that is, it oscillates between the capacitance and inductance. The *figure of merit, energy factor,* or *quality Q* of a circuit is defined by

$$Q = 2\pi\ \frac{\text{peak energy stored}}{\text{energy dissipated per cycle}} \tag{13-12}$$

The quality Q is dimensionless and, as used here, is a measure of the ability of a circuit to produce resonance. Although the symbol Q is also used for reactive power, the two are not equal. One must rely on the units, or lack of units, to distinguish between the two uses of the symbol Q.

In the *RLC* circuit, the peak energy stored at resonance equals $\frac{1}{2}LI_m^2$, whereas the energy dissipated per cycle equals $\frac{1}{2}(I_m^2/R)(1/f)$. Then

$$Q = \frac{2\pi(\frac{1}{2}LI_m^2)}{\frac{1}{2}(RI_m^2)(1/f)} = \frac{2\pi f L}{R} = \frac{\omega L}{R} \tag{13-13}$$

If the ratio of reactance to resistance of a practical coil is calculated by Eq. 13-13, the Q is called the *coil Q*. On the other hand, if the coil is placed in a circuit having additional resistance, the ratio of reactance to total circuit resistance is called the *circuit Q*. Frequently the coil resistance is the only resistance in the resonant circuit; then the circuit Q equals the coil Q.

The Q of a resonant circuit can also be expressed in terms of L and C. At resonance,

$$Q = \frac{\omega_r L}{R} = \frac{2\pi f_r L}{R}$$

Substituting f_r from Eq. 13-10 into the preceding expression, we obtain

$$Q = \frac{1}{R} \sqrt{\frac{L}{C}} \qquad (13\text{-}14)$$

Notice that Eq. 13-14 relates the Q to the L/C ratio and the circuit resistance, both of which determine the sharpness of the resonance curve. It follows that the higher the Q of the circuit, the sharper the response curve.

At resonance, the circuit current is limited only by the resistance R.

$$I = \frac{E}{R}$$

so that the inductive (or capacitive) voltage at resonance equals

$$V_L = I X_L = \frac{E}{R} X_L = QE \qquad (13\text{-}15)$$

With $Q > 1$, it is easy to see why the inductive or capacitive voltage exceeds the applied voltage.

EXAMPLE 13-7

What is the Q of the circuit of Example 13-6?

Solution

$$Q = \frac{X_L}{R} = \frac{400}{50} = 8$$

Alternatively,

$$Q = \frac{V_L}{E} = \frac{200}{25} = 8$$

The *selectivity* of an *RLC* circuit is the ability of the circuit to discriminate among frequencies. Again, the *half-power points* are used to determine which frequencies are selected and which are rejected. Now, however, there are two half-power frequencies, as indicated in Fig. 13-11.

The frequency corresponding to the half-power point below f_r is the *lower half-power frequency f_1*, while the frequency corresponding to the half-power point above

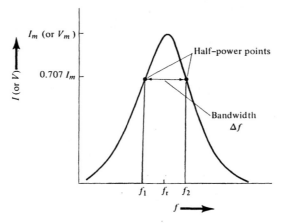

Figure 13-11   Half-power frequencies and bandwidth.

f_r is the *upper half-power frequency* f_2. The spread of frequencies passed by the resonant circuit is the *bandwidth* Δf.

$$\Delta f = f_2 - f_1 \tag{13-16}$$

At the half-power points, the total reactance equals the resistance of the circuit. Using this information, we can derive an expression relating the bandwidth to the resonant frequency. The resulting equation is

$$\Delta f = \frac{f_r}{Q} \tag{13-17}$$

Equation 13-17 emphasizes that with a large Q the bandwidth is small and the selectivity sharp. Coils employed in communication circuits generally have a large Q (sometimes $Q > 100$), while coils employed at low frequencies generally have a low Q (sometimes $Q \approx 1$).

Although the frequency response curve for a resonant circuit is not symmetrical, with a high Q circuit ($Q \geq 10$) the half-power frequencies, for all practical purposes, are separated from f_r by the same frequency difference.

EXAMPLE 13-8

A series resonant circuit has a resonant frequency of 10 kHz. If $Q = 100$, what are the half-power frequencies?

Solution

$$\Delta f = \frac{f_r}{Q} = \frac{10 \times 10^3}{100} = 100 \text{ Hz}$$

$$f_1 = f_r - \frac{\Delta f}{2} = 10\,000 - 50 = 9950 \text{ Hz}$$

$$f_2 = f_r + \frac{\Delta f}{2} = 10\,000 + 50 = 10\,050 \text{ Hz}$$

Drill Problem 13-9 ■

Determine the Q and the bandwidth for a series resonant circuit with the elements $R = 12\ \Omega$, $L = 0.1$ H, and $C = 0.47\ \mu$F.

Answer.   38.4; 19.1 Hz.   ■ ■

Drill Problem 13-10 ■

A circuit resonates at 3.58 MHz with a Q of 150. What are the half-power frequencies?

Answer.   3.568 and 3.592 MHz.   ■ ■

Drill Problem 13-11 ■

A coil with 50 μH inductance and 2 Ω resistance is connected in series with a 100 pF capacitor and a 50 Ω resistor. Determine **(a)** the frequency at which the circuit resonates, **(b)** the coil Q, **(c)** the circuit Q, and **(d)** the cutoff frequencies.

Answer.   (a) 2.25 MHz. (b) 353. (c) 13.6. (d) 2.167 MHz; 2.333 MHz.   ■ ■

13.5 PARALLEL RESONANCE

Parallel resonance is the condition that exists in an ac circuit containing inductive and capacitive branches when the total current is in phase with the voltage across the circuit. This, of course, implies that the total impedance, or conversely, the total admittance, is real.

Consider the general parallel circuit of Fig. 13-12, which depicts resistances R_L and R_C in the inductive and capacitive branches, respectively. The total admittance for the circuit is equal to

$$\mathbf{Y} = \frac{1}{R_L + jX_L} + \frac{1}{R_C - jX_C} \tag{13-18}$$

Rationalizing the denominators, we obtain

$$\begin{aligned}
\mathbf{Y} &= \frac{R_L - jX_L}{R_L^2 + X_L^2} + \frac{R_C + jX_C}{R_C^2 + X_C^2} \\
&= \frac{R_L}{R_L^2 + X_L^2} + \frac{R_C}{R_C^2 + X_C^2} + j\left(\frac{X_C}{R_C^2 + X_C^2} - \frac{X_L}{R_L^2 + X_L^2}\right)
\end{aligned} \tag{13-19}$$

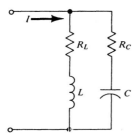

Figure 13-12   General parallel resonant circuit.

For resonance, the j term of Eq. 13-19 is zero. Then

$$\frac{X_C}{R_C^2 + X_C^2} = \frac{X_L}{R_L^2 + X_L^2} \tag{13-20}$$

and

$$\frac{1}{\omega C}(R_L^2 + \omega^2 L^2) = \omega L\left(R_C^2 + \frac{1}{\omega^2 C^2}\right) \tag{13-21}$$

Solving Eq. 13-21 for ω and subsequently for the frequency f_r at which resonance occurs, we find that

$$f_r = \frac{1}{2\pi \sqrt{LC}} \sqrt{\frac{R_L^2 - L/C}{R_C^2 - L/C}} \tag{13-22}$$

It is apparent from Eq. 13-22 that any of the five quantities R_L, R_C, L, C, f may be varied to produce resonance. However, in contrast to the series circuit, where there is always some real resonant frequency for any value of R, L, and C, in a parallel circuit there are some values of R_L, R_C, L, and C for which resonance is impossible. In particular, if the square-root quantity containing the resistances is negative, resonance does not occur. In another extreme, if $R_L^2 = R_C^2 = L/C$, the circuit is resonant at all frequencies and acts as a pure resistance at all frequencies.

If the frequency is the variable quantity in the parallel circuit, we can plot current and impedance vs. frequency. These response curves are typified in Fig. 13-13.

Depending on the values of resistance in the parallel branches, minimum current and maximum impedance occur either close to or exactly at the frequency where the power factor is unity (see Fig. 13-13). Because these characteristics are inverted compared to those for series resonance, parallel resonance is sometimes referred to as *antiresonance*.

If the R_C term in Eq. 13-22 is negligible, the resonant frequency formula reduces to

$$f_r = \frac{1}{2\pi\sqrt{LC}} \sqrt{\frac{R_L^2 - L/C}{-L/C}}$$

$$= \frac{1}{2\pi\sqrt{LC}} \sqrt{1 - \frac{CR_L^2}{L}} \tag{13-23}$$

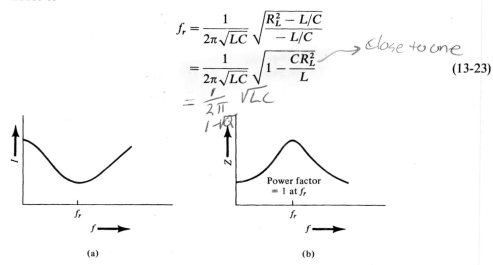

Figure 13-13   **(a)** Current and **(b)** impedance response for the parallel resonant circuit.

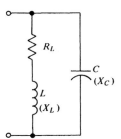

Figure 13-14 Practical parallel resonant circuit.

The corresponding circuit, shown in Fig. 13-14, may be identified as a practical coil placed in parallel with a capacitor. Hence, the name *practical parallel resonant circuit* is used.

EXAMPLE 13-9

A coil having a resistance of 6 Ω and an inductance of 10 mH is placed in parallel with a capacitance of 10 μF. Show that resonance can occur and determine **(a)** the f_r and **(b)** the coil Q at f_r.

Solution. For resonance to occur, the quantity under the square-root sign containing R_L must not be negative.

$$\sqrt{1 - \frac{CR_L^2}{L}} = \sqrt{1 - \frac{(10 \times 10^{-6})(36)}{10 \times 10^{-3}}} = \sqrt{0.964} = 0.982$$

(a)
$$f_r = \frac{1}{2\pi\sqrt{LC}}(0.982) = \frac{1}{2\pi\sqrt{(10^{-2})(10 \times 10^{-6})}}(0.982)$$

$$= \frac{1}{2\pi(3.16 \times 10^{-4})}(0.982) = 495 \text{ Hz}$$

(b) At f_r,

$$X_L = 2\pi f L = 2\pi(495)(10^{-2})$$
$$= 31.1 \ \Omega$$

$$Q = \frac{X_L}{R} = \frac{31.1}{6} = 5.2$$

Drill Problem 13-12 ■

Refer to the general parallel circuit of Fig. 13-12 and consider that $L = 0.1$ H, $C = 20$ μF, $R_L = 120$ Ω, and $R_C = 100$ Ω. Determine the resonant frequency.

Answer. 154 Hz   ■ ■

Drill Problem 13-13 ■

A coil with an inductance of 200 mH and a resistance of 10 Ω is placed in parallel with a capacitance of 0.1 μF. Determine **(a)** the resonant frequency and **(b)** the impedance of the circuit at 1, 2, and 10 kHz. Comment on the inductive or capacitive nature of the impedance.

Answer.   (a) 1125 Hz. (b) 5973 $\angle\ 90°\ \Omega$, 1165 $\angle\ -90°\ \Omega$, 161 $\angle -90°\ \Omega$. Below the resonant frequency, the impedance is inductive; above the resonant frequency, it is capacitive.   ■ ■

13.6 THE EQUIVALENT PARALLEL RESONANT CIRCUIT

The practical parallel resonant circuit of Fig. 13-14 is equivalent to the circuit of Fig. 13-15. The equivalent resistance R'_L and the equivalent inductive reactance X'_L are obtained by finding the admittance of the practical coil and separating this into an equivalent conductance and susceptance. Thus,

$$\mathbf{Y}_L = G - jB = \frac{1}{R_L + jX_L} = \frac{R_L}{R_L^2 + X_L^2} - j\frac{X_L}{R_L^2 + X_L^2}$$

so that

$$R'_L = \frac{1}{G} = \frac{R_L^2 + X_L^2}{R_L} \tag{13-24}$$

and

$$X'_L = \frac{1}{B} = \frac{R_L^2 + X_L^2}{X_L} \tag{13-25}$$

It is interesting to note that for a large coil Q, R'_L, X'_L can be approximated by

$$R'_L = R_L + \frac{X_L^2}{R_L} = R_L + QX_L \approx QX_L \tag{13-26}$$

and

$$X'_L = \frac{R_L^2}{X_L} + X_L = \frac{R_L}{Q} + X_L \approx X_L \tag{13-27}$$

The advantage of using an equivalent circuit as shown in Fig. 13-15 is that a load resistor connected across the parallel circuit can be combined with R'_L. This situation arises in some electronic circuits.

Using the definition for Q, we obtain the ratio of energy stored to the energy dissipated per cycle:

$$Q = \frac{V^2/X'_L}{V^2/R'_L} = \frac{R'_L}{X'_L} \tag{13-28}$$

Even though Eq. 13-28 is an inverted relationship with regard to the previous definition of Q, one must remember that R'_L, is now a large resistance. In fact, Q *of the equivalent parallel circuit equals the coil* Q, as shown by a substitution of Eqs. 13-24

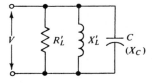

Figure 13-15   Equivalent parallel resonant circuit.

and 13-25 into Eq. 13-28. Then

$$Q = \frac{R'_L}{X'_L} = \frac{X_L}{R} \qquad (13\text{-}29)$$

EXAMPLE 13-10

A coil having a resistance of 1 Ω and an inductance of 100 mH is placed in parallel with a capacitance of 1 μF. Determine **(a)** the f_r, **(b)** Q at f_r, and **(c)** values of the equivalent circuit of Fig. 13-15.

Solution

$$f_r = \frac{1}{2\pi\sqrt{LC}} \sqrt{1 - \frac{CR_L^2}{L}}$$

$$= \frac{1}{2\pi\sqrt{(10^{-1})(10^{-6})}} \sqrt{1 - \frac{10^{-6}}{10^{-3}}}$$

$$\approx \frac{1}{2\pi\sqrt{10 \times 10^{-8}}} = 503 \text{ Hz}$$

(b) At f_r,

$$X_L = 2\pi f L = 2\pi(503)(10^{-1})$$
$$= 316 \text{ Ω}$$

$$Q = \frac{X_L}{R} = \frac{316}{1} = 316$$

(c) Since Q is high,

$$R'_L \approx QX_L = (316)(316) = 100 \text{ kΩ}$$
$$X'_L \approx X_L = 316 \text{ Ω}$$

Therefore,

$$L' = L$$

Hence the two equivalent circuits of Fig. 13-16.

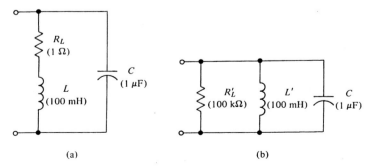

(a)   (b)

Figure 13-16 Example 13-10: **(a)** original circuit; **(b)** its equivalent.

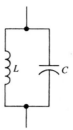

Figure 13-17 Ideal parallel resonant circuit.

Notice in Example 13-10 that $R'_L \gg R_L$. In fact, for a smaller coil resistance, the R'_L becomes even larger and, in the limiting case of $R_L = 0$, $R'_L = \infty$. One then has the *ideal parallel circuit* of Fig. 13-17. It is apparent that for the ideal parallel circuit, the resonant frequency is obtained from the expression

$$f_r = \frac{1}{2\pi\sqrt{LC}}$$

(13-30)

When $Q \geq 10$ for a parallel circuit, Eq. 13-30 may be used to determine the resonant frequency with little error. Equations 3-16 and 3-17 are also applicable to parallel resonant circuits. Thus, if one knows the resonant frequency and Q, one can easily determine the bandwidth and half-power frequencies.

Drill Problem 13-14 ■

A 100 pF capacitor is in parallel with a coil having an inductance of 10 μH and a resistance of 3 Ω. Determine **(a)** the resonant frequency, **(b)** Q, and **(c)** the bandwidth.

Answer. (a) 5.03 MHz. (b) 105. (c) 47.9 kHz.   ■ ■

Drill Problem 13-15 ■

Determine the equivalent parallel resistance R'_L for the circuit described in Drill Problem 13-14.

Answer. 33.2 kΩ.   ■ ■

13.7 MULTIRESONANT CIRCUITS

In summary, the series resonant circuit allows maximum current flow at its resonant frequency, whereas the parallel resonant circuit allows only minimum current flow at its resonant frequency. By a combination of the two, it is possible to accentuate the passage or rejection of a particular frequency. (Actually, a range or band of frequencies near the desired frequency is passed or rejected. The bandwidth, of course, depends on Q.)

Two circuits having both series and parallel resonant combinations for the passage or rejection of a band of frequencies are shown in Fig. 13-18. Ideal resonant

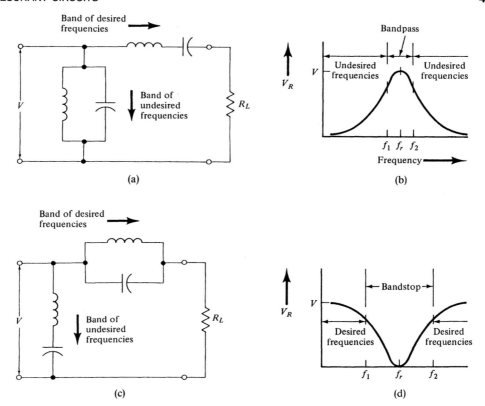

Figure 13-18 **(a)** Bandpass filter; **(b)** bandpass filter response; **(c)** bandstop filter; **(d)** bandstop filter response.

circuits are used in Fig. 13-18 to allow for a simpler qualitative analysis. A voltage V is applied to the filter circuit, and it is assumed that this voltage consists of the superimposed voltages of many frequencies. (As will be shown in Chapter 16, any nonsinusoidal voltage is made of many sinusoidal components of different frequencies.)

Let us consider that both resonant combinations in Fig. 13-18(a) resonate or are *tuned* at the same frequency. The series resonant circuit appears as a short for the resonant frequency but appears as a high impedance to other frequencies. Simultaneously, the parallel resonant circuit appears as an open circuit for the resonant frequency and a low-impedance circuit to other frequencies. The filter acts as a *bandpass filter*; that is, only the range or band of frequencies near the resonant frequency passes from the filter input to the load resistor V_R. Hence, we have the bandpass filter response indicated in Fig. 13-18(b).

We may contrast the bandpass filter to the bandstop filter of Fig. 13-18(c). In the bandstop filter the frequencies near the resonant frequency are deliberately shorted at the input by the series resonant circuit and further attenuated by the parallel resonant circuit. Hence, we have the bandstop response of Fig. 13-18(d).

Sometimes it is desirable to accentuate one frequency while rejecting another. This can be done by combining impedances so that series resonance occurs at one frequency as parallel resonance occurs at another. Circuits with this property are called *multiple resonant* or *multiresonant circuits*.

One type of multiresonant circuit is shown in Fig. 13-19. By selecting L and C_1 properly, we can provide rejection of a specified frequency. At a lower frequency, the parallel combination appears inductive so that C_2 can be chosen for series resonance and accentuation of a lower frequency.

EXAMPLE 13-11

A circuit such as Fig. 13-19 is used to accentuate a 1 kHz signal voltage and reject a 5 kHz signal. If a 10 mH coil is available, determine the capacitors C_1 and C_2 needed.

Solution

At 5 kHz,

$$f_r = 5 \times 10^3 = \frac{1}{2\pi\sqrt{LC_1}}$$

$$LC_1 = \frac{1}{(2\pi \times 5 \times 10^3)^2} = 1.015 \times 10^{-9}$$

Therefore,

$$C_1 = \frac{1.015 \times 10^{-9}}{10^{-2}} = 1.015 \times 10^{-7} = 0.102 \ \mu\text{F}$$

At 1 kHz,

$$X_L = 2\pi f L = 2\pi(10^3)(10^{-2}) = 62.8 \ \Omega$$

$$X_{C1} = \frac{1}{2\pi f C_1} = \frac{1}{2\pi(10^3)(1.02 \times 10^{-7})} = 1560 \ \Omega$$

The impedance of the parallel circuit is

$$\mathbf{Z}_P = \frac{(jX_L)(-jX_C)}{j(X_L - X_C)} = \frac{(j62.8)(-j1560)}{j(62.8 - 1560)}$$

$$= j62.8 \ \frac{-j1560}{-j1497} = j65.5 \ \Omega$$

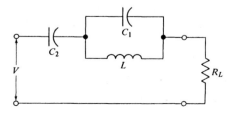

Figure 13-19 A multiresonant circuit.

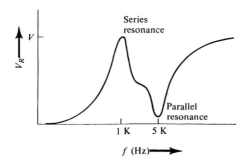

Figure 13-20 Voltage response across R_L for Example 13-11.

For series resonance with Z_p at 1 kHz, $X_{C_2} = 65.5 \ \Omega$ or

$$C_2 = \frac{1}{2\pi f(65.5)} = \frac{1}{2\pi(10^3)(65.5)} = 2.42 \ \mu F$$

The voltage response measured across R_L for this example is shown in Fig. 13-20.

Many different varieties of multiresonant filters are used in practice and are listed in electrical engineering handbooks.

Drill Problem 13-16 ■

Describe the behavior of the filter circuit of Fig. 13-21.

Answer. The parallel resonant circuit stops the band of frequencies 7.296–7.387 MHz; hence, the circuit is a bandstop filter ■ ■

Drill Problem 13-17 ■

The tuned circuit of a radio-frequency amplifier consists of a parallel resonant wavetrap and a series-tuning element $\mathbf{Z}$, as shown in Fig. 13-22. The wavetrap blocks a 1.55 MHz signal. However, the overall circuit including $\mathbf{Z}$ is series tuned to 1.45 MHz. Determine **(a)** the value of C_1 needed, **(b)** the value of capacitor or inductor needed for $\mathbf{Z}$.

Answer. (a) 0.0105 μF. (b) Capacitor; $C = 0.00155 \ \mu F$. ■ ■

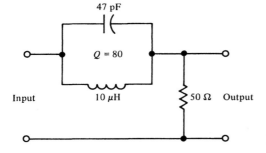

Figure 13-21 Drill Problem 13-16.

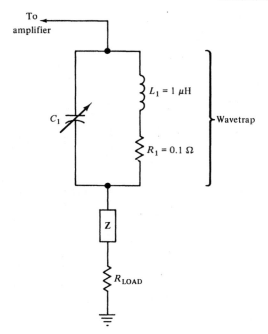

Figure 13-22   Drill Problem 13-17.

13.8 THE PIEZOELECTRIC CRYSTAL

We first considered the *piezoelectric effect* in Chapter 2. At that time we defined it as the ability of certain crystalline substances to transform mechanical strain into electric charge and vice versa. Materials usually associated with piezoelectricity are Rochelle salts, tourmaline, and quartz. Each of these materials has a particular strength and piezoelectric activity; quartz's characteristics are midway between those of the other two.

The natural shape of quartz is a hexagonal prism with pyramidal ends. For electronic and filter circuit use, the prismatic quartz is cut into thin rectangular slabs, the orientation and thickness of which determine the physical properties of the slab. The slab is then mounted between two metal plates and is usually enclosed in a metal case. Hence, it is easy to see why we use the schematic symbol shown in Fig. 13-23

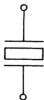

Figure 13-23   Schematic symbol for the piezoelectric crystal.

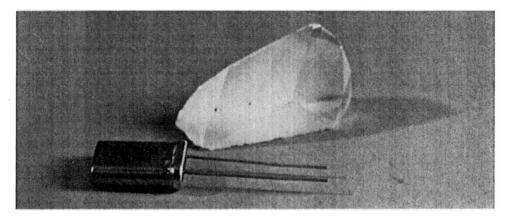

Figure 13-24 A piece of natural quartz crystal and an encased quartz crystal.

to represent the piezoelectric crystal. A piece of natural quartz crystal and an encased quartz crystal are shown in Fig. 13-24.

When a small alternating voltage is applied to a crystal, mechanical vibrations develop. In turn, the vibrations result in a voltage of a particular frequency at the crystal faces. In this manner, the crystal exhibits electromechanical resonance. The resonant crystal frequency depends on the crystal cut and ranges from approximately 1 kHz to about 100 MHz. However, by using electronic circuits that divide or multiply frequencies, it is possible to obtain crystal resonators for almost any frequency.

If we compare the electromechanical properties of the piezoelectric crystal to the electrical characteristics of a tuned circuit, we obtain the equivalent circuit of Fig. 13-25. L_x, C_x, and R_x represent the equivalent circuit elements of the crystal itself, and C_m represents the capacitance of the mounting plates that hold the crystal.

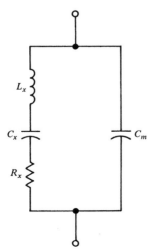

Figure 13-25 Electrical equivalent of the piezoelectric crystal.

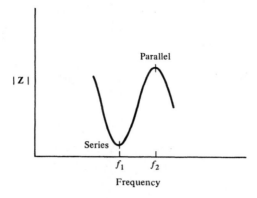

Figure 13-26   Impedance characteristic of a quartz crystal.

Notice that since the crystal equivalent circuit has two reactive branches, it is possible to obtain both a series and a parallel resonant condition. The series resonant frequency is determined by L_x and C_x, whereas the parallel resonant frequency is determined by L_x and the total combination of C_x and C_m. Of course, any external capacitance that shunts the crystal will slightly alter the parallel resonant frequency.

If we plot a response or impedance curve for the crystal circuit, we obtain the graph of Fig. 13-26. Although the graph accents the different values of resonant frequency f_1 and f_2, they are usually only a few hundreds of a percent away from each other. The series resonant mode is usually used for filter designs, whereas, the parallel resonant mode is usually used for designing *oscillators*. Oscillators are amplifiers that produce an alternating voltage or current on their own.

Why are crystals so important as circuit elements when we can obtain resonance with discrete LC elements? The answer is that a discrete LC circuit can achieve a maximum Q of approximately 100. In comparison, a crystal can exhibit a Q of approximately 100 000. Thus, crystals are used in circuits requiring accurate and precise frequency control. Many radio, microprocessor, and measurement circuits use crystals.

QUESTIONS

1. What characterizes a frequency selective circuit?
2. What is a filter?
3. What is a frequency response curve and how is it drawn?
4. Define the cutoff or half-power frequency.
5. With regard to frequency selective circuits, what do the terms *passed* and *rejected* mean?
6. What significance does the 3 dB point have to filter circuits?
7. What is resonance?
8. What are some conditions existing for series resonance? Sketch the response curves.

9. How do the L/C ratio and resistance affect the frequency response curve of a resonant circuit?

10. Why do the capacitive and inductive voltage peaks occur just slightly off the resonant frequency?

11. What is meant by the quality of a resonant circuit?

12. How might the circuit and coil Q differ?

13. How is Q related to the L/C ratio, the resistance, and the shape of the response curve?

14. What is bandwidth? How does it relate to the Q of a circuit?

15. What is meant by the term *high Q*?

16. What is parallel resonance?

17. Under what conditions will a parallel circuit containing an L and a C in separate branches not resonate?

18. Why is parallel resonance sometimes called antiresonance?

19. How does parallel Q differ from series Q?

20. What characterizes bandpass and bandstop filters?

21. What is a multiresonant circuit?

PROBLEMS

1. The resistance of Fig. 13-27 is 1 MΩ. If the capacitance is 0.005 μF, at what frequency will the voltage across the resistance equal 0.707 × the applied voltage?

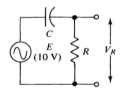

Figure 13-27   Problem 13-1.

2. When the measured voltage in the RC network is the voltage across the capacitance, the filter network functions as a low-pass filter. In Fig. 13-28, $R = 100$ kΩ, $C = 640$ pF, and $E = 10$ V. Plot a frequency response curve of V_C for the audio range of frequencies. At what frequency does cutoff occur?

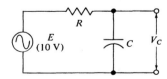

Figure 13-28   Problem 13-2.

3. If, in Fig. 13-27, $R = 500$ kΩ, $C = 0.004$ μF, and $E = 10$ V, determine the cutoff frequency. How might the cutoff frequency be lowered? What additional capacitance is needed for $f_c = 33$ Hz?

4. Frequency response curves are often idealized by straight lines as shown in Fig. 13-29. Consider a filter network with the characteristic response of Fig. 13-29 and an input voltage of 100 mV. Determine the output voltage at **(a)** 10 kHz, **(b)** 200 kHz, and **(c)** 2 MHz.

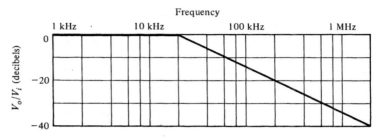

Figure 13-29   Idealized frequency response (Problem 13-4).

5. A power amplifier with an input of 40 μW develops an output of 20 W. What is the power gain in decibels?

6. At a particular frequency the input voltage to a network is 120 mV and the output is 80 mV. Based on the input voltage, what is the output voltage in decibels?

7. A series circuit consists of a 169 pF capacitor, a 0.15 mH coil, and a resistor of 75 Ω. At what frequency does resonance occur?

8. A series circuit with $L = 0.2$ H, $R = 5\ \Omega$, and a variable capacitor of value C resonates at 1 kHz. Determine the value of C.

9. The antenna circuit of a radio receiver consists of a series connection of a 10 mH coil, a variable capacitor, and a 50 Ω resistor. An 880 kHz signal produces a potential difference of 100 μV at resonance. Find **(a)** the capacitance for resonance and **(b)** the current at resonance.

10. A coil with a resistance of 10 Ω and an inductance of 0.1 H is in series with a capacitor and a 100 V, 60 Hz source. The capacitor is adjusted to give resonance. Calculate **(a)** the capacitance needed for resonance and **(b)** the coil and capacitor voltages at resonance.

11. What is Q of the antenna circuit of Problem 13-9?

12. What is Q of the circuit of Problem 13-10?

13. A series circuit consists of a 2.03 pF capacitance, a 5 μH inductance, and a 120 Ω resistance. **(a)** At what frequency does resonance occur? **(b)** What is the circuit bandwidth? **(c)** What are the cutoff frequencies?

14. At 8 MHz, a series circuit consisting of a 35 pF capacitor, a 24 Ω resistor, and a coil with 6 Ω resistance resonates. Determine **(a)** the bandwidth of the circuit and **(b)** the current at resonance if the applied voltage is 10 V.

15. A series circuit consists of an inductance of 0.2 H, a capacitance of 0.127 μF, and a resistance of $R\ \Omega$. **(a)** What is the resonant frequency? **(b)** What is the bandwidth if $R = 100\ \Omega$? **(c)** What is the bandwidth if $R = 2.5\ \Omega$? **(d)** What maximum current flows in each case (b) and (c)? **(e)** Sketch the response curves of (b) and (c) on one graph. ($V = 1\ V.$)

16. A coil and capacitor of a series-tuned circuit resonate with a bandwidth of 8 kHz. If Q of the circuit is 50 and the L/C ratio is 10^5, calculate **(a)** the upper and lower cutoff frequencies and **(b)** the impedance of the coil at resonance.

17. Can resonance occur for the circuit of Fig. 13-14 if $R_L = 10\ \Omega$, $L = 0.1$ mH, and $C = 100$ pF? If so, calculate the frequency and Q at resonance.

18. At what frequency will resonance occur in a parallel circuit having two branches if branch 1 consists of a 4 Ω resistor in series with a 20 μF capacitor and branch 2 consists of a 1 mH coil having a resistance of 6 Ω.

19. A parallel circuit has two branches and an applied voltage of 120 V. Branch 1 is a 159 pF capacitor, and branch 2 is a coil of 50 Ω resistance and 159 μH inductance. (a) What is the resonant frequency? (b) What is the total current at resonance?

20. Given the circuit of Fig. 13-30, compute (a) the resonant frequency, (b) the total current at resonance, and (c) the resistance for which resonance is lost.

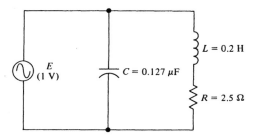

Figure 13-30   Problem 13-20.

21. Determine (a) the frequency of resonance and (b) the total current I at resonance for the circuit of Fig. 13-31.

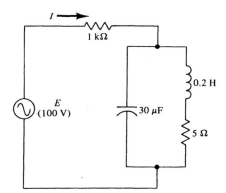

Figure 13-31   Problem 13-21.

22. A parallel resonant circuit, as in Fig. 13-14, resonates at 8 kHz with a Q of 15. Find the impedance of the circuit at resonance if the coil resistance is 10 Ω.

23. It is desired that a coil of 10 mH resonate with a Q of 40, when connected in parallel with a 0.1 μF capacitor. Determine (a) the coil resistance needed and (b) the impedance at resonance.

24. A load resistance of 15 kΩ is connected to the parallel circuit of Problem 13-23. What is the new Q after the load resistance is connected (loaded Q)?

25. Calculate the resonant frequencies for the LC combinations in the filter circuits of Fig. 13-32. Sketch the resistor voltage versus frequency curves and comment on the filter types.

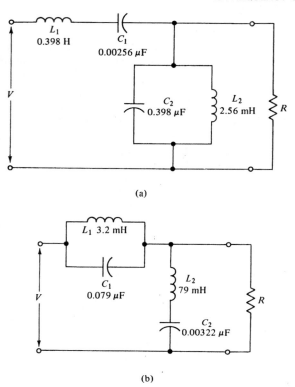

(a)

(b)

Figure 13-32   Problem 13-25.

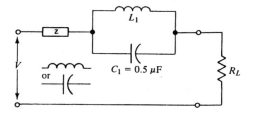

Figure 13-33   Problem 13-26.

26. A circuit such as Fig. 13-33 is to be used to accentuate a 20 kHz signal and reject a 10 kHz signal. **(a)** Sketch a frequency response curve of the load resistance voltage. **(b)** What type of element, capacitor or inductor, is required for **Z**? **(c)** Calculate the L_1 and component value of **Z** needed.

Chapter 14

Alternating Current Circuit Analysis

14.1 INTRODUCTION

The fundamental definitions of a node, a branch, and a complex network (from Chapter 5) depend only on the *topology* or geometry of a network and not on the type of sources (dc or ac) in the network. Thus it is not surprising that the analysis techniques and theorems of Chapter 6 are applicable to sinusoidal ac networks when proper consideration is given to impedance and phasor quantities.

In this chapter, we consider complex ac circuits, that is, circuits that contain two or more ac sources in different branches. To solve for the voltages and currents in such complex circuits, we apply the *loop*, *nodal*, or *superposition* method for analysis, as well as Thevenin's, Norton's, and the maximum power transfer theorems and source and delta–wye conversion techniques.

In the complex ac networks of this chapter, the sources are the same constant frequency. Therefore, X_L and X_C, which normally vary with frequency, are constant. However, we must not forget to associate the operators $+j$ and $-j$ with X_L and X_C, respectively. In order to concentrate on the various circuit analysis techniques, we shall not concern ourselves with changes between the time and frequency domains. We begin our analysis with the circuit quantities specified in the frequency domain and conclude when the desired quantities are determined in the frequency domain.

14.2 SOURCE CONVERSION

In Chapter 5 we studied the effect of internal resistance on the terminal voltage of a practical source. We also studied how the practical voltage source, including the

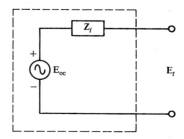

Figure 14-1   Practical ac voltage source.

internal resistance, is represented by a current source equivalent. Similarly, we consider the practical ac voltage source and its current source equivalent.

A practical ac voltage source not only contains a source of alternating emf but a combination of resistance, capacitance, and inductance. From a circuit standpoint, we represent the combination of resistance, inductance, and capacitance within a source as a series impedance called *internal impedance*. The internal impedance, denoted $\mathbf{Z}_i$, is placed in series with the ideal voltage source, denoted $\mathbf{E}_{oc}$, as indicated in Fig. 14-1.

Depending on the internal impedance and any load that may be connected to the terminals of the network of Fig. 14-1, the terminal voltage $\mathbf{E}_t$ may be less than, equal to, or greater than the magnitude of $\mathbf{E}_{oc}$. This contrasts with the terminal voltage of a dc source, which is always equal to or less than the open-circuit voltage.

The practical ac voltage source is equivalent to the current source of Fig. 14-2. Here, a short-circuit current source $\mathbf{I}_{sc}$ is in parallel with the internal impedance $\mathbf{Z}_i$. Alternatively, an internal admittance $\mathbf{Y}_i$ may be specified for the current source equivalent.

As was pointed out earlier, a convenient way to show the phase relationship between voltage and current is the *phasor diagram*, which represents the phasors as stationary—that is, at one instant in time. This type of representation is used to solve for the unknown currents or voltages of a more complex ac circuit. In an ac circuit, the sources are the known quantities; one knows their effective values and phase differences. For example, in Figs. 14-1 and 14-2, the quantities $\mathbf{E}_{oc}$ and $\mathbf{I}_{sc}$ represent the effective values and angular displacement for those quantities. The plus and minus signs of the voltage source and the current direction of the current source represent the instantaneous polarity and direction, respectively, of those phasor quantities.

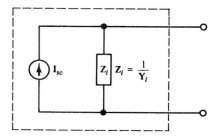

Figure 14-2   Practical ac current source.

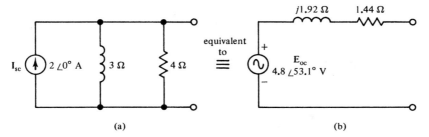

Figure 14-3   Example 14-1: **(a)** A current source; **(b)** the equivalent voltage source.

These marks are vital when writing KVL and KCL equations. For the two sources of Figs. 14-1 and 14-2 to be equivalent, the following relationship must be satisfied:

$$\mathbf{E}_{oc} = \mathbf{I}_{sc}\mathbf{Z}_i \qquad (14\text{-}1)$$

EXAMPLE 14-1

What is the equivalent voltage source representation for the current source of Fig. 14-3(a)?

Solution

$$\mathbf{Z}_i = 4\,\Omega\,\|\,j3\,\Omega = \frac{(4)(j3)}{(4+j3)} = \frac{12\,\angle\,90°}{5\,\angle\,36.9°}$$

$$= 2.4\,\angle\,53.1° = 1.44 + j1.92\,\Omega$$

$$\mathbf{E}_{oc} = \mathbf{I}_{sc}\mathbf{Z}_i = (2\,\angle\,0°)(2.4\,\angle\,53.1°) = 4.8\,\angle\,53.1°\,\mathrm{A}$$

Hence, the equivalent circuit of Fig. 14-3(b).

Drill Problem 14-1 ■

Convert the sources of Fig. 14-4 to their respective equivalent sources.

Answer. (a) $\mathbf{I}_{sc} = 1\,\angle\,6.9°$ A; $\mathbf{Z}_i = 10\,\angle\,53.1°\,\Omega$. (b) $\mathbf{E}_{oc} = 3.54\,\angle\,-45°$ V; $\mathbf{Z}_i = 707\,\angle\,-45°\,\Omega$.   ■ ■

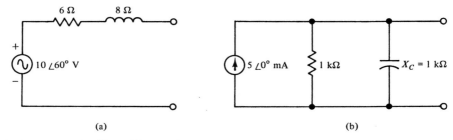

(a)   (b)

Figure 14-4   Drill Problem 14-1.

14.3 LOOP ANALYSIS

In Chapter 6 a *loop* or *circulating current* was defined as a current flowing throughout a closed loop. It was demonstrated that KVL equations can be written in which the loop currents become unknown quantities. Then, in the general analysis of a given complex circuit, if loops are successively chosen so that each new loop contains at least one new element not contained in a previously chosen loop, a sufficient set of independent KVL equations with unknown loop currents can be written.

A more specific approach involves the use of *mesh currents*, the currents chosen to circulate in the "topological windows." If there are only voltage sources in the complex network and the mesh currents are chosen with the same circulating sense, the system determinant is composed of positive terms on the principal diagonal and negative terms symmetrical about the principal diagonal. Then, for an ac circuit containing impedances

$$\Delta = \begin{vmatrix} Z_{11} & -Z_{12} & -Z_{13} \\ -Z_{21} & Z_{22} & -Z_{23} \\ -Z_{31} & -Z_{32} & Z_{33} \end{vmatrix}$$

where Z_{ii} = total *self-impedance* through which I_i flows,

Z_{ij} = the *mutual impedance*, that is, the impedance shared by the meshes i and j.

Of course, we recognize that the impedance elements of the system determinant represent the coefficients of the phasor current variables in the set of simultaneous equations.

The steps used to solve an ac circuit are patterned after the steps in Chapter 6:

1. Determine the independent loops to be used. The meshes constitute a set of independent loops.
2. Assume a phasor current flow in each loop. As was stated earlier, the current directions may be arbitrarily chosen.
3. Note which impedances are shared by two loops. Each of these impedances has a total phasor voltage that is the combination of the two phasor voltages developed when the two loop currents flow through the impedance.
4. Obtain the set of KVL equations by following each loop current around each loop and adding, by phasor algebra, the phasor voltages.
5. Solve the simultaneous equations for the unknown phasor loop currents.

Even though our procedure for solving an ac circuit by loop analysis does not seem any more complicated than the procedure for loop analysis of a dc resistive circuit, the use of phasor quantities complicates our work. In particular, simultaneously solving for the unknown loop currents involves much complex algebra.

A review of the use of determinants for the solution of simultaneous equations appears in the appendix. Specifically, an example is given to illustrate the solution of simultaneous equations with complex coefficients. A BASIC computer program for the solution of simultaneous equations with complex coefficients is also included.

However, in the following examples, the long, tedious work involving phasor manipulation is shown so that you will become thoroughly familiar with the circuit analysis procedures.

EXAMPLE 14-2

Solve for the phasor currents specified in the circuit of Fig. 14-5(a).

Solution.   Loop currents are chosen as in Fig. 14-5(b). Starting at each source and traversing the loops in the directions of the currents, we obtain the loop equations:

$$\text{Loop } a: \quad (+j3)\mathbf{I}_a + 4\mathbf{I}_a + 4\mathbf{I}_b - 20\angle 0° = 0$$
$$\text{Loop } b: \quad (-j5)\mathbf{I}_b + 4\mathbf{I}_a + 4\mathbf{I}_b - 10\angle 90° = 0$$

After rearrangement, the equations are

$$(4 + j3)\mathbf{I}_a + 4\mathbf{I}_b = 20\angle 0° = 20$$
$$4\mathbf{I}_a + (4 - j5)\mathbf{I}_b = 10\angle 90° = +j10$$

By determinants,

$$\mathbf{D} = \begin{vmatrix} 4 + j3 & 4 \\ 4 & 4 - j5 \end{vmatrix} = (4 + j3)(4 - j5) - 16$$
$$= 16 + j12 - j20 - j^2 15 - 16 = 15 - j8 = 17\angle{-28.1°}$$

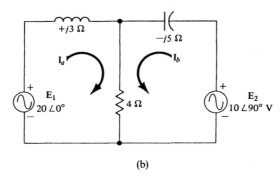

(a)

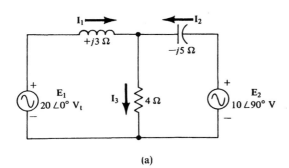

(b)

Figure 14-5   Example 14-2: **(a)** An ac circuit; **(b)** the chosen loop currents.

$$\mathbf{D}_a = \begin{vmatrix} 20 & 4 \\ +j10 & 4-j5 \end{vmatrix} = 80 - j100 - j40 = 80 - j140$$
$$= 161 \angle -60.3$$

$$\mathbf{D}_b = \begin{vmatrix} 4+j3 & 20 \\ 4 & +j10 \end{vmatrix} = j40 + j^2 30 - 80 = -110 + j40$$
$$= 117 \angle 160°$$

Then

$$\mathbf{I}_1 = \mathbf{I}_a = \frac{\mathbf{D}_a}{\mathbf{D}} = \frac{161 \angle -60.3°}{17 \angle -28.1°} = 9.5 \angle -32.2° \text{ A}$$

$$\mathbf{I}_2 = \mathbf{I}_b = \frac{\mathbf{D}_b}{\mathbf{D}} = \frac{117 \angle 160°}{17 \angle -28.1°} = 6.9 \angle 188.1° \text{ A}$$

$$\mathbf{I}_3 = \mathbf{I}_a + \mathbf{I}_b = 9.5 \angle -32.2° + 6.9 \angle 188.1°$$
$$= 8.04 - j5.06 - 6.83 - j0.97 = 1.21 - j6.03$$
$$= 6.15 \angle -78.7° \text{ A}$$

EXAMPLE 14-3

Solve for the current **I** in Fig. 14-6.

Solution. Circulation currents $\mathbf{I}_a$ and $\mathbf{I}_b$ are specified as shown; the respective loop equations are

Loop *a*: $(6 - j8)\mathbf{I}_a - (3 - j4)\mathbf{I}_b - 110 \angle 120° = 0$
Loop *b*: $(6 - j8)\mathbf{I}_b - (3 - j4)\mathbf{I}_a - 110 \angle 0° = 0$

After rearrangement, the equations are

$$(6 - j8)\mathbf{I}_a - (3 - j4)\mathbf{I}_b = 110 \angle 120°$$
$$-(3 - j4)\mathbf{I}_a + (6 - j8)\mathbf{I}_b = 110 \angle 0°$$

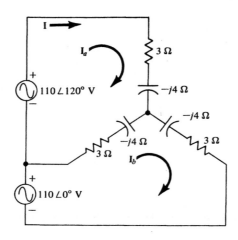

Figure 14-6   Example 14-3.

$$D = \begin{vmatrix} 6 - j8 & -(3 - j4) \\ -(3 - j4) & 6 - j8 \end{vmatrix}$$

$$= 36 - j48 - 64 - j48 - (9 - j24 - 16)$$

$$= -21 - j72 = 75 \angle -106.3°$$

$$D_a = \begin{vmatrix} (-55 + j95.3) & -(3 - j4) \\ 110 + j0 & 6 - j8 \end{vmatrix}$$

$$= -330 + j572 + j440 + 762 + 330 - j440$$

$$= 762 + j572 = 953 \angle 36.9°$$

$$I = I_a = \frac{D_a}{D} = \frac{953 \angle 36.9°}{75 \angle -106.3°} = 12.7 \angle 143.2° \text{ A}$$

EXAMPLE 14-4

Determine the phasor current I for the circuit of Fig. 14-7(a).

Solution. Loop currents are chosen as shown in Fig. 14-7(b). Notice that loop current I_c is chosen counterclockwise; thus

$$I_c = 2 \angle 30° = 1.73 + j1 \text{ A}$$

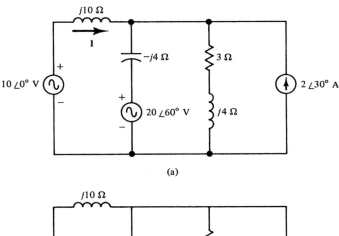

(a)

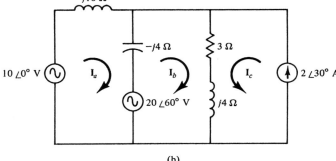

(b)

Figure 14-7 Example 14-4: **(a)** The circuit; **(b)** the circuit with loop currents selected.

The loop equations are

Loop a: $I_a(j10 - j4) - I_b(-j4) + 20 \angle 60° - 10 \angle 0° = 0$

Loop b: $I_b(3 + j4 - j4) + I_c(3 + j4) - I_a(-j4) - 20 \angle 60° = 0$

Rewriting, we have

Loop a: $j6I_a + j4I_b = 10 \angle 0° - 20 \angle 60°$

$= 10 - 10 - j17.3$

$= -j17.3$

Loop b: $j4I_a + 3I_b = 20 \angle 60° - I_c(3 + j4)$

$= 10 + j17.3 - (1.73 + j1)(3 + j4)$

$= 10 + j17.3 - (5.19 + j6.92 + j3 - 4)$

$= 8.81 + j7.38$

Using determinants, we have

$$D = \begin{vmatrix} j6 & j4 \\ j4 & 3 \end{vmatrix} = j18 - j^2 16$$

$$= 16 + j18 = 24.1 \angle 48.4°$$

$$D_a = \begin{vmatrix} -j17.3 & j4 \\ 8.81 + j7.38 & 3 \end{vmatrix} = -j51.9 - j35.2 + 29.5$$

$$= 29.5 - j87.1 = 91.9 \angle -71.3°$$

$$I = I_a = \frac{D_a}{D} = \frac{91.9 \angle -71.3°}{24.1 \angle 48.4°} = 3.81 \angle -119.7° \text{ A}$$

Drill Problem 14-2 ■

Determine the phasor current I in the circuit of Fig. 14-8 using loop analysis.

Answer. $5.48 \angle 122°$ A.   ■ ■

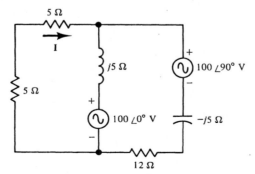

5Ω

I

$j5 \Omega$

$100 \angle 90°$ V

5Ω

$100 \angle 0°$ V $-j5 \Omega$

12Ω

Figure 14-8   Drill Problem 14-2.

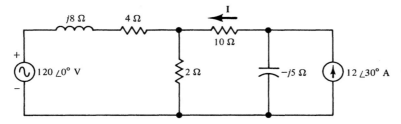

Figure 14-9 Drill Problem 14-4.

Drill Problem 14-3 ■

Convert the current source and parallel impedance of $3 + j4 \, \Omega$ of Fig. 14-7(a) to a voltage source equivalent and solve for the phasor current **I** using loop analysis.

Answer. $3.81 \angle -119.7° \text{ A.}$ ■ ■

Drill Problem 14-4 ■

Determine the current **I** that flows in the 10 Ω resistor of Fig. 14-9. Use loop analysis.

Answer. $2.87 \angle -42.8° \text{ A.}$ ■ ■

14.4 NODAL ANALYSIS

Before considering the nodal method of analysis as applied to ac circuits, the reader should review the method presented in Chapter 6. The following steps, paralleling those of Chapter 6, are used for a nodal solution of an ac circuit:

1. Select one major node as reference and assign each of the $n - 1$ remaining nodes its own unknown phasor voltage.
2. Assign branch current to each branch.
3. Express the branch currents in terms of the node potentials.
4. Write a current equation at each of the $n - 1$ unknown nodes.
5. Substitute the current expressions into the current equations.
6. Solve the resultant equations for the unknown phasor node voltages and, ultimately, the phasor branch currents.

Although the assignment of the phasor current for a given branch is generally arbitrary, the expression for the current in terms of the node voltages is not arbitrary. If the phasor current is designated as flowing from a to b, as in Fig. 14-10(a), the phasor current is

$$\mathbf{I} = \frac{\mathbf{V}_a - \mathbf{V}_b}{\mathbf{Z}}$$

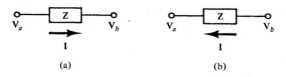

Figure 14-10   Designation of phasor currents **(a)** from *a* to *b* and **(b)** from *b* to *a*.

whereas if the phasor current is specified from *b* to *a* as in Fig. 14-10(b), the phasor current expression is

$$I = \frac{V_b - V_a}{Z}$$

It follows that the two currents of Fig. 14-10 are 180° out of phase.

EXAMPLE 14-5

Determine the phasor node voltages V_a and V_b for the circuit of Fig. 14-11.

Solution.   There are three nodes and the bottom node is selected as the reference. The KCL equations at nodes *a* and *b* are

$$\text{node } a\text{:}\qquad I_1 + I_2 - 5\angle 30° = 0$$
$$\text{node } b\text{:}\qquad -I_1 + I_3 + 4\angle 60° = 0$$

Substituting the expressions for the currents into the KCL equations, we obtain

$$\text{node } a\text{:}\qquad \frac{V_a - V_b}{j5} + \frac{V_a}{-j10} = 5\angle 30°$$
$$\text{node } b\text{:}\qquad -\frac{V_a - V_b}{j5} + \frac{V_b}{8} = -4\angle 60°$$

Rationalizing the denominators, combining coefficients, and converting the source currents from polar to rectangular form, we have

$$\text{node } a\text{:}\qquad -j0.1V_a + j0.2V_b = 4.33 + j2.5$$
$$\text{node } b\text{:}\qquad j0.2V_a + (0.125 - j0.2)V_b = -2 - j3.46$$

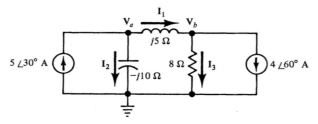

Figure 14-11   Example 14-5.

Then,

$$\mathbf{D} = \begin{vmatrix} -j0.1 & j0.2 \\ j0.2 & 0.125 - j0.2 \end{vmatrix} = 0.02 - j0.0125$$

$$= 0.0236 \angle -32°$$

$$\mathbf{D}_a = \begin{vmatrix} 4.33 + j2.5 & j0.2 \\ -2 - j3.46 & 0.125 - j0.2 \end{vmatrix}$$

$$= 0.349 - j0.153 = 0.381 \angle -23.7°$$

$$\mathbf{D}_b = \begin{vmatrix} -j0.1 & 4.33 + j2.5 \\ j0.2 & -2 - j3.46 \end{vmatrix} = 0.154 - j0.667$$

$$= 0.684 \angle -77°$$

$$\mathbf{V}_a = \frac{\mathbf{D}_a}{\mathbf{D}} = \frac{0.381 \angle -23.7°}{0.0236 \angle -32°} = 16.1 \angle 8.3° \text{ V}$$

$$\mathbf{V}_b = \frac{\mathbf{D}_b}{\mathbf{D}} = \frac{0.684 \angle -77°}{0.0236 \angle -32°} = 29 \angle -45° \text{ V}$$

Notice in Example 14-5 that the coefficient for the voltage term used to define any branch current is really the admittance of that branch. Referring to Fig. 14-10(a) we might instead define the branch current **I** as

$$\mathbf{I} = (\mathbf{V}_a - \mathbf{V}_b)\mathbf{Y}$$

Although circuit analysts might like to have the admittances specified in a problem, impedances are more commonly known. Hence, as in Example 14-5, rationalization of the coefficients is usually involved. Many circuit analysts also prefer that all sources in a nodal problem be specified as current sources. When voltage sources appear in a circuit to be analyzed by nodal analysis, we have the choice of writing the KCL equations in terms of the given voltage sources or converting the voltage sources to equivalent current sources. These alternatives are illustrated by the following examples.

EXAMPLE 14-6
Solve for the branch current specified in the circuit of Fig. 14-12.

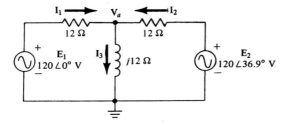

Figure 14-12 Circuit of Example 14-6.

Solution. As shown in Fig. 14-12, the bottom node is selected as the reference node and the top node is assigned an unknown potential $\mathbf{V}_a$.

The node equation is

$$\mathbf{I}_1 + \mathbf{I}_2 - \mathbf{I}_3 = 0$$

where

$$\mathbf{I}_1 = \frac{120 \angle 0° - \mathbf{V}_a}{12} \qquad \mathbf{I}_2 = \frac{120 \angle 36.9° - \mathbf{V}_a}{12}$$

and

$$\mathbf{I}_3 = \frac{\mathbf{V}_a}{j12}$$

The current expressions are substituted into the node equation with the result

$$\left(\frac{120 - \mathbf{V}_a}{12}\right) + \left(\frac{96 + j72 - \mathbf{V}_a}{12}\right) - \left(\frac{\mathbf{V}_a}{j12}\right) = 0$$

Notice that the polar representations for the sources have been converted to rectangular form. Multiplying the preceding equation by $j12$, we can clear the denominators.

$$j120 - j\mathbf{V}_a + j96 + j^2 72 - j\mathbf{V}_a - \mathbf{V}_a = 0$$

or $-72 + j216 = (1 + j2)\mathbf{V}_a$. Then

$$\mathbf{V}_a = \frac{-72 + j216}{1 + j2} = \frac{(-72 + j216)(1 - j2)}{5}$$

$$= \frac{-72 + j216 + j144 + 432}{5} = 72 + j72 = 102 \angle 45° \text{ V}$$

Finally,

$$\mathbf{I}_1 = \frac{120 - \mathbf{V}_a}{12} = \frac{120 - 72 - j72}{12} = 4 - j6 = 7.2 \angle -56.3° \text{ A}$$

$$\mathbf{I}_2 = \frac{96 - j72 - \mathbf{V}_a}{12} = \frac{96 + j72 - 72 - j72}{12} = 2 \angle 0° \text{ A}$$

$$\mathbf{I}_3 = \frac{\mathbf{V}_a}{j12} = \frac{102 \angle 45°}{12 \angle 90°} = 8.5 \angle -45° \text{ A}$$

EXAMPLE 14-7

Refer to Fig. 14-12 of Example 14-6. Convert the voltage sources to current sources and solve for the node voltage $\mathbf{V}_a$.

Solution. For the left source and 12 Ω resistor,

$$\mathbf{I}_1 = \frac{\mathbf{E}_1}{12} = \frac{120 \angle 0°}{12 \angle 0°} = 10 \angle 0° \text{ A}$$

For the right source and 12 Ω resistor,

$$\mathbf{I}_2 = \frac{\mathbf{E}_2}{12} = \frac{120 \angle 36.9°}{12 \angle 0°} = 10 \angle 36.9° \text{ A}$$

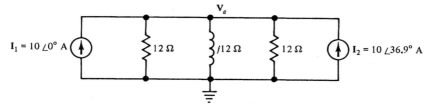

Figure 14-13   Circuit of Example 14-7.

Thus, we have the equivalent single-node pair circuit of Fig. 14-13. We write a single-node equation,

$$\frac{V_a}{12} + \frac{V_a}{12} + \frac{V_a}{j12} - 10 \angle 0° - 10 \angle 36.9° = 0$$

or

$$0.0833V_a + 0.0833V_a - j0.0833V_a = 10 + 8 + j6$$

Combining terms, we have

$$(0.166 - j0.0833)V_a = 18 + j6$$

Finally,

$$V_a = \frac{18.98 \angle 18.4°}{0.186 \angle -26.6} = 102 \angle 45° \text{ V}$$

Again, it must be emphasized that *when voltage sources are replaced by their current source equivalents, the circuit is an equivalent circuit.* Thus, the branch currents I_1 and I_2 in Fig. 14-12 are not the same currents that flow in the 12 Ω resistors of Fig. 14-13. Rather, we should return to the original circuit if we wish to determine the original, specific currents.

Drill Problem 14-5 ■

Using nodal analysis, determine the current **I** for the circuit of Fig. 14-14. Note that the circuit is from Drill Problem 14-2 with current source equivalents.

Answer.   5.48 ∠ 122° A.   ■ ■

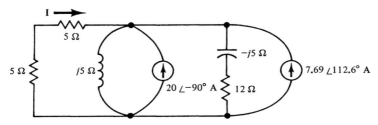

Figure 14-14   Drill Problem 14-5.

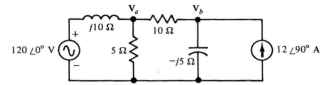

Figure 14-15   Drill Problem 14-6.

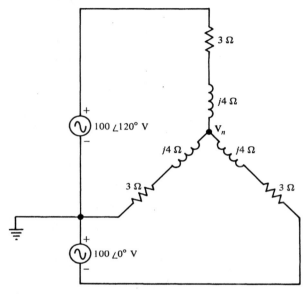

Figure 14-16   Drill Problem 14-7.

Drill Problem 14-6 ■

Solve the node voltages $\mathbf{V}_a$ and $\mathbf{V}_b$ for the circuit of Fig. 14-15. Use nodal analysis.

Answer. $37.5 \angle -51.3°$ V; $41.8 \angle 12.1°$ V.   ■ ■

Drill Problem 14-7 ■

Determine the node voltage $\mathbf{V}_n$ for the circuit of Fig. 14-16. Use nodal analysis.

Answer. $57.7 \angle 150°$ V.   ■ ■

14.5 SUPERPOSITION

When several ac sources are present in a linear, bilateral network, we can apply the superposition principle to find the phasor currents or voltages in the network. As before, we consider the presence of one source at a time and superimpose the phasor currents or voltages to obtain an actual solution.

As the effect of each source is considered, the other sources are removed from the circuit, but their internal impedances must remain. Thus, when a voltage source is removed, a very low internal impedance, ideally a short circuit, remains. When a current source is removed, a high internal impedance remains; ideally, an open circuit results.

The following example illustrates the application of superposition to ac networks.

EXAMPLE 14-8

Using superposition, find the current **I** for the circuit of Fig. 14-17(a).

Solution. The left source is considered first [Fig. 14-17(b)]. Then

$$\mathbf{Z}_1 = (12\ \Omega) \| (j12\ \Omega) = \frac{(12)(j12)}{(12 + j12)} = \frac{j12}{1 + j1} = 6j(1 - j1)$$
$$= 6 + j6 = 8.5\ \angle 45°\ \Omega$$

and by voltage division

$$\mathbf{V}_1 = 120\ \angle 0° \left(\frac{\mathbf{Z}_1}{\mathbf{Z}_1 + 12\ \Omega}\right) = 120\ \angle 0° \left(\frac{8.5\ \angle 45°}{18 + j6}\right) = 120 \left(\frac{8.5\ \angle 45°}{19\ \angle 18.4°}\right)$$
$$= 53.7\ \angle 26.6°\ V$$

Then

$$\mathbf{I}_1 = \frac{\mathbf{V}_1}{j12} = \frac{53.7\ \angle 26.6°}{12\ \angle 90°} = 4.48\ \angle -63.4° = 2.0 - j4.0\ A$$

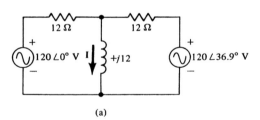

(a)

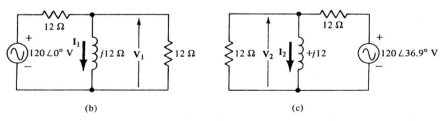

(b)   (c)

Figure 14-17 (a) Circuit of Example 14-8: (b) The left source considered; (c) the right source considered.

Next, the right source is considered [Fig. 14-17(c)]. Then

$$\mathbf{Z}_2 = (12\ \Omega)\,\|\,(j12\ \Omega) = 6 + j6 = 8.5\ \angle\,45°\ \Omega$$

By voltage division,

$$\mathbf{V}_2 = 120\ \angle\,36.9°\left(\frac{\mathbf{Z}_2}{\mathbf{Z}_2 + 12\ \Omega}\right) = 120\ \angle\,36.9°\left(\frac{8.5\ \angle\,45°}{19\ \angle\,18.4°}\right)$$

$$= 53.7\ \angle\,63.5°\ \mathrm{V}$$

Then

$$\mathbf{I}_2 = \frac{\mathbf{V}_2}{j12} = \frac{53.7\ \angle\,63.5°}{12\ \angle\,90°} = 4.48\ \angle\,-26.5° = 4.0 - j2.0\ \mathrm{A}$$

Finally,

$$\mathbf{I} = \mathbf{I}_1 + \mathbf{I}_2 = 6 - j6 = 8.5\ \angle\,-45°\ \mathrm{A}$$

Although, superposition as an analysis technique has the advantage of not requiring simultaneous equations, the amount of work needed to analyze a given problem by superposition is usually the same as in loop or nodal analysis. Its greatest usefulness may be when we have a circuit with two or more sinusoidal sources that are not the same frequency. In the latter case, the impedance varies with the frequency, so the circuit response to each source varies with frequency. Even though each separate response may be sinusoidal, the total response is generally nonsinusoidal, as will be shown in Chapter 16.

A special situation in circuit analysis is the superposition of a sinusoidal response (at a frequency ω) and a dc response (at a frequency $\omega = 0$). This case occurs in electronic circuits when constant voltages and currents are used to define an "operating point" and a sinusoidal voltage or current is used as an input signal.

As a simple example of the superposition of a sinusoidal source and a dc source, consider the circuit of Fig. 14-18(a). We are interested in describing the capacitor voltage. When the switch is closed at $t = 0$, the capacitor charges exponentially in response to the dc voltage. Thus, we have the voltage v_{C1} as defined by Fig. 14-18(b). In response to the sinusoidal voltage, the capacitor voltage is the sinusoidal voltage v_{C2} depicted by Fig. 14-18(c). When the two voltages are superimposed ($v_{C_t} = v_{C1} + v_{C2}$), we have the total response depicted by Fig. 14-18(d). In Chapter 16, we study the response of a circuit to which sources of different frequencies or nonsinusoidal sources are applied.

Drill Problem 14-8 ■

Determine by superposition the phasor voltage across the $8\ \Omega$ resistor of Fig. 14-19.

Answer. $18.3\ \angle\,19.1°\ \mathrm{V}$.   ■ ■

Drill Problem 14-9 ■

Use superposition to determine the currents $\mathbf{I}_1$, $\mathbf{I}_2$, and $\mathbf{I}_3$ of Fig. 14-20.

Answer. $6.93\ \angle\,-30°$, $6.93\ \angle\,-150°$, and $6.93\ \angle\,90°\ \mathrm{A}$.   ■ ■

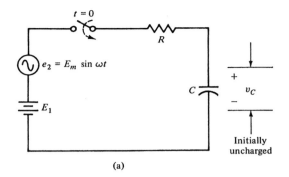

(a)

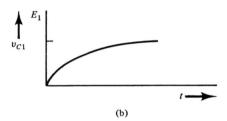

(b)

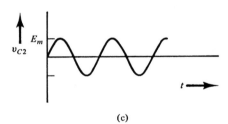

(c)

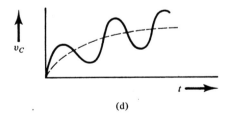

(d)

Figure 14-18 Superposition of dc and sinusoidal voltage: **(a)** The circuit; **(b)** its dc response; **(c)** its sinusoidal response; **(d)** the total response of the capacitor voltage.

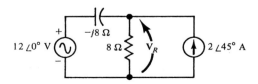

Figure 14-19 Drill Problem 14-8.

433

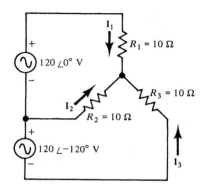

Figure 14-20   Drill Problem 14-9.

14.6 THEVENIN'S AND NORTON'S THEOREMS

In Chapter 5, the practical voltage source was described, electrically, as an ideal voltage source in series with an internal resistance, and the practical current source is described as an ideal current source in parallel with an internal resistance. This concept of an equivalent circuit for a source was extended in Chapter 6 by Thevenin's and Norton's theorems, which allow one to replace not only practical sources, but any two-terminal network, by a simple equivalent electrical circuit.

As with dc circuits, the simplification of an ac network or source to a Thevenin or Norton equivalent circuit is often desirable. It follows that one considers an *internal* or *Thevenin impedance* as opposed to an internal or Thevenin resistance. Thus Thevenin's theorem as applied to an ac network is as follows:

> **Any two-terminal network containing voltage or current sources can be replaced by an equivalent circuit consisting of a voltage equal to the open-circuit voltage of the original circuit in series with the impedance measured back into the original circuit.**

As before, the open circuit or Thevenin voltage is identified as E_{oc} or E_{th}. The internal or Thevenin impedance is identified as Z_{th} and is found by *looking back* into the network with the sources removed but with their internal impedances remaining.

The companion theorem, Norton's theorem, is as follows:

> **Any two-terminal network containing voltage and/or current sources can be replaced by an equivalent circuit consisting of a current source equal to the short-circuit current from the original network in parallel with the impedance measured back into the original circuit.**

The Norton or short-circuit current is identified as I_n or I_{sc}, and the Norton impedance equals the Thevenin impedance and is identified as Z_{th}. Often, the Norton admittance Y_n is specified and it equals the reciprocal of Z_{th}. Both the Thevenin and Norton equivalent circuits are shown in Fig. 14-21. We can convert from the

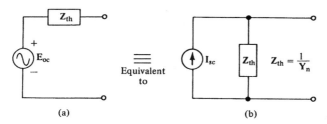

Figure 14-21 Equivalent circuits: **(a)** Thevenin; **(b)** Norton.

Thevenin to the Norton equivalent or from the Norton to the Thevenin equivalent by the relationship

$$\mathbf{E}_{oc} = \mathbf{I}_{sc}\mathbf{Z}_{th} \tag{14-2}$$

EXAMPLE 14-9

Find the Thevenin equivalent circuit between points a and b for the circuit of Fig. 14-22(a).

Solution. $\mathbf{E}_{oc}$ is measured across the 5 Ω resistor. By voltage division,

$$\mathbf{E}_{oc} = 40 \angle 0° \left(\frac{5}{5 + j20}\right) = \left(\frac{200 \angle 0°}{20.6 \angle 76°}\right) = 9.71 \angle -76° \text{ V}$$

From Fig. 14-22(b), the $\mathbf{Z}_{th}$ is

$$\mathbf{Z}_{th} = (5 \ \Omega) \| (j20) = \frac{5(j20)}{5 + j20} = \frac{100 \angle 90°}{20.6 \angle 76°} = 4.85 \angle 14° \ \Omega$$

EXAMPLE 14-10

Find the Norton equivalent circuit for Fig. 14-22(a).

Solution. By the previous analysis,

$$\mathbf{E}_{oc} = 9.71 \angle -76° \text{ V} \quad \text{and} \quad \mathbf{Z}_{th} = 4.85 \angle 14° \ \Omega$$

Then

$$\mathbf{I}_{sc} = \frac{\mathbf{E}_{oc}}{\mathbf{Z}_{th}} = \frac{9.71 \angle -76°}{4.85 \angle 14°} = 2 \angle -90° \text{ A}$$

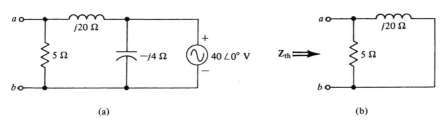

(a)   (b)

Figure 14-22 **(a)** Circuit of Example 14-9; **(b)** finding $\mathbf{Z}_{th}$.

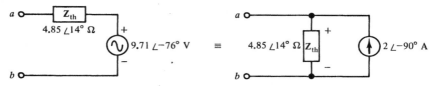

Figure 14-23 Equivalent circuits for Examples 14-9 and 14-10.

This value of current is also found by shorting terminals a and b of Fig. 14-22(a). Then $\mathbf{I}_{sc}$ equals the current flowing through the inductor:

$$\mathbf{I}_{sc} = \frac{40 \angle 0°}{j20} = 2 \angle -90° \text{ A}$$

Thus the equivalent circuits are those in Fig. 14-23.

EXAMPLE 14-11

Find the current $\mathbf{I}$ for the circuit of Fig. 14-24(a) by the use of Norton's theorem.

Solution. The center branch is removed, the circuit is redrawn as in Fig. 14-24(b), and $\mathbf{I}_{sc}$ is calculated:

$$\mathbf{I}_{sc} = \mathbf{I}_1 + \mathbf{I}_2$$

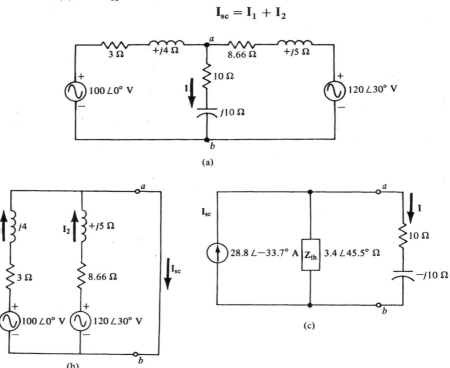

Figure 14-24 (a) Circuit of Example 14-11; (b) the same circuit with the load removed; (c) the Norton equivalent with the load reconnected.

where

$$I_1 = \frac{100 \angle 0°}{3 + j4} = \frac{100 \angle 0°}{5 \angle 53.1°} = 20 \angle -53.1° = 12 - j16 \text{ A}$$

and

$$I_2 = \frac{120 \angle 30°}{8.66 + j5} = \frac{120 \angle 30°}{10 \angle 30°} = 12 \angle 0° = 12 + j0 \text{ A}$$

Then

$$I_{sc} = 12 - j16 + 12 + j0 = 24 - j16 = 28.8 \angle -33.7° \text{ A}$$

Looking back into terminals a and b with the sources shorted, we find

$$Z_{th} = (3 + j4) \| (8.66 + j5)$$

$$= \frac{(5 \angle 53.1°)(10 \angle 30°)}{(3 + j4) + (8.66 + j5)} = \frac{50 \angle 83.1°}{11.66 + j9}$$

$$= \frac{50 \angle 83.1°}{14.7 \angle 37.7°} = 3.4 \angle 45.4° = 2.38 + j2.42 \ \Omega$$

The load is connected to the Norton circuit as in Fig. 14-24(c) so that by current division

$$I = I_{sc}\left(\frac{Z_{th}}{Z_{th} + 10 - j10}\right) = 28.8 \angle -33.7\left(\frac{3.4 \angle 45.4}{2.38 + j2.42 + 10 - j10}\right)$$

$$= 28.8 \angle -33.7\left(\frac{3.4 \angle 45.4}{14.5 \angle -31.5}\right) = 6.75 \angle 43.2° \text{ A}$$

Drill Problem 14-10 ■

Determine the Thevenin equivalents for the circuits of Fig. 14-25.

Answer. (a) $8.94 \angle -26.6° \ \Omega$; $44.7 \angle -116.6°$ V.
   (b) $10 \angle -36.9° \ \Omega$; $20 \angle 30°$ V.   ■ ■

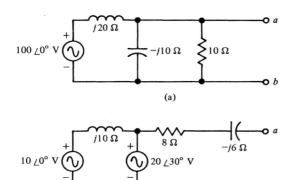

(a)

(b)

Figure 14-25   Drill Problems 14-10 and 14-11.

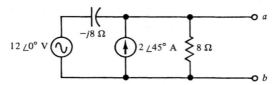

Figure 14-26   Drill Problem 14-12.

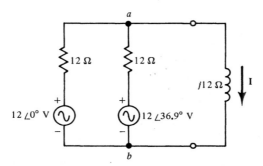

Figure 14-27   Drill Problem 14-13.

Drill Problem 14-11 ■

Determine the Norton equivalents for the circuits of Fig. 14-25.

Answer.   (a) $8.94 \angle -26.6° \ \Omega$; $5 \angle -90° $ A. (b) $10 \angle -36.9° \ \Omega$; $2 \angle 66.9° $ A.

■ ■

Drill Problem 14-12 ■

Determine the Norton equivalent of the circuit of Fig. 14-26.

Answer. $5.66 \angle -45° \ \Omega$; $3.24 \angle 64.1° $ A.

■ ■

Drill Problem 14-13 ■

Determine the current $\mathbf{I}$ that flows in the inductor of Fig. 14-27 by replacing the circuit portion to the left of terminals a and b by a Thevenin equivalent.

Answer. $0.85 \angle -45° $ A.

■ ■

14.7 MAXIMUM POWER TRANSFER

In Chapter 6, it was demonstrated that maximum power is obtained from a dc source with internal resistance R_i when the load resistance R_L equals R_i. In general, an ac source has an *internal impedance* $\mathbf{Z}_i$ (or a Thevenin impedance $\mathbf{Z}_{th}$) that includes a reactance term. However, if $\mathbf{Z}_i$ is purely resistive, then maximum power is delivered to the load if the load is purely resistive and equal to R_i; that is, $R_L = R_i$. In this situation, the load voltage is, by voltage division, one-half of the open-circuit voltage as in the dc case.

On the other hand, when the source internal impedance is complex ($\mathbf{Z}_i = R_i + jX_i$), it is not sufficient to let the load equal only the source resistance if maximum power transfer is desired.

Consider the circuit of Fig. 14-28, where a load impedance $\mathbf{Z}_L = R_L + jX_L$ is connected to a network or source with an open-circuit voltage $\mathbf{E}$ and an internal impedance $\mathbf{Z}_i$. The reactance X_L used here does not necessarily mean inductive reactance; rather, it symbolizes *load reactance* (inductive or capacitive). It is evident that the total impedance of the circuit is

$$\mathbf{Z}_t = (R_i + R_L) + j(X_i + X_L)$$

so that the magnitude of impedance Z_t is

$$Z_t = \sqrt{(R_i + R_L)^2 + (X_i + X_L)^2} \tag{14-3}$$

It follows that the magnitude of the current is

$$I = \frac{E}{Z_t} = \frac{E}{\sqrt{(R_i + R_L)^2 + (X_i + X_L)^2}} \tag{14-4}$$

Then the load power P_L is

$$P_L = I^2 R_L = \frac{E^2 R_L}{(R_i + R_L)^2 + (X_i + X_L)^2} \tag{14-5}$$

With a constant load resistance R_L, the circuit impedance (Eq. 14-3) is minimum and the load power (Eq. 14-5) is maximum if the reactive terms cancel; that is, $X_L = -X_i$. The effect of using a load reactance opposite in sign to the source reactance is that the circuit behaves as a purely resistive circuit. The power expression (Eq. 14-5) then reduces to

$$P_L = \frac{E^2 R_L}{(R_i + R_L)^2}$$

which is shown in Chapter 6 to be maximum when $R_L = R_i$. Hence,

maximum power is transferred from a source when the load impedance equals the complex conjugate of the source impedance.

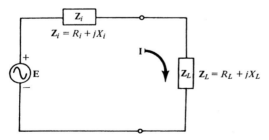

Figure 14-28   Alternating voltage source with load.

Mathematically, the maximum power transfer occurs when

$$\mathbf{Z}_L = \mathbf{Z}_i^*$$

(14-6)

where the asterisk means *complex conjugate*.

If the load is not matched to the source on a conjugate basis, maximum power transfer does not occur. Yet there are times when only a resistive load is used and optimum transfer is desired. It can be shown by differentiation that the best power transfer results, in this case, when the load resistance is matched on a magnitude basis,

$$R_L = Z_i = \sqrt{R_i^2 + X_i^2}$$

(14-7)

The following example emphasizes the maximum power transfer concept by providing a comparison of three load values.

EXAMPLE 14-12

Refer to Fig. 14-29. Find the power transferred to the load if **(a)** the load is a resistance of $5\,\Omega$, **(b)** the load is a resistance matched on a magnitude basis, and **(c)** the load is an impedance matched on a conjugate basis.

Solution

$$\mathbf{Z}_i = 5 + j10 = 11.2 \angle 63.4° \ \Omega$$

(a) $\mathbf{Z}_L = R_L = 5\,\Omega$.

$$\mathbf{I} = \frac{\mathbf{E}}{\mathbf{Z}_i + 5} = \frac{100 \angle 0°}{10 + j10} = \frac{100 \angle 0°}{14.1 \angle 45°} = 7.07 \angle -45° \ \text{A}$$

$$P_L = I^2 R_L = (7.07)^2(5) = 250 \ \text{W}$$

(b) On a magnitude basis, $\mathbf{Z}_L = R_L = Z_i = 11.2\,\Omega$.

$$\mathbf{I} = \frac{\mathbf{E}}{\mathbf{Z}_i + 11.2} = \frac{100 \angle 0°}{16.2 + j10} = \frac{100 \angle 0°}{19 \angle 31.7°} = 5.26 \angle -31.7° \ \text{A}$$

$$P_L = I^2 R_L = (5.26)^2(11.2) = 310 \ \text{W}$$

(c) On a conjugate basis $\mathbf{Z}_L = \mathbf{Z}_i^* = 5 - j10\,\Omega$.

$$\mathbf{I} = \frac{\mathbf{E}}{\mathbf{Z}_i + \mathbf{Z}_L} = \frac{100 \angle 0°}{10 + j0} = 10 \angle 0° \ \text{A}$$

$$P_L = I^2 R_L = (10)^2(5) = 500 \ \text{W}$$

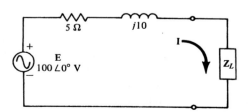

Figure 14-29   Circuit of Example 14-12.

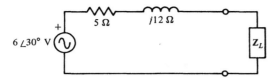

Figure 14-30　Drill Problems 14-14 and 14-15.

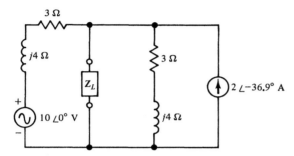

Figure 14-31　Drill Problem 14-16.

Drill Problem 14-14 ■

Determine the load impedance in Fig. 14-30 for which maximum power is delivered to the load. What is the maximum power delivered to the load?

Answer.　$13 \angle -67.4°$ Ω; 1.8 W.　　　　　　　　　　　■ ■

Drill Problem 14-15 ■

If the load impedance in Fig. 14-30 is matched on a magnitude basis, what power is delivered to the load? Compare this to the maximum power that can be delivered.

Answer.　1 W; it is less than the maximum power.　　　　　■ ■

Drill Problem 14-16 ■

Determine for Fig. 14-31 **(a)** the value of Z_L for maximum power and **(b)** the maximum power delivered to Z_L.

Answer.　(a) $2.5 \angle -53.1°$ Ω. (b) 16.3 W.　　　　　　　■ ■

14.8 DELTA AND WYE CONVERSIONS

The delta (Δ) or pi (π) and the wye (Y) or tee (T) configurations are introduced in Chapter 6 for resistance networks. Using the same approach as in Chapter 6, we can obtain expressions for converting a Δ impedance network into an equivalent Y impedance network and vice versa. However, the derivations are so similar that it suffices to restate the conversion formulas with impedance symbols rather than resistance symbols. With reference to Fig. 14-32, the conversion formulas are as follows.

Y *to* Δ *Conversion*:

$$Z_A = \frac{Z_1Z_2 + Z_2Z_3 + Z_3Z_1}{Z_2} \tag{14-8}$$

$$Z_B = \frac{Z_1Z_2 + Z_2Z_3 + Z_3Z_1}{Z_1} \tag{14-9}$$

$$Z_C = \frac{Z_1Z_2 + Z_2Z_3 + Z_3Z_1}{Z_3} \tag{14-10}$$

Δ *to* Y *Conversion*:

$$Z_1 = \frac{Z_AZ_C}{Z_A + Z_B + Z_C} \tag{14-11}$$

$$Z_2 = \frac{Z_BZ_C}{Z_A + Z_B + Z_C} \tag{14-12}$$

$$Z_3 = \frac{Z_AZ_B}{Z_A + Z_B + Z_C} \tag{14-13}$$

Notice that any impedance of the Δ circuit is equal to the sum of the products of all possible pairs of the Y impedances divided by the opposite Y impedance. Also, any impedance of the Y circuit is equal to the product of the two adjacent Δ impedances divided by the sum of the three Δ impedances.

Often in power circuits, a given Δ or Y will have equal impedances in all three branches. In this case, the preceding equations reduce to

$$Z_\Delta = 3Z_Y \tag{14-14}$$

and

$$Z_Y = \frac{Z_\Delta}{3} \tag{14-15}$$

where Z_Δ is one of the Δ impedances and Z_Y is one of the Y impedances.

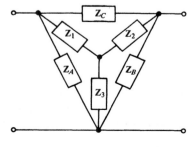

Figure 14-32 Superimposed Δ and Y networks.

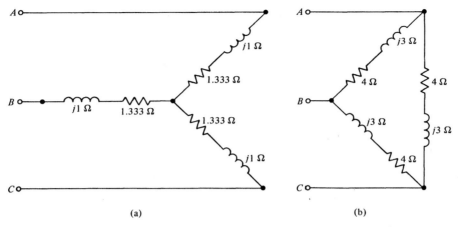

Figure 14-33 Example 14-13: **(a)** Y network; **(b)** the equivalent Δ network.

EXAMPLE 14-13

Find the equivalent Δ network for the Y network of Fig. 14-33.

Solution. From Eq. 14-14, $\mathbf{Z}_\Delta = 3\mathbf{Z}_Y = 3(1.333 + j1) = 4 + j3$ Ω. Hence the equivalent Δ is that of Fig. 14-33(b).

If, as in the preceding example, each of the impedances of a Δ or a Y are equal, we say that the Δ or Y is *balanced*.

One of the applications of Δ–Y conversion formulas is the reduction of the impedance bridge circuit. An ac impedance bridge with general impedance notation is depicted in Fig. 14-34. When such a circuit is used as a measurement circuit, the "bridging" impedance $\mathbf{Z}_5$ represents a "detector" element.

Depending on the voltages and currents we desire, we may apply a variety of circuit analysis techniques to the impedance bridge circuit. For example, if we desire voltages and currents for all the branches of the bridge, we can use loop or nodal analysis. If we desire the voltage or current for the "bridging" element $\mathbf{Z}_5$, we can use loop or nodal analysis, or we can consider the temporary removal of $\mathbf{Z}_5$ and the determination of the Thevenin or Norton equivalent for the remainder of the circuit. If we are interested in voltages and currents external to the bridge circuit, we can reduce the bridge circuit to a simpler circuit by Δ–Y conversion formulas.

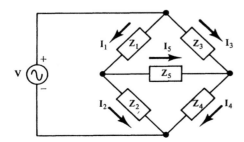

Figure 14-34 Impedance bridge configuration.

Although the impedance bridge is easily recognized when it is drawn in the form of Fig. 14-34, there are times when it is not recognized because it is either drawn in another form or is part of a larger circuit and thus is "disguised." The following example illustrates a circuit that is not drawn in the standard impedance bridge form of Fig. 14-34 yet can be redrawn in the standard form. Once we have the circuit in standard bridge form, we can proceed with the analysis in an appropriate manner.

EXAMPLE 14-14

Determine the current $\mathbf{I}$ flowing to the network of Fig. 14-35.

Solution. When redrawn as in Fig. 14-36(a) the circuit is seen as an impedance bridge circuit. Using a Δ–Y conversion with $\mathbf{Z}_A = -j10\ \Omega$, $\mathbf{Z}_B = j10\ \Omega$, and $\mathbf{Z}_C = -j10\ \Omega$, we have

$$\mathbf{Z}_1 = \frac{\mathbf{Z}_A \mathbf{Z}_C}{\mathbf{Z}_A + \mathbf{Z}_B + \mathbf{Z}_C} = \frac{(-j10)(-j10)}{-j10 + j10 - j10} = -j10\ \Omega$$

$$\mathbf{Z}_2 = \frac{\mathbf{Z}_B \mathbf{Z}_C}{\mathbf{Z}_A + \mathbf{Z}_B + \mathbf{Z}_C} = \frac{(j10)(-j10)}{-j10} = j10\ \Omega$$

$$\mathbf{Z}_3 = \frac{\mathbf{Z}_A \mathbf{Z}_B}{\mathbf{Z}_A + \mathbf{Z}_B + \mathbf{Z}_C} = \frac{(-j10)(j10)}{-j10} = j10\ \Omega$$

With the Y replacing the Δ as indicated by the broken lines, we obtain a series–parallel reducible circuit. Then the total impedance $\mathbf{Z}_t$ is

$$\mathbf{Z}_t = j10 + (15 - j10)\,\|\,(15 + j10)$$

$$= j10 + \frac{(15 - j10)(15 + j10)}{(15 - j10 + 15 + j10)}$$

$$= j10 + \frac{225 - j150 + j150 + 100}{30}$$

$$= 10.83 + j10 = 14.7\ \angle\,42.7^\circ\ \Omega$$

$$\mathbf{I} = \frac{\mathbf{V}}{\mathbf{Z}_t} = \frac{1\ \angle\,0^\circ}{14.7\ \angle\,42.7^\circ} = 0.068\ \angle\,-42.7^\circ\ \text{A}$$

The impedance bridge circuit is said to be *balanced* when the "detector" or bridge current ($\mathbf{I}_5$ in Fig. 14-34) equals zero. For that condition to exist, the following

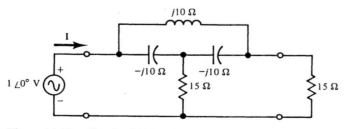

Figure 14-35 Circuit of Example 14-14.

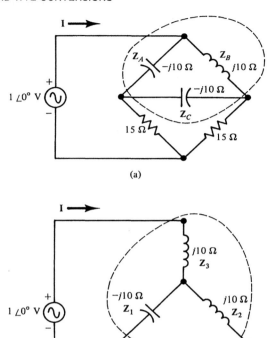

(a)

(b)

Figure 14-36 Circuit of Example 14-14 **(a)** redrawn as a bridge and **(b)** converted to a series–parallel equivalent.

are true:

$$\mathbf{V}_{ac} = \mathbf{V}_{ad}$$

and
$$\mathbf{V}_{cb} = \mathbf{V}_{db}$$

It follows that with $\mathbf{I}_5 = 0$,

$$\mathbf{I}_1\mathbf{Z}_1 = \mathbf{I}_3\mathbf{Z}_3$$

and
$$\mathbf{I}_1\mathbf{Z}_2 = \mathbf{I}_3\mathbf{Z}_4$$

By substituting one of the preceding equations into the other, we can eliminate the currents and obtain the condition of balance in terms of the impedances. Thus, for balance,

$$\frac{\mathbf{Z}_1}{\mathbf{Z}_2} = \frac{\mathbf{Z}_3}{\mathbf{Z}_4} \tag{14-16}$$

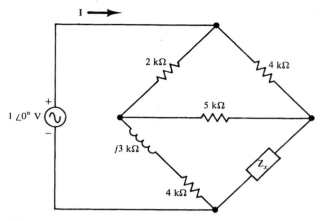

Figure 14-37   Drill Problems 14-17 and 14-18.

In Chapter 17 we shall study the types of impedance bridges used in impedance measuring circuits. Although conversions are often useful in analyzing bridge circuits, the conversions are also useful in three-phase circuits. In the next chapter we study three-phase circuits and find that the load impedances for such circuits are connected in Δ or Y configurations.

Drill Problem 14-17 ■

If the bridge circuit of Fig. 14-37 is balanced, what is the value of Z_x?

Answer. $8 + j6$ kΩ   ■ ■

Drill Problem 14-18 ■

If the impedance Z_x of the bridge circuit of Fig. 14-37 equals $4 + j3$ kΩ, what total current **I** flows to the bridge circuit?

Answer. $0.269 \angle -24°$ mA.   ■ ■

14.9 DEPENDENT SOURCES

The *dependent* or *controlled source* was defined in Section 6.9 as a source for which the voltage or current depends on a voltage or current in some other part of a circuit. As was explained then, dependent sources appear in the electric circuit models of many electronic devices.

Many of the electric signals involved in electronic circuits are nonsinusoidal. Thus, the equivalent models for the electronic devices often have voltages and currents, including dependent source values, expressed in terms of instantaneous values. As in Chapter 6, diamond-shaped source symbols are used for dependent sources (Fig. 14-38). When values are specified in the time domain, small letters indicating instantaneous values are used. When source values are specified in the frequency

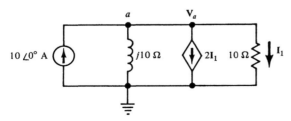

Figure 14-38 Dependent ac sources: (a) Voltage; (b) current.

(a)   (b)

domain, phasor notations are used. In either case, the voltage or current values are expressed as functions of some other voltage or current in the circuit. The dependent source values are included when KVL and KCL equations are written. However, the control variable and dependent source values must be related in turn to the unknown voltages or currents in the Kirchhoff equations.

EXAMPLE 14-15

Determine the phasor voltage $\mathbf{V}_a$ for the circuit of Fig. 14-39.

Solution. A KCL equation is written for node a:

$$\frac{\mathbf{V}_a}{j10} + 2\mathbf{I}_1 + \mathbf{I}_1 = 10 \angle 0°$$

The control variable $\mathbf{I}_1$ is related to $\mathbf{V}_a$ by

$$\mathbf{I}_1 = \frac{\mathbf{V}_a}{10} = 0.1\mathbf{V}_a$$

Substituting $\mathbf{I}_1$ into the KCL equation, we have

$$-j0.1\mathbf{V}_a + 0.2\mathbf{V}_a + 0.1\mathbf{V}_a = 10 \angle 0°$$

or

$$\mathbf{V}_a = \frac{10 \angle 0°}{0.3 - j0.1} = \frac{10 \angle 0°}{0.316 \angle -18.4°} = 31.6 \angle 18.4° \text{ V}$$

Two equivalent circuit models, the transistor "h" parameter and op-amp models, were introduced in Chapter 6. Similar models are used for ac electronic circuit analysis. For ac circuit analysis, the Chapter 6 models are modified by general or sinusoidal dependent voltage and current sources instead of dc sources. Also, any capacitances, such as interelectrode capacitances, or inductances are included if their effects are considered significant.

Figure 14-39 Circuit for Example 14-15.

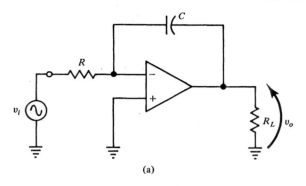

(a)

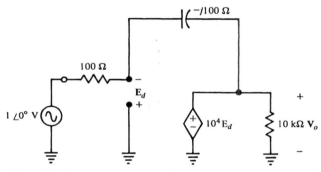

Figure 14-40   Example 14-16: **(a)** An op-amp circuit; **(b)** an equivalent frequency domain circuit.

As in Chapter 6, we shall analyze a few circuits involving equivalent models of electronic devices. A further study of variations to the models and why we use such models is covered in most electronics courses.

EXAMPLE 14-16

An op-amp circuit is shown in Fig. 14-40(a). When the op-amp model is used, some typical circuit values chosen, and the values expressed in the frequency domain, Fig. 14-40(b) is the result. Determine the voltage V_o for the circuit of Fig. 14-40(b).

Solution.   If we convert the input voltage source to a current equivalent, we obtain the two node circuit of Fig. 14-41:

$$\text{At node } b: \quad V_b = V_o = 10^4 E_d$$
$$\text{At node } a: \quad V_a = -E_d$$

We write a node equation at node a:

$$\frac{V_a}{100} + \frac{V_a - V_b}{-j100} = 0.01 \angle 0°$$

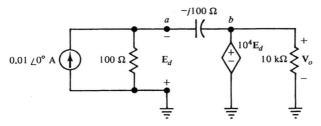

Figure 14-41 Op-amp circuit of Example 14-16.

When this last equation is rationalized and the V_a and V_b values (in terms of E_d) are substituted, we have

$$-0.01E_d - j0.01E_d - j0.01(10^4)E_d = 0.01 \angle 0°$$

or

$$(-0.01 - j0.01 - j10^2)E_d = 0.01 \angle 0°$$

Then

$$E_d \approx \frac{0.01 \angle 0°}{-j10^2} = \frac{0.01 \angle 0°}{100 \angle -90°} = 10^{-4} \angle 90°$$

Finally, $V_o = 10^4 E_d = 1 \angle 90°$ V.

The circuit in Example 14-16 has particular significance in the study of op-amps. Notice that the output is a phasor voltage equal in magnitude to the phasor input voltage and leading it by 90°. That is, if a signal $v_i = 1 \sin \omega t$ is applied at the input, the output voltage is $v_o = 1 \sin(\omega t + 90°) = 1 \cos \omega t$. Since the integral of a sine wave is a cosine wave, the results give us a hint as to the type of circuit in Fig. 14-40(a). That circuit performs integration and is called an *integrator*.

Some other interesting circuit analysis aspects accompany the use of dependent sources. The first of these involves the superposition theorem and the second involves the determination of the impedance "measured back" into a set of network terminals (such as a Thevenin impedance).

To use superposition, we first remove those independent sources that we are not considering at that time. *Dependent sources remain in the circuit.* The effect of a dependent source depends on the excitation that a source receives from independent sources.

Similarly, when we analyze the impedance of a network, we require that the dependent sources remain, although we remove the independent sources. In order to activate the dependent sources, we excite the set of terminals with an independent voltage as we analytically determine the current flowing into the network. The impedance is the ratio of the exciting voltage to the current being driven into the network. This technique is illustrated by the following example.

EXAMPLE 14-17
Determine the Thevenin equivalent for the network of Fig. 14-42(a).

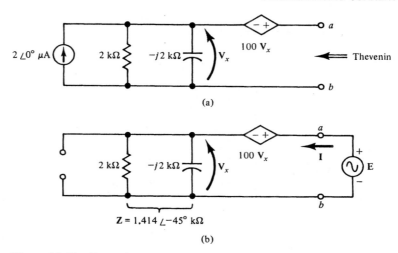

(a)

(b)

Figure 14-42   Example 14-17: **(a)** the original circuit; **(b)** determining Z_{th} with a voltage driving source.

Solution.   We find E_{oc} by finding V_x with terminals a and b open. Then

$$V_x = (2 \times 10^{-6} \angle 0° \text{ A})(2 \text{ k}\Omega \| -j2 \text{ k}\Omega)$$

$$= 2 \times 10^{-6} \angle 0° \left[\frac{(2k)(-j2k)}{2k - j2k} \right]$$

$$= 2 \times 10^{-6} \angle 0° \left[\frac{(2k)(-j2k)}{2.828 \angle -45°k} \right]$$

$$= 2 \times 10^{-6} \angle 0°(1.414 \times 10^3 \angle -45°)$$

$$= 2.828 \times 10^{-3} \angle -45° \text{ V}$$

and

$$E_{oc} = V_x + 100V_x = 101V_x$$

$$= 0.286 \angle -45° \text{ V}$$

We find Z_{th} by removing the independent current source and "driving" the network from the a–b terminals so that a current I flows into the network

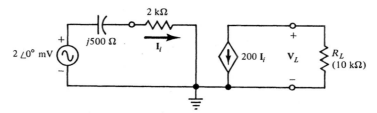

Figure 14-43   Drill Problem 14-19.

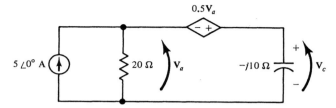

Figure 14-44 Drill Problem 14-20.

as in Fig. 14-42(b). The dependent source remains in the circuit. A KVL equation is $E = 100V_x + V_x = 101V_x$ but $V_x = I(1.414 \angle -45°\text{ k}\Omega)$. Substituting the V_x expression into the E relationship, we have an equation in terms of E and I:

$$E = 101V_x = 101I(1.414 \angle -45°\text{ k}\Omega)$$
$$= (142.8 \angle -45°\text{ k}\Omega)I$$

or
$$Z_{th} = E/I = 142.8 \angle -45°\text{ k}\Omega$$

Alternatively, I_{sc} can be calculated and Z_{th} determined by Eq. 14.2. That is, $Z_{th} = E_{oc}/I_{sc}$.

Drill Problem 14-19 ■

Given a capacitor-coupled transistor amplifier with the frequency domain circuit shown in Fig. 14-43, determine the phasor load voltage V_L.

Answer. $1.94 \angle 194°$ V. ■ ■

Drill Problem 14-20 ■

Determine the voltage across the capacitor in the frequency domain circuit of Fig. 14-44.

Answer. $47.5 \angle -71.6°$ V. ■ ■

Drill Problem 14-21 ■

Determine the impedance "looking into" terminals a–b of Fig. 14-45.

Answer. $11.1 \angle -56.3°\ \Omega$. ■ ■

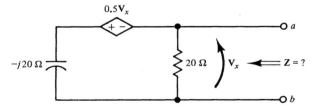

Figure 14-45 Drill Problem 14-21.

QUESTIONS

1. What is meant by internal impedance?
2. What is meant by the topology of a network?
3. How does the topology of a network change with ac rather than dc sources?
4. How must Kirchhoff's laws be modified for ac circuit analysis?
5. What are the steps in solving an ac network by either loop or nodal analysis?
6. How does one handle ac sources when solving a network by superposition?
7. State Thevenin's theorem.
8. State Norton's theorem.
9. What is the maximum power transfer theorem?
10. What is the difference between a conjugate and magnitude match?
11. What are the two statements assisting in the memorization or utilization of the Δ–Y conversion formulas?
12. When is a Y or Δ balanced?
13. When is an impedance bridge balanced?
14. How does a dependent source differ from an independent source?
15. Why are dependent sources not shorted or opened when superposition is applied to a circuit?

PROBLEMS

1. Convert the sources of Fig. 14-46 to their respective equivalent sources.
2. Using loop analysis, solve for the currents specified in the circuit of Fig. 14-47.

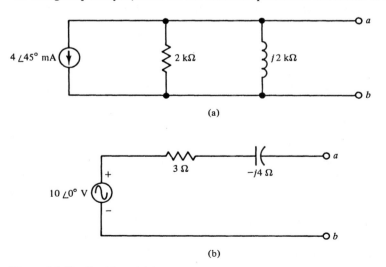

(a)

(b)

Figure 14-46   Problem 14-1.

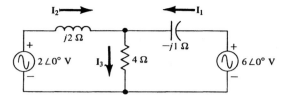

Figure 14-47   Problems 14-2, 14-8, and 14-17.

3. Determine I_1 and I_2 for the circuit of Fig. 14-48.

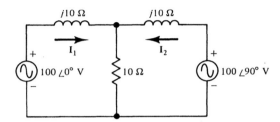

Figure 14-48   Problem 14-3.

4. Using loop analysis, solve for the specified currents for the circuits of Fig. 14-49.

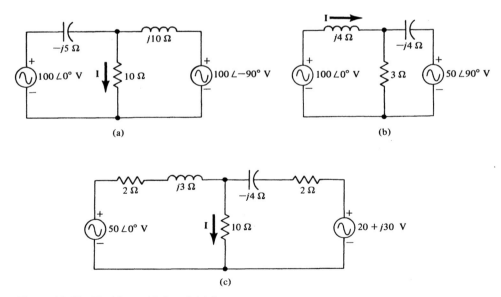

Figure 14-49   Problems 14-4 and 14-9.

5. Using loop analysis, solve for the current **I** in the circuit of Fig. 14-50.

6. Solve for the current **I** in each circuit of Fig. 14-51. Use loop analysis.

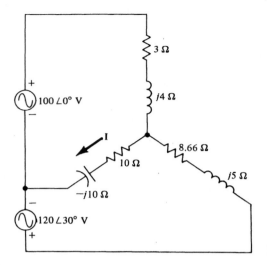

Figure 14-50   Problem 14-5.

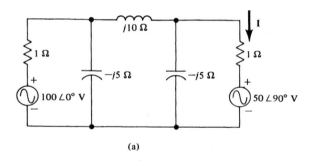

(a)

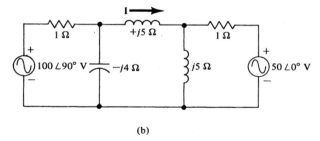

(b)

Figure 14-51   Problems 14-6 and 14-10.

7. Determine the source voltage E_2 in order for the current in the center branch of Fig. 14-52 to be zero.

8. Referring to Fig. 14-47, convert the series branches to current source equivalents and use nodal analysis to determine the voltage across the 4 Ω resistor.

9. Use nodal analysis to solve for the specified currents in Fig. 14-49.

10. Use nodal analysis to solve for the currents specified in Fig. 14-51.

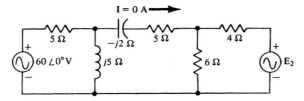

Figure 14-52 Problem 14-7.

11. Use nodal analysis to determine the voltage V_x in Fig. 14-53.

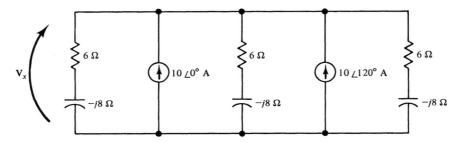

Figure 14-53 Problem 14-11.

12. Solve for the branch currents of Fig. 14-54 by nodal analysis.

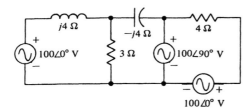

Figure 14-54 Problem 14-12.

13. Using nodal analysis, determine the inductor voltage V_L in Fig. 14-55.

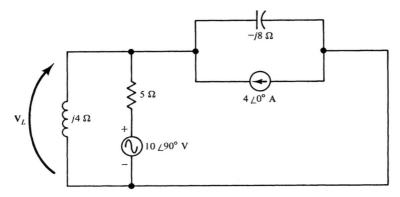

Figure 14-55 Problem 14-13.

14. Use nodal analysis to determine the node voltages V_a and V_b of Fig. 14-56.

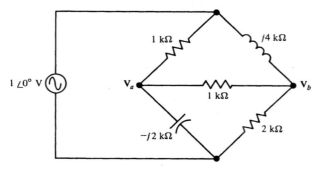

Figure 14-56   Problem 14-14.

15. Using nodal analysis, determine the current I in Fig. 14-57.

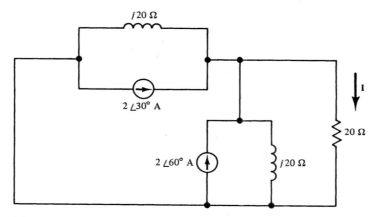

Figure 14-57   Problems 14-15 and 14-16.

16. Use superposition to determine the current I in Fig. 14-57.

17. Use superposition to solve for the currents in the circuit of Fig. 14-47.

18. Solve for the current I in Fig. 14-58 using the superposition method.

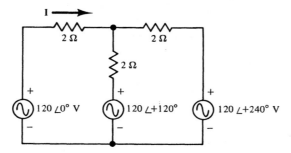

Figure 14-58   Problem 14-18.

19. Find the Thevenin and Norton equivalent circuits for the circuits of Fig. 14-59.

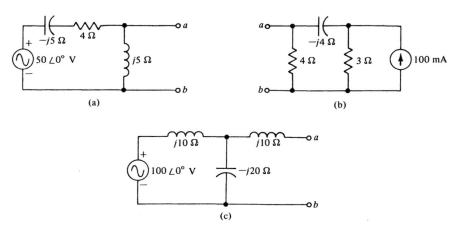

(a)

(b)

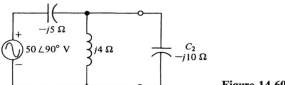

(c)

Figure 14-59 Problem 14-19.

20. Using Thevenin's theorem, solve for the current in capacitor C_2 in Fig. 14-60.

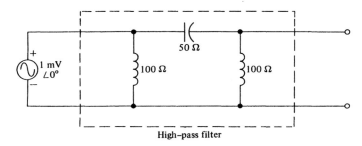

Figure 14-60 Problem 14-20.

21. Refer to Fig. 14-51(b). Use Thevenin's theorem to replace the portions of the circuit to the right and left of the $j5\ \Omega$ inductor in which I flows. Solve for I with the use of these Thevenin equivalents.

22. The circuit of Fig. 14-61 represents a high-pass filter at a particular frequency. Find the Thevenin circuit when a 1 mV source is connected to the filter as shown.

Figure 14-61 Problem 14-22.

23. Determine the Thevenin and Norton equivalent circuits for the circuit of Fig. 14-62.

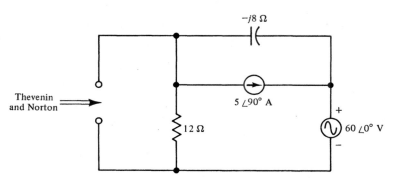

Figure 14-62   Problem 14-23.

24. Determine the maximum power that each load Z_L can receive for the circuits of Fig. 14-63.

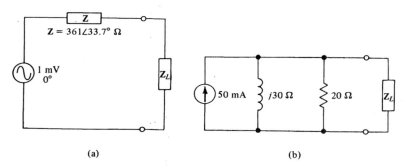

(a)   (b)

Figure 14-63   Problem 14-24.

25. Refer to Fig. 14-64. **(a)** What value must Z_L be in order to receive maximum power? **(b)** What is the maximum power received by Z_L?

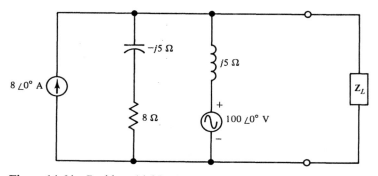

Figure 14-64   Problem 14-25.

26. Determine what value the impedance Z_x of Fig. 14-65 must be in order to receive maximum power.

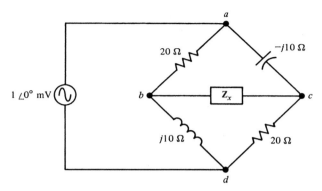

Figure 14-65 Problem 14-26.

27. For Fig. 14-65, what is the maximum power deliverable to Z_x?

28. If a load Z_L is connected to the network of Fig. 14-66, what value must it be to receive maximum power? What is this power? What is the maximum power that is delivered to the load if Z_L is only a resistive load?

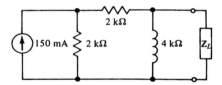

Figure 14-66 Problem 14-28.

29. Find the Y network equivalent to the Δ network of Fig. 14-67.

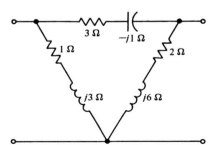

Figure 14-67 Problem 14-29.

30. Convert the high-pass π filter of Fig. 14-61 to a high-pass T filter.

31. Determine the currents I_1, I_2, and I_t for the circuit of Fig. 14-68.

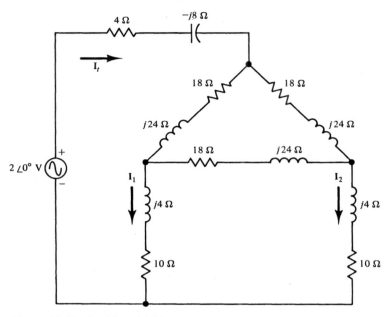

Figure 14-68 Problem 14-31.

32. Refer to the impedance bridge circuit of Fig. 14-65. To what inductive reactance value would the inductive portion have to be changed in order to achieve bridge balance?

33. Using Δ–Y conversions, reduce the circuit of Fig. 14-69 to a two-mesh problem and solve for the source currents I_1 and I_2.

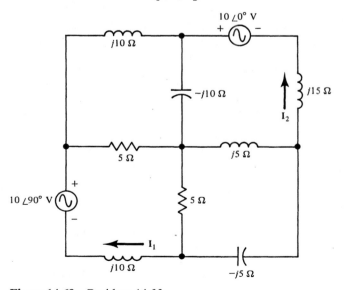

Figure 14-69 Problem 14-33.

34. Determine the voltage V_L for the circuit of Fig. 14-70.

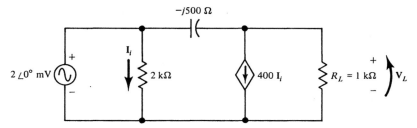

Figure 14-70   Problem 14-34.

35. Determine the voltage V_L for the circuit of Fig. 14-71. What is the total phase shift from the source to the output at V_L?

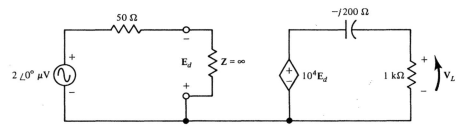

Figure 14-71   Problem 14-35.

36. Determine the open-circuit voltage V_a (also the control voltage) for the circuit of Fig. 14-72.

37. Determine the impedance looking back into the terminals of the circuit of Fig. 14-72.

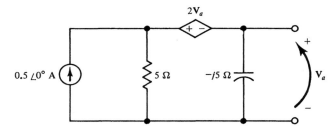

Figure 14-72   Problems 14-36 and 14-37.

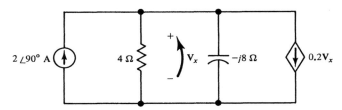

Figure 14-73   Problem 14-38.

38. Determine V_x in Fig. 14-73.

Chapter 15

Polyphase Circuits

15.1 INTRODUCTION

In Chapter 12 a single-phase circuit is defined as an ac circuit with only one source of emf or current. This chapter is concerned with ac circuits with two or more sources of emf interrelated in both phase and magnitude. Because the source voltages are phase-displaced from each other, the circuits of which they are part are called *polyphase circuits* or systems. In the polyphase system, the individual emfs will be of the same frequency and usually the same magnitude. If, in addition, the phasors that represent the emfs are uniformly distributed about a circle of 360 electrical degrees, the system is said to be *symmetrical*. One of the most important types of polyphase systems is the three-phase system used for electric-power generation, transmission, and distribution. The emphasis of this chapter is on the three-phase system, although in some cases the theory is expressed in general for the *n*-phase system, where *n* is the number of phases.

In an earlier chapter it was shown that an emf is developed in a conductor whenever there is a change in magnetic flux lines per unit time. So it is when a conductor moves perpendicularly to a stationary and constant magnetic field. An elementary two-phase generator is shown in Fig. 15-1. Notice that the magnetic field rotates inside a stationary set of conductors, which is common in an ac machine. Thus the relative motion of the conductors to the magnetic field is opposite to the motion of the magnetic field.

When the magnetic poles are in the position shown, the coil *aa'* is cutting a maximum number of flux lines. Therefore, a maximum emf is present across this coil and the application of the right-hand rule indicates that for conductor *a*, the direction of the emf is away from the reader. A second coil represented by conductors *b*

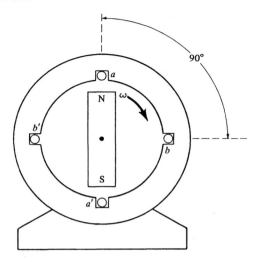

Figure 15-1 Elementary two-phase generator.

and b' is placed 90° physically from coil aa'. At the instant of time shown in Fig. 15-1, no voltage is developed in the b and b' conductors, since the magnetic flux cut by these conductors is zero. Ninety mechanical degrees later, the magnetic field develops a maximum emf across the bb' coil, whereas the aa' coil emf is zero.

Since the single pair of poles in Fig. 15-1 generates a complete cycle of emf in either coil during one complete revolution, the electrical and mechanical degrees are numerically equal. Thus if the emf of coil aa', called $e_{aa'}$, is used as a reference, the emf of coil bb', called $e_{bb'}$, lags by 90°. This relationship is shown by the graph and phasor diagram of Fig. 15-2. The two emfs generated are out of phase by 90° and make up a two-phase nonsymmetrical system of voltages.

A simplified three-phase generator is shown in Fig. 15-3. Notice that three coils a, b, and c are present in this elementary generator and are represented, respectively, by conductors a and a', b and b', and c and c'. At the instant of time indicated in Fig. 15-3, the voltage in coil a, $e_{aa'}$, is a maximum. When the magnetic poles are rotated through 120° with reference to the instantaneous position shown, the emf of coil b, $e_{bb'}$, is maximum, and when the rotation is 240° with reference to the position shown, the emf of coil c, $e_{cc'}$, is maximum. Thus $e_{cc'}$ lags $e_{bb'}$ by 120° and $e_{bb'}$ lags $e_{aa'}$

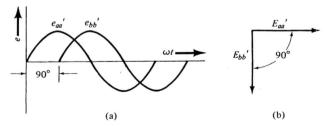

Figure 15-2 (a) Graphical and (b) phasor representation of a set of two-phase voltages.

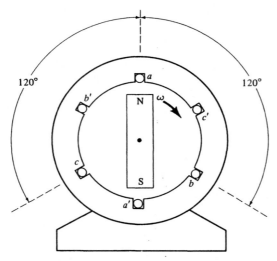

Figure 15-3   Elementary three-phase generator.

by 120°, thereby forming a symmetrical system. These relationships with $e_{aa'}$ as reference are shown in the graph and phasor diagram of Fig. 15-4.

In general, an elementary polyphase generator of n phases has n coils per pair of magnetic poles spaced at intervals of $360/n$ electrical degrees. *Thus the general n-phase system is a symmetrical set of n voltages, the two-phase system being an exception.*

Drill Problem 15-1 ■

Determine the unspecified angle θ in each of the following sets of phasor voltages so that each set is a symmetrical set of polyphase voltages:

(a) $120 \angle 30°$, $120 \angle \theta$, and $120 \angle -90°$ V;

(b) $28 \angle 0°$, $28 \angle 90°$, $28 \angle 180°$, and $28 \angle \theta$ V;

(c) $220 \angle 60°$, $220 \angle -60°$, and $220 \angle \theta$ V.

Answer.   (a) 150°. (b) −90°. (c) 180°.   ■ ■

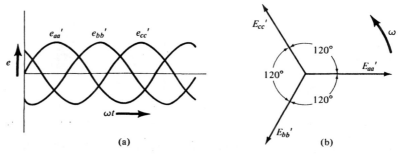

Figure 15-4   **(a)** Graphical and **(b)** phasor representation of a set of three-phase voltages.

15.2 ADVANTAGES OF A POLYPHASE SYSTEM

In practice, the n number of single-phase circuits that comprise a polyphase system are interconnected, as will be shown in Section 15.5, so that more than one phase utilizes a particular transmission line. This sharing of transmission lines enables electric energy to be transmitted from the generator to the load more efficiently than if separate lines were used. An advantage that results from the sharing of transmission lines by the n-phase system is a decrease in the required amount of copper under fixed conditions compared to a single-phase system. For example, a three-phase balanced system requires only $\frac{3}{4}$ the amount of copper of a single-phase system supplying the same amount of power over the same line distances with the same magnitude of voltages between conductors.

A second advantage of the polyphase system compared to a single-phase system is a relatively uniform power transmission. As was shown in Chapter 12, the instantaneous power of a single-phase circuit is a sinusoidally varying quantity of twice the frequency of the applied sine wave voltage. On the other hand, the instantaneous power of a balanced three-phase system is constant even though the instantaneous power in each phase of the system is varying. The comparison between the instantaneous single-phase and balanced three-phase powers is shown in Fig. 15-5.

A third advantage of polyphase voltages is that, with proper connections of the field coils of a polyphase motor, the resultant currents in the field coils create a rotating magnetic field, a concept that is necessary to the study of synchronous and induction motors.

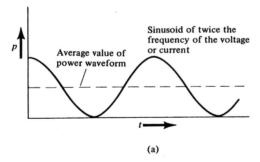

(a)

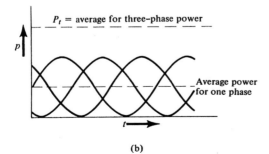

(b)

Figure 15-5 Instantaneous power in **(a)** a single-phase circuit and **(b)** a balanced three-phase circuit.

In making a final comparison between single-phase and polyphase circuits, it must be stated that a single-phase load may be energized by a polyphase circuit simply by connecting the load to any one of the single-phase circuits making up the polyphase circuit. In contrast, it is not possible to energize a three-phase load from a single-phase source.

15.3 DOUBLE-SUBSCRIPT NOTATION

It is imperative in polyphase circuits that phasor voltages and loop tracing directions be properly noted and labeled. Particularly useful is the *double-subscript notation* used in Chapter 5. Recall that in the double-subscript notation for dc voltages, a voltage rise or emf designated by the symbol E or a voltage drop designated by the symbol V has two subscripts attached. The subscripts indicate that the voltage rise or drop is measured from the first subscript to the second subscript, where each subscript refers to a circuit point.

The double-subscript notation is used for polyphase circuits, but since ac voltages and currents periodically reverse direction, the double subscripts can have meaning only *if the order in which the subscripts are written denotes an instantaneous rise or drop or the direction in which the circuit is being traced.* Thus a voltage rise traced from a to b is represented by the symbol $\mathbf{E}_{ab}$ and is equal to the voltage drop from b to a. That is, phasors $\mathbf{E}_{ab}$ and $\mathbf{V}_{ba}$ are coincident (correspond exactly):

$$\mathbf{E}_{ab} = \mathbf{V}_{ba} \tag{15-1}$$

Since displacement of a phasor quantity by 180° is represented by a change in the order of subscripts or by the use of a negative sign, it follows that

$$\mathbf{E}_{ab} = -\mathbf{E}_{ba} \tag{15-2}$$

To illustrate the use of the double-subscript notation, consider the two-phase systems of voltages indicated in Fig. 15-6. Notice first of all that for polyphase circuits the sources of emf are usually represented schematically by an *alternator winding symbol* (the inductor symbol) rather than by the general ac generator symbol. The emf from a to d is found when the coils are connected in series, as in Fig. 15-7(a). The voltage rise $\mathbf{E}_{ad}$ can be specified by the phasor relationship

$$\mathbf{E}_{ad} = \mathbf{E}_{ab} + \mathbf{E}_{cd} \tag{15-3}$$

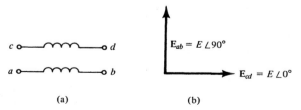

(a)   (b)

Figure 15-6 **(a)** Schematic and **(b)** phasor diagram for a two-phase system.

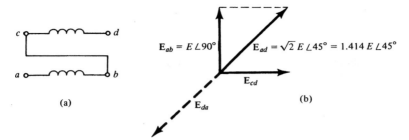

Figure 15-7 **(a)** Schematic and **(b)** phasor addition of polyphase voltages.

With the values from Fig. 15-6(b), E_{ad} is evaluated as

$$E_{ad} = E \angle 90° + E \angle 0° = E + jE = 1.414E \angle 45°$$

This phasor result is shown in Fig. 15-7(b). Also shown is the voltage rise from d to a, E_{da}, which is equal to $-E_{ad}$.

Drill Problem 15-2 ■

Terminals b and d of the coils of Fig. 15-6 are connected together. What is the phasor voltage E_{ac} measured from terminal a to c? Use a phasor diagram to help determine the voltage.

Answer. $1.414E \angle 135°$. ■ ■

15.4 PHASE SEQUENCE AND ITS MEASUREMENT

The phasors used to represent sinusoidal voltages and currents rotate in a counter-clockwise direction with an angular velocity of ω. Remembering this, we find that it is a simple matter to determine the succession of polyphase voltages, such as those of Fig. 15-8. When the phasor voltages are rotated counterclockwise, a voltage E_{ab} first passes through the reference line and is followed by E_{bc} and E_{ca}, in that order. This information is called *phase sequence*, which is defined *as the order in which poly-phase voltages pass through their respective maximum values.* Thus the phase sequence

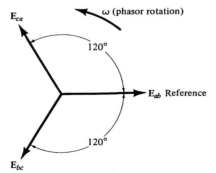

Figure 15-8 Three-phase emfs of sequence *abc.*

of the emfs is E_{ab}, E_{bc}, E_{ca} or, specifying only the subscripts, the sequence is ab, bc, ca. Each of the letters used for the emf subscripts of Fig. 15-8 is common to two voltages. Because of this particular subscripting, it is also possible to describe the given phase sequence as an abc sequence, since either all the first subscript letters or all the second subscript letters form a continuous sequence $abcabcab$, and so forth, as the phasors are rotated counterclockwise. Because of this continuous sequence of letters, the abc sequence describes the same voltage sequence as the notation bca since each is part of the same continuous sequence.

Interchanging the position of any two of the phasors in Fig. 15-8, for example, E_{bc} and E_{ca}, results in the only other three-phase continuous sequence $acbacbac$ and so forth, described as the acb sequence. This change in sequence is called a *phase reversal*, since by this change a three-phase motor will reverse its direction of rotation.

Thus one method of "measuring" the phase sequence of a three-phase system is to observe the direction of rotation of a small three-phase motor connected to the system. By comparing the motor rotation with that for a known phase sequence, we can infer the sequence of the system being "measured."

A second method of checking phase sequence is by the use of a *phase sequence indicator*, a device employing two lamps whose brightness indicates the sequence of voltages. One type of phase sequence indicator is analyzed in Section 15-11.

EXAMPLE 15-1

Determine the phase sequence of the voltages $E_{ab} = 120 \angle 30°$ V, $E_{ca} = 120 \angle -90°$ V, and $E_{bc} = 120 \angle 150°$ V.

Solution. Notice that E_{ab} leads E_{ca}. In turn, E_{ca} leads E_{bc}. Using the first subscripts of the voltage terms as they follow E_{ab}, we have a, c, and b. Hence, we have a phase sequence acb.

Drill Problem 15-3 ■

Determine the phase sequence of the following:

(a) $E_{ab} = 100 \angle 0°$ V; $E_{bc} = 100 \angle 120°$ V; $E_{ca} = 100 \angle -120°$ V.

(b) $E_{na} = 69.3 \angle -30°$ V; $E_{nc} = 69.3 \angle 90°$ V; $E_{nb} = 69.3 \angle -150°$ V.

Answer. (a) acb. (b) abc.   ■ ■

15.5 POLYPHASE CONNECTIONS

One of the two ways that the n coils of the n-phase generator are connected is the *mesh connection*, which is formed when the separate phase windings are connected to form a closed path, as shown for the six-phase system in Fig. 15-9. Notice that lines a–f, connecting the source to the load, begin at the junction point of two neighboring windings. It should be clear from the schematic and phasor diagrams that no relationship exists between the geometric position of the coil in the schematic and the phasor position of the coil voltage; that is, a coil lying schematically in a horizontal position does not necessarily have a horizontal phasor representing its voltage.

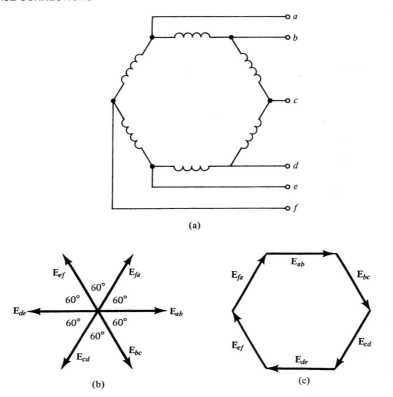

Figure 15-9 (a) Six-phase mesh-connected emfs; (b) the related phasor diagram; (c) the tip to tail diagram.

In general, the coil voltages are symmetrical and *balanced,* or *equal in magnitude,* so that when their phasors are connected as in Fig. 15-9(c) the phasor diagram closes upon itself. Thus, even though it appears that the mesh connection creates a short for the generator windings, the circulating current is zero unless the voltages are unsymmetrical or *unbalanced.*

The mesh connection for three-phase circuits is the *delta* (Δ) shown in Fig. 15-10(a). The delta connects to three lines arbitrarily called *a, b,* and *c* so that the winding voltages are specified as E_{ab}, E_{bc}, and E_{ca}. Again, it should be noted that for a symmetrical, balanced set of voltages a circulating current is not present.

The phasor diagram for the generated emfs is shown in Fig. 15-10(b), where the voltage sequence *abc* with E_{ab} as reference is depicted. Just as the generator windings are closed into a delta, the phasors may also be closed into a delta as in the tip to tail diagram of Fig. 15-10(c).

It is often more desirable to work with the voltage drops at the load rather than the rises at the source. Since V_{ba} must be in the same position as E_{ab}, V_{cb} must be in the same position as E_{bc}, and V_{ac} must be in the same position as E_{ca}, the load voltage phasor diagram of Fig. 15-10(d) identically defines the voltages specified in Fig. 15-10(c).

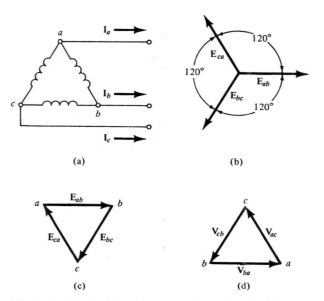

Figure 15-10   **(a)** The delta connection; **(b)**, **(c)**, **(d)** the associated phasor diagrams.

To ensure that an *n*-phase mesh connection or, in particular, a delta connection closes properly, a voltage measurement should be obtained to ensure that the last two points to be connected are at the same potential. For example, since the delta of Fig. 15-11(a) is to be completed, a voltmeter reading is obtained for the voltage between points *a* and *c'*. If the test voltage is zero, the phasor voltages close properly as in Fig. 15-11(b). If the test voltage is as in Fig. 15-11(c), the closure cannot be safely completed until, in this case, coil *c* is reversed.

In actual practice, the test voltage measurement may not be quite zero, since residual voltages and slight voltage unbalancings may produce a small potential difference. These residual voltages are generally not harmful to the successful completion of the delta connection.

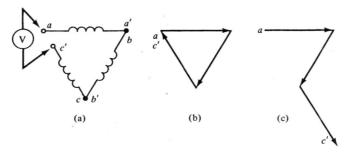

Figure 15-11   **(a)** Delta to be closed; **(b)** proper closure of phasors; **(c)** improper closure of phasors.

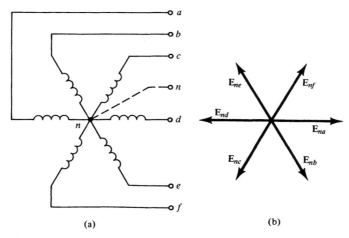

Figure 15-12 **(a)** Six-phase star-connected emfs; **(b)** a phasor diagram.

The second type of polyphase connection is the *star connection*, which results when the individual phase windings are joined at a common junction point as shown for the six-phase system in Fig. 15-12. Notice that an external line wire is connected to each phase and, in addition, a *neutral* or *common line* is available and may be connected externally, as shown by the dotted line.

The three-phase star connection is the wye (Y) shown in Fig. 15-13. If a neutral wire is used, the connection is referred to as a four-wire wye; otherwise, it is a three-wire wye.

Some important definitions related to polyphase connections can now be presented. A *phase voltage* or *current* (E_P or I_P) is the voltage across a phase winding, or the current through a phase winding, respectively. A *line voltage* or *current* (E_L or I_L) is the voltage between two line wires or the current in a line wire, respectively.

Thus, for the delta of Fig. 15-10, it is obvious that the phase voltages equal the line voltages, but the phase currents do not equal the line currents. Conversely, for the wye of Fig. 15-13, the phase voltages are not equal to the line voltages (the neutral

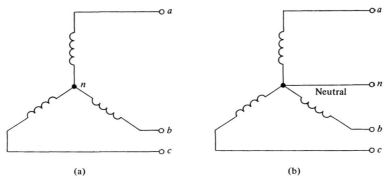

Figure 15-13 Wye connection: **(a)** Three wire and **(b)** four wire.

is not considered a line for definition purposes), but the line currents are the same currents flowing in the respective phases. These relationships are expanded in the next two sections.

15.6 THE BALANCED DELTA SYSTEM

Consider the balanced set of voltages of the delta-connected source of Fig. 15-10. For this properly closed delta, the sum of the emfs around the closed delta equals zero:

$$\mathbf{E}_{ab} + \mathbf{E}_{bc} + \mathbf{E}_{ca} = 0 \tag{15-4}$$

In addition, as was stated earlier, the magnitude of line voltage equals the magnitude of phase voltage,

$$E_L = E_P \tag{15-5}$$

If a balanced delta load (equal impedances) is connected to a delta set of emfs as in Fig. 15-14, the relationship between the phase and line currents can be obtained. The phase current $\mathbf{I}_{ab}$ in Fig. 15-14 is the current flowing from a to b due to a voltage rise from b to a or, alternatively, a voltage drop from a to b. Thus,

$$\mathbf{I}_{ab} = \frac{\mathbf{V}_{ab}}{\mathbf{Z}_{ab}} = \frac{\mathbf{E}_{ba}}{\mathbf{Z}_{ab}} \tag{15-6}$$

In a similar manner,

$$\mathbf{I}_{bc} = \frac{\mathbf{V}_{bc}}{\mathbf{Z}_{bc}} = \frac{\mathbf{E}_{cb}}{\mathbf{Z}_{bc}} \tag{15-7}$$

and

$$\mathbf{I}_{ca} = \frac{\mathbf{V}_{ca}}{\mathbf{Z}_{ca}} = \frac{\mathbf{E}_{ac}}{\mathbf{Z}_{ca}} \tag{15-8}$$

To avoid confusion with the use of voltage rises and drops, it is best *to consider voltage drops at the load, since this permits writing the voltage, current, and impedance subscripts with the same order.* It is evident that the impedance is the same looking from either direction:

$$\mathbf{Z}_{ab} = \mathbf{Z}_{ba} \tag{15-9}$$

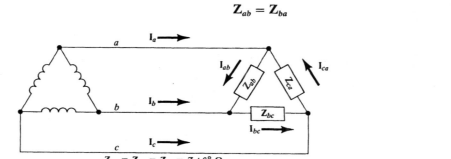

$$Z_{ab} = Z_{bc} = Z_{ca} = Z\,\angle\theta°\ \Omega$$

Figure 15-14   Balanced three-phase delta system.

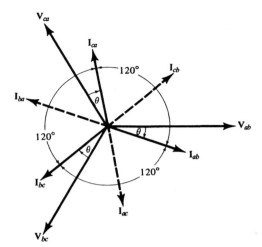

Figure 15-15 Phase currents and voltages of a balanced delta system.

It is conventional to consider the tracing direction of line currents from source to load; therefore, it is sufficient to specify these currents by single subscripts, as was done with currents I_a, I_b, and I_c of Fig. 15-14.

Any line current in Fig. 15-14 does not equal a phase current, but the phasor sum of two phase currents. Thus,

$$I_a = I_{ab} - I_{ca} = I_{ab} + I_{ac} \qquad (15\text{-}10)$$

$$I_b = I_{bc} - I_{ab} = I_{bc} + I_{ba} \qquad (15\text{-}11)$$

$$I_c = I_{ca} - I_{bc} = I_{ca} + I_{cb} \qquad (15\text{-}12)$$

Since each phase has an impedance of $Z \angle \theta°$ Ω, the phase currents will lag the phase voltages by an angle of θ degrees, as shown in Fig. 15-15. The voltage drops V_{ab}, V_{bc}, and V_{ca} are assumed to be those measured at the load of Fig. 15-14. The sequence is *abc* with V_{ab} as reference. Notice that the phase currents are equal in magnitude and displaced from each other by 120°.

From Eq. 15-10, current I_a equals the phasor sum of I_{ab} and I_{ac}. The phasor I_{ac} is $-I_{ca}$ and is obtained geometrically by reversing the direction of I_{ca}. In turn, the phasor I_{ac} bisects the 120° angle between I_{ab} and I_{bc}, as indicated in Fig. 15-16.

Since $I_{ac} = I_{ab}$, their phasor sum I_a bisects the 60° angle between the two component currents, thereby creating geometrically an isosceles triangle with the smaller sides equal to the magnitude of phase current I_P and the larger side equal to the magnitude of line current I_L. If the isosceles triangle is bisected as shown, then by trigonometry

$$\tfrac{1}{2}I_L = I_P \cos 30° \qquad (15\text{-}13)$$

or

$$I_L = 2I_P\left(\frac{\sqrt{3}}{2}\right)$$

Thus

$$I_L = \sqrt{3}\,I_P \qquad (15\text{-}14)$$

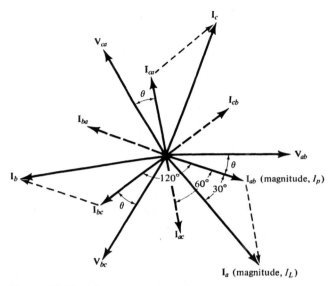

Figure 15-16 Line currents in the balanced delta.

For the balanced delta system, the magnitude of line current is $\sqrt{3}$ × the magnitude of phase current.

EXAMPLE 15-2

A three-phase balanced delta load connected to 220 V lines draws line currents of 135 A. Determine the magnitudes of the phase voltages and phase currents.

Solution. Since $E_P = E_L$, then $E_P = 220$ V. From Eq. 15-14,

$$I_L = \sqrt{3}\,I_P \quad \text{or} \quad I_P = \frac{I_L}{\sqrt{3}} = \frac{135}{\sqrt{3}} = 78 \text{ A}$$

If Fig. 15-14 represents an actual circuit to be analyzed, the circuit analyst needs to know the line voltage magnitude and the phase sequence. As a practical matter, both the line voltage and the phase sequence are easily measured. With this information, the circuit analyst selects one of the line voltages as a reference and the other two must follow in proper sequence. The voltage $\mathbf{V}_{ab}$ is usually taken as the reference.

EXAMPLE 15-3

Each impedance of a balanced three-phase delta load consists of a 20 Ω resistance in series with a 15 Ω reactance. The load is connected to 120 V lines with sequence *abc*. What phasor line currents are expected? Use $\mathbf{V}_{ab}$ as reference.

Solution. For $\mathbf{V}_{ab} = 120 \angle 0°$ V and the sequence *abc*,

$$\mathbf{V}_{bc} = 120 \angle -120° \text{ V} \quad \text{and} \quad \mathbf{V}_{ca} = 120 \angle +120° \text{ V}$$

Then

$$Z_{ab} = Z_{bc} = Z_{ca} = 20 + j15 = 25 \angle 36.9° \ \Omega$$

$$I_P = \frac{V_P}{Z_P} = \frac{120}{25} = 4.8 \text{ A}$$

and

$$I_L = \sqrt{3} I_P = \sqrt{3}(4.8) = 8.31 \text{ A}$$

The line currents can be obtained by adding the phase currents in accordance with Eqs. 15-10–15-12. Alternatively, as noted from Fig. 15-16, the line currents are displaced from the line voltages by the angle $30° + \theta$, which, in this case, equals 66.9°. Therefore,

$$\mathbf{I}_a = 8.31 \angle (0° - 66.9°) = 8.31 \angle -66.9° \text{ A}$$
$$\mathbf{I}_b = 8.31 \angle (-120° - 66.9°) = 8.31 \angle 173.1° \text{ A}$$
$$\mathbf{I}_c = 8.31 \angle (+120° - 66.9°) = 8.31 \angle 53.1° \text{ A}$$

Drill Problem 15-4 ■

A three-phase balanced delta load is a motor operating at 85% lagging power factor. At what angle does the line current (such as $\mathbf{I}_a$) lag the line voltage (such as $\mathbf{V}_{ab}$)?

Answer. 61.8°. ■ ■

Drill Problem 15-5 ■

Balanced delta impedances of $30 \angle -15° \ \Omega$ are connected to 220 V lines. What are the phase and line current magnitudes?

Answer. 7.33 and 12.7 A. ■ ■

Drill Problem 15-6 ■

When a balanced set of resistances are delta connected to 120 V lines, line currents of 5 A are measured. What are the resistance values?

Answer. 41.6 Ω. ■ ■

15.7 THE BALANCED WYE SYSTEM

An example of voltage and current relationships for a balanced wye system is the four-wire circuit; this system degenerates to a three-wire wye, since no current flows in the neutral wire. Consider the four-wire wye system of Fig. 15-17.

The impedances of the load are connected between lines a, b, or c and the neutral point; thus they are called Z_{an}, Z_{bn}, and Z_{cn}. The voltages appearing between lines a, b, or c and neutral are phase voltages, while those voltages appearing between lines are line voltages.

The phase voltages appearing at the load are balanced and are designated with an *abc* sequence as $\mathbf{V}_{an}$, $\mathbf{V}_{bn}$, $\mathbf{V}_{cn}$. The line voltages must be of the same sequence and

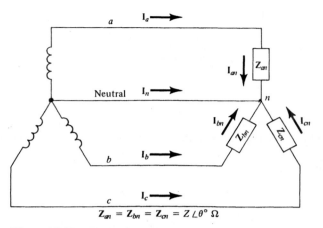

Figure 15-17   Four-wire balanced Y system.

are therefore designated $\mathbf{V}_{ab}$, $\mathbf{V}_{bc}$, and $\mathbf{V}_{ca}$. From Fig. 15-17,

$$\mathbf{V}_{ab} = \mathbf{V}_{an} + \mathbf{V}_{nb} \qquad (15\text{-}15)$$

$$\mathbf{V}_{bc} = \mathbf{V}_{bn} + \mathbf{V}_{nc} \qquad (15\text{-}16)$$

$$\mathbf{V}_{ca} = \mathbf{V}_{cn} + \mathbf{V}_{na} \qquad (15\text{-}17)$$

The phasor diagram of Fig. 15-18 shows these relationships. Formation of the phasor line voltage $\mathbf{V}_{ab}$ is similar to that for the other two phasor line voltages. Geometrically, phasor $\mathbf{V}_{ab}$ is the larger side of the isosceles triangle formed by two phase voltages of magnitudes V_P. Then by trigonometry

$$\tfrac{1}{2}V_L = V_P \cos 30° \qquad (15\text{-}18)$$

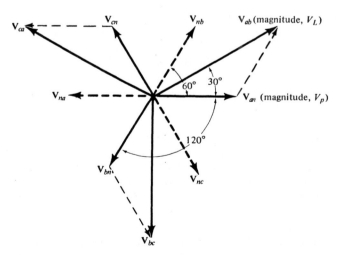

Figure 15-18   Voltage relationships of the balanced Y.

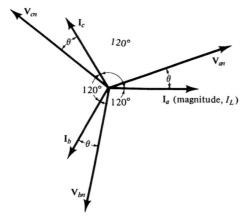

Figure 15-19 Line currents in the balanced Y.

or
$$V_L = \sqrt{3}\, V_P \qquad (15\text{-}19)$$

Thus for the balanced three-phase wye system, the magnitude of line voltage is $\sqrt{3} \times$ the magnitude of the phase voltage. Since the phase current equals the line current, as is evident from Fig. 15-17,

$$I_L = I_P \qquad (15\text{-}20)$$

Notice the 30° angle between the line and phase voltages.

Since the impedances $\mathbf{Z}_{an}$, $\mathbf{Z}_{bn}$, and $\mathbf{Z}_{cn}$ are equal and since the applied voltages are balanced, the currents $\mathbf{I}_a$, $\mathbf{I}_b$, and $\mathbf{I}_c$ form the balanced set of currents shown in Fig. 15-19. For convenience, the current $\mathbf{I}_a$ is chosen as reference and, for completeness, the phase voltages are included.

Referring to Fig. 15-17, a current equation at node n provides

$$\mathbf{I}_a + \mathbf{I}_b + \mathbf{I}_c + \mathbf{I}_n = 0 \qquad (15\text{-}21)$$

or
$$\mathbf{I}_n = -(\mathbf{I}_a + \mathbf{I}_b + \mathbf{I}_c) \qquad (15\text{-}22)$$

Then the phasor currents are obtained by referring to Fig. 15-19 and are substituted into Eq. 15-22 with the results

$$\mathbf{I}_n = -(I_L - I_L \cos 60° - jI_L \sin 60° - I_L \cos 60° + jI_L \sin 60°)$$
$$= 0 \qquad (15\text{-}23)$$

As Eq. 15-23 shows, no current flows in the neutral wire if the wye system is balanced. The neutral wire can even be removed without affecting the system operation. With the unbalanced wye system, however, current does flow in the neutral wire, and removal of the wire drastically affects the circuit operation.

EXAMPLE 15-4

Given the line voltages V_L in Fig. 15-17 as 220 V and $\mathbf{Z}_{an} = \mathbf{Z}_{bn} = \mathbf{Z}_{cn} = 8 + j6\ \Omega$, find the magnitude of the line currents.

Solution. From Eq. 15-19,

$$V_L = \sqrt{3}\, V_P \quad \text{or} \quad V_P = \frac{V_L}{\sqrt{3}} = \frac{220}{\sqrt{3}} = 127 \text{ V}$$

Also,

$$\mathbf{Z}_P = \mathbf{Z}_{an} = \mathbf{Z}_{bn} = \mathbf{Z}_{cn} = 8 + j6 = 10 \angle 36.9° \ \Omega$$

Then

$$I_L = I_P = \frac{V_P}{Z_P} = \frac{127}{10} = 12.7 \text{ A}$$

Example 15-4 can also be worked by using complex analysis for the phasor voltages and currents. If the complex currents are desired, the sequence of voltages must be established and, furthermore, one of the line or phase voltages must be selected as a reference.

EXAMPLE 15-5

Given the line voltages V_L in Fig. 15-17 as 220 V (sequence *abc*) and $\mathbf{Z}_{an} = \mathbf{Z}_{bn} = \mathbf{Z}_{cn} = 8 + j6 \ \Omega$, find the phasor currents, using $\mathbf{V}_{an}$ as reference.

Solution. As in Example 15-4, $V_P = 127$ V and $\mathbf{Z}_{an} = \mathbf{Z}_{bn} = \mathbf{Z}_{cn} = 10 \angle 36.9° \ \Omega$. With $\mathbf{V}_{an}$ as reference and *abc* as the sequence,

$$\mathbf{V}_{an} = 127 \angle 0° \text{ V}$$
$$\mathbf{V}_{bn} = 127 \angle -120° \text{ V}$$
$$\mathbf{V}_{cn} = 127 \angle +120° \text{ V}$$

Then

$$\mathbf{I}_a = \mathbf{I}_{an} = \frac{\mathbf{V}_{an}}{\mathbf{Z}_{an}} = \frac{127 \angle 0°}{10 \angle 36.9°} = 12.7 \angle -36.9° \text{ A}$$

$$\mathbf{I}_b = \mathbf{I}_{bn} = \frac{\mathbf{V}_{bn}}{\mathbf{Z}_{bn}} = \frac{127 \angle -120°}{10 \angle 36.9°} = 12.7 \angle -156.9° \text{ A}$$

$$\mathbf{I}_c = \mathbf{I}_{cn} = \frac{\mathbf{V}_{cn}}{\mathbf{Z}_{cn}} = \frac{127 \angle +120°}{10 \angle 36.9°} = 12.7 \angle +83.1° \text{ A}$$

The line voltages, although not required, are obtained by Eqs. 15-15–15-17 and are included in the phasor diagram of Fig. 15-20.

Drill Problem 15-7 ■

A balanced wye load is connected to lines having voltages of 127 V referenced to the neutral or ground. **(a)** What is the line voltage magnitude? **(b)** What magnitude of line currents flow if each leg of the load impedance is $10 - j4 \ \Omega$?

Answer. (a) 220 V. (b) 11.8 A.   ■ ■

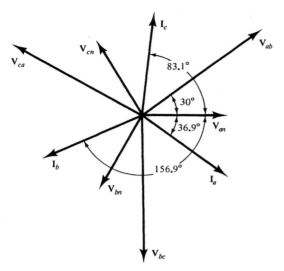

Figure 15-20 Phasor relationships for Example 15-5.

Drill Problem 15-8 ■

Balanced delta impedances of $30 \angle -15°$ Ω are connected to 220 V lines. Convert the delta load to a wye load and calculate the magnitude of the line currents. Compare this to the value obtained in Drill Problem 15-5.

Answer. 12.7 A. The line current is the same value since the Y is equivalent to the Δ. ■ ■

15.8 POWER IN THE BALANCED SYSTEM

The total power in an *n*-phase system, whether balanced or unbalanced, is the sum of the single-phase powers in the respective phases; that is,

$$P_t = P_1 + P_2 + P_3 + \cdots + P_n \tag{15-24}$$

where P_n is the phase power of the *n*th phase.

Under balanced conditions,

$$P_1 = P_2 = P_3 = P_n$$

so Eq. 15-24 becomes

$$P_t = nP_P \tag{15-25}$$

where P_P is the power per phase given by the single-phase power expression

$$P_P = V_P I_P \cos \theta$$

It is important to remember that V_P is the voltage per phase, I_P the phase current, and θ the angle between the phase voltage and current. For the three-phase system, then,

$$P_t = 3V_P I_P \cos \theta \tag{15-26}$$

The development of this last equation in terms of line voltages and currents is desirable and follows from the relationships between phase and line values. For the wye connection

$$P_t = 3V_P I_P \cos \theta = 3\left(\frac{V_L}{\sqrt{3}}\right) I_L \cos \theta$$

$$= \sqrt{3} \, V_L I_L \cos \theta$$

and for the delta connection

$$P_t = 3V_P I_P \cos \theta = 3V_L \left(\frac{I_L}{\sqrt{3}}\right) \cos \theta$$

$$= \sqrt{3} \, V_L I_L \cos \theta$$

Hence the equation for three-phase power in a balanced system, whether wye or delta, is

$$P_t = \sqrt{3} \, V_L I_L \cos \theta \tag{15-27}$$

The right side of Eq. 5-27 consists of two terms, $\sqrt{3} \, V_L I_L$ and $\cos \theta$. The product term $\sqrt{3} \, V_L I_L$ represents the *apparent power of a balanced three-phase circuit* in *volt-amperes* and is denoted by the letter symbol S_t. Thus,

$$S_t = \sqrt{3} \, V_L I_L \tag{15-28}$$

$\cos \theta$ is defined here as *the power factor of the balanced three-phase circuit*. It equals the power factor for an individual phase of the balanced system. Notice that *the angles between line voltages and line currents are not power factor angles*, since they involve angles such as $\theta + 30°$.

The *three-phase reactive power* Q_t is the product of the apparent power and the sine of the phase angle θ.

$$Q_t = S_t \sin \theta = \sqrt{3} \, V_L I_L \sin \theta \tag{15-29}$$

As with a single-phase circuit, a power triangle can be constructed for the three-phase circuit. The real power is along the horizontal axis, the reactive power is along the vertical axis, and the apparent power is the hypotenuse. By the Pythagorean relationship,

$$S_t = \sqrt{P_t^2 + Q_t^2} \tag{15-30}$$

EXAMPLE 15-6

Using the data from Example 15-4 or 15-5, calculate the **(a)** apparent power, **(b)** dissipated power, and **(c)** reactive power.

Solution. From the examples, $V_L = 220$ V, $I_L = I_P = 12.7$ A, and $\theta = 36.9°$. Therefore,

(a) $S_t = \sqrt{3}\, V_L I_L = \sqrt{3}\,(220)(12.7) = 4839$ VA;

(b) $P_t = S_t \cos \theta = 4839 \cos 36.9° = 3870$ W;

(c) $Q_t = S_t \sin \theta = 4839 \sin 36.9° = 2905$ var.

Drill Problem 15-9 ■

The power to a three-phase load is 20 kW. What is the power factor of the load if the line voltage is 440 V and the line current is 32.8 A?

Answer. 0.8. ■ ■

Drill Problem 15-10 ■

A balanced delta load with phase impedances of $10 \angle 30°$ Ω is connected to 220 V lines. Determine the **(a)** apparent power, **(b)** real power, and **(c)** reactive power.

Answer. **(a)** 14.52 kVA. **(b)** 12.58 kW. **(c)** 7.26 kvar. ■ ■

15.9 POLYPHASE POWER MEASUREMENT

According to Eq. 15-24, the total power delivered to a polyphase load is the sum of the powers taken by the individual phase loads. This relationship suggests that n wattmeters be placed in an n-phase circuit in such a way that each wattmeter reads the power of one of the n-phases. In particular, the wattmeter connections of Fig. 15-21 are used to measure the total power of a three-phase wye or delta load.

The system of power measurement shown in Fig. 15-21 applies to either balanced or unbalanced loads. In fact, if the system is balanced, all of the wattmeters will read the same power ($\frac{1}{3}P_t$). Then we need use only one wattmeter and obtain the total power as $3 \times$ the single wattmeter reading.

As simple as the idea of measuring each phase power is, it is often difficult to achieve in practice. Sometimes, the neutral point of a wye is inaccessible, the phase connections of the delta cannot be broken open for insertion of the wattmeters, or only the three load terminals are available and it is not known whether the load is a wye or a delta. In this case, the total power can be measured by the *three-wattmeter method* shown in Fig. 15-22. Although the proof is beyond the scope of this book, it can be shown that the three-wattmeter method indicates the total power, regardless of whether the load is balanced or unbalanced or is a wye or delta.

The three wattmeters in Fig. 15-22 need not have potential coils of equal impedance. Without equal potential coil impedances, the meter readings will not be equal for a balanced load, yet the sum of these readings equals the total power. By an extension of this idea, if one of the potential coils is shorted, that related wattmeter reading becomes zero and the other two readings equal the total power. For example, if the potential coil of the wattmeter measuring P_c in Fig. 15-22 is shorted, the neutral

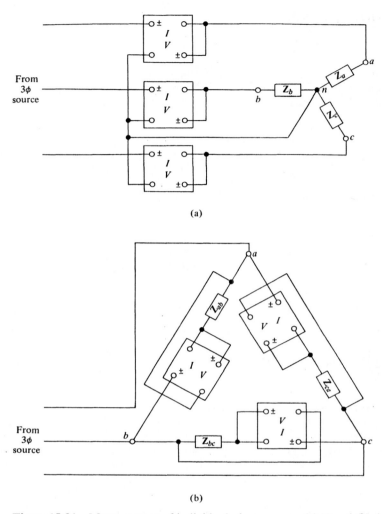

(a)

(b)

Figure 15-21   Measurement of individual phase power: **(a)** Y and **(b)** Δ.

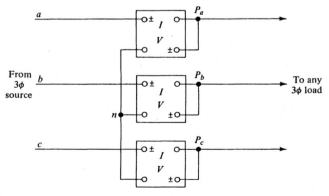

Figure 15-22   Three-wattmeter method of measuring the three-phase power.

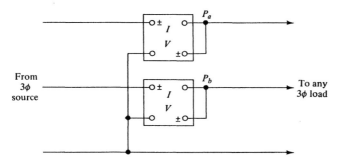

Figure 15-23 Two-wattmeter method of measuring the three-phase power.

point n of Fig. 15-22 then connects to line c. This measurement situation, the *two-wattmeter method*, is illustrated in Fig. 15-23.

Although the individual wattmeters in the two-wattmeter method no longer indicate the power taken by any particular phase, their algebraic sum does equal the total power taken by either a balanced or unbalanced load. A general proof that the method works even for an unbalanced load is beyond the scope of this text. It is sufficient here to show that the method works for the balanced case, in particular, a delta load.

Consider the two-wattmeter method applied to the delta load of Fig. 15-24. Notice that the wattmeter indications P_1 and P_2 depend on the line current and potential drop measured by the respective wattmeter. Thus,

$$P_1 = V_{ac}I_a \cos(\text{angle between } \mathbf{V}_{ac} \text{ and } \mathbf{I}_a)$$

and $$P_2 = V_{bc}I_b \cos(\text{angle between } \mathbf{V}_{bc} \text{ and } \mathbf{I}_b)$$

The line (and phase) voltages $\mathbf{V}_{ab}$, $\mathbf{V}_{bc}$, and $\mathbf{V}_{ca}$ are assumed to be as shown in the phasor diagram of Fig. 15-25. It follows that the respective phase currents lag their respective phase voltages by the impedance angle θ. In turn, the line currents $\mathbf{I}_a$ and $\mathbf{I}_b$ are found by the phasor addition of the appropriate phase currents.

By reversing the phasor voltage $\mathbf{V}_{ca}$, one obtains the angle between $\mathbf{V}_{ac}$ and $\mathbf{I}_a$ as $30° - \theta$. It is also seen that the angle between $\mathbf{V}_{bc}$ and $\mathbf{I}_b$ is $30° + \theta$. Therefore,

$$P_1 = V_{ac}I_a \cos(30° - \theta) \quad \text{and} \quad P_2 = V_{bc}I_b \cos(30° + \theta)$$

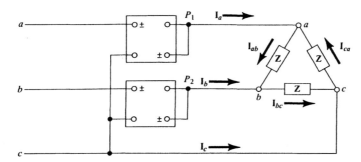

Figure 15-24 Two-wattmeter method applied to a balanced delta circuit.

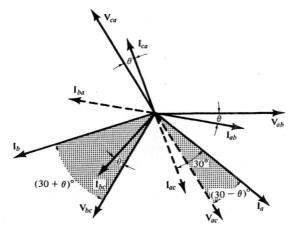

Figure 15-25 Phasor diagram for circuit of Fig. 15-24.

In each of the two preceding equations, the product of voltage and current represents a magnitude only, since the phase relationship is contained in the cosine term. Thus,

$$P_1 = V_L I_L \cos(30° − \theta) \qquad (15\text{-}31)$$

and

$$P_2 = V_L I_L \cos(30° + \theta) \qquad (15\text{-}32)$$

Identical equations result for a similar study of the two-wattmeter method and a balanced wye load.

Combining Eqs. 15-31 and 15-32, we obtain

$$P_1 + P_2 = V_L I_L \cos(30° − \theta) + V_L I_L \cos(30° + \theta)$$
$$= V_L I_L [\cos(30° − \theta) + \cos(30° + \theta)]$$

From trigonometry,

$$\cos(x \pm y) = \cos x \cos y \mp \sin x \sin y$$

so that by substitution

$$P_1 + P_2 = V_L I_L (\cos 30° \cos \theta + \sin 30° \sin \theta + \cos 30° \cos \theta − \sin 30° \sin \theta)$$
$$= 2 V_L I_L \cos 30° \cos \theta = 2 V_L I_L \left(\frac{\sqrt{3}}{2}\right) \cos \theta$$
$$= \sqrt{3} \, V_L I_L \cos \theta$$

Hence,

$$P_t = P_1 + P_2 \qquad (15\text{-}33)$$

Notice in Fig. 15-24 that the $\pm$ terminal of the potential coil of each wattmeter is connected to the line wire in which the current coil of that particular wattmeter is connected. This is the standard connection for obtaining an upscale needle deflection, provided that the measured voltage and current are within 90° of each other in phase. If the measured voltage and current are out of phase by an angle between 90° and 180°, the wattmeter has a downscale deflection. In such a situation, the potential coil is reversed in order to get an upscale reading. In turn, the reading is considered *algebraically negative* and is subtracted from the other wattmeter reading.

Analyzing Eqs. 15-31 and 15-32, we find the two wattmeter readings to be equal for $\theta = 0°$:

$$P_1 = P_2 = \frac{P_t}{2}$$

With $0° < \theta < 60°$, both readings as predicted by Eqs. 15-31 and 15-32 are positive and $P_2 < P_1$. For $\theta = 60°$, one wattmeter reads zero and the other reads the total power. Finally, for $60° < \theta \le 90°$, P_2 is negative, so

$$P_t = P_1 + (-P_2) = P_1 - P_2$$

Note that even though the impedance angle θ is lagging (inductive) in the preceding proof and analysis, the same arguments exist for a leading (capacitive) phase angle.

EXAMPLE 15-7

A balanced three-phase load draws a line current of 40 A when connected to 440 V lines. If the power factor is 0.8 lagging, what power indications would be noted by the two-wattmeter method? What is the total power drawn by the load?

Solution. $\cos \theta = 0.8$; therefore, $\theta = 36.9°$.

$$P_1 = V_L I_L \cos(30° - \theta) = (440)(40)\cos(-6.9°) = 17\,472 \text{ W}$$
$$P_2 = V_L I_L \cos(30 + \theta) = (440)(40)\cos(66.9°) = 6905 \text{ W}$$

and $\quad P_t = P_1 + P_2 = 24\,377 \text{ W}$

A definite ratio between P_1 and P_2 exists for any particular power factor of a balanced load. This ratio, from Eqs. 15-31 and 15-32, is

$$\frac{P_2}{P_1} = \frac{\cos(30° + \theta)}{\cos(30° - \theta)}$$

By trigonometric substitution and subsequent algebraic manipulation, it is possible to solve the preceding equation for the angle in terms of P_1 and P_2. The result is

$$\tan \theta = \sqrt{3}\left(\frac{P_1 - P_2}{P_1 + P_2}\right) \tag{15-34}$$

Equation 15-34 allows one to determine the phase angle of a balanced load from the two wattmeter readings. In using Eq. 15-34, we let P_1 represent the larger of the two wattmeter readings. Since the ratio of P_2 to P_1 has a definite relationship to the power factor, it is also possible to plot a curve of power factor versus P_2/P_1. This curve is shown in Fig. 15-26.

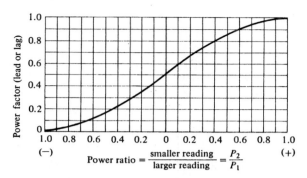

Figure 15-26   Power factor versus wattmeter ratio for the two-wattmeter method.

EXAMPLE 15-8

A three-phase motor is connected to 120 V lines. The two-wattmeter method produces wattmeter readings of 3210 and -1710 W. What is the phase angle of the load?

Solution

$$\tan \theta = \sqrt{3}\left(\frac{P_1 - P_2}{P_1 + P_2}\right) = \frac{3210 - (-1710)}{3210 - 1710}$$

$$= \sqrt{3}\left(\frac{4920}{1500}\right) = 5.68$$

Therefore,

$$\theta = \tan^{-1}(5.68) = 80° \text{ (inductive)}$$

or, using Fig. 15-26,

$$\frac{P_2}{P_1} = \frac{-1710}{3210} = -0.533$$

$$\text{Pf} = \cos \theta = 0.175$$

Therefore,

$$\theta = \cos^{-1}(0.175) \approx 80°$$

Drill Problem 15-11 ■

The power readings for a three-phase inductive load as measured by the two-wattmeter method are 5 and 3 kW. Determine **(a)** the phase angle of the load by use of both Eq. 15-34 and Fig. 15-26 and **(b)** the total power.

Answer.   (a) 23.4°. (b) 8 kW.   ■ ■

Drill Problem 15-12 ■

A balanced wye load operating at a power factor of 93% draws line currents of 12 A when connected to 220 V lines. If the two-wattmeter method is used to measure the power, what does each wattmeter indicate?

Answer.   1641 W; 2611 W.   ■ ■

Drill Problem 15-13 ■

The two-wattmeter method is used to measure the power to a three-phase balanced load. The measurements indicate readings of 4 and -1.5 kW. **(a)** What is the total power? **(b)** What is the total reactive power? **(c)** What is the apparent power?

Answer. (a) 2.5 kW. (b) 9.53 kvar. (c) 9.85 kVA. ■ ■

15.10 THE UNBALANCED DELTA SYSTEM

Any polyphase load that consists of impedances that are not all equal is said to be part of an unbalanced system. Likewise, if the voltages impressed on the load are unequal and differ in phase by angles that are not equal, the system they are part of is also unbalanced. In this and the succeeding section it is assumed that the applied voltages are balanced and any system unbalancing results from an unbalanced load.

For the unbalanced delta circuit, the voltage drops across the respective load impedances are known, thereby enabling the determination of phase currents. The line currents are then found by phasor addition of the phase currents, as illustrated by the following example.

EXAMPLE 15-9

The unbalanced delta load of Fig. 15-27 has balanced voltages of 120 V and the sequence *acb*. Calculate the line currents and the total power dissipated.

Solution. Since the voltage sequence is *acb*, the voltages $\mathbf{V}_{ab}$, $\mathbf{V}_{ca}$, and $\mathbf{V}_{bc}$ are (with $\mathbf{V}_{ab}$ as reference)

$$\mathbf{V}_{ab} = 120 \angle 0° \text{ V}$$
$$\mathbf{V}_{ca} = 120 \angle -120 \text{ V}$$
$$\mathbf{V}_{bc} = 120 \angle +120 \text{ V}$$

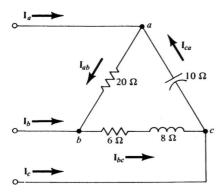

Figure 15-27 Circuit of Example 15-9.

The phase currents are

$$\mathbf{I}_{ab} = \frac{\mathbf{V}_{ab}}{\mathbf{Z}_{ab}} = \frac{120 \angle 0°}{20 \angle 0°} = 6 \angle 0° = 6 + j0 \text{ A}$$

$$\mathbf{I}_{ca} = \frac{\mathbf{V}_{ca}}{\mathbf{Z}_{ca}} = \frac{120 \angle -120°}{10 \angle -90°} = 12 \angle -30° = 10.4 - j6 \text{ A}$$

$$\mathbf{I}_{bc} = \frac{\mathbf{V}_{bc}}{\mathbf{Z}_{bc}} = \frac{120 \angle +120°}{6 + j8} = \frac{120 \angle +120°}{10 \angle 53.2°} = 12 \angle 66.8°$$
$$= 4.73 + j11.0 \text{ A}$$

The line currents are then

$$\mathbf{I}_a = \mathbf{I}_{ab} - \mathbf{I}_{ca} = 6 + j0 - (10.4 - j6) = -4.4 + j6$$
$$= 7.45 \angle 126.2° \text{ A}$$

$$\mathbf{I}_b = \mathbf{I}_{bc} - \mathbf{I}_{ab} = 4.73 + j11 - 6 = -1.27 + j11$$
$$= 11.1 \angle 96.6° \text{ A}$$

$$\mathbf{I}_c = \mathbf{I}_{ca} - \mathbf{I}_{bc} = 10.4 - j6 - (4.73 + j11)$$
$$= 5.67 - j17 = 17.9 \angle -71.5° \text{ A}$$

The total power equals the sum of the individual phase powers P_{ab}, P_{ca}, and P_{bc}.

$$P_{ab} = V_{ab}I_{ab} \cos \theta_{ab} = (120)(6) \cos 0° = 720 \text{ W}$$
$$P_{ca} = 0 \text{ W}$$
$$P_{bc} = V_{bc}I_{bc} \cos \theta_{bc} = (120)(12) \cos 53.2° = 864 \text{ W}$$
$$P_t = P_{ab} + P_{ca} + P_{bc} = 720 + 864 = 1584 \text{ W}$$

Drill Problem 15-14 ∎

An unbalanced delta load of $\mathbf{Z}_{ab} = \mathbf{Z}_{ca} = 6 + j8 \; \Omega$ and $\mathbf{Z}_{bc} = 10 \; \Omega$ is connected to 120 V lines of sequence *abc*. Determine the phasor line currents if $\mathbf{V}_{ab}$ is taken as a reference (120 $\angle 0°$).

Answer.   20.8 $\angle -83.1°$, 13.2 $\angle -176.5°$, and 23.9 $\angle 63.4°$ A.   ∎ ∎

15.11 THE UNBALANCED WYE SYSTEM

The unbalanced three-phase wye system may be either of the four- or three-wire type. In the four-wire wye system, shown in Fig. 15-17, the neutral point *n* is fixed in potential by connection to the source neutral. It follows that balanced and known potentials $\mathbf{V}_{an}$, $\mathbf{V}_{bn}$, and $\mathbf{V}_{cn}$ appear across the load impedances. However, if the load impedances are not equal, the resulting currents $\mathbf{I}_{an}$, $\mathbf{I}_{bn}$, and $\mathbf{I}_{cn}$ are not equal in magnitude.

The solution of an unbalanced four-wire wye system closely follows that of Example 15-5. However, the line currents are not equal in magnitude and the neutral current as given by Eq. 15-22 is not zero.

If the neutral wire of the unbalanced four-wire wye is removed, the neutral point of the resultant three-wire wye is no longer fixed but is free to *float*, that is, to assume a potential determined by the values of load impedance. It now becomes necessary to use loop or nodal analysis to determine the phase currents and voltages. The following example makes use of loop analysis.

EXAMPLE 15-10

The unbalanced wye of Fig. 15-28 is connected to a 100 V, three-phase source of phase sequence *abc*. Find the line currents and phase voltages if Z_{an} and Z_{cn} are lamps each having 100 Ω resistance and Z_{bn} is a capacitor having 100 Ω reactance.

Solution.   Since the phase sequence is *abc*, the line voltages may be assumed to be $V_{ab} = 100 \angle 0°$ V, $V_{bc} = 100 \angle -120°$ V, and $V_{ca} = 100 \angle +120°$ V.

Writing two loop equations with loop currents I_1 and I_2 as shown in Fig. 15-28 results in

$$V_{ab} = I_1(Z_{an} + Z_{bn}) + I_2 Z_{bn} \qquad (15\text{-}35)$$

$$V_{cb} = I_1 Z_{bn} + I_2(Z_{bn} + Z_{cn}) \qquad (15\text{-}36)$$

Substituting values into Eqs. 15-35 and 15-36 and noting that $I_1 = I_{an}$, $I_2 = I_{cn}$, and $V_{cb} = -V_{bc}$, we obtain

$$100 \angle 0° = I_{an}(100 - j100) + I_{cn}(-j100)$$

and
$$100 \angle 60° = I_{an}(-j100) + I_{cn}(100 - j100)$$

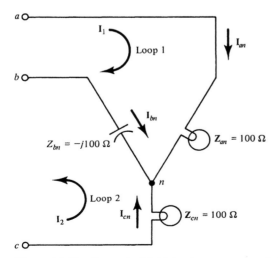

Figure 15-28   Example 15-10 (a phase-sequence indicator).

The equations are modified by dividing by 100 and changing $1 \angle 60°$ to its rectangular form. Thus,

$$1 = \mathbf{I}_{an}(1 - j1) + \mathbf{I}_{cn}(-j1)$$

and

$$0.5 + j0.866 = \mathbf{I}_{an}(-j1) + \mathbf{I}_{cn}(1 - j1)$$

Solving for $\mathbf{I}_{an}$, we obtain

$$
\mathbf{I}_{an} = \frac{\begin{vmatrix} 1 & -j1 \\ 0.5 + j0.866 & 1 - j1 \end{vmatrix}}{\begin{vmatrix} 1 - j1 & -j1 \\ -j1 & 1 - j1 \end{vmatrix}} = \frac{1 - j1 + j0.5 + j^2 0.866}{1 - j2 - 1 + 1}
$$

$$
= \frac{0.134 - j0.5}{1 - j2} = \frac{0.518 \angle -75°}{2.24 \angle -63.4°} = 0.231 \angle -12° \text{ A}
$$

and for $\mathbf{I}_{cn}$ and $\mathbf{I}_{bn}$,

$$
\mathbf{I}_{cn} = \frac{\begin{vmatrix} 1 - j1 & 1 \\ -j1 & 0.5 + j0.866 \end{vmatrix}}{2.24 \angle -63.4°} = \frac{1.366 + j1.366}{2.24 \angle -63.4°}
$$

$$
= \frac{1.93 \angle +45°}{2.24 \angle -63.4°} = 0.863 \angle 108° \text{ A}
$$

$$
\begin{aligned}
\mathbf{I}_{bn} = -(\mathbf{I}_{an} + \mathbf{I}_{cn}) &= -(0.231 \angle -12° + 0.863 \angle 108°) \\
&= -(0.226 - j0.048 - 0.267 + j0.82) \\
&= -(-0.041 + j0.77) = -(0.77 \angle 93°) = 0.77 \angle -87° \text{ A}
\end{aligned}
$$

Finally,

$$
\begin{aligned}
\mathbf{V}_{an} = \mathbf{I}_{an}\mathbf{Z}_{an} &= (0.231 \angle -12°)(100) = 23.1 \angle -12° \text{ V} \\
\mathbf{V}_{bn} = \mathbf{I}_{bn}\mathbf{Z}_{bn} &= (0.77 \angle -87°)(100 \angle -90°) = 77 \angle -187° \text{ V} \\
\mathbf{V}_{cn} = \mathbf{I}_{cn}\mathbf{Z}_{cn} &= (0.863 \angle 108°)(100) = 86.3 \angle 108° \text{ V}
\end{aligned}
$$

The circuit used in Example 15-10 is known as a *phase-sequence indicator*. Notice that, for phase sequence *abc*, lamp *c* has a larger voltage drop across it than lamp *a*; hence lamp *c* is brighter than lamp *a*. On the other hand, for sequence *acb*, lamp *a* is brighter than lamp *c*. This case is presented as a problem at the end of the chapter. Thus, *for the phase-sequence indicator shown, the phase rotation is specified by the lines taken in the order bright lamp, dim lamp, capacitor.*

It should be pointed out that the line currents can be found for Example 15-10 by converting the wye impedances to equivalent delta impedances. The analysis then follows the method of Section 15-10.

Drill Problem 15-15 ■

You are given an unbalanced four-wire wye system for which $\mathbf{V}_{an} = 70 \angle 0°$ V, $\mathbf{V}_{bn} = 70 \angle -120°$ V, and $\mathbf{V}_{cn} = 70 \angle +120°$ V. The phase impedances are $\mathbf{Z}_{an} = 10$ Ω, $\mathbf{Z}_{bn} = 7.07 + j7.07$ Ω, and $\mathbf{Z}_{cn} = 7.07 + j7.07$ Ω. Determine the phasor line and neutral currents.

Answer. $7 \angle 0°$ A; $7 \angle -165°$ A; $7 \angle 75°$ A; $5.36 \angle -112.5°$ A. ■ ■

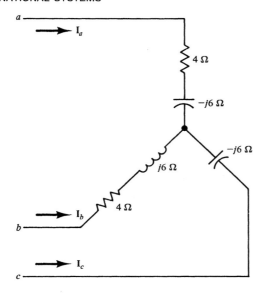

Figure 15-29 Drill Problem 15-16.

Drill Problem 15-16 ■

Determine the line currents for the circuit of Fig. 15-29. The line voltages measure 100 V and are of sequence *abc*. Use $\mathbf{V}_{ab}$ as reference.

Answer. 13.7 $\angle 0.8°$ A; 1.32 $\angle 132.7°$ A; 12.8 $\angle -174.8°$ A. ■ ■

15.12 COMBINATIONAL SYSTEMS

Often a three-phase source delivers power to line wires that are connected to more than one type of load. For instance, a delta and a wye load may both be connected to the lines as shown in Fig. 15-30(a). In addition to a wye or delta load, one or more single-phase loads (similar to the legs of a delta) may be connected between two line wires as in Fig. 15-30(b). When loads of different types are connected to three-phase lines as in Fig. 15-30, the system is called a *combinational system*.

Rather than loop or nodal analysis, which lead to a large number of simultaneous equations, two other methods may be used to solve the line currents of a combinational system.

One approach is to convert any wye loads into equivalent delta loads. The corresponding legs of all deltas, equivalent deltas, and single-phase loads are then combined in parallel to form a single delta. Line currents are then found by the method of Section 15-10.

Another approach is to consider the loads separately and then combine the results by way of superposition. The following example illustrates this latter method.

EXAMPLE 15-11

The delta in Fig. 15-30(b) has a phase impedance of 16 Ω and represents a motor operating at an 86.7% power factor. The single-phase load $\mathbf{Z}'_{ca}$ is a resistive

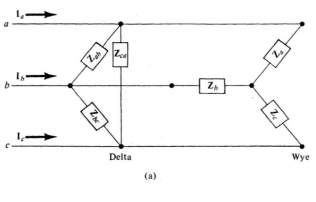

Delta   Wye

(a)

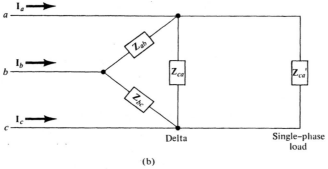

Delta   Single–phase
                              load

(b)

Figure 15-30   Two examples of combinational systems.

heating element of 10 Ω. Calculate the line currents if the line voltage is 230 V for sequence *abc*.

Solution.   With $\mathbf{V}_{ab}$ as reference, $\mathbf{V}_{ab} = 230 \angle 0°$, $\mathbf{V}_{bc} = 230 \angle -120°$, $\mathbf{V}_{ca} = 230 \angle +120°$ V.

Motor: $\theta = \cos^{-1} 0.867 = 30°$.

$$\mathbf{Z}_{ab} = \mathbf{Z}_{bc} = \mathbf{Z}_{ca}$$

Therefore,

$$\mathbf{I}_{ab} = \mathbf{I}_{bc} = \mathbf{I}_{ca} = \frac{230}{16} = 14.38 \text{ A}$$

Since each phase current lags the corresponding phase voltage by 30°

$$\mathbf{I}_{ab} = 14.38 \angle -30° = 12.45 - j7.18 \text{ A}$$
$$\mathbf{I}_{bc} = 14.38 \angle -150° = -12.45 - j7.18 \text{ A}$$
$$\mathbf{I}_{ca} = 14.38 \angle +90° = 0 + j14.38 \text{ A}$$

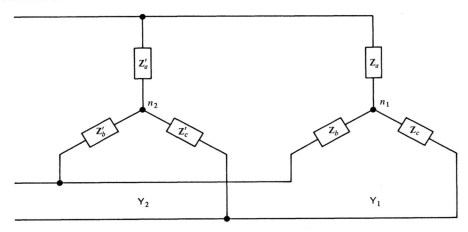

Figure 15-31   A combinational system with two wyes.

Heating Element:

$$I'_{ca} = \frac{V_{ca}}{Z'_{ca}} = \frac{230 \angle 120°}{10 \angle 0°} = 23 \angle 120° = -11.5 + j19.9 \text{ A}$$

$$I_a = I_{ab} - I_{ca} - I'_{ca} = 12.45 - j7.18 - j14.38$$
$$+ 11.5 - j19.9 = 23.9 - j41.4 = 47.8 \angle -60° \text{ A}$$

$$I_b = I_{bc} - I_{ab} = -12.45 - j7.18 - 12.45 + j7.18$$
$$= -24.9 + j0 = 24.9 \angle +180° \text{ A}$$

$$I_c = I_{ca} + I'_{ca} - I_{bc} = j14.38 - 11.5 + j19.9$$
$$+ 12.45 + j7.18 = 0.95 + j41.4 \approx 41.4 \angle 88.7° \text{ A}$$

Although it might be more advantageous in the analysis of combinational systems to convert the wye loads to delta loads, it is, of course, possible to convert the delta loads to wye loads. The problem with this approach is that the elements of the wye loads may not be balanced, and therefore the various branches of the wye cannot always be combined in parallel.

Consider the combinational system of Fig. 15-30(a). If the delta of that figure is converted to a wye, the circuit of Fig. 15-31 is the result. The unwary circuit analyst might consider the two wyes in parallel and proceed to combine the Y_2 values in parallel with the Y_1 values. However, that procedure is correct *only if* the two wyes are each balanced.

The vector voltages for a two-wye system are shown in Fig. 15-32. With a balanced Y_1, the line voltages form an equilateral triangle and the potential for node n_1 is at the "geometric center" of the triangle, so that the phase voltages form a balanced set, as in Fig. 15-32(a). If the second wye, Y_2, is balanced, it too has phase

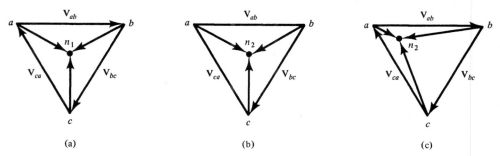

Figure 15-32 Voltages for the two-wye system of Fig. 15-31: **(a)** For Y_1: **(b)** for a balanced Y_2; **(c)** for an unbalanced Y_2.

voltages that form a balanced set of voltages as in Fig. 15-32(b). Clearly, then, node n_2 is at the same potential as node n_1, and the corresponding phase impedances can be considered in parallel. However, if the second wye, Y_2, is not balanced, the potential of node n_2 will not be at the geometric center of the triangle formed by the line voltages. This condition is indicated in Fig. 15-32(c). Nodes n_1 and n_2 are not at the same potential, and it is incorrect to consider the corresponding phase impedances in parallel. Nevertheless, there are times when two or more loads are balanced and it is convenient to consider them as parallel wye loads. As demonstrated in the next section of this chapter, we can even analyze this system by considering only one of the phases.

Drill Problem 15-17 ■

Determine the line currents I_a, I_b, and I_c for the combined system of Fig. 15-33. The line voltages are 120 V and of sequence *abc*. Use V_{ab} as reference.

Answer. 17.3 $\angle -30°$ A; 33.9 $\angle -153°$ A; 28.4 $\angle 57.3°$ A. ■ ■

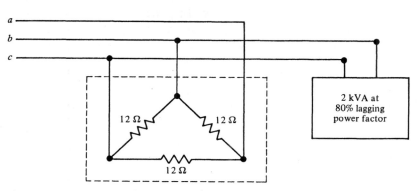

Figure 15-33 Drill Problem 15-17.

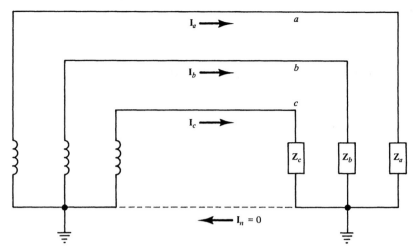

Figure 15-34   The balanced Y circuit redrawn.

15.13 ONE-LINE EQUIVALENT CIRCUITS

A balanced system results in line currents of equal magnitude. Furthermore, the line currents form a balanced set of phasors. If one of the line currents is known, then, the other two are easily derived.

Consider the balanced wye system studied earlier and redrawn in a slightly different form in Fig. 15-34. Usually the line voltages are known, but it is a simple matter to determine the phase voltages from them. It follows that if a phase impedance and a phase voltage are known, the phase current can be found. If one of the three phases is drawn as shown in Fig. 15-35, the circuit is called a *one-line equivalent circuit*. Although only the magnitudes are indicated in Fig. 15-35, a more rigorous approach uses the phase voltage as a phasor quantity and obtains the phase current, also the line current in this case, as a phasor quantity. It is immaterial whether a neutral or ground wire exists because the neutral current is zero for the balanced system. Often, in a power system, that neutral wire is the Earth.

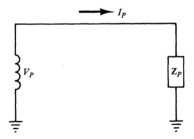

Figure 15-35   A one-line equivalent circuit of the balanced system.

EXAMPLE 15-12

Using a one-line equivalent circuit, determine the line currents for the circuit of Fig. 15-36(a). The line voltages are 120 V and of sequence *abc*.

Solution. The phase voltage magnitude is

$$V_P = \frac{V_L}{\sqrt{3}} = \frac{120}{\sqrt{3}} = 69.3 \text{ V}$$

With $\mathbf{V}_a$ as reference, $\mathbf{V}_a = 69.3 \angle 0°$ V. The one-line equivalent circuit is drawn as in Fig. 15-36(b). Then

$$\mathbf{I}_a = \frac{\mathbf{V}_a}{\mathbf{Z}_a} = \frac{69.3 \angle 0°}{10 \angle 60°} = 6.93 \angle -60° \text{ A}$$

With a sequence *abc*, the currents $\mathbf{I}_b$ and $\mathbf{I}_c$ each successively lag by 120°.

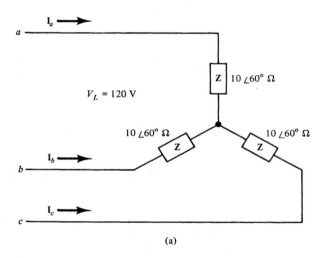

(a)

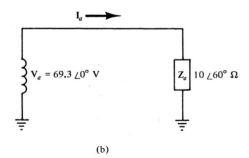

(b)

Figure 15-36 Example 15-12: **(a)** A balanced Y load; **(b)** the one-line equivalent circuit.

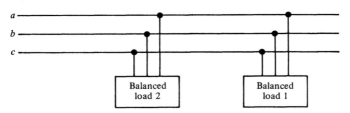

Figure 15-37 A combinational system with balanced loads.

Thus,

$$\mathbf{I}_b = 6.93 \angle -180° \text{ A} \quad \text{and} \quad \mathbf{I}_c = 6.93 \angle 60° \text{ A}$$

The usefulness of one-line equivalent circuits is evident when one analyzes the line currents to a combinational system that consists of only balanced loads. For example, in Fig. 15-37, two general, balanced three-phase loads are depicted. Regardless of the actual internal connections for those loads, each may be considered a wye load. If each of the loads is replaced by an equivalent wye, a circuit such as Fig. 15-32 results. Since the loads are balanced, nodes n_1 and n_2 of Fig. 15-32 are, for this case, at the same potential and the two wyes can be combined in parallel. Again, only one phase need be analyzed for the necessary information. Alternatively, each load is independent of the other, and therefore all phase currents can be superimposed to give a total phase current.

EXAMPLE 15-13

The loads in Fig. 15-37 consist of the following:

Load 1: Wye connection. Each impedance is $20 + j10 \ \Omega$.

Load 2: 3 kVA motor operating at an 85% lagging power factor. If the line voltage is 220 V, sequence acb, determine the line currents.

Solution. The phase voltage is

$$V_P = \frac{V_L}{\sqrt{3}} = \frac{220}{\sqrt{3}} = 127 \text{ V}$$

A representative one-line equivalent circuit is depicted in Fig. 15-38.

Load 1:

$$\mathbf{Z}_a' = 20 + j10 = 22.36 \angle 26.6° \ \Omega$$

$$\mathbf{I}_a' = \frac{\mathbf{V}_a}{\mathbf{Z}_a'} = \frac{127 \angle 0°}{22.36 \angle 26.6°} = 5.68 \angle -26.6° \text{ A}$$

Load 2:

$$3V_P I_P = 3 \text{ kVA}$$

or

$$I_P = \frac{3000}{3(127)} = 7.87 \text{ A}$$

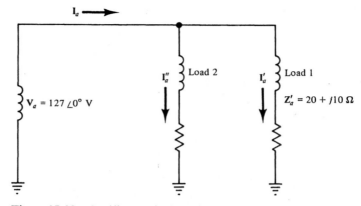

Figure 15-38   One-line equivalent circuit for Example 15-13.

Since I_P lags V_P by $\theta = \cos^{-1}(0.85) = 31.8°$,

$$\mathbf{I}_a'' = 7.87 \angle -31.8° \text{ A}$$

Then, by superposition, $\mathbf{I}_a = \mathbf{I}_a' + \mathbf{I}_a''$.

$$\mathbf{I}_a = 5.68 \angle -26.6° + 7.87 \angle -31.8°$$
$$= 5.08 - j2.54 + 6.69 - j4.15$$
$$= 11.77 - j6.69 = 13.5 \angle -29.6° \text{ A}$$

It follows that

$$\mathbf{I}_c = 13.5 \angle -149.6° \text{ A} \quad \text{and} \quad \mathbf{I}_b = 13.5 \angle 90.4° \text{ A}$$

Drill Problem 15-18 ■

Determine the phasor line currents for the combination circuit of Fig. 15-39 using a one-line equivalent circuit. The line voltage is 120 V and of sequence *abc*.

Answer. $23.5 \angle -18.6°$, $23.5 \angle -138.6°$, and $23.5 \angle 101.4°$ A.   ■ ■

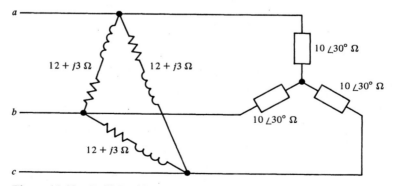

Figure 15-39   Drill Problem 15-18.

QUESTIONS

1. What is a polyphase circuit?
2. When are polyphase voltages said to be symmetrical?
3. What is an *n*-phase system?
4. What are some of the advantages of a polyphase system?
5. How are double subscripts used in three-phase circuits?
6. What is meant by phase sequence? How is it determined?
7. How many possible sequences are there for a three-phase system? Describe them.
8. What are the two types of interconnections used for a general *n*-phase system?
9. How do the line and phase values of voltage of a balanced delta relate to each other? How do the current values relate?
10. How does a four-wire wye differ from a three-wire wye?
11. How do the line and phase values of voltage of a balanced wye relate to each other? How do the current values relate?
12. Why can the neutral wire of a four-wire balanced wye system be removed without a change in load values?
13. How is three-phase apparent power defined?
14. What are the equations for three-phase real power and reactive power?
15. What is meant by the three-wattmeter method of power measurement? Describe the method.
16. Describe the two-wattmeter method of power measurement. When do both meters indicate the same power reading?
17. What is the significance of an algebraically negative power reading for the two-wattmeter method?
18. How does one solve an unbalanced wye system?
19. What is the significance of a floating neutral point?
20. Describe one type of phase sequence indicator utilizing an unbalanced wye load.
21. What is a combinational system?

PROBLEMS

1. Consider three alternator coils for which $\mathbf{E}_{oa} = 120 \angle 0°$, $\mathbf{E}_{ob} = 120 \angle -120°$, and $\mathbf{E}_{oc} = 120 \angle +120°$. Find the voltages corresponding to the various coil connections:
   (a) $\mathbf{E}_{ab} = \mathbf{E}_{ao} + \mathbf{E}_{ob}$.   (b) $\mathbf{E}_{bc} = \mathbf{E}_{bo} + \mathbf{E}_{oc}$.   (c) $\mathbf{E}_{ca} = \mathbf{E}_{co} + \mathbf{E}_{oa}$.
   Sketch all the phasor voltages.
2. Identify the sequences for the phasor voltages of Fig. 15-40.
3. Sketch phasor diagrams for the following sets of balanced voltages having an *abc* sequence:
   (a) $E_{ba} = E_{cb} = E_{ac} = 120$ V. Use $\mathbf{E}_{ba}$ as reference.
   (b) $E_{nb} = E_{na} = E_{nc} = 78$ V. Use $\mathbf{E}_{nb}$ as reference.
4. A balanced delta load has phase impedances of $20 \angle +30°$ Ω. If the line voltage is 208 V, what phase and line currents flow?

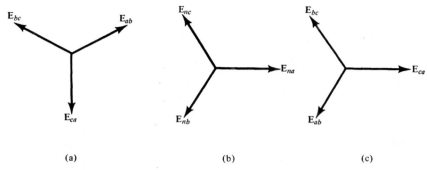

(a)   (b)   (c)

Figure 15-40   Problem 15-2.

5. A balanced delta load with $50 \angle +15°\ \Omega$ in each leg is connected to a three-phase set of voltages with $V_L = 440$ V. What magnitude of line currents flow?

6. A balanced delta has phase impedances of $\mathbf{Z} = 10 + j2\ \Omega$ and line voltages of 120 V. Voltage $\mathbf{V}_{ab}$ is taken as a reference and the sequence is *abc*. Calculate all phase and line currents, and sketch a phasor diagram with all currents and voltages.

7. A delta-connected load of 11 Ω resistors is connected to a three-phase source of voltages; $\mathbf{V}_{ab} = 110 \angle 0°$, $\mathbf{V}_{bc} = 110 \angle 120°$, and $\mathbf{V}_{ca} = 110 \angle -120°$ V. Calculate all phase and line currents and sketch them on a phasor diagram.

8. The line voltage of a balanced wye system is 480 V. Determine the magnitude of the phase voltage.

9. Three-phase line voltages of 230 V are impressed on a balanced wye load having phase impedances of $\mathbf{Z} = 16 + j12\ \Omega$. What magnitude of line currents flow?

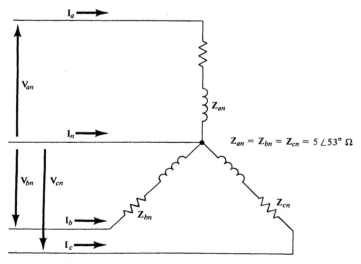

Figure 15-41   Problem 15-10.

10. The balanced wye load of Fig. 15-41 has impedances of $5 \angle 53°$ Ω. The phase voltages are $\mathbf{V}_{an} = 120 \angle 0°$ V, $\mathbf{V}_{bn} = 120 \angle -120°$ V, and $\mathbf{V}_{cn} = 120 \angle +120$ V. **(a)** Solve for the line and neutral currents. **(b)** Sketch a phasor diagram with all voltages and currents.

11. If the impedances of a four-wire wye load are not equal, unequal line currents flow and the neutral current is not zero. What current is measured by an ammeter in the neutral line if $\mathbf{I}_{an} = 60 \angle -30°$ A, $\mathbf{I}_{bn} = 52 \angle -135°$ A, and $\mathbf{I}_{cn} = 35 \angle 75°$ A?

12. What are the phase power and total power for the delta load of Problem 4?

13. What are the dissipated and reactive powers for the delta system of Fig. 15-42?

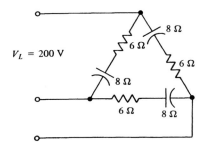

$V_L = 200$ V

8 Ω

6 Ω

6 Ω

8 Ω

6 Ω   8 Ω

Figure 15-42   Problem 15-13.

14. What are the total dissipated and reactive powers for the balanced wye load of Problem 9?

15. A delta circuit has phase voltages of $\mathbf{V}_{ab} = 100 \angle 0°$ V, $\mathbf{V}_{bc} = 100 \angle -120°$ V, and $\mathbf{V}_{ca} = 100 \angle +120°$ V. If the phase currents are $\mathbf{I}_{ab} = 10 \angle -30°$ A, $\mathbf{I}_{bc} = 20 \angle -173.1$ A, and $\mathbf{I}_{ca} = 14.14 \angle 165°$ A, what are the phase powers and the total power?

16. A three-phase balanced load has line currents of 1.75 A when line voltages of 240 V are applied. Each phase has a lagging power factor of 50%. What will each wattmeter indicate if the two-wattmeter method is used?

17. A three-phase balanced load has currents in each line of 2.5 A, when a three-phase source of 120 V is applied. The load is inductive and operates at a power factor of 35%. If the two-wattmeter method is used, what does each meter indicate?

18. A three-phase motor draws 7864 W when the line voltage is 220 V and the line current is 24 A. **(a)** If the power is measured by the three-wattmeter method, what value does each wattmeter indicate? **(b)** If the power is measured by the two-wattmeter method, what value does each wattmeter indicate?

19. The power to a three-phase motor is 76.2 kW. The line voltage is 2250 V and the line current is 26.1 A. **(a)** Determine the power factor of the motor. **(b)** Determine what each wattmeter of the two-wattmeter method indicates.

20. Determine the power factor of a three-phase balanced load when the two-wattmeter method produces indications of -300 and $+1680$ W.

21. Calculate the line currents for the delta of Fig. 15-43. The line voltage is 100 V, of sequence abc, with $\mathbf{V}_{ab}$ as reference.

22. Calculate the line currents for the delta of Fig. 15-44 if $\mathbf{V}_{ab} = 220 \angle 0°$ V, $\mathbf{V}_{bc} = 220 \angle -120°$ V, and $\mathbf{V}_{ca} = 220 \angle +120°$ V.

23. A four-wire wye circuit has impedances of $\mathbf{Z}_{an} = 60 \angle -30°$ Ω, $\mathbf{Z}_{bn} = 40 \angle 0°$ Ω, and $\mathbf{Z}_{cn} = 30 \angle +15°$ Ω. The voltages are $\mathbf{V}_{an} = 120 \angle +120°$ V, $\mathbf{V}_{bn} = 120 \angle 0°$ V, and $\mathbf{V}_{cn} = 120 \angle -120°$ V. What current does an ammeter in the neutral line indicate?

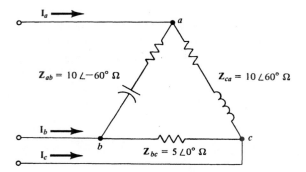

Figure 15-43   Problem 15-21.

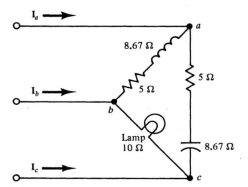

Figure 15-44   Problem 15-22.

24. Solve for the lamp voltages of the phase sequence indicator of Fig. 15-28 for a voltage sequence of *acb*. Compare the results with those for the *abc* sequence.

25. Solve for the line currents in Fig. 15-45. The line voltages are $V_{ab} = 100 \angle 0°$ V, $V_{bc} = 100 \angle -120°$ V, and $V_{ca} = 100 \angle +120°$ V.

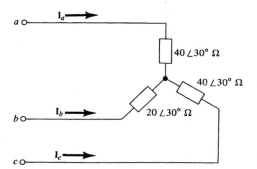

Figure 15-45   Problem 15-25.

26. Find the magnitude of line currents for the combinational system of Fig. 15-46.

27. What phasor current I_b flows to the combinational system of Fig. 15-47?

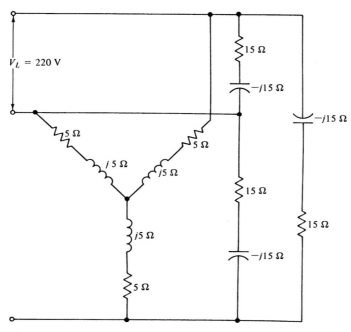

Figure 15-46 Problem 15-26.

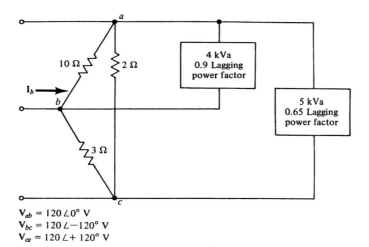

$V_{ab} = 120 \angle 0° \text{ V}$
$V_{bc} = 120 \angle -120° \text{ V}$
$V_{ca} = 120 \angle +120° \text{ V}$

Figure 15-47 Problem 15-27.

28. A combinational circuit consists of a three-phase motor load and a balanced three-phase capacitive load for power factor correction connected across 220 V lines. Without correction, the motor draws line currents of 8.5 A and operates at 90% power factor. (a) What kvar of capacitive load is needed to correct the overall power factor to unity? (b) What are the line currents under those corrected conditions?

29. Using a one-line equivalent circuit, determine the phasor line currents for the combinational circuit of Fig. 15-48. Use $\mathbf{V}_{an}$ as reference and an *abc* sequence.

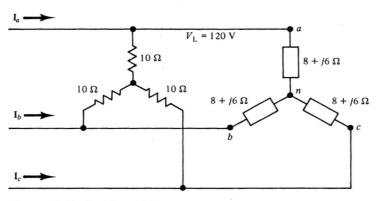

Figure 15-48   Problem 15-29.

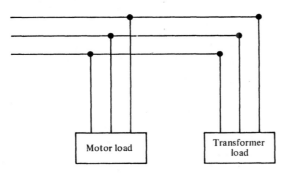

Figure 15-49   Problem 15-30.

30. Refer to Fig. 15-49. One load is a motor load of 10 kVA at 80% lagging power factor and the second load is a transformer load of 8 kVA at 95% lagging power factor. Determine the line current magnitudes if the line voltage is 220 V.

Chapter 16

Response to Nonsinusoidal Voltages

16.1 INTRODUCTION

As was pointed out in Chapter 10, sinusoidal quantities are highly desirable in electrical engineering work because they are easy to manipulate mathematically. Electrical quantities that do not vary sinusoidally are said to be *complex* or *nonsinusoidal*. Although the deviation may be negligible, all sources provide nonsinusoidal waveforms to some extent. In fact, in some electronic circuits, nonsinusoidal waves of such distinctive forms as triangular and square waves are deliberately generated. Also, the concern for nonsinusoidal quantities is not limited to electronics but extends into the power field. For example, it is not uncommon for a three-phase system to have certain peculiar nonsinusoidal effects. In this chapter the effects of nonsinusoidal voltages on linear circuits are investigated. The related nonsinusoidal effects that occur when sinusoidal voltages are applied to nonlinear circuits are also discussed.

Most nonsinusoidal voltages and currents are periodic or, if not periodic, can be considered periodic over short time intervals. In Chapter 10, the *period T* of a periodic waveform was defined as the time interval between successive repetitions. The number of cycles per second $1/T$ defines the *frequency f* in hertz. Thus, any nonsinusoidal waveform has a frequency fundamentally determined by its period.

In the early 1800s, the French mathematician Jean Baptiste Fourier showed that any mathematical function that is periodic, single-valued, finite, and possessed of a finite number of discontinuities may be represented by the sum of a number of sinusoidal components of different frequencies. The sinusoidal components of different frequencies are called the harmonics of the wave. One of the distinct sinusoidal components is the component of the same frequency as the nonsinusoidal waveform itself. This sinusoidal component is the *fundamental* or *first harmonic*. The other sinusoidal components are at frequencies that are integer multiples of the fundamental and are called harmonics. The term at a frequency of twice the fundamental is

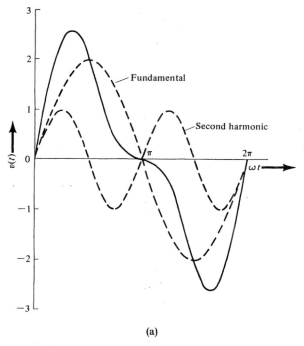

(a)

Figure 16-1   Plot of the complex wave $v(t) = 2 \sin \omega t + 1 \sin 2\omega t$: **(a)** Continuous curve; **(b)** computer-generated curve.

the *second harmonic* and, in general, the term at n times the fundamental frequency is the *nth harmonic* (n is any integer).

Before proceeding with a formal mathematical statement for the Fourier representation, it is instructive to see how two simple harmonically related sinusoidal voltages when added together produce a nonsinusoidal wave. The following examples not only show the development of nonsinusoidal waves, but show that any change in phase of a harmonic results in an entirely new waveform.

In the following examples a continuous curve represents actual point-by-point addition of the sinusoids. Obviously, adding sinusoids point-by-point is very tedious. The personal computer provides the circuit analyst with a tool that eliminates such work. Some BASIC computer programs for adding sinusoids appear in the Appendix. The supplementary solutions to the following examples were generated by those programs. Although the computer-generated solutions are used here only as a check, in actual practice the circuit analyst might generate only the computer solution.

EXAMPLE 16-1

Determine the shape of the complex wave $v(t) = 2 \sin \omega t + 1 \sin 2\omega t$.

Solution.   The fundamental and second harmonic are plotted on the same graph in Fig. 16-1(a). Notice that the horizontal axis has units of ωt so that in an interval $\omega t = 2\pi$, the fundamental passes through one cycle while the second harmonic passes through two cycles. (In general, n cycles of the nth

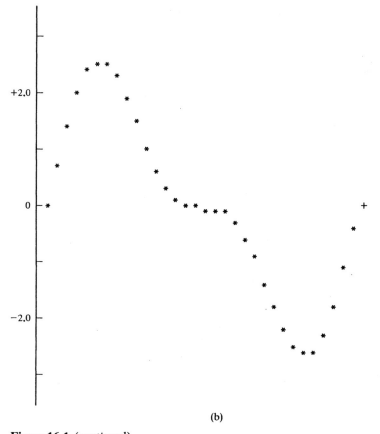

(b)

Figure 16-1 (*continued*)

harmonic occur in the period of one cycle of the fundamental.) Point-by-point addition results in the continuous curve $v(t)$ shown in Fig. 16-1(a). The computer-generated solution is shown in Fig. 16-1(b).

EXAMPLE 16-2

Determine the shape of the complex wave $v(t) = 2 \sin \omega t + 1 \sin(2\omega t + 90°) = 2 \sin \omega t + 1 \cos 2\omega t$.

Solution. The fundamental and second harmonic, each with proper phase, are plotted on the same graph in Fig. 16-2(a). Point-by-point addition results in the continuous curve for $v(t)$. The computer-generated curve for $v(t)$ is shown in Fig. 16-2(b).

Drill Problem 16-1 ■

Determine the shape of the complex wave $v(t) = 2 \cos \omega t - 1 \cos 2\omega t$.

Answer. The curve is the same as for Fig. 16-2(a) but is advanced by 90° of the fundamental. ■ ■

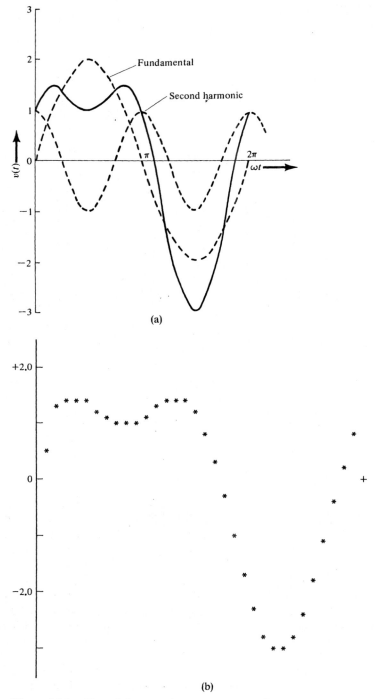

Figure 16-2 Plot of the complex wave $v(t) = 2 \sin \omega t + 1 \cos 2\omega t$: **(a)** Continuous curve; **(b)** computer-generated curve.

16.2 FOURIER REPRESENTATION

The Fourier representation of a function of time $f(t)$ having the previously mentioned properties is a series of the form,

$$f(t) = A_0 + A_1 \sin \omega t + A_2 \sin 2\omega t + A_3 \sin 3\omega t + \cdots + A_n \sin n\omega t$$
$$+ B_1 \cos \omega t + B_2 \cos 2\omega t + B_3 \cos 3\omega t + \cdots + B_n \cos n\omega t \quad (16\text{-}1)$$

or
$$f(t) = A_0 + \sum_{n=1}^{n=\infty} A_n \sin n\omega t + \sum_{n=1}^{n=\infty} B_n \cos n\omega t \quad (16\text{-}2)$$

where A_0 is a constant term, A_n and B_n are amplitudes of the different harmonics, and n is an integer such as 1, 2, 3, 4, ...

From Eqs. 16-1 and 16-2, it is seen that an infinite number of terms are theoretically required to represent an arbitrary function exactly. As n becomes greater, the corresponding amplitudes become smaller. Therefore, only a few terms are often necessary to practically represent a function. Of course, in certain cases, a function may be represented exactly by a series containing a finite number of terms. For example, the complex waveforms of the preceding examples are represented exactly by two terms.

One of the ways the Fourier coefficients, As and Bs, are determined for a given waveform is by a set of integration formulas, the *Euler formulas*:

$$A_0 = \frac{1}{T} \int_0^T f(t)\, dt \quad (16\text{-}3)$$

$$A_n = \frac{2}{T} \int_0^T f(t) \sin \frac{2n\pi t}{T}\, dt \quad (16\text{-}4)$$

$$B_n = \frac{2}{T} \int_0^T f(t) \cos \frac{2n\pi t}{T}\, dt \quad (16\text{-}5)$$

where $n = 1, 2, 3, \ldots$

The formula for A_0 is the formula introduced in Chapter 10 for the evaluation of the average value of a waveform. Since A_0 is the average value of the waveform, one usually obtains this term by inspection or by a simple calculation utilizing the techniques of Chapter 10. The formulas for A_n and B_n are a bit more involved and are provided here for completeness.

Often the nonsinusoidal wave appears on an oscilloscope and it is not possible to express the wave by mathematical equations. The coefficients can then be obtained by graphical techniques, which are usually presented in advanced circuit analysis textbooks. The coefficients that represent the amplitudes of the harmonics can also be obtained directly by specialized instruments such as a wave analyzer or a spectrum analyzer.

If the waveform or function of time is symmetrical about the vertical axis, that is, has the same value for $+t$ and $-t$, it is called an *even function*:

$$\text{even function:} \quad f(t) = f(-t) \quad (16\text{-}6)$$

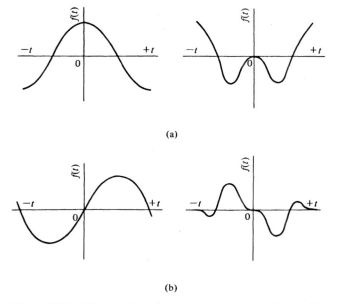

Figure 16-3 **(a)** Even functions: a cosine wave and an arbitrary function. **(b)** Odd functions: a sine wave and an arbitrary function.

However, if the waveform or function of time has a value for $+t$ that is the negative of its value for $-t$, it is called an *odd function*:

$$\text{odd function:} \quad f(t) = -f(-t) \tag{16-7}$$

Examples of even and odd functions are shown in Fig. 16-3. Notice in particular that the cosine function is even, whereas the sine function is odd.

    When a function is even, every term in its Fourier representation must also be even; otherwise the even symmetry is destroyed. Thus, for an even function, all of the sine coefficients equal zero ($A_n = 0$ for all n). However, for an odd function, all of the cosine coefficients equal zero as does the constant term ($A_0 = 0$ and $B_n = 0$ for all n).

    Certain waveforms can be even or odd, depending on the location of the vertical axis. In Fig. 16-4(a), a square wave is shown as an even function so that representation

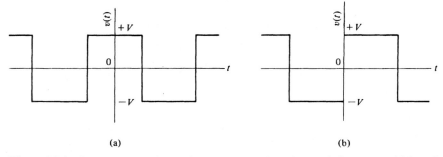

         (a)   (b)

Figure 16-4 Square wave, shown **(a)** as an even function and **(b)** as an odd function.

by a cosine series results. With a shift in the vertical axis to the position shown in Fig. 16-4(b), the square wave is represented as an odd function so that a sine series results. Obviously, the waveforms have an average value of zero; therefore, $A_0 = 0$.

If we evaluate the coefficients for the square wave shown as an even function, we obtain

$$A_0 = 0$$

$$A_n = 0 \qquad \text{for all } n$$

$$B_n = \frac{4V}{n\pi} \sin \frac{n\pi}{2}$$

$$= \begin{cases} \dfrac{+4V}{n\pi} & n = 1, 5, 9, \ldots \\ \dfrac{-4V}{n\pi} & n = 3, 7, 11, \ldots \\ 0 & n = \text{even integers} \end{cases}$$

so that

$$v(t) = \frac{4V}{\pi}(\cos \omega t - \tfrac{1}{3} \cos 3\omega t + \tfrac{1}{5} \cos 5\omega t - \cdots) \qquad (16\text{-}8)$$

Alternatively, if we evaluate the coefficients for the square wave shown as an odd function, we obtain

$$A_0 = 0$$

$$B_n = 0$$

$$A_n = \frac{2V}{n\pi}(1 - \cos n\pi)$$

$$= \begin{cases} \dfrac{+4V}{n\pi} & n = 1, 3, 5, 7, \ldots \\ 0 & n = \text{even integers} \end{cases}$$

Therefore,

$$v(t) = \frac{4V}{\pi}(\sin \omega t + \tfrac{1}{3} \sin 3\omega t + \tfrac{1}{5} \sin 5\omega t + \cdots) \qquad (16\text{-}9)$$

The angular frequency in Eqs. 16-8 and 16-9 equals $2\pi f$, where f is the reciprocal of the period T. Notice that in both expressions, only the odd harmonics are present in the series. This results whenever the function of time has *half-wave symmetry*, a condition for which the function at any time $+t$ is the negative of the value of the function one-half period later in time, $t + T/2$.

$$\text{half-wave symmetry:} \quad f(t) = -f(t + T/2) \qquad (16\text{-}10)$$

With the vertical axis shifted to points other than those shown in Fig. 16-4, the square wave is neither even nor odd and its series contains both sine and cosine terms. However, the sine and cosine terms of the same frequency can be combined

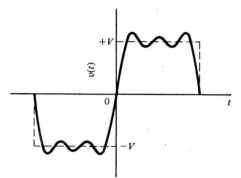

Figure 16-5 Addition of the first three terms of the square wave expression.

into one sine or cosine term with a displacement angle. Regardless of the type of series, the total magnitude of a particular harmonic does not change.

Theoretically an infinite number of terms is needed to represent the square wave. Yet, as shown in Fig. 16-5, a fair approximation results from only the first three terms. Another waveform frequently encountered is the sawtooth waveform of Fig. 16-6. Obviously the average value of the voltage is the area under the triangle formed by the sawtooth; that is, $A_0 = V/2$. Subtracting the average value from the waveform for analysis purposes, we are left with an odd function so that a sine series represents this case. Then $B_n = 0$ and by calculation $A_n = -V/n\pi$.

Therefore, the sawtooth has the Fourier series

$$v(t) = \frac{V}{2} - \frac{V}{\pi} \sin \omega t - \frac{V}{2\pi} \sin 2\omega t - \frac{V}{3\pi} \sin 3\omega t - \cdots \tag{16-11}$$

Many nonsinusoidal voltages result from or are generated by the application of a sinusoidal voltage to a nonlinear or unilateral device. A good example of this is the case where a sine wave of voltage is applied to a load resistor via an ideal diode [Fig. 16-7(a)].

Since the diode is ideal, current flows only during the positive half of the ac voltage cycle. No voltage is developed across the diode during this time period so the resistor voltage equals the applied voltage. During the negative half of the ac voltage cycle, the ideal diode does not conduct current, all of the voltage is developed across the diode, and the resistor voltage equals zero.

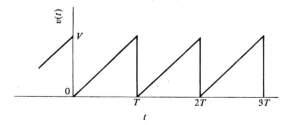

Figure 16-6 Sawtooth waveform.

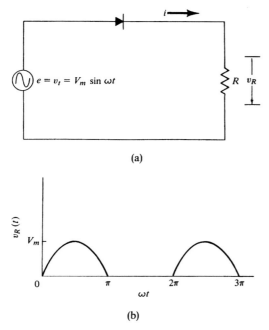

Figure 16-7 (a) Circuit with a unilateral element (diode); (b) the resistor voltage v_R.

The particular voltage waveform for the resistor is shown in Fig. 16-7(b). The waveform is called a *half-wave rectified sine wave* and has the Fourier series

$$v_R(t) = \frac{V_m}{\pi}\left(1 - \frac{\pi}{2}\sin\omega t - \frac{2}{3}\cos 2\omega t - \cdots - \frac{2\cos n\omega t}{n^2 - 1}\right) \tag{16-12}$$

where n is an even number.

When a sinusoidal voltage is applied to an iron-core inductor or transformer, magnetic saturation of the core may occur at some point of current excitation. The net result is a nonsinusoidal current and, in the case of a transformer, a nonsinusoidal voltage at the secondary.

Finally, any arbitrary function of time may contain both sine and cosine terms of the same frequency. Remembering that the sum of two sinusoids of the same frequency is a displaced sinusoid of the original frequency, we may combine terms so that Eq. 16-2 becomes

$$f(t) = A_0 + \sum_{n=1}^{n=\infty} C_n \sin(n\omega t + \theta_n) \tag{16-13}$$

where

$$C_n = \sqrt{A_n^2 + B_n^2} \tag{16-14}$$

and

$$\theta_n = \tan^{-1}\frac{B_n}{A_n} \tag{16-15}$$

EXAMPLE 16-3

Express the current $i = 41 \sin \omega t + 5.38 \sin 5\omega t - 11 \cos \omega t - 3.65 \cos 5\omega t$ A in a more concise form.

Solution

$$C_1 = \sqrt{A_1^2 + B_1^2} = \sqrt{(41)^2 + (-11)^2} = 42.4$$

$$\theta_1 = \tan^{-1} \frac{B_1}{A_1} = \tan^{-1}\left(\frac{-11}{41}\right) = -15°$$

$$C_5 = \sqrt{A_5^2 + B_5^2} = \sqrt{(5.38)^2 + (-3.65)^2} = 6.5$$

$$\theta_5 = \tan^{-1} \frac{B_5}{A_5} = \tan^{-1}\left(\frac{-3.65}{5.38}\right) = -34.2°$$

Therefore

$$i = 42.4 \sin(\omega t - 15°) + 6.5 \sin(5\omega t - 34.2°) \text{ A}$$

Drill Problem 16-2 ■

Using either the methods of Chapter 10 or inspection, determine the average value for the waveform of Fig. 16-8.

Answer.   6.5 V.   ▪ ■

Drill Problem 16-3 ■

Determine whether the following periodic waveforms are even or odd: **(a)** Fig. 16-1 for Example 16-1, **(b)** Fig. 16-2 for Example 16-2, and **(c)** Fig. 16-8.

Answer.   (a) Odd. (b) Neither even nor odd. (c) Even.   ■ ■

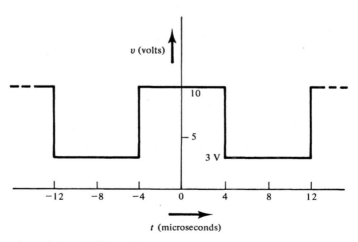

Figure 16-8   Drill Problems 16-2 and 16-3.

Drill Problem 16-4 ■

The following series mathematically represents a wave that is neither odd nor even:

$$i = 3 \sin \omega t - 4 \cos \omega t + 3 \sin 3\omega t + 8 \cos 3\omega t \text{ A}$$

Determine a simplified expression using only displaced sine terms.

Answer. $i = 5 \sin(\omega t - 53.1°) + 8.54 \sin(3\omega t + 69.4°)$ A. ■ ■

16.3 EFFECTIVE VALUES

The effective or rms value for any waveform representing electric current is, from Chapter 10,

$$I = \left(\frac{1}{T} \int_0^T i^2 \, dt \right)^{1/2} \tag{16-16}$$

Similarly, the effective value of voltage for any voltage waveform is

$$V = \left(\frac{1}{T} \int_0^T v^2 \, dt \right)^{1/2} \tag{16-17}$$

The general Fourier representation of a nonsinusoidal alternating current is, by Eq. 16-13,

$$i = I_0 + I_{m1} \sin(\omega t + \theta_1) + I_{m2} \sin(2\omega t + \theta_2) + \cdots \tag{16-18}$$

where $I_0 = A_0$ = the average value of current,
$I_{mn} = C_n$ = maximum value or amplitude of the nth harmonic.

Substituting Eq. 16-18 into Eq. 16-16 and integrating term by term, we obtain

$$I = \sqrt{I_0^2 + \tfrac{1}{2}(I_{m1}^2 + I_{m2}^2 + \cdots + I_{mn}^2)} \tag{16-19}$$

Remembering that the effective and maximum values for any harmonic are related by $\sqrt{2}$, we may write Eq. 16-19 in terms of the effective values as

$$I = \sqrt{I_0^2 + I_1^2 + I_2^2 + \cdots I_n^2} \tag{16-20}$$

Similarly,

$$V = \sqrt{V_0^2 + \tfrac{1}{2}(V_{m1}^2 + V_{m2}^2 + \cdots + V_{mn}^2)} \tag{16-21}$$

or

$$V = \sqrt{V_0^2 + V_1^2 + V_2^2 + \cdots + V_n^2} \tag{16-22}$$

EXAMPLE 16-4

Determine the effective value for the current

$$i = 42.6 \sin(\omega t - 15°) + 6.5 \sin(5\omega t - 34.2) \text{ A}$$

Solution. By Eq. 16-19,

$$I = \sqrt{[(42.6)^2 + (6.5)^2]/2} = \sqrt{928} = 30.5 \text{ A}$$

EXAMPLE 16-5

The square wave of Fig. 16-4 has a magnitude of 1 V. Using the first three terms of its Fourier series (Eq. 16-9), evaluate the effective value and compare the result with the actual effective value of 1 V.

Solution

$$v(t) = \frac{4(1)}{\pi}\left(\sin \omega t + \frac{1}{3}\sin 3\omega t + \frac{1}{5}\sin 5\omega t + \cdots\right)\text{V}$$
$$= 1.27(\sin \omega t + 0.333 \sin 3\omega t + 0.2 \sin 5\omega t + \cdots)\text{V}$$

Then, by Eq. 16-21,

$$V = \sqrt{[(1.27)^2 + (0.42)^2 + (0.254)^2]/2}$$
$$= 0.92^{1/2} = 0.96 \text{ V}$$

The calculated value is within 4% of the actual value.

Drill Problem 16-5 ■

A wave analyzer indicates the following effective values for a voltage measured across a 300 Ω resistor:

Fundamental: 5 V

Second harmonic: 3 V

Third harmonic: 1 V

What power is developed in the resistor?

Answer.   0.117 W.   ■ ■

Drill Problem 16-6 ■

Determine the effective value of current for the following nonsinusoidal current:

$$i = 2 + 10 \sin \omega t + 8 \sin(3\omega t + 30°)$$
$$+ 3 \sin(5\omega t + 60°) - 1.5 \sin 7\omega t \text{ mA}$$

Answer.   9.57 mA.   ■ ■

16.4 CIRCUIT RESPONSE TO NONSINUSOIDAL VOLTAGES

From our study of single-phase circuits, we know that the response of a circuit to a sinusoidal source depends on the impedance that the circuit exhibits at the source frequency. Then, if we apply a nonsinusoidal voltage and mathematically express it as a sum of sinusoids, it follows that the circuit response depends on the impedance of the circuit at each of the harmonic frequencies representing the nonsinusoidal voltage.

In the analysis of any network problem, we invariably depend on Kirchhoff's voltage and current laws. Certainly, these two laws must be satisfied when consider-

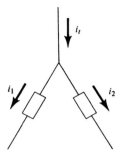

Figure 16-9   Example 16-6.

ing nonsinusoidal voltages and currents. With nonsinusoidal waves, however, the circuit laws must be satisfied for each separate harmonic.

The addition or subtraction of two nonsinusoidal waves in series form is accomplished by adding or subtracting, respectively, the components of the same frequency and then writing the Fourier expression for the resultant. In performing the addition or subtraction, it is convenient to use phasor algebra. Because the maximum values or amplitudes of the harmonics are specified in the Fourier series, the phasor algebra is best accomplished by using these values rather than the effective values, but *only those phasors of the same frequency are combined*, as emphasized by the following example.

EXAMPLE 16-6
Find the total current i_t of Fig. 16-9 if

$$i_1 = 10 \sin \omega t + 5 \sin(3\omega t - 30°) - 3 \sin(5\omega t + 60°) \text{ A}$$

and $\qquad i_2 = 20 \sin(\omega t - 30°) + 10 \sin(5\omega t + 45°) \text{ A}$

Solution.   Phasor diagrams for each harmonic are drawn as in Fig. 16-10. A second set of subscripts, 1, 2, and t, is added to denote branch 1, branch 2, and the total currents, respectively.

Fundamental:

$$\mathbf{I}_{m1_t} = \mathbf{I}_{m1_1} + \mathbf{I}_{m1_2} = 10 \angle 0° + 20 \angle -30°$$
$$= 10 + j0 + 17.3 - j10 = 27.3 - j10$$
$$= 29.1 \angle -20.1°$$

Third harmonic:

$$\mathbf{I}_{m3_t} = \mathbf{I}_{m3_1} + \mathbf{I}_{m3_2} = 5 \angle -30° + 0 = 5 \angle -30°$$

Fifth harmonic:

$$\mathbf{I}_{m5_t} = \mathbf{I}_{m5_1} + \mathbf{I}_{m5_2} = -3 \angle 60° + 10 \angle 45°$$
$$= 3 \angle -120° + 10 \angle 45° = -1.5 - j2.6 + 7.07 + j7.07$$
$$= 5.57 + j4.47 = 7.14 \angle 38.7°$$

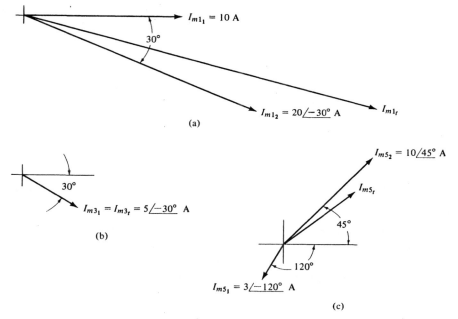

Figure 16-10   Phasor diagrams for Example 16-6 (the phasor magnitudes are maximum and not effective values): **(a)** The fundamental; **(b)** the third harmonic; **(c)** the fifth harmonic.

Finally,

$$i_t = 29.1 \sin(\omega t - 20.1°) + 5 \sin(3\omega t - 30°)$$
$$+ 7.14 \sin(5\omega t + 38.7°) \text{ A}$$

All of the general network theorems developed in earlier chapters are applicable to circuits to which periodic nonsinusoidal voltages are applied. In particular, by superposition, the effects of each harmonic can be evaluated independently and the results combined to yield the resultant nonsinusoidal voltages and currents. In accordance with the superposition theorem, the nonsinusoidal voltage source is replaced by a series connection of harmonically related voltage sources as in Fig. 16-11.

Although capacitance and inductance are considered constant or independent of frequency, the capacitive and inductive reactance are not independent of frequency. Rather, the reactances are different for the different harmonics. The result is that a nonsinusoidal current may differ widely in waveform from the applied nonsinusoidal voltage.

EXAMPLE 16-7

A voltage of $v(t) = 2 \sin 377t + 1 \sin 754t$ V is applied to the circuit of Fig. 16-12, where $R = 6 \ \Omega$ and $X_C = 8 \ \Omega$ at 60 Hz. Determine **(a)** the current expression, **(b)** the current waveform, and **(c)** the effective value of current.

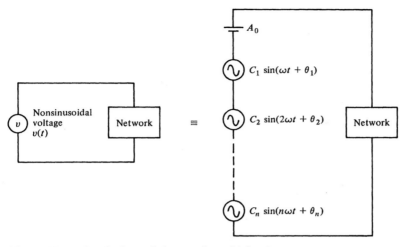

Figure 16-11   Equivalent of the nonsinusoidal voltage source.

Solution.   For convenience, the maximum values of the harmonics are used.

Fundamental ($\omega = 377$ rad/s):

$$\mathbf{Z}_1 = R - jX_C = 6 - j8 = 10 \angle -53.1° \, \Omega$$

$$\mathbf{I}_{m1} = \frac{\mathbf{V}_{m1}}{\mathbf{Z}_1} = \frac{2 \angle 0°}{10 \angle -53.1°} = 0.2 \angle +53.1° \, \text{A}$$

Second harmonic ($\omega = 754$ rad/s):

$$X_C = \frac{1}{2(377)C} = \tfrac{1}{2}(8 \, \Omega) = 4 \, \Omega$$

$$\mathbf{Z}_2 = R - jX_C = 6 - j4 = 7.2 \angle -33.7° \, \Omega$$

$$\mathbf{I}_{m2} = \frac{\mathbf{V}_{m2}}{\mathbf{Z}_2} = \frac{1 \angle 0°}{7.2 \angle -33.7°} = 0.139 \angle +33.7° \, \text{A}$$

(a) $i = 0.2 \sin(377t + 53.1°) + 0.139 \sin(754t + 33.7°)$ A.

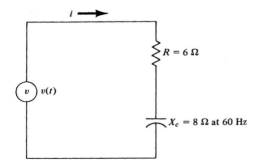

Figure 16-12   Example 16-7.

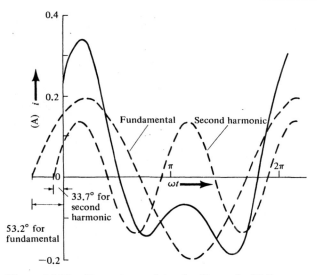

Figure 16-13   Current waveform for Example 16-7.

(b) The current waveform is sketched in Fig. 16-13.

(c)
$$I = \sqrt{(I_{m1}^2 + I_{m2}^2)/2} = \sqrt{(0.2)^2 + (0.139)^2/2}$$
$$= \sqrt{0.0296} = 0.172 \text{ A}$$

The applied voltage in the preceding example is defined by the waveform of Fig. 16-1. Notice that the current waveform for the example differs markedly from the voltage waveform.

EXAMPLE 16-8

A voltage of $v(t) = 100 \sin 1000t + 50 \sin(3000t - 30°)$ V is applied to the circuit of Fig. 16-14 for which $R_1 = 5\ \Omega$, $R_2 = 10\ \Omega$, and $X_L = 5\ \Omega$ at the fundamental frequency. Determine **(a)** the expression for the total current and **(b)** the power developed in R_1.

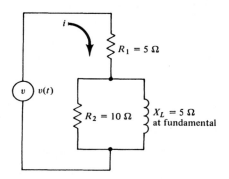

Figure 16-14   Example 16-8.

Solution. The maximum values of the harmonics are used.

Fundamental:

$$R_2 \| jX_L = \frac{(10)(j5)}{10 + j5} = \frac{j10}{2 + j1}\left(\frac{2 - j1}{2 - j1}\right)$$

$$= \frac{10 + j20}{5} = 2 + j4\ \Omega$$

$$\mathbf{Z}_1 = R_1 + 2 + j4 = 5 + 2 + j4 = 7 + j4$$
$$= 8.06\ \angle\,29.7°\ \Omega$$

$$\mathbf{I}_{m1} = \frac{\mathbf{V}_{m1}}{\mathbf{Z}_1} = \frac{100\ \angle\,0°}{8.06\ \angle\,29.7°} = 12.4\ \angle\,-29.7°\ \text{A}$$

Third harmonic:

$$X_L = 3(5) = 15\ \Omega$$

$$R_2 \| jX_L = \frac{(10)(j15)}{10 + j15} = \frac{j30}{2 + j3}\left(\frac{2 - j3}{2 - j3}\right)$$

$$= \frac{90 + j60}{13} = 6.93 + j4.62\ \Omega$$

$$\mathbf{Z}_3 = 5 + 6.93 + j4.62 = 11.93 + j4.62$$
$$= 12.8\ \angle\,21.1°\ \Omega$$

$$\mathbf{I}_{m3} = \frac{\mathbf{V}_{m3}}{\mathbf{Z}_3} = \frac{50\ \angle\,0°}{12.8\ \angle\,21.1°} = 3.91\ \angle\,-21.1°\ \text{A}$$

(a) $\quad i = 12.4 \sin(1000t - 29.7°) + 3.91 \sin(3000t - 51.1°)\ \text{A}$

(b) $\quad I = \sqrt{(I_{m1}^2 + I_{m3}^2)/2} = \sqrt{[(12.4)^2 + (3.91)^2]/2}$
$$= \sqrt{84.5} = 9.2\ \text{A}$$
$$P_1 = I^2 R_1 = (84.6)(5) = 423\ \text{W}$$

If an average or dc value is present in the applied nonsinusoidal voltage, an exponential transient component is part of the network solution. In general, this transient period is very short with respect to the period of the fundamental frequency.

Another consideration with respect to nonsinusoidal response is resonance. Resonance may occur for one of the harmonics if the harmonic voltage and current are in phase. If series resonance occurs and the circuit resistance is small relative to the reactance at resonance, the component of current for the resonant harmonic becomes very prominent. Alternatively, if parallel resonance occurs for one of the harmonic components of voltage, the associated current harmonic is diminished.

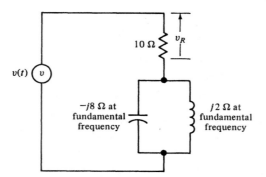

Figure 16-15   Drill Problem 16-7.

Drill Problem 16-7

The nonsinusoidal voltage $v(t) = 1 \sin \omega t + 0.25 \sin 2\omega t$ is applied to the circuit of Fig. 16-15. Determine the voltage expression for v_R.

Answer. $v_R = 0.97 \sin(\omega t - 14.9°)$ V.   ■ ■

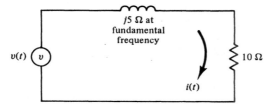

Figure 16-16   Drill Problem 16-8.

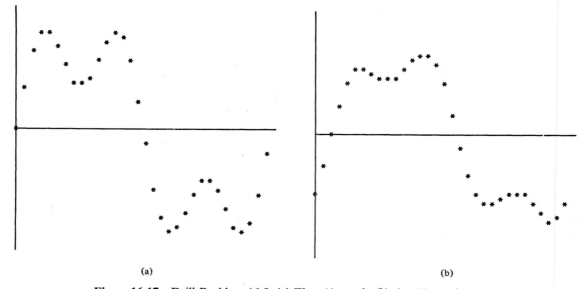

(a)   (b)

Figure 16-17   Drill Problem 16-8: **(a)** The $v(t)$ graph; **(b)** the $i(t)$ graph.

Drill Problem 16-8

A voltage of $v(t) = 1 \sin \omega t + 0.5 \sin 3\omega t$ is applied to the circuit of Fig. 16-16.
(a) Determine the expression for the current $i(t)$.
(b) Sketch the waveforms for $v(t)$ and $i(t)$.

Answer. (a) $i = 0.0893 \sin(\omega t - 26.6°) + 0.028 \sin(3\omega t - 56.3°)$ A.
(b) See Fig. 16-17. ■ ■

QUESTIONS

1. What defines a nonsinusoidal quantity?
2. Define period and frequency for a nonsinusoidal waveform.
3. What is a harmonic?
4. What is meant by a Fourier representation of a function of time?
5. How are Fourier series obtained?
6. What is the significance of the A_0 term in a Fourier series?
7. What is (a) an even function and (b) an odd function?
8. What is meant by half-wave symmetry?
9. What is the effective value of a waveform?
10. If one has the Fourier series for a nonsinusoidal waveform, how does one obtain the effective value?
11. How are nonsinusoidal waves added and subtracted?
12. How does one solve for the various voltages and currents for a circuit to which a nonsinusoidal voltage is applied?
13. Can resonance occur as a response to nonsinusoidal voltages? Explain.

PROBLEMS

1. Determine the shape of the complex wave $v(t) = 10 \sin \omega t + 2.5 \sin(3\omega t - 90°)$ V.
2. Determine the shape of the complex wave $i(t) = 5 \sin \omega t - 1.5 \sin 5\omega t$ A.
3. Determine the shape of the complex wave $v(t) = \cos \omega t - \sin 2\omega t$ V.
4. Find the A_0 term for the waveform of Fig. 16-18.

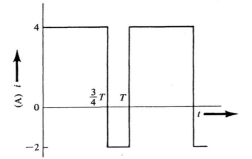

Figure 16-18 Problem 16-4.

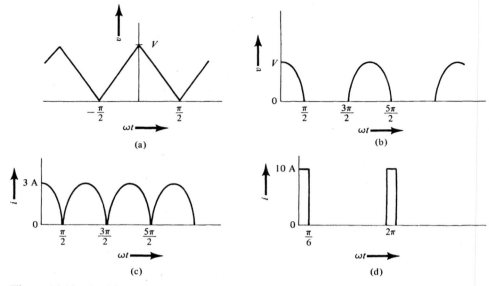

Figure 16-19   Problem 16-5.

5. Given the waveforms of Fig. 16-19, determine the A_0 term and whether the respective Fourier series consist of sine terms, cosine terms, or both.

6. Express the nonsinusoidal voltage, $v = 10 \sin \omega t - 2 \cos \omega t + 2 \cos 3\omega t + 1 \sin 5\omega t$ V as a series of displaced sine waves.

7. The voltage across a resistance of 15 Ω is $v = 100 + 22.4 \sin(\omega t - 45°) + 4.11 \sin(3\omega t - 67°)$ V.
   (a) What is the effective value of the voltage?
   (b) What power is developed in the resistor?

8. Determine the effective values for the following quantities:
   (a) $v = 156 \sin(\omega t + 30°) + 50 \sin(3\omega t - 60°) + 2 \sin(5\omega t + 30°)$ V.
   (b) $i = 31.1 \sin(\omega t + 15°) + 5 \sin 5\omega t$ A.
   (c) $i = 2 + \cos(500t + 30°) + 0.75 \cos(1500t - 15°)$ A.

9. The sawtooth voltage of Fig. 16-6 has a peak value of $V = 10$ V. Using the first five terms of the series expansion, find the effective value.

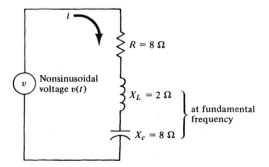

Figure 16-20   Problem 16-11.

10. Determine $i_t = i_1 - i_2$ if $i_1 = 10 \sin \omega t + 5 \sin(3\omega t - 30°) - 3 \sin(5\omega t + 60°)$ A and $i_2 = 20 \sin(\omega t - 30°) + 10 \sin(5\omega t + 45°)$ A.

11. The voltage applied to the circuit of Fig. 16-20 is $v = 50 \sin \omega t + 25 \sin(3\omega t + 60°)$ V. If $R = 8\ \Omega$, $X_L = 2\ \Omega$, and $X_C = 8\ \Omega$ at the fundamental frequency, what is the equation for the current?

12. A voltage of $v = 10 + 141 \sin 1000t$ V appears across the resistor of Fig. 16-21. **(a)** What is the equation for the total current? (Assume that the dc transient period is completed.) **(b)** What power is developed in R?

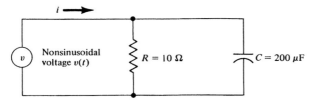

Figure 16-21 Problem 16-12.

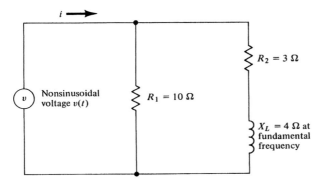

Figure 16-22 Problem 16-13.

13. The voltage applied to the circuit of Fig. 16-22 is $v = 200 \sin 377t + 100 \sin(1131t + 45°)$. If $R_1 = 10\ \Omega$, $R_2 = 3\ \Omega$, and $X_L = 4\ \Omega$ at the fundamental frequency, determine **(a)** the total current and **(b)** the effective value of the total current.

Chapter 17

Electrical Measurements

17.1 INTRODUCTION

Electrical measurement is the actual determination of the extent of an electrical quantity such as voltage, current, resistance, or power. Any device used for measuring an electrical quantity is called an electrical instrument and is generally either an analog or a digital instrument. An *analog instrument* is characterized by an output that is continuously variable. Such a device may have a pointer, the deflection or position of which is a function of the electrical quantity; that is, the deflection is *analogous* to the measured electrical quantity. A *digital instrument* is characterized by an output that is displayed directly in decimal numeric values. When using either type of instrument, certain aspects of measurements must be kept in mind.

The *true* or *actual value* is the actual magnitude of the quantity at the input of the instrument. The true value may be only approached by measurement because the process of measurement alters the measured quantity.

The *measured* or *indicated value* is the magnitude of the quantity indicated or displayed at the output of the instrument. The difference between the measured and true value is called the *error*. The error is often expressed as a percentage of the true or standard value.

$$\% \text{ error} = \frac{\text{error or deviation}}{\text{true value}} \times 100 \qquad (17\text{-}1)$$

Alternatively, the amount within which a measured value agrees with its actual value is called the *accuracy*. Accuracy is often expressed as a percentage of the *full-scale value*, that is, the maximum value of the particular range of the instrument being used.

Manufacturers calibrate an instrument in terms of standards derived from values established and maintained by the National Bureau of Standards. Any instrument has inaccuracies, however, and these inaccuracies are, in general, inversely related to the quality of components used in the instrument. Because of this, manufacturers state the accuracy of a particular instrument on specification sheets and sometimes provide a *calibration curve* or *correction chart* that indicates the actual values by which the instrument readings differ from the standard. As the original accuracy of an instrument is not necessarily maintained over the useful life of the instrument, recalibrations become necessary.

Errors are not limited to those inherent in the instrument itself but are often a result of procedural and personnel performance. For example, failure to adjust the readout of an instrument initially to zero leads to consistently higher or lower measured values than normal. In addition, personnel interpolation errors are always possible with analog instruments. One such error is *parallax error*, which is the error introduced when the operator's line of vision is not perpendicular to the indicating scale. This error is reduced by the proper use of a mirror scale, which provides for alignment of the operator's eyes with the pointer and its reflection.

As was pointed out in Chapter 2, an electric current may be a direct current or an alternating current; the effect that each has on a circuit is unique. Thus it is not surprising that certain instruments are designed for use with direct current, while others are designed for alternating current. Still other instruments are modified for use with both direct and alternating current.

EXAMPLE 17-1

A line voltage has a true frequency of 60 Hz. A frequency meter indicates a frequency of 59.2 Hz. What is the error introduced by the meter?

Solution

$$\text{Error} = \frac{\text{deviation}}{\text{true value}} = \frac{60 - 59.2}{60} = \frac{0.8}{60}$$
$$= 0.0133 = 1.33\%$$

Drill Problem 17-1 ■

A certain meter has an accuracy of 0.5% of its full-scale value when used in the 0–10 V range. What is the maximum voltage error possible in that meter range?

Answer. 0.05 V. ■ ■

Drill Problem 17-2 ■

A certain voltage as measured with a calibrated voltmeter of negligible error is 12.5 V. The same voltage when measured with a voltmeter in need of calibration is 11.8 V. Determine the percentage of error of the second measurement.

Answer. 5.6%. ■ ■

17.2 BASIC ANALOG METER MECHANISMS

The distinction is sometimes made that a *meter* is a device that in a single transformation converts a quantity into output information, whereas an *instrument* is a complex device that may provide many intermediate transformations in producing

Figure 17-1 An analog meter with a blank scale.

output information. In this text, that distinction will be made only in the present section.

An analog meter with a blank scale and a moving pointer is shown in Fig. 17-1. The analog deflection of the pointer or needle is obtained by electrostatic, electrothermal, or electromagnetic conversion of a current. Like most analog meters, the one in Fig. 17-1 uses an *electromagnetic movement*. In particular, the meter of Fig. 17-1 has a mechanism like that depicted in Fig. 17-2.

The basic meter mechanism of Fig. 17-2 is referred to as the *permanent-magnet moving-coil or Weston mechanism* (Edward Weston, 1850–1936). Because the Weston mechanism was developed as a more rugged, but less sensitive version of the *D'Arsonval galvanometer*, an instrument used for detecting very small electric currents, it is often called the D'Arsonval mechanism.

In the permanent-magnet moving-coil mechanism, a coil of fine wire is allowed to pivot or rotate between the poles of a permanent magnet. When current flows in the coil, a magnetic field is created and this field interacts with the field supplied by the permanent magnet. The result is a *torque* or force of rotation that tends to rotate the coil of wire. Because the magnetic field provided by the permanent magnet

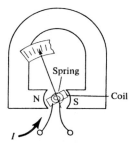

Figure 17-2 A basic electromagnetic mechanism.

does not vary, any change in torque results from a change in the magnitude of current flowing in the coil. A pointer is attached to the coil, and springs are added to provide a mechanical countertorque so that the pointer deflection is directly proportional to the coil current.

An actual permanent-magnet moving-coil mechanism is shown in Fig. 17-3. The moving coil is wound on an aluminum bobbin attached to a pointer. The bobbin is suspended by two short shafts that rest in jewel bearings. Flat coil springs attached to the shaft near each bearing provide the necessary countertorque and serve as electrical connections to the coil. Typically, a meter mechanism such as that shown produces a full-scale deflection with a current somewhere in the microampere or milliampere range. Because the pointer deflection is directly proportional to the current, a linear scale results.

The permanent-magnet moving-coil mechanism is basically a dc instrument in that an upscale torque results only when current flows through the coil in a particular direction. When the meter is used to measure a changing current (such as alternating current), the moving coil in most cases cannot follow the variations and settles in a position determined by the average torque. The reading obtained is an average value of the current. In the case of sinusoidal alternating current, the average values of the current and torque are zero and no deflection results.

Many variations of the mechanism of Fig. 17-3 exist. One of these is a system where the permanent magnet serves as a core about which the moving coil rotates. Another variation is a system in which the moving coil is held in place by *taut bands* or ribbons of metal that not only provide the restoring torque and carry the

Figure 17-3 An actual permanent-magnet, moving-coil mechanism (that of the meter of Fig. 17-1).

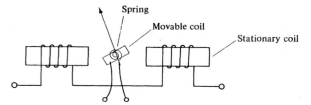

Figure 17-4   Basic electrodynamic mechanism.

current to the coil, but also eliminate the friction associated with jewel bearing movements. The taut band mechanism is sensitive to small currents but is also easily damaged by vibrational stresses. The taut band suspension is similar to the coil suspension used in the D'Arsonval galvanometer.

The second type of meter mechanism is the *electrodynamometer* or *electrodynamic mechanism*. In this mechanism a set of stationary field coils provides the magnetic field in which the moving coil rotates (see Fig. 17-4). Depending on certain design factors, the movable and stationary coils are connected, within the meter housing, in series or parallel. The result is that if a current reversal occurs, it occurs for both coils simultaneously so that the direction of torque remains unchanged. Since the torque is proportional to both coil currents, a nonlinear scale results. In fact, with no magnetic material in the system, the deflection, and hence the scale, has a squared relationship.

The electrodynamic mechanism is versatile, since it can be used for both dc and ac measurements. Not only can it be adapted to the measurement of current and voltage, but also to the measurement of power. However, it cannot compete in cost with either the D'Arsonval mechanism for direct current or the *moving iron vane mechanism* (discussed next) for alternating current.

If two strips of magnetic material are placed in parallel in a magnetic field, the strips become similarly magnetized and a repelling force develops between them. This principle is used in a third type of meter mechanism, the *moving iron vane mechanism* shown in Fig. 17-5. Two thin, curved strips of soft iron called *vanes* are located within a stationary coil. One of the vanes is stationary, whereas the other is mounted with jewel bearings and is free to rotate. When the coil is energized, the stationary and moving vanes are magnetized alike; this results in a corresponding

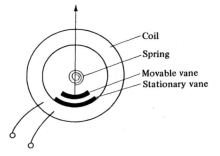

Figure 17-5   Concentric iron vane mechanism.

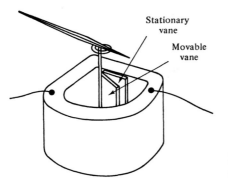

Figure 17-6   Book-type iron vane mechanism.

amount of repulsion of the movable vane. As in other mechanisms, the pointer moves upscale against a spring tension.

Like poles are always induced in the adjacent ends of the two vanes, regardless of the direction of the coil current. It follows that the mechanism is useful for either dc or ac measurements. However, if one uses an iron vane meter for dc measurements, it is necessary to account for residual magnetic effects in the iron vanes by reversing the meter connections and averaging the two readings.

Because of its circular vanes, the iron vane mechanism of Fig. 17-5 is known as a *concentric-type mechanism*. Another variation is the *book-type mechanism* (of Fig. 17-6), which uses two rectangular vanes; the movable vane moves like the cover of an opening book.

The permanent-magnet moving-coil mechanism is relatively inexpensive, but is basically a dc mechanism. A *rectifier*, such as a diode, is a device that changes alternating current to direct current by providing a low resistance (ideally zero) to current flow in one direction and a high resistance (ideally infinite) in the opposite direction (see Chapter 4). Diodes or rectifiers, which are characterized by this unilateral electrical behavior, permit alternating current to flow through a dc meter in one direction only. One such meter rectifier circuit is shown in Fig. 17-7. Notice that a sinusoidal alternating voltage v is applied as the measured quantity and a series resistance R_S is used to limit the current flow to the meter.

With a single diode connected in series with the meter as shown in Fig. 17-7(a), current flows to the meter when the diode is forward-biased. Thus, as indicated in

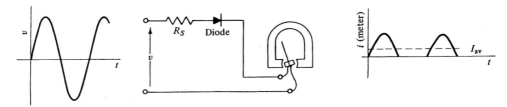

Figure 17-7   **(a)** A half-wave rectifier circuit used with a dc meter; **(b)** the associated meter current.

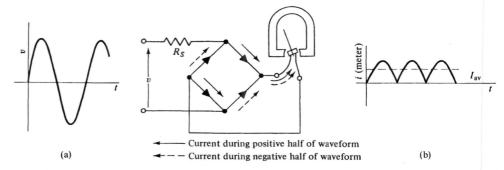

(a)

← ——— Current during positive half of waveform
← – – – Current during negative half of waveform

(b)

Figure 17-8 **(a)** A full-wave bridge rectifier circuit used with a dc meter; **(b)** the associated meter current.

Fig. 17-7(b), current flows only during the positive half of the alternating voltage. This configuration is therefore called a *half-wave rectifier circuit*.

A second type of meter rectifier circuit is shown in Fig. 17-8(a). This rectifier circuit, in contrast to the half-wave type, allows current to flow through the meter during both the positive and negative halves of the applied voltage. Because of this and the bridge configuration that the four diodes and meter form, the circuit is called a *full-wave bridge rectifier circuit*. The solid arrows indicate the path of conventional current during the positive half of the voltage, while the dotted arrows indicate the current path during the negative half of the voltage. The current is described by Fig. 17-8(b).

In both rectifier cases, the meter mechanism responds to the *average value* of the meter current, I_{av}. But, as was shown in Chapter 10, an *effective value* of current relates any varying current to the actual power developed by that current. For this reason and because most alternating currents are sinusoidal in nature, the scales of rectifying instruments are calibrated in terms of the effective value of a sine wave of current. However, if these scales are used to measure nonsinusoidal currents, appreciable error may result.

17.3 CURRENT MEASUREMENT: THE AMMETER

The basic device used to measure current in amperes is the *ammeter*. When the current in a circuit is substantially less than an ampere, its value is determined by the use of a *milliammeter* or *microammeter*, which measures full-scale currents in milliamperes or microamperes, respectively.

The symbol for the ammeter is a circle with the enclosed letter "A," as in Fig. 17-9. In this figure, *an ammeter is inserted into the path of current, that is, in series to carry the current being measured*. If the meter is a dc type, it has polarity marks to ensure proper connection for an upscale indication. In particular, *conventional current flow into the positive terminal* for an upscale meter deflection. If the meter is an ac type, current direction is of no significance and no polarity marks appear on the

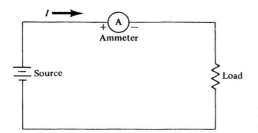

Figure 17-9 Ammeter: Its symbol and its circuit connection.

instrument. However, the meter is still connected in series with the part of the circuit for which the current is being determined.

It is obvious that a coil used in a basic meter mechanism has a certain resistance R_m, called meter resistance. Thus one may think of a meter mechanism in terms of the circuit of Fig. 17-10. By Ohm's law, when a current I_m sufficient to produce a full-scale deflection flows through the mechanism, a voltage drop V_m is produced across the meter mechanism:

$$V_m = I_m R_m \qquad (17\text{-}2)$$

The resistance of a moving-coil mechanism is relatively low; hence, the voltage drop across the mechanism at full scale is low.

EXAMPLE 17-2

A dc milliammeter has a full-scale rating of 1 mA and a meter resistance of 25 Ω. What is the voltage across the meter when maximum current flows in the meter?

Solution

$$V_m = I_m R_m = (1 \times 10^{-3})(25) = 25 \text{ mV}$$

In order to use a meter mechanism to measure a current larger than the rating of the mechanism, a large part of the current being monitored must be bypassed or shunted around the mechanism as in Fig. 17-11. The shunt resistance R_{sh} needed to increase the current range is usually calculated using the full-scale values. It is realized that, with constant resistances, other currents and scale positions are directly proportional to the full-scale values.

With reference to Fig. 17-11, if a meter with a full-scale rating of I_m indicates a full-scale deflection for a current $I_t (I_t > I_m)$, then the shunted value of current is

$$I_{\text{sh}} = I_t - I_m \qquad (17\text{-}3)$$

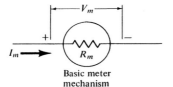

Basic meter mechanism

Figure 17-10 Resistance of a meter mechanism.

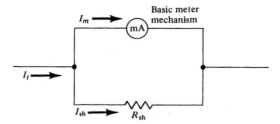

Figure 17-11   Circuit for increasing the current range of a meter.

Since the shunt and meter form a parallel circuit, the voltages across both are equal and in terms of the resistances

$$I_{sh}R_{sh} = I_m R_m \tag{17-4}$$

EXAMPLE 17-3

We wish to use the meter mechanism of Example 17-2 (1 mA, 25 Ω) to measure 50 mA at full scale. What shunt resistance is needed?

Solution

$$I_{sh} = 50 - 1 = 49 \text{ mA}$$
$$I_{sh}R_{sh} = I_m R_m = 25 \text{ mV}$$

Therefore

$$R_{sh} = \frac{25 \times 10^{-3}}{49 \times 10^{-3}} = 0.51 \ \Omega$$

Multiple-scale ammeters (or milliammeters) are constructed using either switching arrangements or multiple terminals as in Fig. 17-12. In each part of this figure, an ammeter (or milliammeter) with three ranges is shown.

In Fig. 17-12(a), each of the shunt resistors is separately calculated for each range and a simple switching arrangement allows the selection of a particular range. In order to prevent excessive current from passing through the meter mechanism during range switching, the switch must be of the *shorting* or *make-before-break type*.

The ammeter in Fig. 17-12(b) has a shunt made of series resistances. In the selection of progressively higher current ranges, more of the shunt resistance is shorted by the switch. Compared to the first type, the series shunt type provides a margin of safety due to switch failure, because the shunt is always connected across the meter mechanism. The required resistances are calculated by first evaluating the resistance for the highest range and then progressing to each lower range to find the additional resistance needed.

The third type of ammeter, with multiple terminals, is shown in Fig. 17-12(c). The shunt arrangement is known as the *Aryton shunt*. In operating this type of ammeter, one uses the positive terminal and the appropriate negative terminal (in the dc case). This arrangement avoids the use of a switch and its variable contact resis-

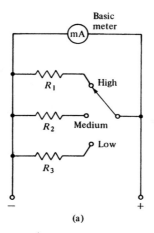

(a)

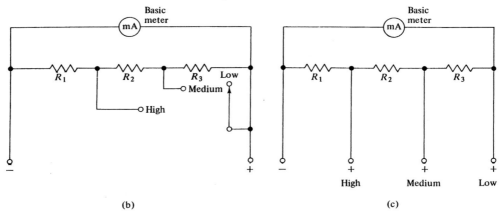

(b)

(c)

Figure 17-12 Multiple-range ammeter: **(a)**, **(b)** Two switching arrangements; **(c)** Ayrton shunt, multiple terminals.

tance. However, the resistance calculations are more involved and require simultaneous equations.

Drill Problem 17-3 ■

A voltage of 2 mV is measured across a meter when a full-scale current of 1 μA flows through the meter. What is the meter resistance?

Answer.   2 kΩ.   ■ ■

Drill Problem 17-4 ■

A dc milliameter has a full-scale rating of 10 mA and a meter resistance of 2 Ω. **(a)** What is the voltage across the meter when maximum current flows in the meter? **(b)** How much power does the meter dissipate at full-scale deflection?

Answer.   (a) 20 mV. (b) 200 μW.   ■ ■

Drill Problem 17-5 ■

If the meter of Drill Problem 17-4 is shunted so that the full-scale current occurs at 200 mA, what is the shunting resistance?

Answer.   0.105 Ω.   ■ ■

Drill Problem 17-6 ■

Refer to the multirange ammeter of Fig. 17-12(b). Determine the shunting resistances if a basic meter of 10 Ω resistance and 5 mA full-scale current is used and ranges of 500, 200, and 100 mA are desired.

Answer. $R_1 = 0.101\ \Omega$; $R_2 = 0.155\ \Omega$; $R_3 = 0.27\ \Omega$.   ■ ■

17.4 VOLTAGE MEASUREMENT: THE VOLTMETER

The basic device used to measure electric potential or voltage in volts is the *voltmeter*. As was discussed in the preceding section, any basic meter mechanism has a voltage V_m across its terminals when a full-scale current I_m flows through the meter. It follows that the basic meter mechanism can be calibrated in microvolts, millivolts, or volts, depending on the voltage needed to produce a full-scale deflection. Since the moving-coil mechanism needs a voltage only in the microvolt or millivolt range for full-scale deflection, it is essentially a *microvoltmeter* or *millivoltmeter*.

The symbol for a voltmeter is a circle with the enclosed letter "V," as indicated in Fig. 17-13. A voltmeter is connected *across or in parallel with the element or circuit portion for which the voltage is being measured.* As with a dc ammeter, a dc voltmeter has polarity marks; therefore, one has to connect the plus (+) terminal of the voltmeter to the higher point of potential and the minus (−) terminal to the lower point of potential in order to obtain an upscale meter deflection. An ac voltmeter does not have polarity marks, but the instrument is still connected across the element for which the voltage is desired. Because a voltage measurement does not involve breaking the circuit, the voltmeter need be connected only when a reading is desired. Hence, in Fig. 17-13, arrowheads denote the temporary placement of the voltmeter leads.

A voltmeter with a higher voltage range V is constructed by connecting a resistance R_{se} in series with a meter mechanism having a full-scale voltage capability of V_m (see Fig. 17-14). The series resistance R_{se} is called a multiplier; its value is determined from the voltage equation,

$$V = I_m R_{se} + V_m \tag{17-5}$$

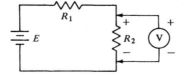

Figure 17-13   Voltmeter: Its symbol and its circuit connection.

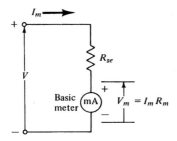

Figure 17-14 Circuit for increasing the voltage range of a meter.

EXAMPLE 17-4

A voltmeter with a full-scale value of 10 V is constructed using a 1 mA, 25 Ω mechanism. **(a)** What multiplier resistance is needed? **(b)** What current flows through the voltmeter with only 8 V applied?

Solution. For full scale, $V_m = 25$ mV. **(a)**

$$R_{se} = \frac{V - V_m}{I_m} = \frac{10 - 0.025}{1 \times 10^{-3}} = \frac{9.975}{1 \times 10^{-3}}$$
$$= 9975 \ \Omega$$

(b) The total resistance of the meter R_t is

$$R_t = R_m + R_{se} = 25 + 9975 = 10 \ \text{k}\Omega$$

With 8 V applied, the total meter current I_t is

$$I_t = \frac{8}{R_t} = \frac{8}{10 \times 10^3} = 0.8 \times 10^{-3} \ \text{A}$$

Notice that I_t is 0.8 of the full-scale current so that the meter deflection is also 0.8 of full scale.

Multiple-scale voltmeters (or millivoltmeters) are constructed either using switching arrangements or multiple terminals, as in Fig. 17-15.

In Fig. 17-15(a), each of the multiplier resistors are separately calculated and are then selected by a *nonshorting switch*. The multiplier resistors for Fig. 17-15(b) and (c) are calculated by first evaluating the resistance needed for the lowest range. The additional resistance needed for each higher range is then determined.

Drill Problem 17-7 ■

A meter with a resistance of 2 Ω and a full-scale current of 10 mA is used with a multiplier resistor in order to measure voltages to a full-scale value of 15 V. What is the multiplier resistance?

Answer. 1.498 kΩ. ■ ■

Drill Problem 17-8 ■

A multiple-range voltmeter with multiple terminals as in Fig. 17-15(c) is constructed with a basic meter having a resistance of 4 Ω and a full-scale current

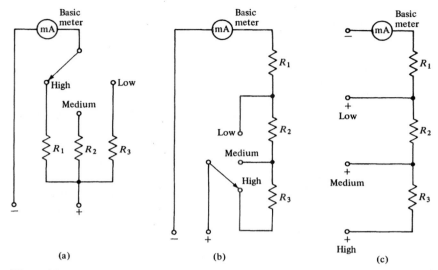

Figure 17-15 Multiple-range voltmeter: **(a)**, **(b)** Two switching arrangements; **(c)** multiple terminals.

of 10 mA. Determine the resistance values of R_1, R_2, and R_3 for voltage ranges of 10 V, 50 V, and 100 V.

Answer. $R_1 = 996 \ \Omega$; $R_2 = 4 \ \text{k}\Omega$; $R_3 = 5 \ \text{k}\Omega$.   ■ ■

17.5 VOLTMETER SENSITIVITY; LOADING EFFECTS

A voltmeter has a resistance for any range that equals the sum of the basic meter resistance and any multipliers used in that range. It follows that when one connects a voltmeter to a circuit as in Fig. 17-13, a current, determined by the voltmeter resistance, flows through the voltmeter. So as not to change the original circuit conditions, that is, *load* the original circuit, the voltmeter current should be negligible. A measure of the loading effect produced by a voltmeter is the *sensitivity*, defined as the reciprocal of the current necessary for full-scale deflection:

$$\text{sensitivity} = \frac{1}{I_m} \quad \frac{\text{ohms}}{\text{volt}} \tag{17-6}$$

where I_m is the full-scale current.

Relatively speaking, a smaller meter current results in a larger voltage sensitivity, indicative of less loading. As indicated in Eq. 17-6, the unit of sensitivity is *ohms per volt*, which is the reciprocal of the current unit of volts per ohm.

The actual voltmeter resistance equals the sensitivity × the full-scale voltage. It is important to remember that the voltmeter resistance remains constant throughout a range, even though the voltage reading may not be a full-scale reading.

EXAMPLE 17-5

A multirange voltmeter with 50 and 250 V ranges uses a 50 μA meter mechanism. **(a)** What is the sensitivity? **(b)** What resistance does the voltmeter present on each range?

Solution

(a)
$$\text{sensitivity} = \frac{1}{I_m} = \frac{1}{50 \times 10^{-6}} = 20 \text{ k}\Omega/\text{V}$$

(b) R = sensitivity × range.

50 V range.

$$R = (20 \times 10^3)(50) = 1 \times 10^6 = 1 \text{ M}\Omega$$

250 V range.

$$R = (20 \times 10^3)(250) = 5 \times 10^6 = 5 \text{ M}\Omega$$

The sensitivity of a voltmeter is often printed on the voltmeter scale. Unless one selects the proper sensitivity voltmeter, serious measurement errors may result. For high-current, low-resistance power applications, a low-sensitivity voltmeter may be used. Although the low-sensitivity meter presents a large loading effect, it is inexpensive and mechanically rugged. However, for low-current, high-resistance electronic applications, a high-sensitivity voltmeter should be used. The loading effect and error introduced during voltage measurement is shown in the following example.

EXAMPLE 17-6

A voltmeter with a 50 V scale and a sensitivity of 20 kΩ/V is used to measure the voltage across one of two 500 kΩ resistors connected in series to a 100 V source. (The circuit diagram of Fig. 17-13 is applicable here.) **(a)** What is the actual voltage across each resistor? **(b)** When the voltmeter is connected, what voltage will it indicate? **(c)** What error is made in measurement?

Solution. **(a)** By voltage division, the voltages V_1 and V_2 across R_1 and R_2, respectively, equal

$$V_1 = V_2 = E\left(\frac{R_2}{R_1 + R_2}\right) = 100\left(\frac{5 \times 10^5}{1 \times 10^6}\right) = 50 \text{ V}$$

(b) The voltmeter resistance R_m on the 50 V range is

$$R_m = (20 \times 10^3)(50) = 1 \times 10^6 = 1 \text{ M}\Omega$$

With the meter connected to R_2, a parallel circuit with resistance R_{eq} is formed:

$$R_{eq} = \frac{R_m R_2}{R_m + R_2} = \frac{(1 \times 10^6)(0.5 \times 10^6)}{1.5 \times 10^6} = 333 \text{ k}\Omega$$

The voltage measured by the meter is that across R_{eq}, which by voltage division is

$$V'_2 = E\left(\frac{R_{eq}}{R_1 + R_{eq}}\right) = 100\left(\frac{3.33 \times 10^5}{8.33 \times 10^5}\right) = 40 \text{ V}$$

(c) By Eq. 17-1, the error is

$$\text{error} = \frac{\text{deviation}}{\text{true value}} = \frac{50 - 40}{50} = 0.2 = 20\%$$

Drill Problem 17-9 ■

A basic meter with a resistance of $2 \, \Omega$ and a full-scale current of 10 mA needs a multiplying resistor of $1498 \, \Omega$ in order to form a voltmeter of 15 V full scale. **(a)** What is the sensitivity of the meter? **(b)** What is the resistance of the voltmeter?

Answer. (a) $100 \, \Omega/\text{V}$. (b) $1500 \, \Omega$. ■ ■

Drill Problem 17-10 ■

A multirange voltmeter with a $25 \, \mu\text{A}$ meter mechanism has ranges of 5 V, 10 V, and 50 V. **(a)** What is the meter sensitivity? **(b)** What is the voltmeter resistance on each range?

Answer. (a) $40 \, \text{k}\Omega/\text{V}$. (b) $200 \, \text{k}\Omega$; $400 \, \text{k}\Omega$; $2 \, \text{M}\Omega$. ■ ■

Drill Problem 17-11 ■

Refer to Fig. 17-13. $R_1 = R_2 = 100 \, \text{k}\Omega$ and $E = 20$ V. As shown, a voltmeter is momentarily connected across R_2 so as to permit measurement of the resistor voltage. **(a)** What voltage does the meter indicate? **(b)** What error results from the loading effect of the meter? The meter resistance is $100 \, \text{k}\Omega$.

Answer. (a) 6.67 V. (b) 33.3%. ■ ■

17.6 RESISTANCE MEASUREMENT

Voltmeter–Ammeter Method

There are many ways of measuring or determining the value of a resistance. One of the most obvious ways is to measure the voltage and current for the resistance and then to apply Ohm's law. This method is known as the *voltmeter–ammeter method*. Two possible voltmeter–ammeter connections used to measure an unknown resistance R_x are shown in Fig. 17-16. The connection in which the voltmeter is across, or *shunts*, only R_x is called the *short-shunt connection*, whereas the connection in which the voltmeter shunts the combination of R_x and the ammeter is called the *long-shunt connection*.

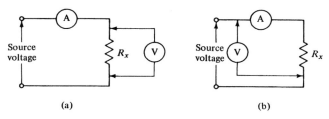

Figure 17-16 Voltmeter–ammeter method: **(a)** Short-shunt connection; **(b)** long-shunt connection.

Errors are introduced with both voltmeter–ammeter connections, since in the case of the short-shunt connection, the ammeter measures the sum of the resistor and voltmeter currents; in the case of the long-shunt connection, the voltmeter measures the sum of the resistor and ammeter voltages. It follows that the voltmeter reading divided by the ammeter reading may not be the true value of R_x.

If the unknown resistance has a value comparable to the voltmeter resistance, the short-shunt method introduces a large ammeter error. It follows that if R_x is relatively high, the long-shunt method is preferable. If the unknown resistance has a value comparable to the ammeter resistance, the long-shunt method introduces a large voltmeter error. Then, if R_x is relatively low, the short-shunt method is preferable. In determining which method to use, one may momentarily connect the voltmeter leads in short shunt and compare the ammeter readings with that obtained prior to connection of the voltmeter. If the readings indicate a loading effect by the voltmeter connection, the connection should be changed to long shunt. Regardless of the method used, it is always possible to correct for the higher ammeter or voltmeter reading, as indicated by the following example.

EXAMPLE 17-7

A short-shunt connection is used to measure an unknown resistance (refer to Fig. 17-16). The current, as indicated by a milliammeter, is 125 mA; the voltage, as indicated by a 20 V meter with a sensitivity of 100 Ω/V, is 10 V. Determine the resistance from **(a)** uncorrected readings and **(b)** corrected readings.

Solution. (a) $R_x = V/I = 10/0.125 = 80 \ \Omega$ (uncorrected).
(b) The voltmeter resistance R_m and current I_m are

$$R_m = (100)(20) = 2 \ \text{k}\Omega$$

$$I_m = \frac{10}{R_m} = \frac{10}{2 \times 10^3} = 5 \ \text{mA}$$

Then

$$R_x = \frac{V}{I - I_m} = \frac{10}{0.125 - 0.005} = \frac{10}{0.120} = 83.3 \ \Omega \qquad \text{(corrected)}$$

Therefore the corrected value is $R_x = 83.3 \ \Omega$. A 4.0% error is made if the milliammeter reading is not corrected.

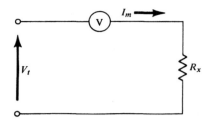

Figure 17-17   Series voltmeter method.

Series Voltmeter Method

A second method of measuring an unknown resistance is the series voltmeter method, described by the circuit of Fig. 17-17. In this method, a voltmeter with a resistance of R_m measures its own voltage drop V. The current through the meter is then

$$I = \frac{V}{R_m} \tag{17-7}$$

Since the current through the unknown resistance R_x also equals I

$$I = \frac{V_t - V}{R_x} \tag{17-8}$$

where V_t is the applied voltage. Combining the preceding equations and solving for R_x, we obtain

$$R_x = R_m\left(\frac{V_t - V}{V}\right) \tag{17-9}$$

By knowing the voltmeter resistance and measuring V_t, we can solve for R_x by Eq. 17-9. The best results are obtained when the voltmeter resistance is close to the unknown resistance value.

Ohmmeter Method

A third method of measuring resistance uses the *ohmmeter*, an instrument that contains a voltage source and a meter directly calibrated in ohms. One type of ohmmeter is the *series ohmmeter*, so-called because the meter movement is in series with the source of emf and the unknown resistance. The circuit diagram and basic ohmmeter scale are shown in Fig. 17-18.

If the terminals of the ohmmeter are left open, $R_x = \infty$ and no current flows. As is shown on the basic scale, zero meter deflection corresponds to $R_x = \infty$. However, when the terminals or leads are shorted, $R_x = 0$ and maximum current flows. So that exactly full-scale current flows when $R_x = 0$, a *zero adjust* control is provided. The zero adjust allows one to calibrate the ohmmeter with the test leads shorted, thus compensating for lead resistance and battery aging.

If the sum of all the resistances internal to the ohmmeter is called R_i,

$$R_i = R_1 + R_2 + R_m \tag{17-10}$$

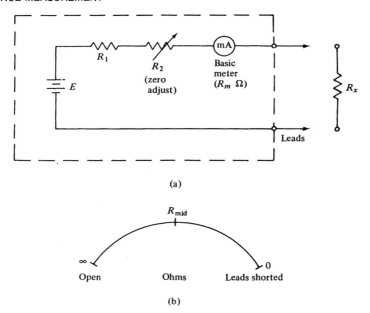

(a)

(b)

Figure 17-18   Series ohmmeter: **(a)** Circuit; **(b)** basic scale.

and the meter current for any value of R_x is

$$I = \frac{E}{R_i + R_x} \qquad (17\text{-}11)$$

where E is the internal ohmmeter voltage. It follows from Eq. 17-11 that if full-scale current flows when $R_x = 0 \ \Omega$, half-scale current flows when $R_x = R_i$. Within the limitations of the emf and meter mechanism, the internal resistances R_1 and R_2 are selected to provide a particular mid-scale resistance R_{mid}.

Notice that, for the left half of the ohmmeter scale, resistances between ∞ and R_{mid} are indicated; for the right-half, resistances between R_{mid} and 0 are indicated. The resulting scale is nonlinear, and at either scale end the accuracy is poor. Therefore, multirange ohmmeters having different mid-scale resistance values are desirable. The different scales are obtained through range switching, meter mechanism shunts, and different potential sources.

A second type of ohmmeter is the *shunt ohmmeter*, so called because the meter movement is in parallel with the unknown resistance. The basic shunt ohmmeter circuit and scale are shown in Fig. 17-19. Notice that a switch S is necessary to prevent current flow from the source of emf when the ohmmeter is not in use.

If the terminals of the shunt ohmmeter are shorted, $R_x = 0$ and all current is shunted away from the meter mechanism. However, when the terminals are open, $R_x = \infty$ and maximum meter current flows. As before, a control is provided for the adjustment of full-scale deflection, but it is now an *infinity adjust* (∞ adjust). As in

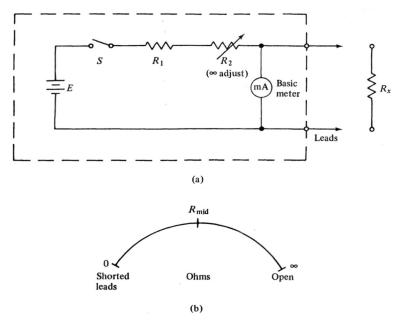

(a)

(b)

Figure 17-19   Shunt ohmmeter: **(a)** Circuit; **(b)** basic scale.

the series-type ohmmeter, when the unknown resistance equals the meter resistance, the meter reading is at half-scale. Compared to the series type, however, the shunt ohmmeter has a low meter resistance, making it particularly useful for unknown resistances that are relatively low. Regardless of the type of ohmmeter, one must be certain that *it is not connected to an energized or active circuit.*

Wheatstone Bridge Method

A fourth method of measuring resistance is by means of the *Wheatstone bridge*, which is defined by the circuit of Fig. 17-20, where R_1, R_2, and R_3 are known resistances and R_x is the resistance whose value is to be determined. At least one of the known resistances is variable. With the switch S closed, the variable resistance is adjusted

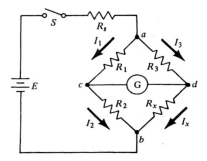

Figure 17-20   Wheatstone bridge.

until there is no current in the zero center galvanometer G. When no current flows in the galvanometer, the circuit is in a *null* or *balanced* condition.

For the bridge to be balanced, points c and d must be at the same potential; for this to be true, the voltage from a to c must equal the voltage from a to d. It follows that

$$I_1 R_1 = I_3 R_3 \tag{17-12}$$

and similarly

$$I_2 R_2 = I_x R_x \tag{17-13}$$

Dividing one equation by the other, we obtain

$$\frac{I_1 R_1}{I_2 R_2} = \frac{I_3 R_3}{I_x R_x} \tag{17-14}$$

However, with no current in the galvanometer,

$$I_1 = I_2 \quad \text{and} \quad I_3 = I_x$$

so that Eq. 17-14 reduces to

$$\frac{R_1}{R_2} = \frac{R_3}{R_x}$$

or

$$R_x = R_3 \left(\frac{R_2}{R_1} \right) \tag{17-15}$$

Equation 17-15 suggests that only the ratio of R_2 to R_1 is required, not the individual values. Conveniently, the ratio may be made equal to a power of 10. Notice, too, that bridge balance and the determination of R_x are completely independent of the magnitude of source voltage. Commercially available bridges use a switching arrangement for the multiplying ratio. Resistance can also be measured with an impedance bridge circuit, which is discussed in Section 17.11.

As with the ohmmeter, one must not use the bridge to measure the resistance of a component that is a part of another energized circuit. The component for which the resistance is desired must first be isolated from the original circuit and then connected to the bridge circuit. Lead resistance should then be found and subtracted from the subsequent resistance measurements.

In this chapter the bridge circuit is described from a measurement perspective so that a balanced condition is dictated. However, the bridge circuit may also be encountered in a nonmeasurement role. Then the various circuit analysis techniques of Chapter 6 are used to solve for the voltages and currents. In particular, the Δ–Y conversion formulas of Chapter 6 are very useful.

Drill Problem 17-12 ■

Consider the measurement circuit of Fig. 17-21. The milliammeter indicates 10 mA, and the voltmeter, with a resistance of 20 kΩ, indicates 20 V. Determine the resistance R_x from (a) the uncorrected readings and (b) the corrected readings.

Answer. (a) 2 kΩ. (b) 2.22 kΩ. ■ ■

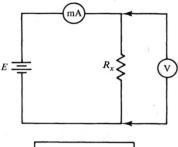

Figure 17-21   Drill Problem 17-12.

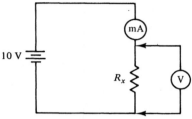

Figure 17-22   Drill Problem 17-13.

Drill Problem 17-13 ■

Consider the measurement circuit of Fig. 17-22. The voltmeter indicates 10 V and has a resistance of 20 kΩ. The milliammeter indicates 200 mA and has a resistance of 2 Ω. Determine the resistance R_x from **(a)** the uncorrected readings and **(b)** the corrected readings.

Answer.   (a) 50 Ω. (b) 48 Ω.   ■ ■

Drill Problem 17-14 ■

A bridge circuit like that shown in Fig. 17-23 is used to measure the resistance of a strain gauge. What is the resistance of the strain gauge if the bridge is balanced with R_3 adjusted to 30.62 Ω?

Answer.   122.5 Ω.   ■ ■

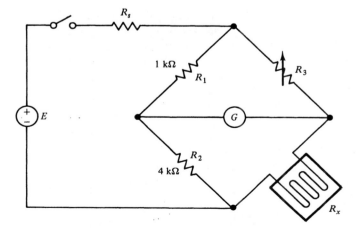

Figure 17-23   Drill Problem 17-14.

17.7 DIGITAL INSTRUMENTS

A digital instrument is a highly sophisticated electronic device that measures electrical quantities and displays the measured value in decimal numeric form. In almost all cases, the input quantity is a voltage or is converted into a voltage. Thus, all digital instruments, including small panel-mounted instruments, are fundamentally *digital voltmeters* (DVM). A DVM converts the analog input to a digital output, a process known as *analog to digital* (A/D) *conversion*. Although a number of different A/D conversion processes are used in instrumentation, two of the most frequently used processes are the *comparison* and *dual-slope integration methods*.

The comparison method, sometimes called the *successive approximation method*, is depicted in block diagram form in Fig. 17-24. Although it may at first seem contradictory, this A/D conversion process utilizes *digital to analog* (D/A) *conversion*. A D/A converter is a network of precision resistors, operational amplifiers, and electronic switches that add, or sum, a series of binary digits representing a numerical magnitude.

In the block diagram of Fig. 17-24, the analog voltage to be measured, V_i, is applied to a comparator circuit. Notice that the D/A converter sends the analog of what is considered the binary or digital equivalent of the input back to the comparator. The comparator circuit detects any difference or error between the input signal and the signal generated by the D/A converter. If there is a difference, the comparator signals the group of logic circuits to minimize that difference by altering the binary signal that represents the digital equivalent. This signal is then converted by the D/A converter, again compared to the input, and changed if necessary. Thus, by successive approximation, the binary signal follows the analog input voltage.

Figure 17-24 also shows *latch, decoder, driver,* and *control circuits*. A *latch* is a digital circuit that holds data after having had data applied earlier in time. A *decoder* is a digital circuit that generates an output signal to identify which one of a number of possible binary combinations is present. The *driver circuit* is an amplifier circuit

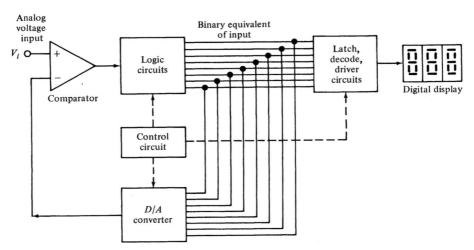

Figure 17-24 Comparison or successive-approximation-type DVM.

Figure 17-25   Two LED display units: (left) Ten-segment bar graph display; (right) seven-segment display.

that increases the voltage or current to a level sufficient to operate a particular digital display. The *control* circuit is a circuit that provides overall timing and determines the start and finish of the conversion process, the conversion repetition rate, and the latching of the binary signal for the output.

The output is indicated in Fig. 17-24 by *seven-segment displays,* so called because each of the display units has seven segments, each of which can be selectively operated so that groups can be formed to represent the decimal characters (0–9).

Various types of display technology are used in digital instruments. Some of the more popular types are the *light emitting diode* (LED), the *liquid crystal display* (LCD), and the *electroluminescent display* (ELD). Each type of display has its advantages. For example, the LED emits light and is useful in a dark environment but is not very visible in bright sunlight. The LCD does not emit light, but rather through rear projection or reflection alters the transmission of light. It is useful in bright sunlight, but depends on auxiliary lighting in dark environments.

Two examples of digital LEDs appear in Fig. 17-25. The display on the left is a ten-segment display that can be used as a bar graph. The display on the right is a seven-segment display. Although it is usually used to display a decimal character, a limited number of other characters can be formed with the seven segments.

Another A/D conversion method used in digital instruments is the *dual-slope integration method,* illustrated by Fig. 17-26. This method uses an *integrator,* a circuit that performs the mathematical function of integration, as well as the subsequent amplification, of an input signal. Notice in particular that the input signal to the integrator in Fig. 17-26 is either the voltage being observed, V_i, or an internal reference voltage V_{REF}.

The actual conversion process begins with the control circuit setting the counter to zero, the integrating capacitor charge to zero, and the electronic switch at the input to the input voltage V_i. Then the counter begins its count and the input signal is

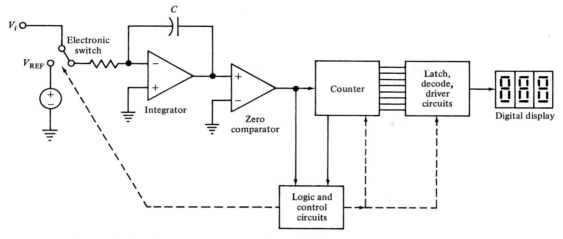

Figure 17-26 Dual-slope integrating type DVM.

integrated. As the input signal V_i is integrated, the output voltage of the integrator circuit continues to rise, as illustrated by the left-hand portion of the voltage graph of Fig. 17-27. When the counter reaches a certain count (selected by design), the counter resets and the logic circuits switch the integrator input to the reference voltage.

A reference voltage opposite in polarity to V_i is used so that as the reference voltage is integrated, the capacitor begins to discharge and the integrator output voltage begins to fall, as indicated in the right-hand portion of Fig. 17-27. The zero comparator detects when the output of the integrator reaches zero. At that point the counter stops at a count N. Since the count N is proportional to the input voltage V_i, the count can then be latched, decoded, and displayed.

Digital panel meters are available in many functions and ranges. Digital panel meters offer many advantages over the analog type. Not only are they more accurate and faster, but they generally have the capability of being connected directly to larger digital systems. A typical digital panel meter is shown in Fig. 17-28.

The digital panel meter of Fig. 17-28 typifies a family of $3\frac{1}{2}$-digit process monitors for measuring and displaying a number of different quantities including temperature, pressure, level, flow, voltage, and current. Each of the members of the

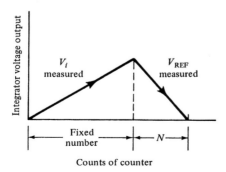

Figure 17-27 Output for the dual-slope integrator.

Figure 17-28 The MEASUROMETER™ II digital process monitor. (Courtesy of Analogic Corporation.)

MEASUROMETER™ II family features a modular design that comprises a signal conditioning circuit, an A/D converter, and a seven-segment LED display. The exact performance characteristics depend on the monitor function. For example, the dc voltage or current monitor measures voltages from 0–200 V dc and is capable of measuring currents to 200 mA dc, each at an accuracy of $\pm 0.05\%$ of reading ± 1 count.

In contrast to the analog panel meter, which derives all its operating power from the same voltage or current that is to be measured, the digital panel meter derives almost all its operating power from an internal power supply and only negligible power from the measured circuit. The internal power supply of the typical digital panel meter is connected to an ac power line and draws approximately 5 W.

17.8 MULTIMETERS

An instrument that uses switching arrangements to combine the ammeter, voltmeter, and ohmmeter functions is called a multimeter. Because multimeters are designed for general-purpose or service-type measurements, provisions are made for both dc and ac measurements. Some multimeters use self-contained batteries, whereas others require connection to external 115 V, 60 Hz lines.

A commercially available volt–ohm–milliammeter, commonly known as a *VOM*, is shown in Fig. 17-29. This instrument has a taut-band, permanent-magnet moving-coil mechanism and a $4\frac{1}{2}$-in. analog scale. The schematic diagram for the instrument is shown in Fig. 17-30. The instrument provides nine dc voltage ranges,

Figure 17-29 Model 260®-6XL volt–ohm–milliammeter (VOM). (Courtesy of Simpson Electric Company.)

each accurate to 2% of full scale, and a full-wave rectifier circuit affording seven ac voltage ranges, each accurate to 3% of full scale. Although the instrument responds to the average rectified value, the ac scales are calibrated in terms of the rms or effective value of a sine wave. The meter sensitivity is 20 000 Ω/V for direct current and 5000 Ω/V for ac use.

Additionally, the 260®-6 XL VOM provides six current ranges, four conventional resistance ranges, and two low-power resistance ranges. The low-power resistance ranges utilize an open-circuit voltage of only 100 mV, thereby allowing safe resistance measurements of semiconductors and integrated circuits.

Some multimeters utilize solid-state circuits, which load a measured circuit less than a conventional VOM. In particular, an electronic device known as the *field effect transistor* (FET) is used in the FET VOM, with a resulting instrument resistance in the 10 MΩ range. Since multimeters are used for alternating current, the term "input impedance," the total opposition in ohms that the meter offers to the ac flow, is used instead of "input resistance."

Although used in a manner similar to the conventional analog multimeter, the *digital multimeter* (DMM) utilizes complex electronic circuits and digital display. A distinct advantage of a DMM is its inherent accuracy. A disadvantage is that variations are not as readily recognizable as with an analog display.

One of the descriptive features of a digital meter, including the DMM, is the number of digits in the display. However, manufacturers often use ranges that do not utilize all the possible decimal characters for the most significant digit. In such

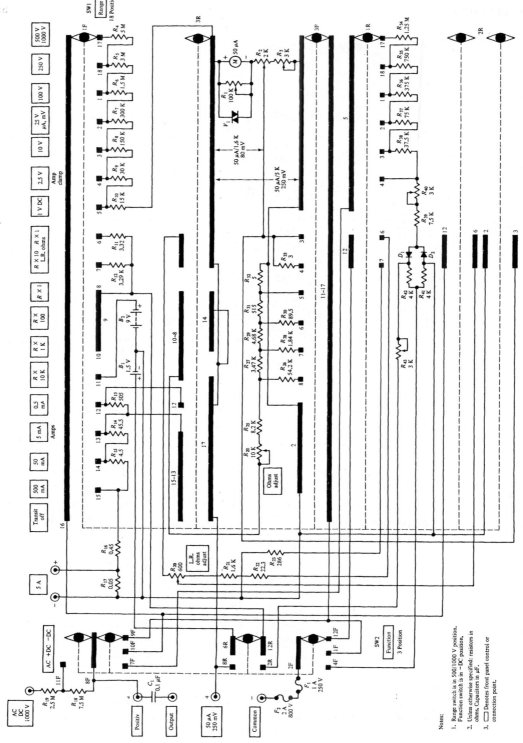

Figure 17-30 Schematic for the 260®-6XL VOM. (Courtesy of Simpson Electric Company.)

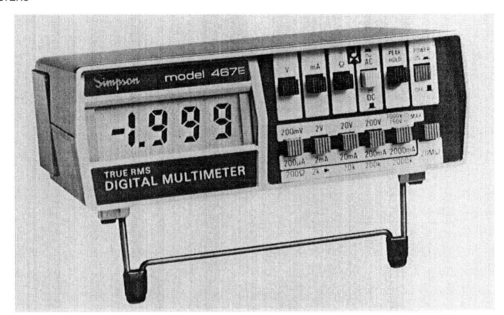

Figure 17-31   Model 467 E digital multimeter (DMM). (Courtesy of Simpson Electric Company.)

a case, the number of display digits is specified as an integer *plus* a fraction, usually $\frac{1}{2}$ digit, regardless of the fractional part of the display. For example, if the decimal point is neglected, a three-digit display normally goes up to a maximum count of 999. Suppose now that a manufacturer desires an instrument that goes up to a full count of 1999; a fourth digit must be used. Because 0 and 1 are the only characters used on the most significant display, the display is called a $3\frac{1}{2}$-digit display.

Figure 17-31 is one example of a commercially available DMM. This instrument is a $3\frac{1}{2}$-digit DMM with a LCD. It features five dc voltage and five ac voltage ranges with an input resistance or impedance, respectively, of 10 MΩ. It has six resistance, five dc current, and five ac current ranges. In addition, it provides an audible indication of continuity and logic levels, and "captures" and "holds" the maximum value of a complex waveform.

As a second example of a DMM, consider the hand-held model in Fig. 17-32. It uses a $3\frac{1}{2}$-digit LCD and features five ranges for each of the functions of dc voltage, ac voltage, resistance, dc current, and ac current. It provides an audible tone for continuity checks and measures diode resistance at the 1 mA level.

It should be pointed out that a number of optional accessories can be used with a multimeter in order to extend its range and versatility. Some of these accessories extend the voltage or current ranges while others measure nonelectrical effects such as temperature via transducing probes. The AMP–CLAMP® shown in Fig. 17-33 is an example of an optional accessory. It allows one to measure relatively high alternating currents without opening the circuit. The clamp is connected to an appropriate multimeter, the jaws are opened, and the clamp is placed around the

Figure 17-32 Model 470™ handheld DMM. (Courtesy of Simpson Electric Company.)

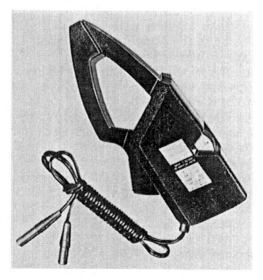

Figure 17-33 Model 153 AMP-CLAMP®. (Courtesy of Simpson Electric Company.)

Figure 17-34 Touch Test model 21 DMM. (Courtesy of Non-Linear Systems, Inc.)

current-carrying conductor. The current is reduced through transformer action to a level that can be measured by the basic multimeter.

Figure 17-34 shows a final example of a DMM model. This instrument tests and measures ten electrical parameters with 44 ranges. In particular, it features a $3\frac{1}{2}$-digit, LCD, and touch switch selection and control of function ranges and power. In addition to 24 ranges for the basic electrical quantities of voltage, current, and resistance, it provides for measurements of temperature in two ranges and capacitance in six ranges. Conductance measurements and diode and continuity testing are also possible.

17.9 POWER MEASUREMENT: THE WATTMETER

In Chapter 4, power, the rate of doing work, was described mathematically for a dc circuit as the product of voltage and current. It follows that one may determine the power in a dc circuit from voltage and current measurements. Although the preceding method is perhaps preferred for dc power measurements, dc power can be measured with a *wattmeter*, an instrument that directly measures power.

The wattmeter is essentially an electrodynamic mechanism in which the stationary and movable coils separately measure the current and voltage of the circuit (see Fig. 17-35). The stationary coil is designed to handle the current of the circuit, whereas the moving coil is connected in series with a multiplying resistor and is designed to respond to the voltage of the circuit. Because the torque of the moving coil depends directly on the magnetic fields produced by both coils, the moving-coil deflection is proportional to the product of voltage and current. Hence, the instrument is calibrated directly in watts.

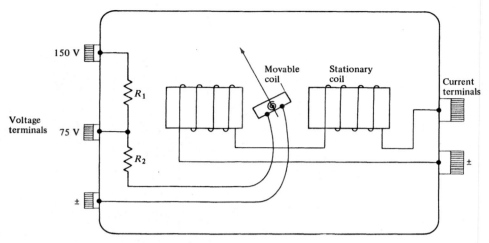

Figure 17-35   Wattmeter.

In an ac circuit, the voltage and current are generally not in phase so that their product does not equal the power. Rather, the power equals the product of voltage and current, or the *apparent power*, and a *power factor*. Often one does not know the power factor of an ac circuit and cannot determine the power from separate voltage and current measurements. Yet the electrodynamic wattmeter allows one to determine directly the power in an ac circuit.

Wattmeters have one voltage and one current terminal signed with a ± symbol. For an upscale deflection, current must simultaneously enter both signed terminals or leave both signed terminals. A correct wattmeter connection is shown in Fig. 17-36. Reversal of either the voltage or current element results in a downscale deflection.

The wattmeter connection shown in Fig. 17-36 is a *short-shunt connection*, since the voltage element of the wattmeter is across the load only. Obviously, the wattmeter reads not only the load power, but the power lost in the voltage element. When one uses a *long-shunt connection*, the wattmeter reads the load power plus that

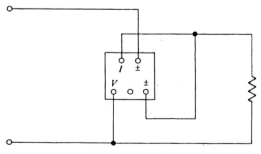

Figure 17-36   Wattmeter connection for upscale deflection (short-shunt connection).

lost in the current element. Knowing the resistance of the appropriate element, one can correct for the instrument power loss.

EXAMPLE 17-8

The wattmeter in Fig. 17-36 reads 200 W when the line voltage is 150 V. If the resistance of the voltage element is 4 kΩ, what is the true load power?

Solution

$$P = \text{reading} - \text{meter loss} = 200 - \frac{V^2}{R}$$

$$= 200 - \frac{(150)^2}{4 \times 10^3} = 200 - 5.6 = 194.4 \text{ W}$$

Both the voltage and current elements of a wattmeter have ratings that should not be exceeded. Unfortunately, it is possible to exceed either the voltage or current rating even should the wattmeter pointer not go off scale. For example, a wattmeter with a 750 W range may have a voltage rating of 150 V and a current rating of 5 A. If the wattmeter is connected so that a voltage of 100 V and a current of 7 A are measured, the wattmeter range is not exceeded, yet the current rating is. Therefore, one should have an idea of the voltage and current levels being applied to a wattmeter. A shunt can be used so as not to exceed a current rating and a voltage divider or a series resistor can be used so as not to exceed a voltage rating. The wattmeter reading is then multiplied by a scale factor.

Drill Problem 17-15 ■

The wattmeter of Fig. 17-37 reads 420 W. The resistances of the voltage and current coils of the wattmeter are 5 kΩ and 1 Ω, respectively. What power is actually absorbed by the load resistance *R*?

Answer.   407.8 W.   ■ ■

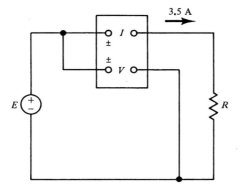

Figure 17-37   Drill Problem 17-15.

17.10 ENERGY MEASUREMENT: THE WATTHOUR METER

Energy is the product of power and time and is measured in *wattseconds* or *joules*. Since the voltage and current in a dc circuit are constant values, the energy is easily calculated from a measurement of power and time,

$$W = VIt = Pt \tag{17-16}$$

where W is in wattseconds, P is in watts, and t is in seconds. As was pointed out in Chapter 4, the wattsecond is too small an electrical unit and hence the larger unit *kilowatt-hour* (kWh) is preferred.

If the current and voltage are not constant, one measures energy directly by the use of a *watthour meter*. Both dc and ac watthour meters are available; both types are *summing* or *integrating* instruments. The basic assembly of an ac induction-type watthour meter is shown in Fig. 17-38.

The ac induction watthour meter has potential and current coils, but unlike in the wattmeter, all the coils are stationary. The potential coil is connected across the source lines, whereas the current coils are connected in series with the load. The combination of stationary coils is called the stator.

A disk or *rotor* mounted on a shaft receives a torque through an electromagnetic induction process whenever the two sets of coils are energized. The rotor, in turn, is mechanically connected to a meter register via a gear train (not shown). The register then provides a record of the number of rotor shaft revolutions and is calibrated in kilowatt-hours. A constant of proportionality, the *watthour constant* K_h, is the number of watthours corresponding to one disk revolution. Thus the energy in watthours is

$$W = K_h \times \text{disk revolutions} \tag{17-17}$$

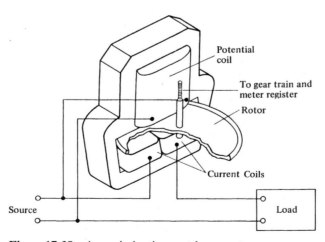

Figure 17-38 An ac induction watthour meter.

Figure 17-39 A single-phase watthour meter. (Courtesy of General Electric Co.)

In kilowatt-hours,

$$W = \frac{K_h \times \text{disk revolutions}}{1000} \qquad (17\text{-}18)$$

EXAMPLE 17-9

A kilowatt-hour meter indicates a 10 kWh change for 347 revolutions of the rotor. Determine the constant K_h for the meter.

Solution. From Eq. 17-18,

$$K_h = \frac{W \times 1000}{\text{disk revolutions}} = \frac{10 \times 1000}{347} = 28.8 \text{ Wh/rev}$$

A single-phase watthour meter with a conventional register and analog dials is shown in Fig. 17-39. This self-contained meter has a potential coil wound for 240 V and a current coil wound for 30 A.

The analog, dial and pointer display is read from left to right; the left dial and pointer indicate the most significant digit. To read the meter one takes the number just passed by the pointer on each scale. Because of the nature of the internal gearing, the dials have alternate clockwise and counterclockwise scales, as noted in Fig. 17-39. In order to determine the amount of energy used over a certain period of time, the reading at the beginning of the time period is subtracted from the reading at the end of the time period.

EXAMPLE 17-10

Determine the kilowatt-hour reading for the meter display of Fig. 17-40.

Solution. By reading the dials from left to right, we obtain a reading of 48 204 kWh.

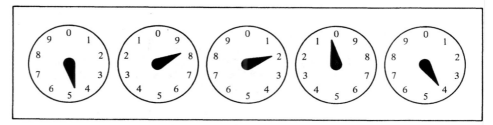

Figure 17-40   Kilowatt–hour meter display for Example 17-10.

A digital TOU (time of use) register is shown in Fig. 17-41. This digital register, the GE TM-80, includes an optional analog display and allows power companies to monitor and better manage their electric energy demands. The register can also be retrofitted to analog registers already in use. It is microprocessor-controlled and contains a nine-digit vacuum fluorescent display. The register records a number of selectable energy and demand measurements and automatically adjusts its timing for holidays, weekends, daylight savings time, and leap year. It contains a lithium battery that maintains the time-keeping and data storage for up to 40 days of outage during the five-year rated life of the battery.

Notice the optical port coupler on the glass meter enclosure to the right and slightly below the digital display. This through-the-cover coupler is called OPTOCOM™ and allows access to another instrument, a register programmer, while maintaining the security of the meter enclosure.

The register programmer is shown in Fig. 17-42. Notice the optical coupler that magnetically latches to the optical port coupler for the TM-80 register. The programmer is microprocessor based and uses a rechargeable battery for 24 h of field

Figure 17-41   TM-80, Time of Use, digital register. (Courtesy of General Electric Co.)

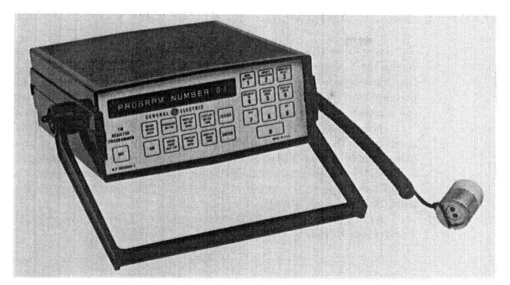

Figure 17-42   The TM Register Programmer. (Courtesy of General Electric Co.)

operation. The programmer can store a number of programs, load them into the
TOU register, and extract all metered data and register contents.

Drill Problem 17-16 ■

How many revolutions of the rotor are necessary for a kilowatt-hour meter
with a constant of 3.6 Wh/rev to register a change of 1 kWh?

Answer.   278 rev.   ■ ■

Drill Problem 17-17 ■

After a period of 20 days, a kilowatt-hour meter that originally had a reading
of 48 204 kWh shows the reading in Fig. 17-43. Determine the average kilowatt-
hour use/day.

Answer.   861 kWh/day.   ■ ■

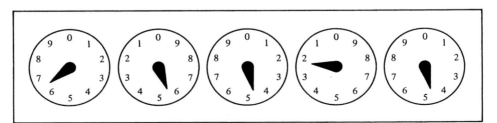

Figure 17-43   Drill Problem 17-17.

17.11 IMPEDANCE MEASUREMENT

The opposition that a circuit offers to the flow of ac is called impedance. By measuring the voltage and current in an ac circuit and utilizing Eq. 12-1, we can obtain the magnitude of circuit impedance. However, it is often desirable to separate impedance into its resistive and reactive components. One instrument used to measure the separate resistive and reactive parts of an impedance is the *ac bridge*.

The circuit of the general ac bridge is shown in Fig. 17-44. The configuration is similar to that of the Wheatstone bridge, yet distinct differences exist between the components. The ac bridge has impedance arms, rather than resistance arms; instead of a battery and galvanometer, an ac signal source and a null detector are used. If the signal voltage is in the audio range, a set of headphones may be used as the null detector; otherwise, a sensitive ac voltmeter is used.

As in the Wheatstone bridge, at balance no current flows through the detector. The voltage from a to c equals that from a to d, so that

$$\mathbf{I}_1\mathbf{Z}_1 = \mathbf{I}_3\mathbf{Z}_3 \tag{17-19}$$

Similarly,
$$\mathbf{I}_2\mathbf{Z}_2 = \mathbf{I}_x\mathbf{Z}_x \tag{17-20}$$

It follows that for balanced conditions

$$\frac{\mathbf{Z}_1}{\mathbf{Z}_2} = \frac{\mathbf{Z}_3}{\mathbf{Z}_x}$$

$$\mathbf{Z}_x = \mathbf{Z}_3\left(\frac{\mathbf{Z}_2}{\mathbf{Z}_1}\right) \tag{17-21}$$

In order to obtain balance, at least one of the known impedances must have a resistive and reactive component. If $\mathbf{Z}_3$ is chosen to be complex, $\mathbf{Z}_1$ and $\mathbf{Z}_2$ can be conveniently chosen to be purely resistive with a ratio that is a power of 10. Then from Eq. 17-21,

$$R_x + jX_x = (R_3 + jX_3)\left(\frac{R_2}{R_1}\right) \tag{17-22}$$

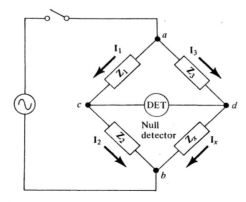

Figure 17-44 AC impedance bridge circuit.

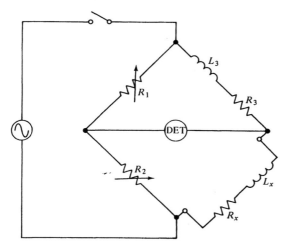

Figure 17-45 Inductance comparison bridge.

For Eq. 17-22 to be satisfied, the real parts of both sides must be equal, and so must the imaginary parts. Equating these parts, we obtain

$$R_x = R_3\left(\frac{R_2}{R_1}\right) \tag{17-23}$$

$$X_x = X_3\left(\frac{R_2}{R_1}\right) \tag{17-24}$$

A simple inductance bridge is shown in Fig. 17-45. When the bridge is balanced, we solve for R_x by using Eq. 17-23. From Eq. 17-24, we then solve for the inductance L_x:

$$\omega L_x = \omega L_3\left(\frac{R_2}{R_1}\right)$$

or

$$L_x = L_3\left(\frac{R_2}{R_1}\right) \tag{17-25}$$

The inductance comparison bridge of Fig. 17-45 depends on a standard inductance L_3. Inductance standards are expensive and not as ideal as capacitance standards. Thus, other bridge configurations that use capacitance standards are often used. One such bridge is the Hay bridge of Fig. 17-46.

When Eq. 17-21 is applied to the Hay bridge, the following formulas for the unknowns R_x and L_x are found to specify the balance conditions:

$$R_x = \frac{R_1 R_2 R_3 \omega^2 C^2}{1 + R_1^2 \omega^2 C^2} \tag{17-26}$$

$$L_x = \frac{R_2 R_3 C}{1 + R_1^2 \omega^2 C^2} \tag{17-27}$$

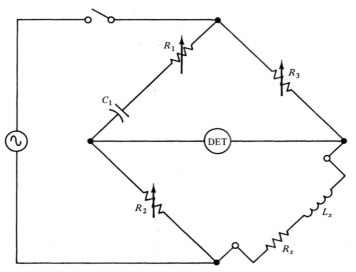

Figure 17-46   Hay bridge for measuring inductance.

For measuring capacitance a simple capacitance comparison bridge is used. Such a bridge is depicted in Fig. 17-47. When the balance conditions are imposed, the unknown capacitance C_x and series resistance R_x are found to be

$$C_x = C_2\left(\frac{R_1}{R_3}\right)$$ (17-28)

$$R_x = R_2\left(\frac{R_3}{R_1}\right)$$ (17-29)

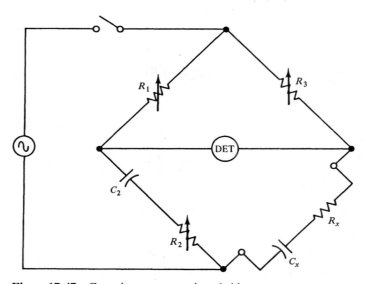

Figure 17-47   Capacitance comparison bridge.

Often the ratio of the series resistance to equivalent reactance is specified for a capacitor. This ratio is called the *dissipation factor D*:

$$D = \frac{R_x}{X_{Cx}} = \omega C_x R_x \qquad (17\text{-}30)$$

The factor, of course, is frequency dependent. A dissipation factor can also be defined for an inductor:

$$D = \frac{R_x}{\omega L_x} \qquad (17\text{-}31)$$

However, the reciprocal of the dissipation factor of the inductor is most commonly used. The reciprocal of Eq. 17-31 should be recognized as the coil Q:

$$Q = \frac{1}{D_x} = \frac{\omega L_x}{R_x} \qquad (17\text{-}32)$$

Commercial impedance bridges and meters often combine the basic bridge functions in one convenient instrument. One such impedance meter is shown in Fig. 17-48. The impedance meter shown is actually a bridge that measures inductance, capacitance, resistance, conductance, and dissipation factor at a basic accuracy of 0.25%. The test frequency for the meter shown is 1 kHz. The display is a $3\frac{1}{2}$-digit LED display that blanks for overload. As noted on the front panel, the meter

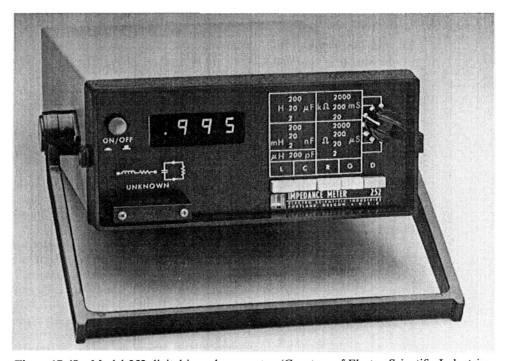

Figure 17-48 Model 252 digital impedance meter. (Courtesy of Electro Scientific Industries, Inc.)

measures inductance to 200 H, capacitance to 200 μF, resistance to 2000 kΩ, and conductance to 2000 mS. The dissipation factor can be measured to 1.999.

Drill Problem 17-18 ■

Refer to the Hay bridge. Using the equation for balance (Eq. 17-21), show all the steps and derive the equations for calculating the values of R_x and L_x in terms of the other bridge components.

Answer. Equations 17-26 and 17-27.   ■ ■

Drill Problem 17-19 ■

Refer to the capacitance comparison bridge. Using the equation for balance (Eq. 17-21), show all the steps and derive the equations for calculating the values of R_x and C_x in terms of the other bridge components.

Answer. Equations 17-29 and 17-28.   ■ ■

Drill Problem 17-20 ■

The capacitance comparison bridge of Fig. 17-47 is balanced at 1 kHz when $R_1 = 200$ kΩ, $R_2 = 200$ Ω, $R_3 = 2$ kΩ, and $C_2 = 1$ μF. Determine **(a)** the values of R_x and C_x and **(b)** the dissipation factor.

Answer. (a) 2 Ω; 100 μF. (b) 1.26.   ■ ■

17.12 FREQUENCY MEASUREMENT

The frequency of a periodic electrical voltage or current can be determined directly by the use of a frequency meter or indirectly through comparison with a known frequency.

One of the many frequency meters that directly indicate frequency is the *reed-type meter* of Fig. 17-49. In this type of meter, many reeds are mounted on a common support with their free ends visible at the meter face. Each reed has its own natural frequency of vibration. When an internal electromagnet is excited by the current of unknown frequency, an alternating magnetic field is produced; if the frequency of the field corresponds to the vibration frequency of a reed, that particular reed vibrates with considerable amplitude. If two adjacent reeds vibrate with the same amplitude, the unknown frequency is halfway between those indicated by the two vibrating reeds. The reed-type meter is useful only at low frequencies and only over a limited range of frequencies (approximately 40–400 Hz).

Another type of frequency meter is the *digital frequency meter*, which measures frequencies up to about 100 MHz and displays a digital readout of the measured frequency. The digital frequency meter is commonly called a counter, since it determines the frequency by electronically counting the number of cycles of the unknown in a standard time interval, usually 1 s. In addition to its basic function of measuring frequency, the counter can count uniform or random pulses or events and display the total.

An example of a commercial digital frequency meter is shown in Fig. 17-50. This counter covers frequencies of 10 Hz–60 MHz in two ranges and has a switchable

Figure 17-49 A reed-type frequency meter.

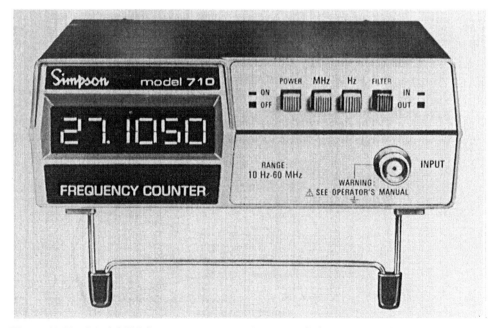

Figure 17-50 Model 710 frequency counter. (Courtesy of Simpson Electric Company.)

low-pass filter that can eliminate noise. The unit features a six-digit LED display and an input impedance of 1 MΩ shunted by 30 pF of capacitance.

Finally, an *oscilloscope* is an electronic instrument that presents a visual indication of instantaneous excursions (the waveform) of a voltage. An unknown frequency can be determined indirectly with an oscilloscope if the waveform of the unknown voltage is compared to the waveform of a known voltage. Furthermore, if the oscilloscope has a calibration feature, an unknown frequency can be easily determined from the display. The oscilloscope is discussed further in the next section.

17.13 THE OSCILLOSCOPE

Perhaps the most versatile electrical instrument is the *cathode-ray oscilloscope*, which permits a visual indication of the instantaneous values of a measured voltage. Therefore, it is indispensable for the analysis of complex waveforms and, when calibrated, can also be used as a dc or ac voltmeter. The frequency and phase of an ac signal can also be determined with a calibrated oscilloscope.

The oscilloscope consists of a number of complex electronic circuits; only a brief description of the operational features is presented here. The main component is a *cathode-ray tube* (CRT), a special type of vacuum tube in which a stream of electrons is emitted from an *electron gun*, deflected, and allowed to strike a fluorescent screen with light emitted at the point of electron impact (see Fig. 17-51).

The electron gun consists of a heated *cathode* that emits the electron stream, a *grid* that controls the stream intensity, and *anodes* that accelerate and focus the stream or *beam*. The electron beam passes between two sets of *deflection plates* which, if charged, electrostatically deflect the beam. The plates used to deflect the beam horizontally are called the *horizontal deflection plates* and those used to deflect the beam vertically are called the *vertical deflection plates*.

A simplified block diagram of the oscilloscope appears in Fig. 17-52. In addition to the CRT, there are four functional blocks: *horizontal deflection, vertical deflection, sweep,* and *trigger circuits.*

The *horizontal deflection circuits* are the amplifiers and level controls that supply a very high voltage to the horizontal plates and cause the electron beam to hori-

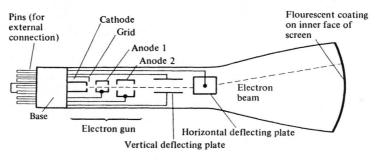

Figure 17-51   Cathode-ray tube (electrostatic deflection type).

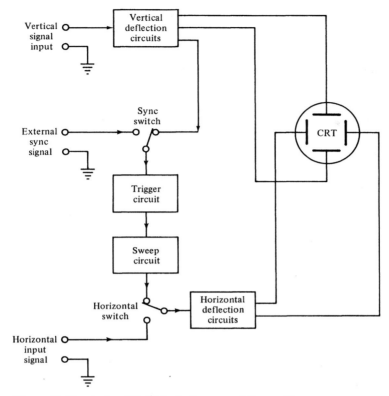

Figure 17-52 A simplified block diagram of the oscilloscope.

zontally sweep across the CRT screen. The horizontal deflection circuits respond to either an external signal applied at the horizontal input terminals or an internal sweep signal.

The *sweep* circuit is basically an electronic oscillator that develops a sawtooth voltage for ultimately deflecting (sweeping) the electron beam uniformly to the right of the CRT and then abruptly bringing it back (retracing) to the left of the CRT. The sawtooth voltage has the form shown in Fig. 17-53. The frequency of the sweep voltage is variable; the controls used to vary it are called the *horizontal frequency* or *time-base controls*.

The *vertical deflection circuits* are the amplifiers and level controls that respond to a given signal on the vertical input terminals and supply a very high voltage to the vertical plates thereby causing the electron beam to move vertically. If a repetitive signal applied to the vertical circuits is properly synchronized with the sweep signal, as the sweep signal deflects the beam across the CRT screen the vertical signal simultaneously deflects the beam vertically. Thus, as illustrated in Fig. 17-54, a graph of the vertical signal is presented on the CRT screen.

Synchronization is the process of starting the sweep of the beam across the CRT screen at the same point for each cycle of a repetitive voltage waveform applied at the

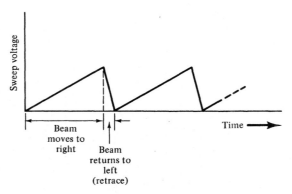

Figure 17-53 Sweep voltage (sawtooth) of an oscilloscope.

vertical input. In Fig. 17-54 the sweep signal begins when the vertical input voltage, which, in this case, is a sine wave, passes through the point P. It follows that each sweep will trace the observed signal over and over starting at the same point P. Hence, a stable $X-Y$ display or graph of the signal appears on the CRT screen.

The circuit that starts the sweep so that proper synchronization of the sweep and observed signals occurs is the *trigger circuit*. As indicated in Figure 17-52, the trigger circuit derives its triggering action internally from the signal applied to the vertical circuits, or externally from a signal that is time related to the displayed signal. Some oscilloscopes even have a *delayed sweep* feature that uses a main sweep and a linearly related secondary sweep called the delayed sweep. This allows for proper synchronization and observation of not only the waveform synchronized to the main sweep, but also that portion traced by the delayed sweep.

Many general-purpose laboratory oscilloscopes have provisions for simultaneously displaying two signal waveforms. Such instruments, called *dual-trace oscilloscopes*, use internal electronic switching to select between two vertical amplifiers as depicted in Fig. 17-55. At least two switching modes are available with dual-trace oscilloscopes: the *alternate* and *chop modes*.

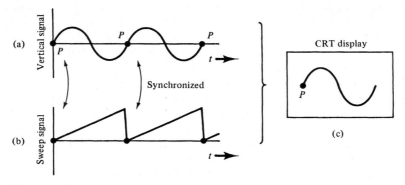

Figure 17-54 **(a)** A vertical signal and **(b)** a property synchronized sweep signal result in **(c)** the vertical signal being displayed on the face of the CRT.

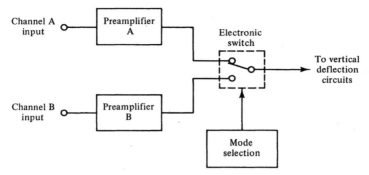

Figure 17-55 Switching circuit for a dual trace oscilloscope.

The *alternate mode* is used to alternately display one channel for a full sweep, then the second channel on the next sweep, and so on. The alternate mode is generally used to view relatively high-frequency signals, for which sweep speeds may be much faster than the screen phosphor decay time.

The *chop mode* is used to alternately display each channel for some fixed interval or fraction of the total sweep time. Therefore, each signal is displayed as a number of small chopped segments that merge and appear continuous to the eye. The chop mode is commonly used when observing low-frequency signals.

A general-purpose laboratory oscilloscope with triggered sweep is shown in Fig. 17-56. Notice that the oscilloscope face has grid lines (spaced 1 cm apart). These grid lines in conjunction with internally generated calibration signals allow the measurement of the voltage level and the time or frequency of an unknown voltage. The oscilloscope shown in Fig. 17-56 provides dual-trace displays of electronic signals from direct current to 60 MHz, with amplitudes from 2 mV to 400 V (dc + peak ac) or 800 V (peak-to-peak at 1 kHz or less). The sweep speeds are from 0.5 s to 0.05 μs and can be extended to 5 ns/div by using a 10 × magnifier.

Figure 17-56 Model 2213 dual trace oscilloscope. (Courtesy of Tektronix, Inc.)

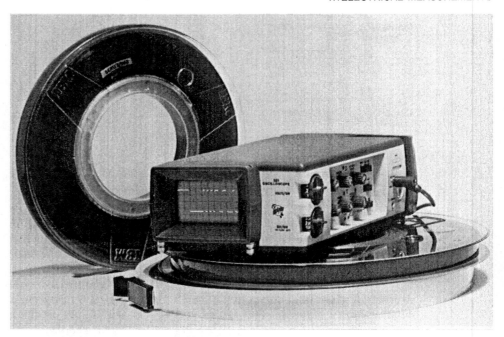

Figure 17-57 Model 221 oscilloscope. (Courtesy of Tektronix, Inc.)

A lightweight oscilloscope designed for use with either 120 V ac power or battery power is shown in Fig. 17-57. This particular unit weighs only 3.5 lbs and is approximately $3 \times 5 \times 10$ in. The vertical sensitivity ranges from 5 mV/div–100 V/div with a bandwidth extending to 5 MHz. The unit is a single-trace oscilloscope with sweep speeds up to 100 ns/div.

The network of grid lines on the front of the CRT is called the *graticule*. When the vertical or time base circuits are calibrated, the graticule is used to determine the voltage amplitude or elapsed time, respectively. Most oscilloscopes provide a means for switching the vertical sensitivity or time-base controls to a calibrated mode. For example, the vertical sensitivity and time-base controls of the oscilloscope in Fig. 17-56 are calibrated when the internal vernier knobs of those respective controls (with

the identification ⊙CAL) are fully clockwise. Under calibrated conditions,

$$\text{Voltage amplitude} = \begin{pmatrix} \text{Vertical distance} \\ \text{on graticule} \end{pmatrix} \times \text{volts/div} \qquad (17\text{-}33)$$

and

$$\text{Elapsed time} = \begin{pmatrix} \text{Horizontal distance} \\ \text{on graticule} \end{pmatrix} \times \text{time/div} \qquad (17\text{-}34)$$

Of course, if the elapsed time for one cycle of a repetitive waveform is determined,

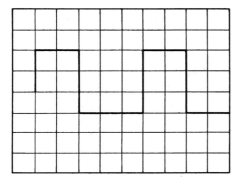

Figure 17-58   Example 17-11.

then the frequency is determined by

$$f = \frac{1}{T} \tag{17-35}$$

where f is the frequency in hertz and T is the period in seconds.

EXAMPLE 17-11

The display of a calibrated oscilloscope is depicted in Fig. 17-58. Determine the amplitude and frequency of the waveform if the vertical control is set for 2 V/div and the time base is set for 5 μs/div.

Solution. The voltage magnitude is equivalent to 3 div and the time base is equivalent to 5 div for one cycle. Then

$$\text{Voltage amplitude} = 3 \text{ div} \times 2 \text{ V/div} = 6 \text{ V}$$
$$T = 5 \text{ div} \times 5 \text{ }\mu\text{s/div} = 25 \text{ }\mu\text{s}$$

and

$$f = \frac{1}{T} = \frac{1}{25 \times 10^{-6}} = 40 \text{ kHz}.$$

Drill Problem 17-21 ■

Two voltages of the same frequency are measured with a calibrated dual-trace oscilloscope. The display is depicted in Fig. 17-59. If the time base is set for

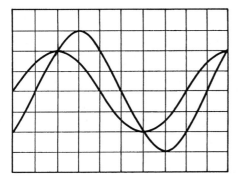

Figure 17-59   Drill Problem 17-21.

2 μs/div and the vertical control for each channel is set for 5 V/div, determine the **(a)** peak–peak amplitudes, **(b)** frequency, and **(c)** phase shift between the two voltages.

Answer.   (a) 20 and 30 V. (b) 62.5 kHz. (c) 45°.   ■ ■

QUESTIONS

1. What is meant by the terms "true value," "measured value," "error," and "accuracy"?
2. What are some of the errors that enter into electrical measurements?
3. Describe the three basic electromagnetic meter mechanisms.
4. What is a rectifier and how is it used in measuring instruments?
5. How does one increase the current capability of an ammeter?
6. How does one make a voltmeter?
7. What is voltmeter sensitivity? How does it relate to the accuracy of a voltage measurement?
8. What are four methods for measuring resistance?
9. What are the advantages and disadvantages of the long- and short-shunt measuring techniques?
10. How are the series and parallel ohmmeters different? Describe their scales.
11. How do the wattmeter and watthour meter differ?
12. How is impedance measured?
13. What are some advantages in using digital meters?
14. How is analog input converted to digital output in a digital meter?
15. Describe the operation of a cathode-ray tube.
16. What are the functions of the deflection and sweep circuits of an oscilloscope?

PROBLEMS

1. A 15 V voltmeter indicates 7.2 V when 7.0 V is applied. What is the error?
2. What is the accuracy of the reading in Problem 1 based on the full-scale value?
3. A 10 A ammeter with an accuracy of $\pm 2\%$ of full scale indicates a current of 4.9 A. What is the range of the true value?
4. What is the resistance of a 50 μA meter if a voltage drop of 25 mV results from full-scale current?
5. A dc milliammeter has a full-scale reading of 5 mA. If the milliammeter has a resistance of 2 Ω, what shunt is needed to extend the range to 50 A?
6. A 1 mA meter movement with a resistance of 150 Ω is used for a multirange milliammeter with scales of 5, 50 and 100 mA. Using the multirange configuration of Fig. 17-12(a), solve for the required shunt resistances.
7. If the meter movement of Problem 6 is used in the configuration of Fig. 17-12(b) and the same scales are desired, what are the values of R_1, R_2, and R_3?

8. A voltmeter with a full-scale value of 15 V is to be constructed using a 5 mA, 25 Ω mechanism. What is the needed multiplier resistance?

9. A dc meter mechanism has full-scale deflection with 100 mV across it. If the meter resistance is 10 Ω, what value of multiplier resistance is needed to measure 1000 V?

10. A 50 μA, 2 kΩ meter movement is used in the multirange voltmeter of Fig. 17-15(b). What values are needed for R_1, R_2, and R_3 if the voltage ranges are 2.5, 10, and 50 V?

11. A voltmeter has a sensitivity of 20 kΩ/V. (a) What is the full-scale meter current? (b) What resistance does the voltmeter have on each range if the ranges are 2.5, 10, 50, and 250 V?

12. A 50 V voltmeter with a sensitivity of 20 kΩ/V is used to measure the voltage across a 50 kΩ resistance. What is the percent of change of circuit resistance following the connection of the voltmeter?

13. Two 50 kΩ resistances are connected in series to a 50 V source. A 50 V voltmeter with a resistance of 2 MΩ is used to measure the voltage across one of the resistances. What is the percent error between the measured and true values?

14. A 150 V voltmeter with a sensitivity of 1 kΩ/V is used in a short-shunt connection to determine the value of an unknown resistance. The voltmeter indicates 135 V and the ammeter indicates 13.5 mA. What percent error is made if one does not correct for meter resistance?

15. A 250 V voltmeter with a sensitivity of 20 kΩ/V is used in the series voltmeter method of resistance measurement. What is the value of the unknown resistance if the source voltage is 120 V and the voltmeter reading is 95 V?

16. A wattmeter is connected short shunt and reads 500 W. If the line voltage is 120 V, the current coil resistance negligible, and the voltage coil resistance equal to 2 kΩ, what error does the short-shunt connection introduce in the wattmeter reading?

17. A lamp having a rating of 100 W is connected to a watthour meter having a constant of $\frac{1}{3}$. If the applied voltage is at rated value, how many revolutions does the disk make in 1 min?

18. The kilowatt-hour reading at a residence is 13 136 at the beginning of a 30-day period. At the end of the period the kilowatt-hour meter indicates the reading of Fig. 17-60. (a) What is the average kilowatt-hour usage per day? (b) What is the total energy charge if each kilowatt-hour costs the resident 6¢?

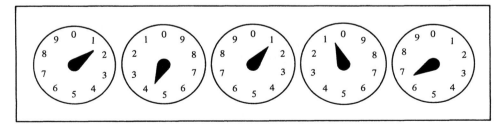

Figure 17-60   Problem 17-18.

19. Determine the unknown capacitance and series resistance for a capacitance comparison bridge when at balance the known components have the following values: $R_1 = 2$ kΩ, $R_2 = 5$ kΩ, $R_3 = 20$ kΩ, and $C_2 = 40$ μF.

20. A coil of unknown resistance and inductance is connected to a Hay bridge that uses a 1 kHz source. Balance is achieved when the known components have the following values:

$R_1 = 10$ kΩ, $R_2 = 1$ kΩ, $R_3 = 100$ Ω, and $C_1 = 2$ μF. Determine the unknown series resistance and inductance.

21. The repetitive display for a calibrated oscilloscope is depicted in Fig. 17-61. For a no-signal condition the oscilloscope trace occurs at the midpoint of the vertical scale. The vertical and time-base controls are set for 5 V/div and 0.1 ms/div, respectively. (a) What are the positive and negative amplitudes of the signal? (b) What are the period and frequency of the signal?

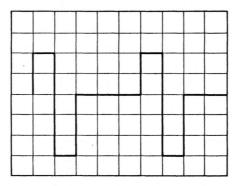

Figure 17-61   Problem 17-21.

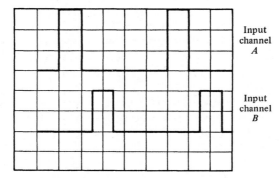

Input channel *A*

Input channel *B*

Figure 17-62   Problem 17-22.

22. A dual-trace oscilloscope is used for measuring the input and output signals of a circuit. Figure 17-62 depicts how the input and output signals appear on the CRT screen. The oscilloscope controls are set as follows:

> channel A (input): 100 mV/div
>
> channel B (output): 2 V/div
>
> time base: 0.2 μs/div

Determine (a) the amplitude of the input and output signals, (b) the frequency of the signals, and (c) the delay between the input and output signals.

Chapter 18

Direct Current Graphical Analysis

18.1 LOAD LINES AND SOURCE LINES

As was pointed out in Chapter 4, a linear resistance is a resistance whose voltage versus current ($V-I$) or current versus voltage ($I-V$) characteristic is a straight line. Conversely, a nonlinear resistance is a resistance characterized by a nonlinear $V-I$ or $I-V$ characteristic.

Because of temperature effects in particular, all resistances are basically non-linear, but those that exhibit only a small resistance change over a range of operating voltage and current are often called linear. The $V-I$ characteristics of a 250 Ω linear resistance, a 100 W carbon filament lamp, and a 100 W incandescent (tungsten filament) lamp are shown in Fig. 18-1. The slope of the linear resistance characteristic is constant so that a change in current ΔI anywhere along the curve results in the same change in voltage ΔV. Although the carbon filament lamp exhibits a linear characteristic at relatively high values of current, it exhibits a nonlinear characteristic at low values of current. In contrast to the other two curves, the curve of the incandescent lamp is completely nonlinear. Notice that since each of the devices is a passive device, no voltage exists across the device unless current flows through the device, that is, each curve passes through the origin. The $V-I$ or $I-V$ characteristic of a device or a group of devices used as an electrical load is called the *load line*.

The devices whose $V-I$ characteristics appear in Fig. 18-1 are *bilateral* since they exhibit the same $V-I$ characteristics regardless of direction of current flow. On the other hand, a *unilateral* device is one whose resistance or $V-I$ characteristic changes markedly with the direction of current through the device. An excellent example of a unilateral device is the semiconductor diode whose characteristic is presented in Chapter 4.

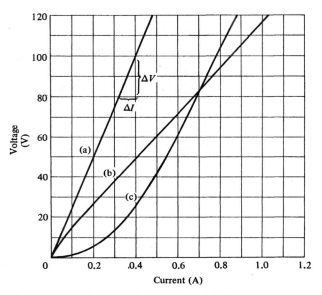

Figure 18-1   *V–I* characteristics: **(a)** 250 Ω resistor; **(b)** carbon filament lamp; **(c)** tungsten filament lamp.

Another nonlinear load device is the *thyrite resistor*, typified by the *I–V* characteristic of Fig. 18-2. Because the thyrite resistor maintains a relatively constant voltage across it at high currents, this instrument is useful as a voltage regulating device.

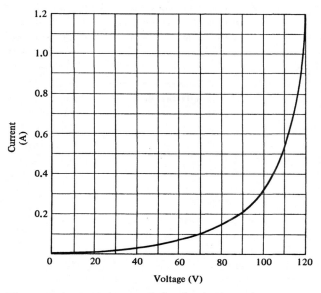

Figure 18-2   *I–V* characteristic of a thyrite resistor.

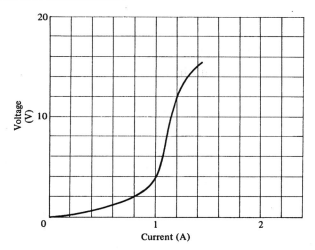

Figure 18-3 *V–I* characteristic of a ballast resistor.

In contrast to the thyrite resistor, the *ballast resistor* passes a relatively constant current over a range of applied voltages, as shown in Fig. 18-3. Therefore, it is useful as a current regulating device.

A particularly interesting characteristic is displayed by the *tunnel diode*, a semiconductor device electrically characterized by the *I–V* characteristic of Fig. 18-4.

The *V–I* or *I–V* characteristic of an active device used as an electrical energy source is called a *source line*. The terminal voltage versus current characteristics of voltage sources were introduced in Chapter 5 and are shown in Fig. 18-5. The ideal voltage source has the curve *a* characteristic as it supplies a constant voltage regardless of the current drawn. However, the practical source, having some internal resistance R_i, exhibits a decrease in voltage as current is drawn. If the internal resistance is constant or linear, the source line is characterized by a linear decrease in voltage, as shown in curve *b*. At any point along the curve, the terminal voltage E_t is

$$E_t = E - IR_i \tag{18-1}$$

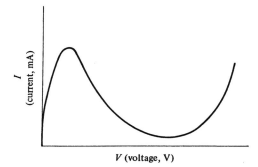

Figure 18-4 Tunnel diode characteristic.

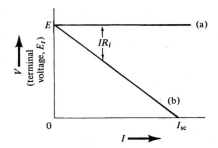

Figure 18-5 Voltage source characteristic:
(a) Ideal source; (b) practical source.

As an increasing amount of current is taken from a voltage source, the terminal voltage decreases until at $E_t = 0$, the maximum current, the short-circuit current I_{sc} flows. For a linear $V-I$ characteristic, this current from Eq. 18-1 is

$$I_{sc} = \frac{E}{R_i} \qquad (18\text{-}2)$$

The second type of electrical source is the current source, which ideally supplies a constant value of current regardless of the voltage developed by the current. Its electrical characteristic is typified by curve a of Fig. 18-6. Of course, a practical source has some internal resistance that shunts part of the developed current so that part does not reach the source terminals—hence the characteristic curve b.

Not all sources have linear $V-I$ characteristics. For example, a study of the various types of dc generators reveals that not only are the $V-I$ characteristics nonlinear, but the terminal voltage at the rated output current may be higher than the open-circuit voltage. The $V-I$ characteristics for four types of dc generators are shown in Fig. 18-7. It is obvious that for the flat and overcompounded generators factors other than internal resistance are present, since the $V-I$ curves indicate an increase in voltage with an increase in current rather than the decrease in voltage with an increase in current associated with internal resistance. In fact, the internal resistance is only one factor affecting the characteristic of each type of generator; other factors are the type of generator windings and the winding connections used.

When an electrical circuit contains nonlinear loads or sources, one solves for the various voltages and currents in the circuit by obtaining algebraic equations, including those for the nonlinear elements, and solving these simultaneously. However, if one has graphs of the electrical characteristic of the nonlinear elements, one

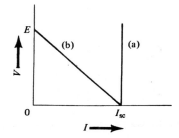

Figure 18-6 Current source characteristic:
(a) Ideal source; (b) practical source.

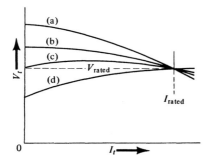

Figure 18-7 Generator V–I characteristic: **(a)** Shunt; **(b)** under-compound; **(c)** flat-compound; **(d)** over-compound types.

can obtain a solution by graphical means. The graphical technique for solving circuits with nonlinear elements is presented in the following sections.

18.2 OPERATING OR Q POINT

Regardless of whether one is considering an electric circuit with linear or nonlinear elements, Kirchhoff's voltage and current laws must be satisfied at all times. With reference to the basic electric circuit of Fig. 18-8, Kirchhoff's current law requires that the current leaving the source terminal, I_t, equal the load current I_L; Kirchhoff's voltage law requires that the terminal voltage of the source, E_t, equal the load voltage V_L.

One solves graphically for the voltage and current of a circuit by superimposing the source line and load line on a common graph (see Fig. 18-9). The point at which

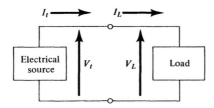

Figure 18-8 Basic electrical circuit.

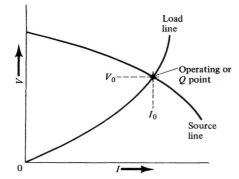

Figure 18-9 Graphical solution.

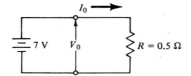

Figure 18-10   Circuit of Example 18-1.

the two curves intersect is the point at which Kirchhoff's current and voltage laws are satisfied for that particular source and load. This point is the *operating point*; the coordinates at this point are V_0 and I_0, for the operating voltage and current, respectively. In electronics work, the operating point is called the Q point, short for the *quiescent point*. The following examples illustrate the graphical solution of some simple circuits.

EXAMPLE 18-1

A 7 V source is connected to a linear resistance of 0.5 Ω as shown in Fig. 18-10. Graphically determine the load current and voltage.

Solution.   The source line is plotted as an ideal voltage characteristic as in Fig. 18-11. The load line passes through the origin and has a slope of

$$R = 0.5 \ \Omega = \frac{\Delta V}{\Delta I}$$

Alternatively,

$$\Delta V = 0.5 \ \Delta I$$

so if $\Delta I = 10$ A, then $\Delta V = 5$ V. The load line has a slope of 5 V for every 10 A and thus passes through the coordinate (10 A, 5 V). Because only two points

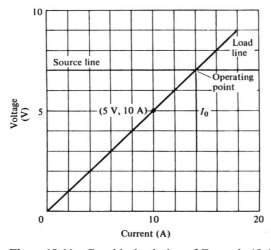

Figure 18-11   Graphical solution of Example 18-1.

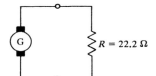

Figure 18-12   Circuit of Example 18-2.

are needed to graphically describe a straight line, the load line is drawn using the coordinates (0, 0) and (10 A, 5 V), as shown in Fig. 18-11. The operating point is at the intersection of the two curves and has the coordinates (14 A, 7 V). Thus $I_0 = 14$ A and $V_0 = 7$ V.

EXAMPLE 18-2

A generator is connected to a 22.2 Ω resistor as shown in Fig. 18-12. The generator characteristic is given in Fig. 18-13. Determine the load voltage and current.

Solution.   For the load line, $\Delta V = 22.2 \, \Delta I$. If the current changes from 0.0 to 0.3 A, the voltage changes from 0.0 to 6.66 V. Hence the load line passes through the coordinate (0.3 A, 6.66 V). The load voltage and current are obtained from the intersection of the curves as $V_0 = 8.9$ V and $I_0 = 0.405$ A.

EXAMPLE 18-3

A 28 V source with an internal resistance of 0.75 Ω is connected to a 1 Ω load resistor, as in Fig. 18-14. Determine the load voltage and current and the voltage across the internal resistance V_i.

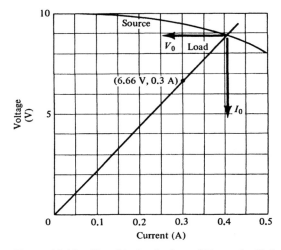

Figure 18-13   Graphical solution of Example 18-2.

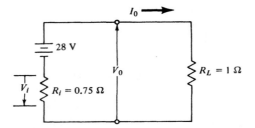

Figure 18-14   Circuit of Example 18-3.

Solution.   The short-circuit current of the source is

$$I_{sc} = \frac{28}{0.75} = 37.3 \text{ A}$$

Using the short-circuit and open-circuit points, the source curve is plotted as in Fig. 18-15. A load line with a slope of 1 Ω passes through the coordinate (30 A, 30 V) and intersects the source line at a coordinate of (16 A, 16 V). Hence $I_0 = 16$ A, $V_0 = 16$ V, and $V_i = 12$ V (the difference between the open-circuit and operating voltages).

Drill Problem 18-1 ■

A 25 Ω resistive load is connected to the generator described in Example 18-2. Using the characteristic source line specified in that example, determine the current and voltage for the 25 Ω load.

Answer. $V_0 \approx 9.2$ V; $I_0 \approx 3.7$ A.   ■ ■

Drill Problem 18-2 ■

A solar cell is connected to a resistive load as shown in Fig. 18-16. If in bright sunlight (approximately 10^5 lumens/m²) the solar cell has the nonlinear V–I characteristic curve of Fig. 18-17, determine the load voltage and current.

Answer. $V_0 \approx 0.35$ V; $I_0 \approx 105$ mA.   ■ ■

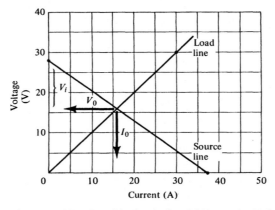

Figure 18-15   Graphical solution of Example 18-3.

Figure 18-16 Circuit of Drill Problem 18-2.

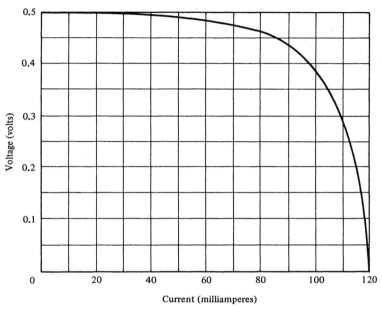

Figure 18-17 Typical solar cell characteristic for bright sunlight (approximately 10^5 lumens/m^2).

18.3 SERIES RESISTANCES

One of the defining properties of a series circuit is that the current is the same in each of the series elements (see Fig. 18-18). For any current, in turn, the total voltage across the series combination, V_t, equals the sum of the voltages across the individual elements. It follows that *the total V–I characteristic for a series combination of*

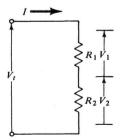

Figure 18-18 Series resistances.

resistances is obtained by graphically adding the voltages of the series elements at various values of current. In the case of linear elements, it is necessary to graphically add the voltages at only one value of current. This follows from the fact that the sum of two straight-line relationships is a third straight-line relationship. On the other hand, when one of the series elements is nonlinear, a graphical addition is performed at many different values of current and the resulting characteristic is obtained as a smooth curve connecting these points of addition.

EXAMPLE 18-4

Graphically determine the total $V–I$ characteristic for two resistors in series if $R_1 = 2\ \Omega$ and $R_2 = 4\ \Omega$.

Solution. The $V–I$ characteristics of R_1 and R_2 are drawn as in Fig. 18-19. At any particular current, the voltage drops across R_1 and R_2 are added. In this case, the voltages are added at a chosen current of 2 A. The result is the coordinate point a through which the total straight-line characteristic passes.

EXAMPLE 18-5

A ballast resistor with the nonlinear characteristic of Fig. 18-3 is connected in series with a resistor of 5 Ω and a 20 V source as shown in Fig. 18-20. Determine the circuit current and the voltages across the ballast and 5 Ω resistors.

Solution. All work is superimposed on the ballast resistor graph. First, the load resistor line (5 Ω) is drawn through the origin and the coordinate point (2 A, 10 V), as in Fig. 18-21. Next, points for the total resistance characteristic $R_t = R_L + R_B$ are determined by adding the voltages of the R_L and R_B curves

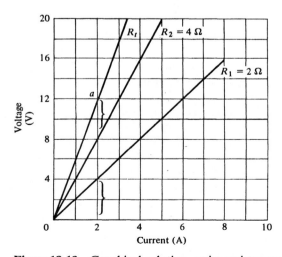

Figure 18-19   Graphical solution, series resistances of Example 18-4.

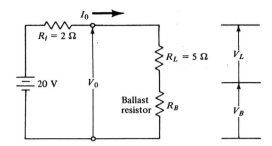

Figure 18-20   Circuit of Example 18-5.

along many selected constant current lines. A smooth curve through these points results in a nonlinear curve for R_t.

Because the source has a short-circuit current of 10 A, which is off the R_B graph, one must use Eq. 18-1 to determine the source line. In particular, for a current of 2 A, the source voltage is 16 V and this point, along with the open-circuit voltage point, determines the source line.

As shown in Fig. 18-21, the total resistance and source lines intersect at the point (1.18 A, 17.5 V), so $I_0 = 1.18$ A and $V_0 = 17.5$ V. Since the operating current I_0 is the same current flowing through R_L and R_B, we determine the associated voltage drops V_L and V_B by moving from the operating point along the constant current line of I_0 to the curves R_L and R_B. The I_0 line intersects the R_L and R_B curves at the respective voltages V_L and V_B. From the graph then, $V_L = 5.9$ V and $V_B = 11.4$ V. Notice that within the accuracy of the graph the sum of these two voltages equals the total operating voltage V_0.

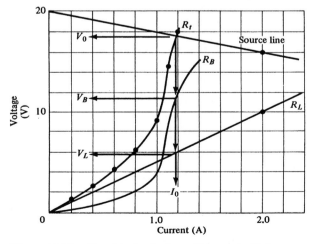

Figure 18-21   Graphical solution of Example 18-5.

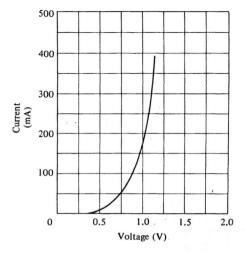

Figure 18-22 Typical semiconductor diode characteristic.

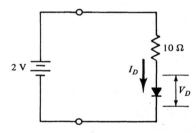

Figure 18-23 Circuit for Drill Problem 18-3.

Drill Problem 18-3 ■

A diode having the $I–V$ characteristic of Fig. 18-22 is connected in series with a 10 Ω and a 2 V source as shown in Fig. 18-23. Determine the diode current and voltage I_D and V_D, respectively.

Answer. ≈ 115 mA; ≈ 0.9 V.   ■ ■

18.4 PARALLEL RESISTANCES

One of the defining properties of a parallel circuit is that the voltages across the parallel elements are equal (see Fig. 18-24). However, for any voltage V_t, the total current to the parallel combination I_t equals the sum of the individual branch

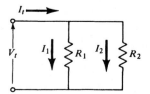

Figure 18-24 Parallel resistances.

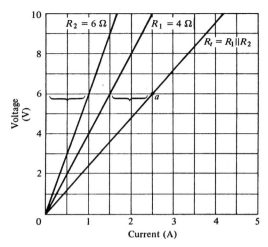

Figure 18-25 Graphical solution, parallel resistances of Example 18-6.

currents. It follows that *the total V–I characteristic for a parallel combination of resistances is obtained by graphically adding the currents through the parallel elements at various values of voltage.* As in the case of series combinations, if linear characteristics are added, the origin and one other coordinate point are sufficient to determine the total characteristic. However, when one of the parallel elements is nonlinear, a graphical addition is performed at many different values of voltage, and the resulting characteristic is obtained as a smooth curve connecting these points.

EXAMPLE 18-6

Graphically determine the total $V–I$ characteristic for two resistors in parallel if $R_1 = 4\ \Omega$ and $R_2 = 6\ \Omega$.

Solution. The $V–I$ characteristics of R_1 and R_2 are drawn as in Fig. 18-25. At any particular voltage, the current values are added. In this case, a voltage of 6 V is considered, and hence the coordinate point through which the total straight-line characteristic passes is *a*.

One should not generalize and say that for series elements the characteristics are added vertically and for parallel elements the characteristics are added horizontally. Although the generalization holds for $V–I$ characteristics, it is just the opposite for $I–V$ characteristics. Instead, one should keep in mind that the characteristics are added along either constant voltage or current lines; whether the particular lines are horizontal or vertical is incidental.

EXAMPLE 18-7

A diode having the $I–V$ characteristic of Fig. 18-22 is connected in parallel with a 10 Ω resistor and a 1.5 V source with an internal resistance of 2 Ω (see Fig. 18-26). Determine the operating voltage and current, and the diode voltage and current.

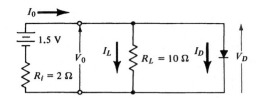

Figure 18-26   Circuit of Example 18-7.

Solution.   The 10 Ω (R_L) load characteristic passes through the coordinate (100 mA, 2 V) and is plotted on the diode graph as in Fig. 18-27. The total load characteristic R_t is obtained by adding the diode and R_L currents at selected voltages. Next, the source line is superimposed on the diode graph. The source has an open-circuit voltage of 1.5 V and, when delivering 500 mA, has a terminal voltage of 0.5 V; hence the source line in Fig. 18-27. The operating point is $V_0 = 0.98$ V, $I_0 = 265$ mA.

Since the operating voltage is the same as that across the diode, $V_D = V_0 = 0.98$ V. To obtain the diode current, one moves from the operating point along the constant voltage line of V_0 to the diode characteristic. The V_0 line intersects the diode characteristic at $I_D = 170$ mA. If one continues along the constant voltage line of V_0, one intersects the R_L characteristic at $I_L = 95$ mA. As expected, $I_0 = I_D + I_L$.

Drill Problem 18-4 ■

A parallel combination consisting of a ballast resistor (V–I characteristic of Fig. 18-3) and a 10 Ω resistor is connected to a 20 V source of 7.5 Ω internal resistance. (See Fig. 18-28.) Determine graphically the following quantities: V_0, I_0, and I_B.

Answer.   6.5 V; 1.75 A; 1.1 A.   ■ ■

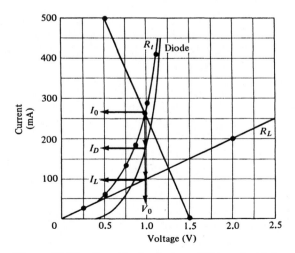

Figure 18-27   Graphical solution of Example 18-7.

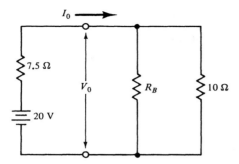

Figure 18-28 Circuit for Drill Problem 18-4.

18.5 SERIES–PARALLEL RESISTANCE

The graphical solution of circuits containing series and parallel combinations is a little more cumbersome. Nevertheless, when the circuit contains nonlinear elements, a graphical solution is particularly useful. As with an analytical solution, we proceed with a graphical solution by reducing the characteristics to one total load line. After the operating point is determined, we move along the appropriate constant voltage or constant current lines to find the individual voltages and currents.

EXAMPLE 18-8

The series–parallel circuit of Fig. 18-29 uses a 250 Ω resistor, a carbon filament lamp, and a tungsten filament lamp, each of which is described by the appropriate characteristic of Fig. 18-1. A source voltage of 100 V is applied. Determine the operating voltage V_0, operating current I_0, tungsten lamp voltage V_{tl}, carbon lamp voltage V_{cl}, resistor current I_R, and tungsten lamp current I_{tl}.

Solution. The graphical solution appears in Fig. 18-30. In obtaining a solution, we first combine the 250 Ω resistor and tungsten lamp characteristics

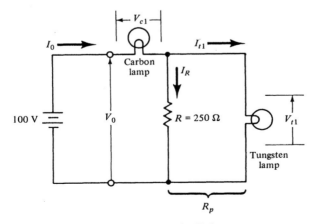

Figure 18-29 Circuit of Example 18-8.

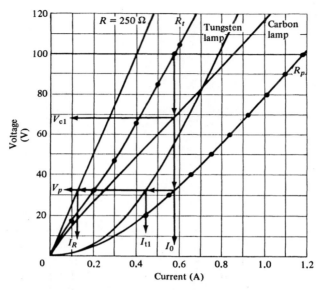

Figure 18-30   Graphical solution of Example 18-8.

along constant voltage lines to form the parallel resistance R_p. In turn, the R_p characteristic is combined along constant current lines with the carbon lamp characteristic. This latter addition results in the total load line R_t. The intersection of the source and total load line is at (100 V, 0.58 A); hence $V_0 = 100$ V and $I_0 = 0.58$ A.

As the operating current is common to the carbon lamp and R_p, we move from the operating point along the constant current line I_0. This current line intersects the carbon lamp and R_p characteristics at 68 and 32 V, respectively; hence $V_{cl} = 68$ V and $V_p = V_{tl} = 32$ V. Notice that $V_{tl} + V_{cl} = V_0$.

Since R_p is constructed from the parallel combination of the 250 Ω resistor and tungsten lamp, we move from the R_p curve along the constant voltage line V_p. The V_p line intersects the 250 Ω and tungsten lamp curves at their respective operating currents. Thus, $I_R = 0.13$ A and $I_{tl} = 0.45$ A. Notice that $I_R + I_{tl} = I_0$.

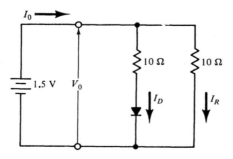

Figure 18-31   Circuit for Drill Problem 18-5.

Drill Problem 18-5 ∎

A diode having the electrical characteristic of Fig. 18-22 is used in the series–parallel circuit of Fig. 18-31. Graphically determine the approximate values of V_0, I_0, I_R, and I_D.

Answer.   1.5 V, 220 mA, 150 mA, and 70 mA.   ∎ ∎

18.6 ELECTRONIC LOAD LINES

In electronic circuits, we encounter many unilateral and nonlinear devices. One of these is the semiconductor diode, which has already been considered. Unlike the semiconductor diode, some other electronic devices have electrical characteristics that make it possible for more than one operating point to exist.

Before considering such a situation, it should be pointed out that electronic devices often are connected to a series circuit consisting of a voltage source and a load resistor R_L (see Fig. 18-32). The circuit, when connected to an electronic device, is called the *load*. Notice, however, that even though the circuit is called the load, it functions as a source; hence one plots the V–I characteristic as a source.

Now consider the graphical analysis of the tunnel diode circuit of Fig. 18-33(a). In the analysis of this circuit, the load line is superimposed on the tunnel diode characteristic, as shown in Fig. 18-33(b). Notice that the open-circuit voltage E and the short-circuit current $I_{sc} = E/R_L$ determine the points of intersection of the load line

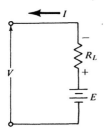

Figure 18-32   Electronic load.

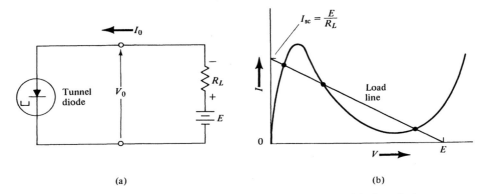

(a)   (b)

Figure 18-33   **(a)** A tunnel diode circuit. **(b)** The associated graphical analysis.

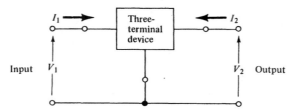

Figure 18-34   Three-terminal device.

with the horizontal and vertical axes. In the case depicted, the load line intersects the tunnel diode characteristic at three points; hence there are three possible operating points. Each of the three operating points has a certain stability associated with it, and any slight change in voltage may trigger a change from one point to another.

Some electronic devices, such as the transistor, are three-terminal devices. As was pointed out in Chapter 6, a three-terminal network has two input terminals and two output terminals, one of which is common to the input (see Fig. 18-34). The common terminal and one other terminal serves as the input pair, and the common terminal and the remaining terminal serves as the output pair. By convention, the currents are defined entering the input and output terminals.

To describe the electrical behavior of a three-terminal device properly, it is necessary to characterize the voltage and current relationships for both the input and output terminal pairs. However, the electrical behavior of the device at one set of terminals usually depends on the voltage or current at the other terminal pair. The result is that instead of one V–I or I–V curve for the input or output, we obtain a family of curves.

Consider the transistor and the associated voltage and current designations of Fig. 18-35. The transistor has three distinct electrical regions: the *base*, the *collector*, and the *emitter*. The terminals connected to these regions are labeled b, c, and e, respectively. In accordance with the double-subscript notation introduced in Chapter 5, the voltages V_{be} and V_{ce} are, respectively, the voltages at the base and collector referenced to the emitter. The base and collector currents I_b and I_c are the input and output currents, respectively.

The input characteristics for the particular transistor connection shown in Fig. 18-35 (known as the common emitter connection) is typified by the set of I_b versus V_{be} curves of Fig. 18-36(a). Notice that a particular input curve is associated with a particular value of voltage V_{ce}.

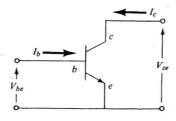

Figure 18-35   The transistor as a three-terminal device.

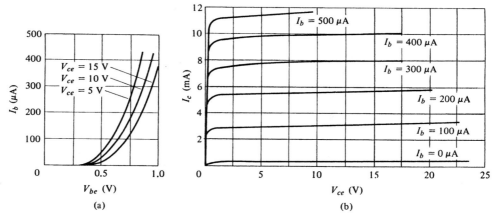

Figure 18-36 Typical transistor characteristics for a common emitter connection: **(a)** input; **(b)** output.

The associated output characteristics are the set of I_c versus V_{ce} curves of Fig. 18-36(b). Notice that a particular output curve is associated with a particular value of input current I_b.

The transistor is considered an active device, since an input signal is amplified, that is, made larger, in passing through the device. This desired amplifying action takes place about a particular dc operating point called the quiescent or Q point. In order to obtain a dc Q point and to provide a load resistance to which the amplified signal is applied, an electronic load characterized by that of Fig. 18-37 is used. The load is essentially the same as that shown in Fig. 18-32, except that here the voltage source is called V_{CC}.

Even though V_{CC} is truly a source supplying the dc quiescent current, the $I-V$ characteristic of V_{CC} and R_L is called the load line. In determining the operating point of a transistor circuit, the load line is superimposed on the output characteristics as in Fig. 18-38. Many operating points are possible, but we select and supply one particular value of I_b (usually with another source). For example, in Fig. 18-38, if $I_b = I_{b2}$, the operating point is completely specified.

Usually, with transistor circuits, one is primarily interested in the output characteristics and the associated operating point. However, graphical analysis can also

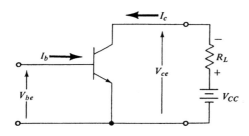

Figure 18-37 Transistor with a load connected.

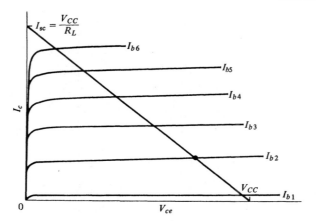

Figure 18-38   Transistor load line analysis.

be applied to the input characteristics. Transistor manufacturers make both input and output characteristic curves available.

Drill Problem 18-6 ■

A diode having the electrical characteristic depicted in Fig. 18-22 is connected in series with an electronic load as shown in Fig. 18-39. **(a)** Determine the diode current and voltage, I_D and V_D, respectively. **(b)** Compare this answer with the results from Drill Problem 18-3.

Answer.   (a) 115 mA; 0.9 V. (b) The results are the same since the two circuits are really the same.   ■ ■

Drill Problem 18-7 ■

Refer to the transistor circuit of Fig. 18-37 and the electrical characteristic of Fig. 18-36. Determine the operating voltage and current for the following conditions: $V_{CC} = 25$ V, $R_L = 2.5$ kΩ, and $I_b = 200$ μA.

Answer. $I_c \approx 5.6$ mA; $V_{ce} \approx 11$ V.   ■ ■

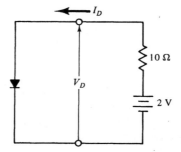

Figure 18-39   Circuit for Drill Problem 18-6.

QUESTIONS

1. What characterizes linear and nonlinear resistances?
2. What is meant by the terms load line and source line?
3. What is meant by an operating point?
4. How does one graphically find an operating point?
5. How does one obtain a composite $V-I$ curve for two or more series elements?
6. How does one obtain a composite $V-I$ curve for two or more parallel elements?
7. Describe the electronic load line.
8. How does the electrical characterization of a three-terminal device differ from that of a two-terminal device?

PROBLEMS

In the following problems, the characteristic curves introduced in the chapter should be used unless otherwise stated.

1. Plot on one graph the $V-I$ curves for the following loads: 50, 100, and 200 Ω.
2. Plot on one graph the $I-V$ curves for the following loads: 1, 2, and 5 Ω.
3. A certain resistance is directly proportional to the current through it. Plot the $V-I$ characteristic of this resistor if it is given by $R = 0.5I$ Ω. (Use maximum scale values of 50 V and 10 A.)
4. A generator has an open-circuit voltage of 120 V and an internal resistance $R_i = 0.4I$ Ω. Plot the $V-I$ characteristic of the generator. (Use maximum scale values of 120 V and 10 A.)
5. Solve graphically for the voltage and current for the circuit of Fig. 18-40. (Use maximum scale values of 100 mA and 200 V.)

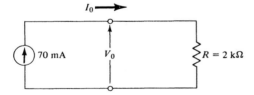

Figure 18-40   Problem 18-5.

6. A practical source has an open-circuit voltage of 6.8 V and a short-circuit current of 24 A. A resistor of 1.4 Ω is placed across the source. Graphically determine the current through the resistor and the voltage across the resistor.
7. Given the circuit of Fig. 18-41, what are the lamp voltage and current (a) if a carbon lamp is used and (b) if a tungsten lamp is used.
8. Solve for the load voltage and current for the circuit of Fig. 18-42. (Use maximum scale values of 10 V and 15 mA.)
9. Using the circuit of Fig. 18-43 and the $I-V$ characteristics for R_A (Fig. 18-44), solve for (a) the voltage across R_A, (b) the current through R_A, and (c) the voltage across R_L.

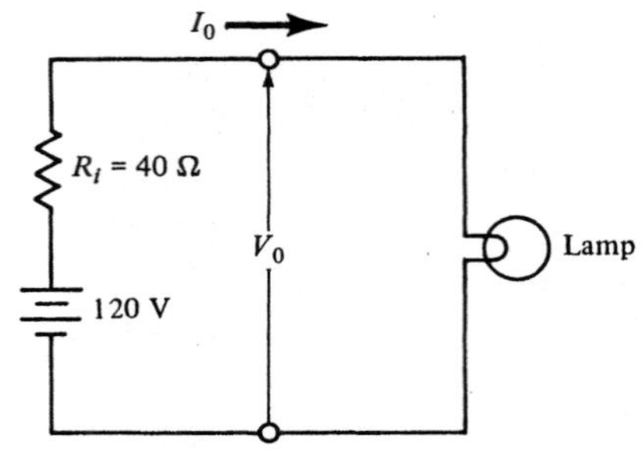

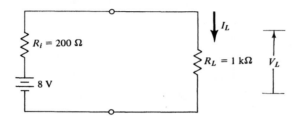

Figure 18-41   Problem 18-7.

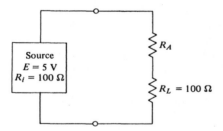

Figure 18-42   Problem 18-8.

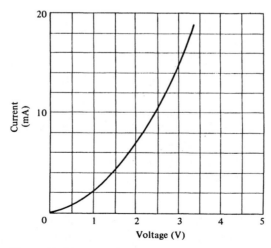

Figure 18-43   Problem 18-9.

Figure 18-44 $I-V$ characteristic for R_A for Problem 18-9.

10. What type of resistance (V–I characteristic) will produce a constant resistance of 200 Ω when connected in series with the incandescent lamp of Fig. 18-1?

11. Given the circuit of Fig. 18-45 and the ballast resistor R_B characteristic of Fig. 18-3, solve for the operating point and the currents in R_B and the 25 Ω resistor.

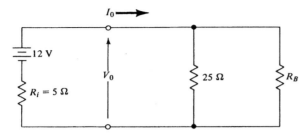

Figure 18-45   Problem 18-11.

12. The generator in Fig. 18-46 has the characteristic shown in Fig. 18-47. Graphically determine the operating voltage and the current in each resistor.

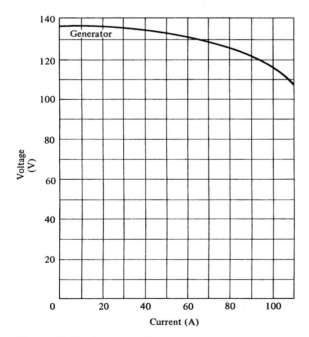

Figure 18-46   Problem 18-12.

Figure 18-47   Problem 18-12.

13. Given the characteristic for R_A (Fig. 18-44) and the circuit of Fig. 18-48, solve for **(a)** the current in R_A, **(b)** the current in R_L, and **(c)** the voltage across R_A.

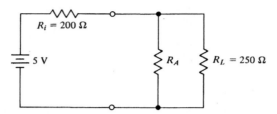

Figure 18-48   Problem 18-13.

14. Using the circuit of Fig. 18-49 and the incandescent lamp characteristic of Fig. 18-1, determine the lamp voltages and currents. (Use a current scale maximum of 2.5 A.)

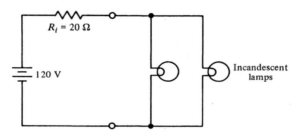

Figure 18-49   Problem 18-14.

15. Given the circuit of Fig. 18-50 and the generator characteristic of Fig. 18-51, solve graphically for the two currents I_1 and I_2.

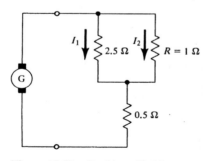

Figure 18-50   Problem 18-15.

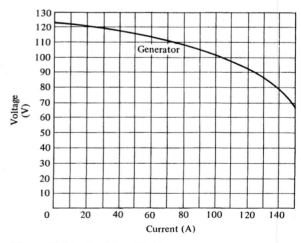

Figure 18-51   Problem 18-15.

16. Using scales of 50 V and 10 A, solve graphically for the voltage and current for each resistor of Fig. 18-52.

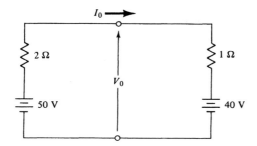

Figure 18-52   Problem 18-16.

17. Given the circuit of Fig. 18-53, solve for the operating voltage, operating current, and diode current. Use the diode curve of Fig. 18-22.

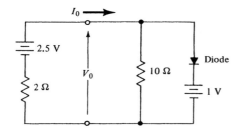

Figure 18-53   Problem 18-17.

18. Sketch load lines on the output curves of the transistor characteristic of Fig. 18-36 if **(a)** $V_{CC} = 20$ V; $R_L = 5$ kΩ and **(b)** $V_{CC} = 10$ V; $R_L = 2$ kΩ.

19. With reference to the circuit of Fig. 18-37 and the transistor characteristic of Fig. 18-36, find the Q point if $I_b = 200$ μA, $V_{CC} = 15$ V, and $R_L = 2$ kΩ.

Transformers

19.1 TRANSFORMER ACTION

The concepts of mutual inductance and magnetic coupling between two coils were introduced in Chapter 9. We said that two coils are mutually coupled if some or all of the magnetic flux produced in one coil passes through the second coil. A *transformer* is an electrical device that uses magnetically coupled coils to transfer energy from one circuit to another.

The transformer is depicted in Fig. 19-1: A mutual flux links two windings, to which, in turn, the circuits are connected. One of the circuits, external to the transformer, serves as a source of electric energy; the winding to which this circuit is con-

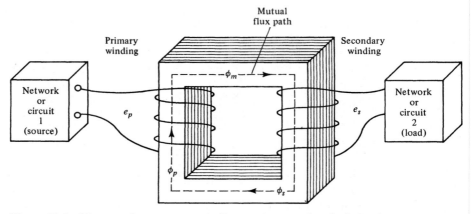

Figure 19-1 The transformer magnetically couples two electrical circuits.

nected is called the *primary winding*. The second circuit or network connected to the transformer serves as the electrical load, and the winding to which it is connected is called the *secondary winding*.

Although transformers can be used in pulsed, unidirectional current circuits, they are primarily used to couple and transform alternating voltages and currents. Thus, in most cases, the source network in Fig. 19-1 is a sinusoidal source. The load network can be a simple resistance or a more complicated impedance network.

Recall from Chapter 9 that whenever changing flux lines link a coil of wire or winding, an emf is developed in that coil in accordance with Faraday's law and Lenz's law:

$$e = N \frac{d\phi}{dt} \tag{19-1}$$

where e = the counter emf, V
   N = the number of turns in the coil
   $d\phi/dt$ = the change of flux per unit time linking the turns N, Wb/s

If negligible resistance losses occur in the transformer windings, and if the leakage flux is negligible, then all the flux generated by the primary winding links the secondary winding. Then

$$\phi_s = \phi_p \tag{19-2}$$

and
$$\frac{d\phi_s}{dt} = \frac{d\phi_p}{dt} \tag{19-3}$$

Furthermore, under the stated conditions, the primary winding counter emf e_p very nearly equals the applied voltage. If the applied voltage is a sinusoid, it follows from Eq. 19-1 that the flux variation is sinusoidal. In turn, this sinusoidal flux links the secondary winding and in accordance with Eq. 19-1 develops a counter emf at the secondary winding. It is this secondary voltage, denoted e_s, and the consequent current that are applied to the load.

19.2 TRANSFORMER CONSTRUCTION

As was noted in Chapter 8, steel, iron, and other ferromagnetic materials are commonly used for magnetic cores. Indicative of a laminated sheet-steel core is the transformer core depicted in Fig. 19-1. For the time being, the transformer core will be considered a ferromagnetic core. However, a transformer core can be constructed from many nonmagnetic materials, such as air or ceramic. In fact, many radiofrequency transformers use air or ceramic cores in order to reduce the hysteresis and eddy current losses that ferromagnetic materials offer at relatively high frequencies (see Section 8.9).

Laminated sheet-steel transformers are generally classified into two general types that depend on how the coils are placed on the core. *Core-type* construction has the coils distributed over a considerable portion of a single-path magnetic circuit. This construction is depicted in Fig. 19-2. Notice that the windings are on the

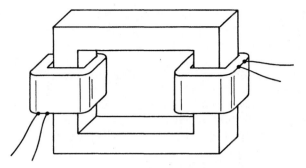

Figure 19-2   Core-type transformer.

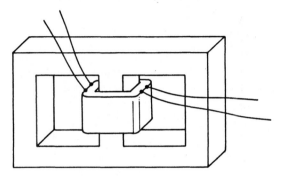

Figure 19-3   Shell-type transformer.

core sides. Compared to core-type construction, the *shell-type* construction has considerably less of the magnetic circuit surrounded by the coils. The construction is typified by Fig. 19-3, in which the primary and secondary windings are on the same magnetic circuit branch and are encircled by the parallel branches of the magnetic circuit. Some shell-type transformers are shown in Fig. 19-4. As is typical with transformer windings, the coils of the transformers of Fig. 19-4 were dipped in varnish and then baked.

Regardless of the construction features, the iron or steel core transformer has the schematic symbol of Fig. 19-5(a). A transformer with a nonmagnetic core material is represented schematically by the air core transformer symbol of Fig. 19-5(b).

19.3 VOLTAGE AND CURRENT RATIOS

Consider a transformer having a primary winding of N_p turns and a secondary winding of N_s turns, as depicted in Fig. 19-6. Let us suppose that the transformer is ideal. An *ideal transformer* is one in which the winding resistances are negligible, the coefficient of coupling between the windings is unity, core losses are negligible, and a negligible amount of current is needed to create the exciting flux ϕ_m. Many actual transformers approach the ideal.

(a)

Figure 19-4 Shell-type transformers for **(a)** chassis mounting; **(b)** circuit board mounting.

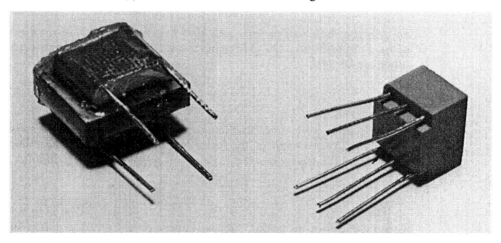

(b)

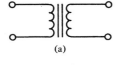

(a)

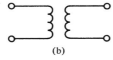

(b)

Figure 19-5 Schematic symbol for a transformer: **(a)** Iron or ferrite core; **(b)** air core.

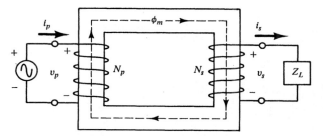

Figure 19-6   Ideal transformer with voltage and current notations.

With a sinusoidal voltage applied to the primary winding, the primary winding voltage v_p is, by Faraday's law,

$$v_p = e_p = N_p \frac{d\phi}{dt} \tag{19-4}$$

Similarly,

$$v_s = e_s = N_s \frac{d\phi}{dt} \tag{19-5}$$

$d\phi/dt$ is the same in Eqs. 19-4 and 19-5 because the coefficient of coupling is taken to be 1. Substituting the expression for $d\phi/dt$ from one of the equations into the other, we obtain

$$\frac{v_p}{v_s} = \frac{N_p}{N_s} \tag{19-6}$$

Equation 19-6 states mathematically that the primary and secondary voltages are related by the ratio of the turns of the respective windings. The turns ratio is often called the *transformation ratio* and is given the letter symbol a:

$$a = \frac{N_p}{N_s} \tag{19-7}$$

Because the transformation ratio is dimensionless and represents the ratio of two similar quantities, we can use effective values of voltage rather than the instantaneous values as in Eq. 19-6. Thus,

$$\frac{V_p}{V_s} = \frac{N_p}{N_s} = a \tag{19-8}$$

For $a > 1$, $V_p > V_s$ and we have a *step down* in voltage from primary to secondary. In this situation the transformer is called a *step-down transformer*. When $a < 1$, $V_p < V_s$. We now have a *step-up transformer* since the voltage is increased from primary to secondary. Transformer voltage ratings are often specified as V_p/V_s. Thus, a 2300/230 V transformer would normally be used with 2300 V on the primary, and 230 V would be induced across the secondary.

EXAMPLE 19-1

A 2300/230 V transformer has 36 turns on the low-voltage side. Determine the number of turns on the high-voltage side.

Solution. From Eq. 19-8, we obtain

$$N_p = N_s \frac{V_p}{V_s} = 36\left(\frac{2300}{230}\right) = 360 \text{ turns}$$

Let us now consider the action of the load current i_s for the transformer of Fig. 19-6. With a load Z_L, a current i_s flows through the N_s turns of the secondary. Hence a magnetomotive force of $N_s i_s$ is developed in the secondary. This mmf must be counteracted in the primary by a primary current i_p flowing through the primary of N_p turns. Thus,

$$N_p i_p = N_s i_s \tag{19-9}$$

Rearranging this last equation and using effective values rather than instantaneous values, we have

$$\frac{I_p}{I_s} = \frac{N_s}{N_p} = \frac{1}{a} \tag{19-10}$$

A comparison of Eq. 19-10 to Eq. 19-8 leads us to conclude that a voltage step up is accompanied by a current step down and vice versa. Furthermore, by substituting Eq. 19-8 into Eq. 19-10, we obtain the relationship

$$V_p I_p = V_s I_s \tag{19-11}$$

The significance of Eq. 19-11 is that the volt-amperes delivered to the primary equal the volt-amperes delivered from the secondary. In fact, the equation is so important that manufacturers provide a volt-ampere rating in addition to the voltage rating in order to specify completely the capability of a transformer.

EXAMPLE 19-2

A transformer is rated at 2300/115 V and 4.6 kVA at 60 Hz. Determine **(a)** the turns ratio and **(b)** the primary and secondary current ratings.

Solution. (a) $a = V_p/V_s = 2300/115 = 20$. (b)

$$I_p = \frac{\text{VA rating}}{V_p} = \frac{4600 \text{ VA}}{2300 \text{ V}}$$

$$= 2 \text{ A}$$

Then $I_s = aI_p = 20(2) = 40$ A.

Drill Problem 19-1 ■

The secondary current to a 2300/230 V transformer is 35 A. Determine **(a)** the turns ratio, **(b)** the primary current, and **(c)** the kilovolt-ampere rating.

Answer. (a) 10. (b) 3.5 A. (c) 8.05 kVA. ■ ■

Drill Problem 19-2 ■

If the voltage on the primary of a 2300/115 V transformer is 2100 V, what is the secondary voltage?

Answer.   105 V.   ■ ■

19.4 REFLECTED IMPEDANCE

It should be apparent that any change in the load impedance in Fig. 19-6 affects the secondary current of the ideal transformer. In turn, any change in the secondary current is reflected as a similar change, modified by the transformation ratio, in primary current. As the following analysis shows, we can even consider the load impedance to be reflected into the primary circuit.

At the secondary winding, a load impedance Z_L across the induced voltage V_s causes a current I_s to flow:

$$I_s = \frac{V_s}{Z_L} \tag{19-12}$$

Relating this current to the primary current by Eq. 19-10, we have

$$I_p = \frac{1}{a} I_s = \frac{1}{a}\left(\frac{V_s}{Z_L}\right) \tag{19-13}$$

If $V_s = (1/a)V_p$ is substituted into Eq. 19-13 and the equation is rearranged, we obtain

$$\frac{V_p}{I_p} = a^2 Z_L \tag{19-14}$$

However, the ratio V_p/I_p represents the input impedance at the primary side Z_p, so that

$$Z_p = a^2 Z_L \tag{19-15}$$

Thus, *any load impedance across the secondary is reflected by the square of the turns ratio to the primary side*; hence the equivalent circuit is that of Fig. 19-7.

Because the transformer allows us to transform a given impedance into another equivalent load, it provides a way of matching to ensure maximum power transfer, as illustrated in the following example.

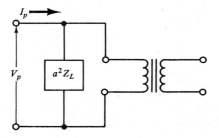

Figure 19-7 An equivalent circuit for the ideal transformer loaded by Z_L.

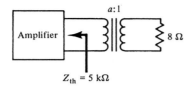

Figure 19-8 Example 19-3.

$Z_{th} = 5 \text{ k}\Omega$

EXAMPLE 19-3

A low-frequency amplifier has an output or Thevenin impedance of 5 kΩ. It is to supply a maximum amount of power to an 8 Ω load, as depicted in Fig. 19-8. What should be the turns ratio of the matching transformer?

Solution. For maximum power,

$$Z_p = Z_{th} = 5 \text{ k}\Omega$$

and from Eq. 19-15

$$a^2 = \frac{Z_p}{Z_L} = \frac{5 \text{ k}\Omega}{8 \text{ }\Omega} = 625$$

Then $a = 25$.

Drill Problem 19-3 ■

An audio transformer has 800 primary turns that are connected to an amplifier with an output impedance of 2 kΩ. If a speaker of 4 Ω is coupled to the amplifier with the transformer, how many turns are needed for the secondary winding? Express the answer in full turns.

Answer. 36 turns.   ■ ■

19.5 TRANSFORMER POLARITY

Transformers are sometimes connected in parallel to supply a single load. Furthermore, they are often connected together to form polyphase systems. Before these connections can be safely made, the phase relationships of the secondary voltages must be proper with respect to each other.

Lenz's law (Chapter 9) briefly states that when flux is increasing in a coil, a counter emf is generated by the coil in opposition to the increasing flux. Furthermore, the right-hand rule allows us to determine the direction of the magnetic flux from the current direction through the coil or vice versa. Let us use both these principles to study the polarity of the voltages for the transformer depicted in Fig. 19-6.

When the primary current in Fig. 19-6 is increasing, application of the right-hand rule to the primary winding determines that the increasing flux ϕ_m is clockwise through the core. In opposition to that increasing flux, each of the coils generates a counter emf and a resultant current that tends to oppose the changing flux. Thus, the primary winding develops a counter emf that attempts to limit the increasing

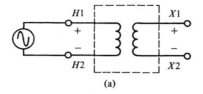

(a)

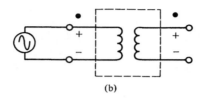

(b)

Figure 19-9 Terminology used to indicate transformer polarity or phase: **(a)** Number; **(b)** dot.

current i_p. This voltage has the indicated instantaneous polarity. The secondary attempts to limit the increasing flux through the generation of a counterclockwise flux. Application of the right-hand rule to the secondary shows that for a counterclockwise flux the secondary current flows from the top terminal of the winding; hence we see the indicated polarity for the induced secondary voltage v_s.

The relative direction of the induced voltages in the primary and secondary windings with respect to the winding terminals of a transformer is called the *transformer polarity*. Winding terminals with the same instantaneous polarity are identified by odd numbers or dots. Thus, in Fig. 19-9(a), the terminals identified with a 1 have the same instantaneous polarity. That is, as $H1$ goes positive compared to $H2$, $X1$ goes positive compared to $X2$. If dots are used, as in Fig. 19-9(b), the terminals with dots each have the same polarity relative to the unmarked terminals and are in phase with each other.

Because the windings of most transformers are sealed and not visible, we must rely on the manufacturer's polarity marks. If a transformer lacks these marks, a simple voltage test can be performed to determine the polarity.

The method for determining the polarity is illustrated in Fig. 19-10 and consists of the following procedure: The high-voltage winding is used as the primary, and one terminal of that winding is arbitrarily marked with a polarity dot. A connection is made between the dotted primary terminal and one of the secondary terminals,

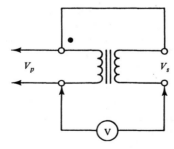

Figure 19-10 Circuit for determining transformer polarity.

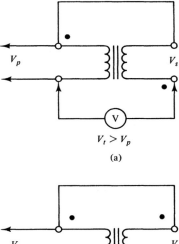

$V_t > V_p$

(a)

$V < V_p$

(b)

Figure 19-11 Terminal connections for
(a) additive and **(b)** subtractive polarity.

as noted in Fig. 19-10. Then a voltmeter is connected between the open secondary terminal and the undotted primary terminal. When a rated voltage V_p is applied to the primary, the voltmeter indicates a test voltage V_t given by

$$V_t = V_p \pm V_s \tag{19-16}$$

If $V_t > V_p$, the winding voltages are in phase, the *polarity is additive,* and a dot is placed on the secondary as shown in Fig. 19-11(a). If $V_t < V_p$, the winding voltages are out of phase, the *polarity is subtractive,* and a dot is placed on the secondary as shown in Fig. 19-11(b).

The importance of transformer polarity becomes evident if we wish to interconnect windings into a multiwinding transformer. A *multiwinding transformer* is a transformer with three or more separate windings. A four-winding transformer with each winding related to the others by a transformation ratio of $1:1 (a = 1)$ is shown in Fig. 19-12. The transformer is rated at 1 kVA at a voltage of 120 V per winding. A pictorial representation of the transformer of Fig. 19-12 appears in Fig. 19-13.

EXAMPLE 19-4

A three-winding transformer has secondary windings of 5 and 10 V. What is the voltage V_x when the windings are connected **(a)** as in Fig. 19-14(a) and **(b)** as in Fig. 19-14(b)?

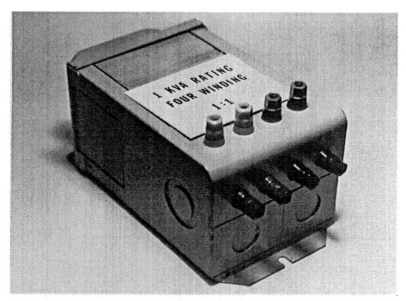

Figure 19-12 A four-winding transformer.

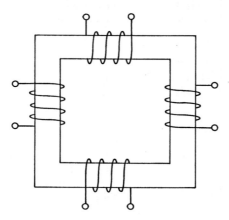

Figure 19-13 Pictorial representation of the transformer of Fig. 19-12.

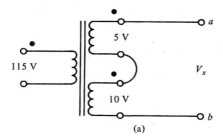

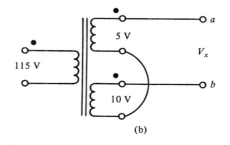

(a)

(b)

Figure 19-14 Example 19-4.

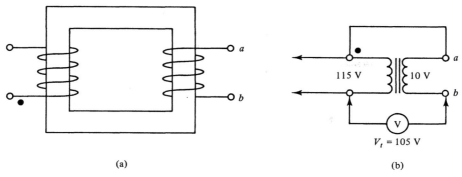

(a)   (b)

Figure 19-15   Drill Problem 19-4.

Solution.   (a) In Fig. 19-14(a), the two secondary voltages are in phase, and

$$V_x = 5 + 10 = 15 \text{ V}$$

(*a* is dotted relative to *b*.)

        (b) In Fig. 19-14(b), the two secondary voltages are out of phase, and

$$V_x = 5 - 10 = -5 \text{ V}$$

(*b* is dotted relative to *a*.)

Drill Problem 19-4 ■

Determine the placement of the polarity dot for the transformers shown in Fig. 19-15(a) and 19-15(b).

Answer.   (a) At *a*. (b) At *a*.   ■ ■

Drill Problem 19-5 ■

Determine the output voltage V_x and the associated polarity dot for the multi-winding transformer of Fig. 19-16.

Answer.   30 V; *a* is dotted.   ■ ■

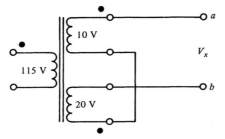

Figure 19-16   Drill Problem 19-5.

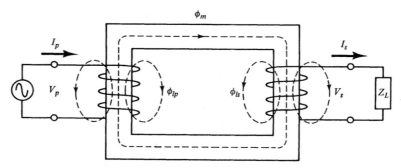

Figure 19-17   Leakage flux in the practical iron core transformer.

19.6 THE PRACTICAL IRON CORE TRANSFORMER

Having developed the transformation relationships among voltage, current, and impedance for the so-called ideal transformer, let us investigate how a practical transformer differs from our ideal model.

First, the practical transformer may have leakage flux occurring at the respective windings as indicated in Fig. 19-17. Primary leakage flux is identified in Fig. 19-17 as ϕ_{lp} and secondary leakage flux is identified as ϕ_{ls}. These leakage fluxes produce primary and secondary inductances L_p, L_s and their respective inductive reactances X_p, X_s.

Additionally, each of the windings has a finite resistance. We may identify the winding resistance of the primary as R_p and that of the secondary as R_s.

Lastly, we shall account for the magnetization reactance of the core and the core losses due to hysteresis and eddy currents. These effects are represented by the inductive reactance X_m and the resistance R_c, respectively. The current to these elements is called the magnetization current I_m.

When we combine all the preceding nonideal effects into one equivalent circuit that represents the practical transformer, we have the equivalent circuit of Fig. 19-18. The various resistances and reactances in the equivalent circuit are called the *transformer constants*. These constants are determined from simple electrical tests, two of which are described in Section 19.7.

Although the equivalent circuit appears fairly complicated, we can complicate it even more by adding shunting capacitances, representing the capacitances of the windings, at the primary and secondary terminals. We shall not add the capacitances to the equivalent circuit, but they must be kept in mind when we consider the frequency response (in Section 19.9).

Notice that an ideal transformer is a part of the equivalent circuit. An ideal transformer provides ideal transformation between the voltages V'_p and V'_s and the currents $I'_p = I_p - I_m$ and I_s. Of course, these latter voltages and currents differ from the terminal voltages and currents because of the nonideal circuit elements.

To simplify the analysis of the transformer circuit, it is customary, at least for the loaded power system transformer, to neglect the core and magnetization con-

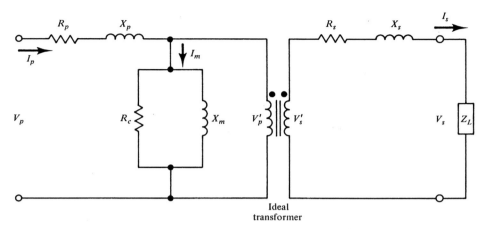

Figure 19-18   Equivalent circuit of the practical iron core transformer.

stants. (They are considered as open circuits with $I_m \approx 0$.) Furthermore, the load impedance and secondary constants can be referred to the primary so that the equivalent of Fig. 19-19 results. (The constants could just as well have been referred to the secondary.)

The circuit of Fig. 19-19 can be simplified further if we "lump" or combine the winding resistances into a total equivalent winding resistance R_e. Then

$$R_e = R_p + a^2 R_s \tag{19-17}$$

When the winding reactances are combined, we obtain a total equivalent winding reactance X_e:

$$X_e = X_p + a^2 X_s \tag{19-18}$$

Hence we obtain the further simplified circuit of Fig. 19-20.

When the winding resistances and reactances are combined, we have the total equivalent winding impedance $\mathbf{Z}_e$, where

$$\mathbf{Z}_e = R_e + jX_e = \sqrt{R_e^2 + X_e^2} \tan^{-1} \frac{X_e}{R_e} \tag{19-19}$$

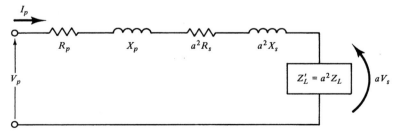

Figure 19-19   A simplified transformer equivalent circuit (negligible I_m; constants and load referred to primary).

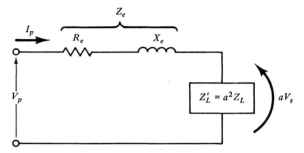

Figure 19-20 A further simplified transformer equivalent circuit.

The following examples illustrate how the transformer equivalent circuit is used in the analysis of transformer circuits.

EXAMPLE 19-5

A 2300/230 V transformer with constants of $R_p = 0.2\ \Omega$, $X_p = 0.5\ \Omega$, $R_s = 0.002\ \Omega$, and $X_s = 0.005\ \Omega$ is loaded by an effective resistive load of 0.5 Ω. Determine **(a)** the total internal impedance referred to the primary and **(b)** the equivalent load resistance referred to the primary.

Solution. The transformation ratio is $a = V_p/V_s = 10$. Using Eqs. 19-17 and 19-18, we have

$$R_e = R_p + a^2 R_s = 0.2 + (10)^2(0.002)$$
$$= 0.2 + 0.2 = 0.4\ \Omega$$

$$X_e = X_p + a^2 X_s = 0.5 + (10)^2(0.005)$$
$$= 0.5 + 0.5 = 1\ \Omega$$

(a) Then $\mathbf{Z}_e = R_e + jX_e = 0.4 + j1 = 1.08\ \angle\ 68.2°\ \Omega$.

(b) $Z'_L = a^2 Z_L = (10)^2(0.5) = 50$. Hence the equivalent circuit is that of Fig. 19-21.

EXAMPLE 19-6

Suppose that the transformer of Example 19-5 is loaded as indicated and 2300 V is applied. **(a)** What primary current flows? **(b)** What is the secondary voltage?

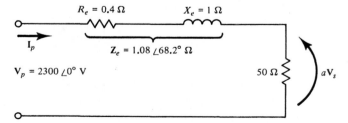

Figure 19-21 Equivalent circuit for Example 19-5.

Solution. (a) Referring to the equivalent circuit of Fig. 19-21, we have

$$\mathbf{Z}_t = R_e + 50 + jX_e = 50.4 + j1 \approx 50.4 \angle 1.14° \ \Omega$$

$$\mathbf{I}_p = \frac{\mathbf{V}_p}{\mathbf{Z}_t} = \frac{2300 \angle 0°}{50.4 \angle 1.14°} = 45.6 \angle -1.14° \ \text{A}$$

(b) A KVL equation is written:

$$\mathbf{V}_p = \mathbf{I}_p \mathbf{Z}_e + a\mathbf{V}_s$$

or

$$\begin{aligned}
a\mathbf{V}_s &= \mathbf{V}_p - \mathbf{I}_p \mathbf{Z}_e \\
&= 2300 \angle 0° - (45.6 \angle -1.14°)(1.08 \angle 68.2°) \\
&= 2300 \angle 0° - 49.3 \angle 67° \\
&= 2300 - 19.2 - j45.4 \\
&= 2281 - j45.4 \approx 2281 \angle -1.14°
\end{aligned}$$

Then

$$\mathbf{V}_s = \frac{2281 \angle -1.14°}{10} = 228 \angle -1.14° \ \text{V}$$

Drill Problem 19-6 ■

A 40 kVA, 2400/240 V transformer has transformer constants of $R_p = 0.8 \ \Omega$, $X_p = 1 \ \Omega$, $R_s = 0.0072 \ \Omega$, and $X_s = 0.009 \ \Omega$. Determine the total equivalent winding impedance referred to the primary (high) side.

Answer.   2.43 $\angle 51.3° \ \Omega$.   ■ ■

Drill Problem 19-7 ■

A load of 1.44 Ω is connected to the secondary of the transformer of Drill Problem 19-6. If the primary voltage is 2400 V, what is (a) the primary current, (b) the secondary voltage, and (c) the secondary current?

Answer.   (a) 16.5 $\angle -0.75°$ A. (b) 237.5 $\angle -0.75°$. (c) 165 $\angle -0.75°$ A.   ■ ■

19.7 SHORT-CIRCUIT AND OPEN-CIRCUIT TESTS

The transformer constants R_e, X_e, and Z_e of the equivalent circuit of Fig. 19-20 are easily determined by a simple laboratory procedure, called the *short-circuit test* because the transformer is actually shorted under controlled test conditions. The circuit used for a short-circuit test is illustrated in Fig. 19-22.

The short-circuit test procedure consists of connecting the circuit shown in Fig. 19-22, setting the potentiometer or variable transformer initially to 0 V, increasing the voltage across the transformer until the rated current flows in the high-voltage side, and finally measuring the power, current, and voltage for these conditions. The measured short-circuit quantities of power, current, and voltage, will be denoted P_{sc}, I_{sc}, and V_{sc}, respectively.

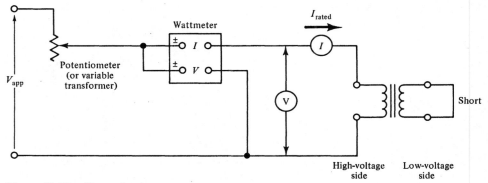

Figure 19-22   Short-circuit test.

Because the short on the secondary is reflected to the primary, the rated primary current flows in the high-voltage side when the transformer voltage is considerably less than the rated value. This voltage is generally about 5% or less of the rated voltage. Since the transformer voltage is low, the core losses are negligible and all the measured power is attributed to the copper loss of the windings, referred to the primary in this case. Thus, the total equivalent resistance R_e is

$$R_e = \frac{P_{sc}}{I_{sc}^2} \qquad (19\text{-}20)$$

The total equivalent impedance Z_e is

$$Z_e = \frac{V_{sc}}{I_{sc}} \qquad (19\text{-}21)$$

It follows that the total equivalent reactance X_e is determined from the relationship

$$X_e = \sqrt{Z_e^2 - R_e^2} \qquad (19\text{-}22)$$

EXAMPLE 19-7

A 2300/230 V, 22 kVA transformer is short-circuited tested. The measured data are $P_{sc} = 280$ W, $V_{sc} = 80$ V, and $I_{sc} = 9.5$ A. Determine the total equivalent transformer constants.

Solution.   Using Eqs. 19-20–19-22, we calculate R_e, Z_e, and X_e to be

$$R_e = \frac{P_{sc}}{I_{sc}^2} = \frac{280}{(9.5)^2} = 3.1 \ \Omega$$

$$Z_e = \frac{V_{sc}}{I_{sc}} = \frac{80}{9.5} = 8.42 \ \Omega$$

$$X_e = \sqrt{Z_e^2 - R_e^2} = \sqrt{(8.42)^2 - (3.1)^2} = 7.83 \ \Omega$$

A second transformer test, the *open-circuit test*, features an unloaded (open-circuit) secondary. With an unloaded condition reflected to the primary, the test enables one to determine the core losses and excitation constants. The procedure consists of connecting the circuit of Fig. 19-23, adjusting the potentiometer or variable transformer for the rated voltage, and measuring the values of power, current, and voltage for the open circuit. These open-circuit measurements are identified as P_{oc}, I_{oc}, and V_{oc}, respectively.

The open-circuit test measurements can be used to determine the constants R_c and X_m, but, as previously stated, these excitation constants have negligible effect under normal load conditions. For our purposes, then, the usefulness of the test lies not in the determination of the excitation constants, but in the measured open-circuit power loss. The open-circuit power loss P_{oc} is nearly equal to the core loss P_c, the value of which is needed for transformer efficiency calculations.

The core loss P_c comprises hysteresis and eddy current losses and, as was explained in Chapter 8, is frequency dependent. However, for a particular frequency, the core loss is constant for all conditions of load, from no load to full load. This is in contrast to the copper loss, which depends on the load current. The total power loss in the transformer is the sum of the copper and core losses:

$$\text{losses} = P_{\text{copper}} + P_{\text{core}} \qquad (19\text{-}23)$$

In terms of the power measurements from the short-circuit and open-circuit test, when the load current and voltage are at the same values as for the short-circuit and open-circuit tests, respectively, the losses are

$$\text{losses} = P_{sc} + P_{oc} \qquad (19\text{-}24)$$

Knowing the transformer losses, we can calculate its efficiency by the following formula, originally presented in Chapter 4:

$$\text{efficiency} = \frac{\text{output}}{\text{output} + \text{losses}} \qquad (19\text{-}25)$$

where the output and losses are in watts. The efficiency of an iron core transformer is relatively high, and for some large power distribution transformers the efficiency is close to 99%.

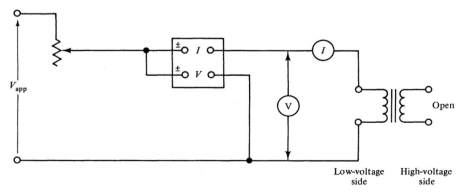

Figure 19-23 Open-circuit test.

EXAMPLE 19-8

The 22 kVA transformer of Example 19-7 is open-circuit tested and P_{oc} is measured as 100 W. If the transformer supplies its rated kilovolt-amperes at unity power factor, what is its efficiency at that load?

Solution. The output and losses are

$$\text{output} = S\cos\theta = (22\text{ kVA})(1) = 22\,000\text{ W}$$
$$\text{losses} = P_{sc} + P_{oc} = 280 + 100 = 380\text{ W}$$

Then

$$\text{efficiency} = \frac{\text{output}}{\text{output} + \text{losses}} = \frac{22\,000}{22\,380}$$
$$= 0.983 = 98.3\%$$

Drill Problem 19-8 ■

A 100 VA transformer is short-circuit tested. The measured data are $P_{sc} = 4$ W, $V_{sc} = 8$ V, and $I_{sc} = 0.8$ A. Determine the transformer constants R_e, Z_e, and X_e.

Answer. 6.25, 10, and 7.8 Ω.   ■ ■

Drill Problem 19-9 ■

A 200 VA power-supply transformer is first short-circuit tested, then open-circuit tested. The measured data are $P_{sc} = 10$ W and $P_{oc} = 2$ W. What is the efficiency at a full volt-ampere load and unity power factor?

Answer. 94.3%.   ■ ■

Drill Problem 19-10 ■

If the transformer described in Example 19-7 is used with a current of 8 A, what is the copper loss?

Answer. 198.4 W.   ■ ■

Drill Problem 19-11 ■

If the transformer described in Drill Problem 19-9 is used at a full volt-ampere load but at 80% power factor, what is its efficiency?

Answer. 93%.   ■ ■

19.8 AUTOTRANSFORMERS

The *autotransformer* is a transformer that consists of one continuous winding. In the conventional two-winding transformer depicted in Fig. 19-24, the two windings are magnetically linked by a common core but are electrically isolated from each other.

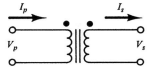

Figure 19-24 Two-winding transformer.

In contrast, an autotransformer, as depicted in Fig. 19-25, has its two windings both magnetically and electrically interconnected.

Although the autotransformer depicted in Fig. 19-25 utilizes a continuous winding with a connection point called a *tap*, a two-winding transformer can be used just as well in constructing an autotransformer. In fact, the polarity test procedure of Section 19-5 utilizes a circuit that we now recognize as an autotransformer circuit. Regardless of whether an autotransformer is constructed of one continuous winding or by electrically connecting the windings of a two-winding transformer, the relationships governing two-winding transformers apply to autotransformers as well.

EXAMPLE 19-9

A 2300/230 V, 46 kVA two-winding transformer is connected as an autotransformer as depicted in Fig. 19-26. Determine **(a)** the voltage rating V_s, **(b)** the current rating I_s, and **(c)** the kilovolt-ampere rating as an autotransformer.

Solution. (a) The secondary rating is the sum of the two separate winding voltages since they are in phase:

$$V_s = 2300 + 230 = 2530 \text{ V}$$

(b) The current rating I_s equals the rating of the individual 230 V winding.

$$I_s = \frac{46 \text{ kVA}}{230 \text{ V}} = 200 \text{ A}$$

(c) As an autotransformer, the transformer has output kilovolt-amperes S_s, where

$$S_s = V_s I_s = (2530)(200) = 506 \text{ kVA}$$

As indicated in the preceding example, the autotransformer has a kilovolt-ampere output that is greater than the rating of its two-winding counterpart. The increased kilovolt-ampere capability results because a large part of the kilovolt-amperes is transferred conductively. In Example 19-9, 46 kVA is actually trans-

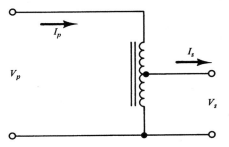

Figure 19-25 Autotransformer.

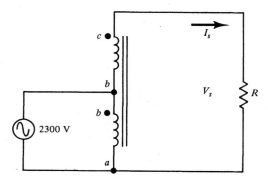

Figure 19-26   Example 19-9.

formed, but 460 kVA is transferred conductively. However, the direct copper connection between the two windings is disadvantageous because it provides no electrical isolation between output and input. The autotransformer also has a higher efficiency than its two-winding counterpart since the same core and copper losses are now based on the higher kilovolt-ampere capability.

Drill Problem 19-12 ■

The low-voltage winding of the autotransformer of Example 19-9 (winding *bc*) is reversed. Determine the new **(a)** secondary voltage and **(b)** kilovolt-ampere rating.

Answer.   (a) 2070 V. (b) 414 kVA.   ■ ■

19.9 FREQUENCY CONSIDERATIONS

When the equivalent circuit of an iron core transformer was described, it was pointed out that stray capacitance exists between the turns of both the primary and secondary windings. The presence of this shunting capacitance along with the inherent inductance of the windings suggests a certain frequency dependence of the transformer. Although we have neglected the effect of frequency, we have done so realizing that the simplified transformer equivalent circuit is valid only over a certain frequency range. This frequency range is from about 50 Hz through, perhaps, 1 kHz and is called the *midband frequencies*. However, transformers are used for coupling a variety of communications, measurement, and control circuits in which the frequency may not always be in the midband range. At frequencies lower or higher than those at midband, an appreciable drop-off in voltage occurs, as indicated by the response curve of Fig. 19-27.

Notice that in the midband range the load voltage is a maximum, denoted V_{Lm}. As the frequency is progressively lowered, the stray capacitances appear as open circuits, and the inductances of the equivalent circuit appear as shorts. In particular, the magnetization reactance X_m progressively shorts out the ideal transformer (see Fig. 19-18), so that less transformation takes place as the frequency is lowered. Eventually, a low enough frequency f_1 is reached so that the voltage has dropped to

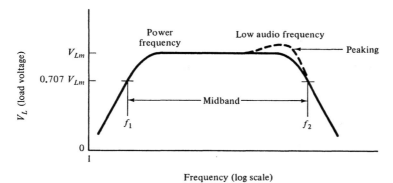

Figure 19-27 Typical response curve of the iron core transformer.

$0.707 V_{Lm}$. This frequency is called the half-power frequency or cutoff frequency (see Chapter 13).

As the frequency increases above midband, the leakage reactances become increasingly important and begin to appear as open circuits. Simultaneously, the stray capacitances behave as short circuits, producing the drop-off at high frequencies seen in Fig. 19-27. As with any frequency response curve, we can define a high-frequency cutoff f_2 at which the voltage drops to 0.707 of its midband value. This is also indicated in Fig. 19-27.

Finally, the combination of stray capacitance and leakage reactances may resonate at frequencies above midband. The peaking resulting from this resonant effect is indicated by the dotted portion of the response curve. The height and width of the peak depends on the values of stray capacitance, leakage inductance, and load impedance.

19.10 THREE-PHASE TRANSFORMER CONNECTIONS

Many industrial operations require large amounts of power to be supplied by three-phase power transmission. To handle the transformation of voltage and current levels, transformer connections compatible with three-phase operation are used. Either a group of similar single transformers, called a *transformer bank,* or a special three-phase transformer with all three sets of windings on one core is used for three-phase transformation. A three-phase core-type transformer and its schematic symbol are depicted in Fig. 19-28.

When single transformers are grouped into banks, it is important that transformers with similar voltage and volt-ampere ratings be used. In addition, it is important that the polarity marks be observed since the windings, both the primary set and the secondary set, must be properly connected into a delta (Δ) or wye (Y) configuration. As was discussed in Section 15-5, a delta-connected set of generator coils— or transformer windings, as in the present case—can only be safely closed if the phasor voltages around the Δ sum to zero. Thus, it is important to observe polarity markings.

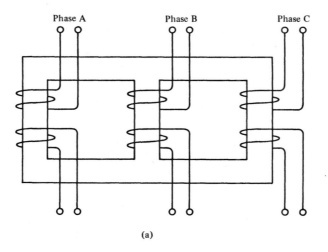

(a)

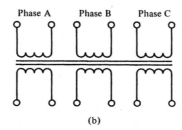

(b)

Figure 19-28 **(a)** Three-phase core-type transformer; **(b)** the associated schematic.

There are four standard ways of connecting a three-phase transformer or a bank of transformers for three-phase operation. The connections are Y–Y, Δ–Δ, Y–Δ, and Δ–Y, where the first character refers to the primary configuration and the second to the secondary. The relationships of the transformed voltages and currents for each of the four connections are discussed in the remainder of this section. Keep in mind that similar transformers, of transformation ratio a, are used so that balanced three-phase system equations apply.

The Y–Y connection and the line voltage and current relationships are depicted in Fig. 19-29. Notice in particular the following notation: V_{Lp} is the line voltage at the primary side, V_{Ls} is the line voltage at the secondary side, I_{Lp} is the line current at the primary side, and I_{Ls} is the line current at the secondary side. It follows from the transformation equations that,

$$V_{Ls} = V_{Lp}/a \qquad\qquad (19\text{-}26)$$

and
$$I_{Ls} = aI_{Lp} \qquad\qquad (19\text{-}27)$$

The Δ–Δ connection is depicted in Fig. 19-30. Here, there is a straight transformation $(a:1)$ between the line voltages and the individual phase currents. Thus,

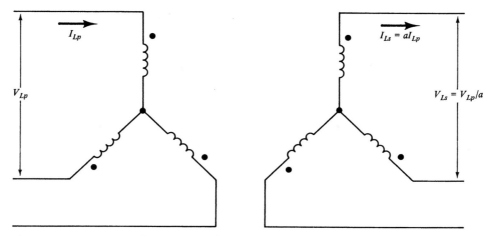

Figure 19-29   Y–Y transformer connection.

Eqs. 19-26 and 19-27 also apply to the Δ–Δ connection. This connection is especially useful because if one of the transformers is removed for repair or replacement, the other two will still provide the proper three-phase voltages and can still handle the load except at a $1/\sqrt{3}$ or 58% level of the original three-phase transformer bank. The two-transformer configuration is called the *open delta* configuration.

The Y–Δ connection is shown in Fig. 19-31. In addition to the transformation ratio a occurring between the primary and secondary phases, there is the additional factor of $\sqrt{3}$ that arises from the line-to-phase values. Hence

$$V_{Ls} = \frac{V_{Lp}}{a\sqrt{3}} \qquad (19\text{-}28)$$

and

$$I_{Ls} = a\sqrt{3}\, I_{Lp} \qquad (19\text{-}29)$$

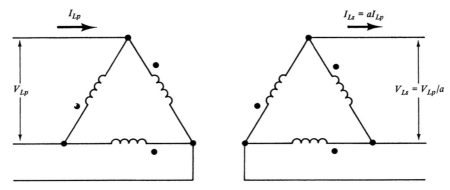

Figure 19-30 Δ–Δ transformer connection.

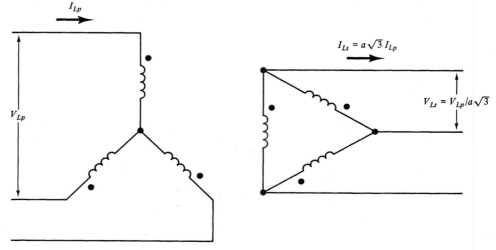

Figure 19-31   Y–Δ transformer connection.

Finally, the Δ–Y connection is shown in Fig. 19-32. Again, the factor of $\sqrt{3}$ occurs in addition to the transformation ratio a. Thus, the equations relating the line voltages and currents for the Δ–Y connection are

$$V_{Ls} = \frac{\sqrt{3}}{a} V_{Lp} \tag{19-30}$$

and

$$I_{Ls} = \frac{a}{\sqrt{3}} I_{Lp} \tag{19-31}$$

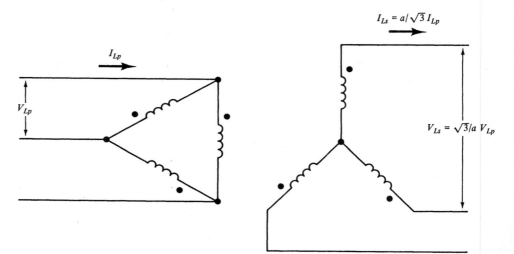

Figure 19-32   Δ–Y transformer connection.

Regardless of the type of connection used, the total three-phase apparent power S_t and the total three-phase power P_t are, from Chapter 15,

$$S_t = \sqrt{3}\, V_L I_L \qquad (19\text{-}32)$$

and
$$P_t = S_t \cos\theta = \sqrt{3}\, V_L I_L \cos\theta \qquad (19\text{-}33)$$

where V_L and I_L are the line values of voltage and current, both taken at the same (primary or secondary) transformer side. Of course, with a three-transformer bank, the apparent power and power are shared by the transformers. When the load is balanced and similar transformers are used, the load sharing is equal.

EXAMPLE 19-10

A three-phase circuit is shown in Fig. 19-33. **(a)** Identify the transformer connection. **(b)** Determine V_{Ls} and I_{Ls}. **(c)** Determine the volt-ampere load for each transformer.

Solution. (a) The primary is connected Δ; the secondary is connected Y. Thus, we have a Δ−Y connection.

(b) From Eqs. 19-30 and 19-31, we have

$$V_{Ls} = \frac{\sqrt{3}}{a} V_{Lp} = \frac{\sqrt{3}}{10}(240) = 41.6 \text{ V}$$

and
$$I_{Ls} = \frac{a}{\sqrt{3}} I_{Lp} = \frac{10}{\sqrt{3}}(5) = 28.9 \text{ A}$$

(c) Using primary values for S_t, we have

$$S_t = \sqrt{3}\, V_L I_L = \sqrt{3}\,(240)(5) = 2078 \text{ VA}$$

Each transformer equally shares the load, so

$$S_1 = S_2 = S_3 = S_t/3 = 2078 \text{ VA}/3 = 693 \text{ VA}$$

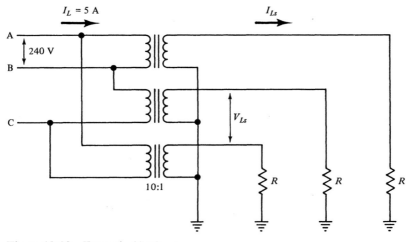

Figure 19-33 Example 19-10.

Drill Problem 19-13 ■

What should the transformation ratio *a* of the transformers of a Y–Δ bank be if the line voltages are 3300 and 220 V at the primary and secondary, respectively?

Answer. 8.66.   ■ ■

Drill Problem 19-14 ■

Three 40 kVA transformers are connected Δ–Δ. **(a)** What is the total kilovolt-amperes available from the transformer bank? **(b)** To what kilovolt-amperes should the load be reduced if one of the transformers becomes open and an open delta results?

Answer. (a) 120 kVA. (b) 69.3 kVA.   ■ ■

QUESTIONS

1. What is a transformer?
2. Define the primary and secondary windings of a transformer.
3. How do the core- and shell-type transformers differ?
4. What is an ideal transformer?
5. What is the transformation ratio?
6. When is a transformer called step down? When is it step up?
7. How does the turns ratio affect transformer voltage and current transformations?
8. What is meant by a reflected impedance?
9. What is transformer polarity?
10. How is transformer polarity determined?
11. Describe an equivalent circuit and a simplified equivalent circuit for an iron core transformer.
12. What information is obtained from the short-circuit test? How is the test performed?
13. What information is obtained from the open-circuit test? How is the test performed?
14. Describe the two types of losses that occur in a transformer.
15. What is an autotransformer?
16. Why is the response of a transformer poor at low frequencies and at high frequencies?
17. What is a transformer bank?
18. What are the four standard ways of connecting three-phase transformers?

PROBLEMS

1. A transformer is tested and found to be able to deliver 60 A at 230 V when the primary current is 2.5 A. Calculate **(a)** the ratio of transformation, **(b)** the primary voltage, and **(c)** the kVA that can be supplied.

2. A primary voltage of 2900 V is applied to a 3000/120 V, 10 kVA transformer. Determine (a) the secondary voltage and (b) the kVA that can be supplied at this lower voltage. (The current cannot exceed the rated value regardless of the voltage.)

3. A transformer with a step-down ratio of 10:1 is followed by a transformer with a step-down ratio of 5:1. What is the voltage at the secondary of the second transformer if the voltage at the primary of the first is 1200 V?

4. A welding transformer operating from 120 V has 5 A of primary current. If the transformer secondary supplies 1 V across the load, what current flows from the secondary?

5. An ac source with a source impedance of 10 Ω is to be matched to a 490 Ω impedance with the use of a transformer. What turns ratio is needed for the transformer?

6. A circuit with an impedance of 200 kΩ is to be matched to an impedance of 80 Ω by a transformer with a 1000-turn primary. What is the number of secondary turns required?

7. Suppose that the transformer in Fig. 19-34 is ideal. Determine (a) the secondary voltage, (b) the secondary current, (c) the primary current, and (d) the power delivered to the load.

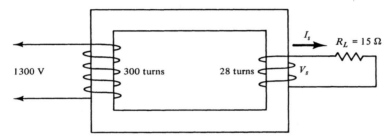

Figure 19-34 Problem 19-7.

8. Determine the placement of the secondary polarity dots for the transformers depicted in Fig. 19-35.

9. Consider Fig. 19-36. Determine the connections between windings and to the load in order to obtain 10 V at the load.

10. Determine the primary current I_p to the transformer of Fig. 19-37 (a) if the transformer is ideal and (b) if the transformer is not ideal but has equivalent constants of $R_e = 5$ Ω and $X_e = 10$ Ω referred to the primary.

11. A 1200/240 V transformer has the constants $R_p = 0.1$ Ω, $X_p = 0.6$ Ω, $R_s = 0.002$ Ω, and $X_s = 0.012$ Ω. If the transformer is loaded with a load having an effective resistance of 1 Ω, determine the equivalent circuit with all impedances referred to the primary.

12. If the transformer of Problem 19-11 is connected to 1200 V, what are the magnitudes of the secondary voltage and current?

13. A 1 kVA, 240/120 V transformer is short-circuit tested; the measurements are $V_{sc} = 20$ V, $P_{sc} = 40$ W, and $I_{sc} = 4.17$ A. Determine the total equivalent transformer constants R_e and X_e.

14. The transformer of Problem 19-13 is open-circuit tested, and the power measurement at rated voltage is 22 W. What is the efficiency of the transformer at a full-load with power factors of (a) unity and (b) 0.7?

15. A 5 kVA transformer has a core loss of 45 W and a full-load copper loss of 120 W. Calculate the efficiency at unity power factor and (a) a full load and (b) a load rated at 80%.

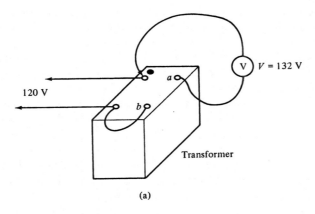

120 V

V = 132 V

Transformer

(a)

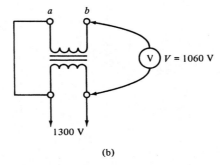

a　b

V = 1060 V

1300 V

(b)

Figure 19-35　Problem 19-8.

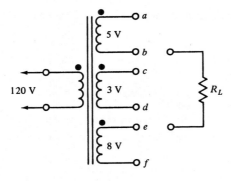

5 V

3 V

8 V

120 V

R_L

a
b
c
d
e
f

Figure 19-36　Problem 19-9.

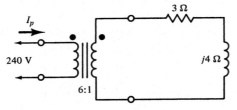

3 Ω

I_p

240 V

$j4$ Ω

6:1

Figure 19-37　Problem 19-10.

16. A 120/12 V two-winding transformer is rated at 120 VA. If the transformer is connected as an autotransformer with a primary of 120 V and a secondary of 132 V, determine the full-load secondary current and the apparent power.

17. A 5 kVA, 2400/240 V distribution transformer is connected as an autotransformer to supply a voltage of 2640 V to a load. Determine the full-load kilovolt-ampere rating for the autotransformer.

18. An industrial plant uses a bank of transformers each rated at 40 kVA and 2300/230 V. The bank is connected in a Y–Δ configuration. For a rated kilovolt-ampere load, determine the line voltages and currents **(a)** on the primary side and **(b)** on the secondary side.

19. The 60 kVA load depicted in Fig. 19-38 is supplied via a Y–Δ transformer bank from 13 200 V lines. **(a)** Specify the voltage ratio and kilovolt-ampere rating of the two-winding transformers used in the bank. **(b)** What are the rated primary and secondary line currents for the bank?

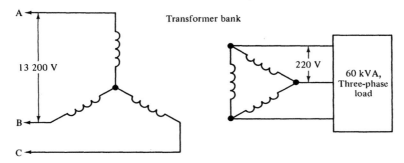

Figure 19-38 Problem 19-19.

20. A transformer bank with Δ–Y connections is used to step up line voltages from 240 to 13 200 V. **(a)** What transformation ratio is needed for each of the transformers? **(b)** What line currents flow at the primary and secondary sides if the total load is 150 kVA?

Chapter 20

Two-Port Networks

20.1 INTRODUCTION

Electrical networks are classified by the number of terminals available for external connections. The simplest network is the two-terminal network depicted in Fig. 20-1. We can completely characterize this two-terminal network by specifying the ratio of voltage V_1 to current I_1:

$$Z_1 = \frac{V_1}{I_1} \qquad (20\text{-}1)$$

We have referred to the impedance Z_1 as the *input impedance*, but it is also commonly called the *driving point impedance*. The inverse ratio is called the *driving point admittance*.

$$Y_1 = \frac{1}{Z_1} = \frac{I_1}{V_1} \qquad (20\text{-}2)$$

Circuit analysts define a pair of terminals at which a current may enter and leave a network as a *port.* Alternatively, the network of Fig. 20-1 is called a *one-port*

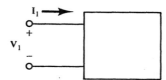

Figure 20-1   Two-terminal network.

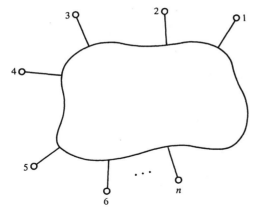

Figure 20-2 An n-terminal network.

network. In general, a network may have any number of externally accessible terminals. A general network with n-terminals is depicted in Fig. 20-2. Of course, the n-terminal network is part of a larger network or circuit as other networks and energy sources are connected to one or more pairs of its terminals.

Although the concept of the n-terminal network is useful from a theoretical standpoint, most applications of n-terminal networks involve a four-terminal network. When a four-terminal network is identified by two distinct pairs of terminals as in Fig. 20-3, we call the network a *two-port network* or simply *two-port.* As was discussed in Chapter 6, we think of one pair of terminals of a four-terminal network as an input pair and the second pair as an output pair. Regardless of which terminal pair is considered the input or output, it is common practice to consider the currents being driven into the network—hence the reference directions specified in Fig. 20-3.

Whereas the electrical behavior of the simple one-port is specified by one equation relating V_1 and I_1, in order to characterize electrically the two-port it is necessary to write two equations relating the quantities V_1, V_2, I_1, and I_2. Circuit analysts use a number of equation sets to describe two-port voltage and current relationships. The various coefficient terms that relate the terminal voltages and currents sets are called *parameters.* In this chapter, three sets of two-port equations are discussed. Each set is identified by a specific parameter that describes the coefficient terms of the particular equations. In each case, the two-port is linear and contains no independent sources. However, to make the material useful for your subsequent study of electronics, we will allow dependent sources within the two-port.

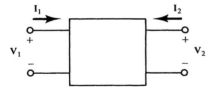

Figure 20-3 Two-port network with standard voltage and current reference directions.

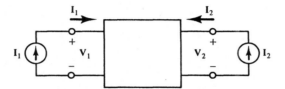

Figure 20-4 Two-port driven by current sources.

20.2 IMPEDANCE PARAMETERS

If current sources are connected to the two-port network as shown in Fig. 20-4, each of the voltages V_1 and V_2 can be related to the source currents I_1 and I_2 through the following linear combinations:

$$V_1 = z_{11}I_1 + z_{12}I_2 \qquad (20\text{-}3)$$

$$V_2 = z_{21}I_1 + z_{22}I_2 \qquad (20\text{-}4)$$

The terms z_{11}, z_{12}, z_{21}, and z_{22} are called the *impedance* (z) *parameters* and have the units of ohms. The parameters can be evaluated by removing the appropriate current source, that is, by setting $I_1 = 0$ or $I_2 = 0$. Thus,

$$z_{11} = \left.\frac{V_1}{I_1}\right|_{I_2=0} \qquad (20\text{-}5)$$

$$z_{12} = \left.\frac{V_1}{I_2}\right|_{I_1=0} \qquad (20\text{-}6)$$

$$z_{21} = \left.\frac{V_2}{I_1}\right|_{I_2=0} \qquad (20\text{-}7)$$

and

$$z_{22} = \left.\frac{V_2}{I_2}\right|_{I_1=0} \qquad (20\text{-}8)$$

We also refer to the z parameters as the *open-circuit impedance parameters* since they are obtained individually by forcing the appropriate current to zero, that is, by opening one side of the network.

If we wish to model the electrical behavior of a two-port network characterized by the z-parameter equations of Eqs. 20-3 and 20-4, we can incorporate dependent sources as shown in the equivalent network of Fig. 20-5. The impedance parameters

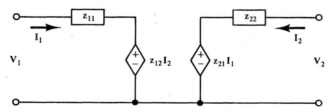

Figure 20-5 An equivalent circuit model for the z parameter equations.

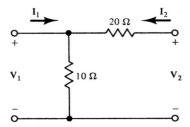

Figure 20-6 Network for Examples 20-1 and 20-3.

z_{11} and z_{22} represent the driving point impedances at those respective ports. When one port is considered the input, the driving point impedance at that port is called the *input impedance*. It follows that the impedance at the other port is called the *output impedance*. The two impedances z_{12} and z_{21} are ratios of both an input and an output quantity and are called *transfer impedances*. Although we used current sources as driving sources in Fig. 20-4, voltage sources can be used instead because it is simply the voltage to current ratio that is important to the determination of a parameter.

EXAMPLE 20-1

Determine the z parameters for the network of Fig. 20-6.

Solution. First, a voltage is applied at the left as shown in Fig. 20-7(a) so that z_{11} and z_{21} can be determined. For the circuit of Fig. 20-7(a),

$$I_1 = \frac{V_1}{10 \ \Omega}$$

and $$V_2 = V_1 = (10 \ \Omega)I_1$$

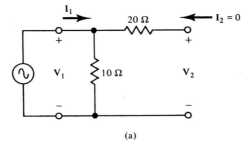

(a)

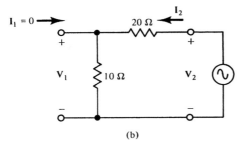

(b)

Figure 20-7 Voltage sources applied to the network of Fig. 20-6 **(a)** at port 1 and **(b)** at port 2.

These relationships for I_1 and V_2 are substituted into Eqs. 20-5 and 20-7, respectively:

$$z_{11} = \left.\frac{V_1}{I_1}\right|_{I_2=0} = \frac{V_1}{V_1/10} = 10\ \Omega$$

and

$$z_{21} = \left.\frac{V_2}{I_1}\right|_{I_2=0} = \frac{I_1(10)}{I_1} = 10\ \Omega$$

With a voltage source at the right as shown in Fig. 20-7(b), V_1 and I_2 are calculated as

$$V_1 = V_2\left(\frac{10\ \Omega}{10\ \Omega + 20\ \Omega}\right) = \frac{V_2}{3}$$

and

$$I_2 = \frac{V_2}{10 + 20} = \frac{V_2}{30}$$

These relationships for V_1 and I_2 are substituted in Eqs. 20-6 and 20-8, so that

$$z_{12} = \left.\frac{V_1}{I_2}\right|_{I_1=0} = \frac{V_2/3}{V_2/30} = 10\ \Omega$$

and

$$z_{22} = \left.\frac{V_2}{I_2}\right|_{I_1=0} = \frac{V_2}{V_2/30} = 30\ \Omega$$

When a two-port is linear and bilateral, the transfer impedance terms are equal ($z_{12} = z_{21}$). In such a case, the equivalent wye (Y) or tee (T) of Fig. 20-8 can be used. However, the network representation contains no dependent sources, only passive elements. The network of Example 20-1 is bilateral and is really a degenerate T network.

Two-port relationships enable the circuit analyst to replace a more complicated part of a larger circuit by an equivalent so that the analysis of the larger circuit is somewhat simplified. For example, the resistive network of Fig. 20-9(a), since it is bilateral, can be replaced by an equivalent T, as shown in Fig. 20-9(b). Then the various analysis methods, such as loop and nodal analysis, can be applied to this

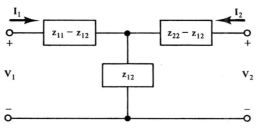

Figure 20-8   An equivalent network (T) for the bilateral network ($z_{21} = z_{12}$).

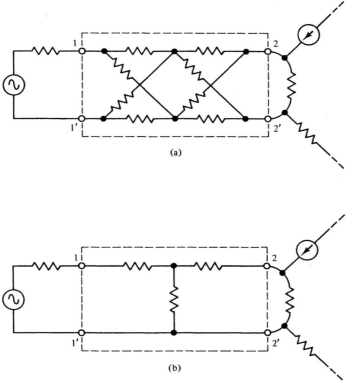

(a)

(b)

Figure 20-9 **(a)** A network within a larger circuit; **(b)** the network of (a) replaced by its equivalent.

simpler circuit. Two-port relationships also permit us to model electronic devices in their simplest form.

EXAMPLE 20-2

Determine the voltage generator or source current I_g and the load voltage V_L for the circuit of Fig. 20-10.

Solution. The **z** parameters can be substituted into the equivalent circuit of Fig. 20-5 and KVL equations then can be written. Alternatively, we can obtain the two-port relationships from Eqs. 20-3 and 20-4. When the **z** parameters are

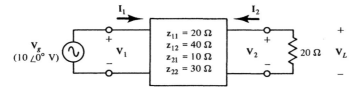

Figure 20-10 Example 20-2.

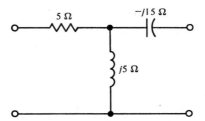

Figure 20-11   Drill Problem 20-1.

substituted into those equations, we obtain

$$\mathbf{V}_1 = 20\mathbf{I}_1 + 40\mathbf{I}_2$$

and
$$\mathbf{V}_2 = 10\mathbf{I}_1 + 30\mathbf{I}_2$$

Furthermore, $\mathbf{I}_1 = \mathbf{I}_g$, $\mathbf{V}_1 = 10 \angle 0°$ V, $\mathbf{V}_2 = \mathbf{V}_L$, and $\mathbf{I}_2 = -\mathbf{V}_L/20$ Ω.

Substituting these latter relationships into the z-parameter equations, we obtain two simultaneous equations with the variables $\mathbf{I}_g$ and $\mathbf{V}_L$:

$$10 \angle 0° = 20\mathbf{I}_g - \mathbf{V}_L\!\left(\frac{40}{20}\right)$$

and
$$\mathbf{V}_L = 10\mathbf{I}_g - \mathbf{V}_L\!\left(\frac{30}{20}\right)$$

That is,

$$20\mathbf{I}_g - 2\mathbf{V}_L = 10$$

and
$$10\mathbf{I}_g - 2.5\mathbf{V}_L = 0$$

When these latter equations are solved simultaneously, we obtain

$$\mathbf{I}_g = 0.833 \text{ A} \qquad \mathbf{V}_L = 3.33 \text{ V}$$

Drill Problem 20-1 ■

Determine the parameters z_{11}, z_{12}, z_{21}, and z_{22} for the circuit of Fig. 20-11.

Answer. $5 + j5$ Ω; $j5$ Ω; $j5$ Ω; $-j10$ Ω.   ■ ■

Drill Problem 20-2 ■

Determine the open-circuit output voltage for the two-port shown in Fig. 20-12.

Answer. $0.8 \angle 90°$ V.   ■ ■

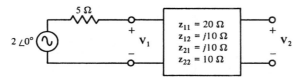

Figure 20-12   Drill Problem 20-2.

20.3 ADMITTANCE PARAMETERS

Another set of equations relating the voltages and currents of the two-port of Fig. 20-4 is based upon the expression of the currents I_1 and I_2 in terms of the voltages V_1 and V_2. These equations are

$$I_1 = y_{11}V_1 + y_{12}V_2 \tag{20-9}$$

$$I_2 = y_{21}V_1 + y_{22}V_2 \tag{20-10}$$

The terms y_{11}, y_{12}, y_{21}, and y_{22} are called the *admittance* (**y**) *parameters* and have the units of siemens. Since the set of **y** parameter equations form a set of nodal-type equations, one interpretation of a circuit that models the equations is the network depicted in Fig. 20-13.

An analysis of Eqs. 20-9 and 20-10 shows that if the various **y** parameters are evaluated for a given network, either V_1 or V_2 must be set equal to zero. Thus the individual parameters are obtained by forcing the appropriate voltage to zero, that is, by shorting the appropriate port. For this reason, the **y** parameters are also called the *short-circuit admittance parameters*.

Notice that if port 2 is shorted, $V_2 = 0$ and the parameters y_{11} and y_{21} are determined as

$$y_{11} = \left.\frac{I_1}{V_1}\right|_{V_2 = 0} \tag{20-11}$$

and

$$y_{21} = \left.\frac{I_2}{V_1}\right|_{V_2 = 0} \tag{20-12}$$

If port 1 is shorted, $V_1 = 0$ and the parameters y_{12} and y_{22} are obtained as

$$y_{12} = \left.\frac{I_1}{V_2}\right|_{V_1 = 0} \tag{20-13}$$

and

$$y_{22} = \left.\frac{I_2}{V_2}\right|_{V_1 = 0} \tag{20-14}$$

The admittance parameters y_{11} and y_{22} represent the driving point admittances at those respective ports. When one of these ports is considered the input port, the driving point admittance at that port is called the *input admittance*; the driving point admittance at the remaining port, the output port, is called the *output admittance*.

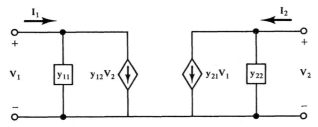

Figure 20-13 A network model for the **y** parameter equations.

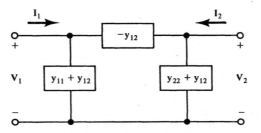

Figure 20-14 A y parameter model (π) for a bilateral two-port.

Because the parameters y_{12} and y_{21} are ratios between the quantities at both ports, those parameters are known as *transfer admittances*. Furthermore, when the two-port is not only linear but also bilateral, the transfer admittances are equal ($y_{12} = y_{21}$) and a network of passive elements can be used to represent the two-port. The delta (Δ) or pi (π) network of Fig. 20-14 is one such network.

EXAMPLE 20-3

Determine the **y** parameters for the two-port of Fig. 20-6.

Solution. First, a voltage is applied at the left and the port 2 terminals are shorted as in Fig. 20-15(a). Then

$$Z_t = 10\ \Omega \| 20\ \Omega = \frac{(10)(20)}{30} = 6.67\ \Omega$$

$$I_1 = \frac{V_1}{Z_t} = \frac{V_1}{6.67\ \Omega}$$

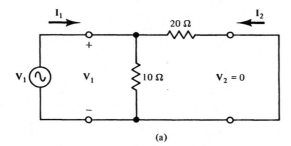

(a)

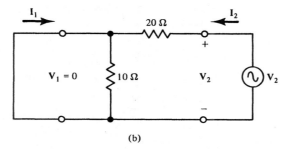

(b)

Figure 20-15 Example 20-3; circuits for determining **(a)** y_{11} and y_{21} and **(b)** y_{12} and y_{22}.

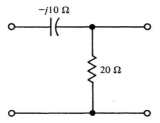

Figure 20-16 Drill Problem 20-3.

and
$$y_{11} = \frac{V_1/6.67}{V_1}\bigg|_{V_2=0} = \frac{1}{6.67} = 0.15 \text{ S}$$

By current division,

$$I_2 = -I_1\left(\frac{10 \,\Omega}{10 \,\Omega + 20 \,\Omega}\right) = \frac{-I_1}{3} = \frac{-V_1}{3(6.67)}$$

Then
$$y_{21} = \frac{I_2}{V_1}\bigg|_{V_2=0} = \frac{-V_1/20}{V_1} = -0.05 \text{ S}$$

If a voltage is applied to port 2 and port 1 is shorted as in Fig. 20-15(b), we obtain

$$I_2 = \frac{V_2}{20} \quad \text{and} \quad I_1 = -I_2$$

Substituting these expressions into Eqs. 20-13 and 20-14, we have

$$y_{12} = \frac{I_1}{V_2}\bigg|_{V_1=0} = \frac{-V_2/20}{V_2} = -\tfrac{1}{20} = -0.05 \text{ S}$$

and
$$y_{22} = \frac{I_2}{V_2}\bigg|_{V_1=0} = \frac{V_2/20}{V_2} = 0.05 \text{ S}$$

Comparing the results of Examples 20-1 and 20-3, we notice that the driving point admittances are not the reciprocals of the driving point impedances. However, there are definite equations that relate the **z** and **y** parameters. These conversion formulas are found in most electrical engineering handbooks.

Drill Problem 20-3 ■

Determine the **y** parameters for the coupling network of Fig. 20-16.

Answer. $y_{11} = j0.1 \text{ S}$; $y_{12} = y_{21} = -j0.1 \text{ S}$; $y_{22} = 0.05 + j0.1 \text{ S}$. ■ ■

Drill Problem 20-4 ■

A circuit model of a field effect transistor (FET) appears in Fig. 20-17. Determine the **y** parameters for the circuit.

Answer. $y_{11} = y_{12} = 0$; $y_{21} = g_m$; $y_{22} = 1/r_{ds}$. ■ ■

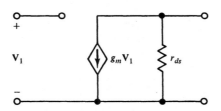

Figure 20-17 Drill Problem 20-4 (FET circuit model).

20.4 HYBRID PARAMETERS

A third set of two-port equations define the input voltage V_1 and the output current I_2 in terms of the input current I_1 and the output voltage V_2:

$$V_1 = h_{11}I_1 + h_{12}V_2 \tag{20-15}$$

$$I_2 = h_{21}I_1 + h_{22}V_2 \tag{20-16}$$

Following the same pattern of defining the **h** parameters in terms of the appropriate voltages and currents as for the other parameter sets, we have

$$h_{11} = \left.\frac{V_1}{I_1}\right|_{V_2=0} \tag{20-17}$$

$$h_{21} = \left.\frac{I_2}{I_1}\right|_{V_2=0} \tag{20-18}$$

$$h_{12} = \left.\frac{V_1}{V_2}\right|_{I_1=0} \tag{20-19}$$

$$h_{22} = \left.\frac{I_2}{V_2}\right|_{I_1=0} \tag{20-20}$$

The reason the parameters are called hybrid becomes apparent on examination of the ratios of Eqs. 20-17–20-20: the h_{11} parameter is dimensionally an impedance and is measured in ohms, the h_{21} and h_{12} parameters are dimensionally unitless, and the h_{22} parameter is dimensionally an admittance measured in siemens. Notice that h_{21} is the ratio of currents and is called the *short-circuit forward current gain*. On the other hand, h_{12} is the ratio of voltages and is called the *open-circuit reverse voltage gain*. For a bilateral network, $h_{12} = -h_{21}$.

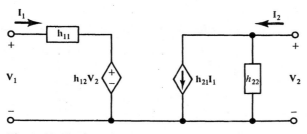

Figure 20-18 **h** parameter model for a two-port.

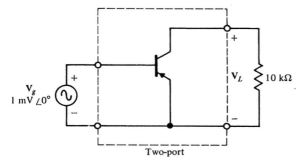

Figure 20-19 Circuit for Example 20-4.

Although the set of parameters is not dimensionally homogeneous, it is very appropriate for describing electronic devices such as the bipolar junction transistor (BJT). Figure 20-18 depicts a two-port representation that uses **h** parameters.

The determination of the **h** parameters follows the format adopted in the study of the **z** and **y** parameters; a source is applied to the appropriate port and the opposite port is either shorted or left open, depending on the parameter we wish to determine.

The following example illustrates the use of circuit analysis techniques with an **h**-parameter model.

EXAMPLE 20-4

A transistor has the **h** parameters $h_{11} = 1$ kΩ. $h_{12} = 1 \times 10^{-4}$, $h_{21} = 50$, and $h_{22} = 1 \times 10^{-4}$ S. Determine the load voltage V_L when the transistor is connected as shown in Fig. 20-19.

Solution. The circuit is redrawn with the **h**-parameter model of the transistor as in Fig. 20-20. A KVL equation at the input yields the expression for I_1:

$$I_1 = \frac{1 \times 10^{-3} - 1 \times 10^{-4} V_L}{1 \text{ k}\Omega}$$

At the output, we have

$$V_L = -50I_1(10 \text{ k}\Omega \| 10 \text{ k}\Omega)$$
$$= -50I_1(5 \text{ k}\Omega)$$

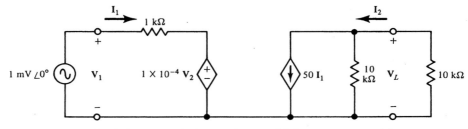

Figure 20-20 Circuit for Example 20-4, redrawn with **h** parameter model.

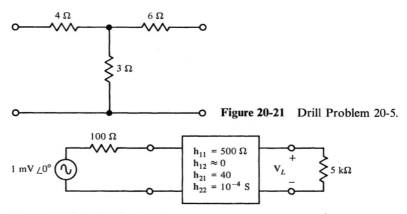

Figure 20-21 Drill Problem 20-5.

Figure 20-22 Drill Problem 20-6.

Substituting the expression for I_1 into the V_L equation, we obtain

$$V_L = -50\left(\frac{1 \times 10^{-3} - 1 \times 10^{-4}\, V_L}{1\ \text{k}\Omega}\right)(5\ \text{k}\Omega)$$

$$= -50(1 \times 10^{-3} - 1 \times 10^{-4}\, V_L)(5)$$

$$= -250 \times 10^{-3} + 250 \times 10^{-4}\, V_L$$

Thus

$$V_L - 250 \times 10^{-4}\, V_L = -250 \times 10^{-3}$$

$$0.975V_L = -250 \times 10^{-3}$$

$$V_L = -256 \times 10^{-3} = 256\ \angle\, 180°\ \text{mV}$$

Drill Problem 20-5 ■

Determine the **h** parameters for the circuit of Fig. 20-21.

Answer. $h_{11} = 6\,\Omega,\ h_{12} = \tfrac{1}{3},\ h_{21} = -\tfrac{1}{3},\ h_{22} = \tfrac{1}{9}\,\text{S}.$

■ ■

Drill Problem 20-6 ■

Determine the load voltage V_L for the circuit of Fig. 20-22.

Answer. $0.222\ \angle\, 180°\ \text{V}.$

■ ■

QUESTIONS

1. What is an electrical port?
2. What is meant by input impedance?
3. What is meant by output impedance?
4. What is a transfer impedance?
5. How are the transfer parameters related for a bilateral network?

6. Why are the impedance parameters called open-circuit parameters?

7. Why are the admittance parameters called short-circuit parameters?

8. What is a transfer admittance?

9. What is meant by the short-circuit forward current gain of a network?

10. What is meant by the open-circuit reverse voltage gain of a network?

11. What are the units of the hybrid parameters?

PROBLEMS

1. Determine the z parameters for the networks of Fig. 20-23.

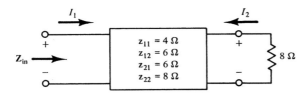

(a)   (b)

Figure 20-23   Problem 20-1.

2. Determine the input impedance to the network of Fig. 20-24.

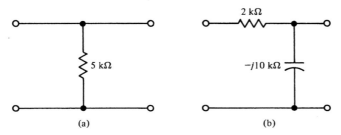

Figure 20-24   Problems 20-2 and 20-3.

3. Determine the magnitude of the voltage across the 8 Ω load resistor of Fig. 20-24 when a 1 V signal is applied to the network input.

4. Determine the y parameters for the networks of Fig. 20-25.

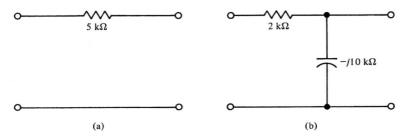

(a)   (b)

Figure 20-25   Problem 20-4.

5. Determine the voltage V_L for the circuit of Fig. 20-26.

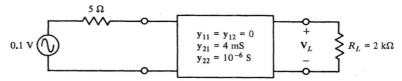

Figure 20-26   Problem 20-5.

6. Determine the ratio V_L/V_g for the circuit of Fig. 20-27.

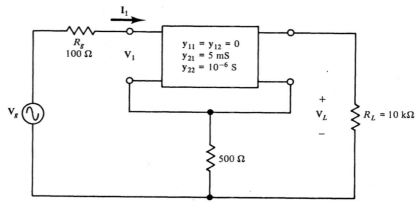

Figure 20-27   Problem 20-6.

7. Determine the **h** parameters for the network shown in Fig. 20-28.

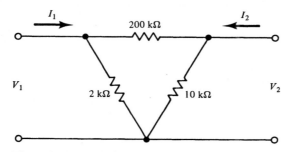

Figure 20-28   Problem 20-7.

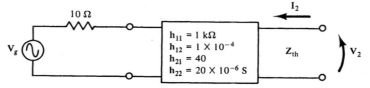

Figure 20-29   Problem 20-8.

8. Determine the Thevenin impedance Z_{th} for the circuit of Fig. 20-29 by exciting the circuit with a voltage V_2 and determining the current I_2.

9. Determine the input impedance Z_{in} for the circuit of Fig. 20-30.

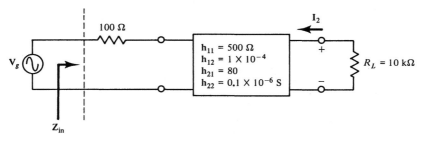

Figure 20-30 Problem 20-9.

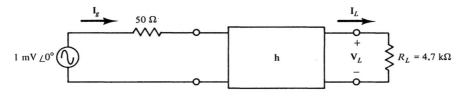

Figure 20-31 Problem 20-10.

10. Determine the voltage gain $A_v = V_L/V_g$ and current gain $A_i = I_L/I_g$ for the circuit shown in Fig. 20-31 if the **h** parameters of the two-port are $h_{11} = 2$ kΩ, $h_{21} = 100$, and $h_{12} = h_{22} = 0$.

Appendix

A.1 DETERMINANTS

A *determinant* is a square array of numbers representing the algebraic sum of a set of possible products. The number of terms in a determinant determines its size or *order*. If a determinant has four terms, it has two rows and two columns and is called a 2×2 or second-order determinant. If it has nine terms, it is called a 3×3 or third-order determinant. It follows that, in general, an $n \times n$ or nth-order determinant contains n^2 terms.

Symbolically, a determinant is denoted when the numbers are arranged in rows and columns and bounded by vertical lines. For example, a second-order determinant D is denoted

$$D = \begin{vmatrix} a_1 & a_2 \\ a_3 & a_4 \end{vmatrix} \tag{A-1}$$

where a_1, a_2, a_3, and a_4 are the *numbers, terms,* or *elements* of the determinant. The numbers from the upper left to lower right, a_1 and a_4, form the *principal diagonal*. The numbers from the upper right to lower left, a_2 and a_3, form the *secondary diagonal*.

By definition, the *value* or *expansion* of the second-order determinant is obtained as *the product of the elements in the principal diagonal minus the product of the elements in the secondary diagonal*; that is,

$$D = \begin{vmatrix} a_1 & a_2 \\ a_3 & a_4 \end{vmatrix} = a_1 a_4 - a_2 a_3 \tag{A-2}$$

If, for example, the terms a_1, a_2, a_3, and a_4 are 2, -1, 9, and -4, respectively, the value of the determinant is

$$D = \begin{vmatrix} 2 & -1 \\ 9 & -4 \end{vmatrix} = -8 - (-9) = 1 \tag{A-3}$$

The third-order determinant has nine terms and is defined by

$$D = \begin{vmatrix} a_1 & a_2 & a_3 \\ a_4 & a_5 & a_6 \\ a_7 & a_8 & a_9 \end{vmatrix} = a_1 a_5 a_9 + a_2 a_6 a_7 + a_3 a_4 a_8 - a_3 a_5 a_7 - a_1 a_6 a_8 - a_2 a_4 a_9 \tag{A-4}$$

where the as again represent the terms of the determinant. Now the numbers a_1, a_5, and a_9 form the principal diagonal and the numbers a_3, a_5, and a_7 form the secondary diagonal.

There are several ways of obtaining the six terms in the expansion defined by Eq. A-4. One of these is the following method:

1. *The first and second columns are rewritten to the right of the determinant.*

$$D = \begin{vmatrix} a_1 & a_2 & a_3 \\ a_4 & a_5 & a_6 \\ a_7 & a_8 & a_9 \end{vmatrix} \begin{matrix} a_1 & a_2 \\ a_4 & a_5 \\ a_7 & a_8 \end{matrix} \tag{A-5}$$

2. *The products of the elements in each of the three diagonals from left to right are multiplied by $+1$, and the products of the elements in each of the three diagonals from right to left are multiplied by -1.*

$$D = \begin{vmatrix} a_1 & a_2 & a_3 \\ a_4 & a_5 & a_6 \\ a_7 & a_8 & a_9 \end{vmatrix} \begin{matrix} a_1 & a_2 \\ a_4 & a_5 \\ a_7 & a_8 \end{matrix} \tag{A-6}$$

$$(-1)(-1)(-1) \quad (+1)(+1)(+1)$$

3. *The algebraic sum of the six products thus formed equals the value of the determinant.*

The method is illustrated by the following expansion:

$$D = \begin{vmatrix} 4 & -3 & 1 \\ 3 & -1 & -2 \\ 1 & -2 & -1 \end{vmatrix} = \begin{vmatrix} 4 & -3 & 1 \\ 3 & -1 & -2 \\ 1 & -2 & -1 \end{vmatrix} \begin{matrix} 4 & -3 \\ 3 & -1 \\ 1 & -2 \end{matrix}$$

$$= +4 + 6 - 6 + 1 - 16 - 9 = -20 \tag{A-7}$$

Determinants are particularly useful in the solution of simultaneous equations. A system of linear independent equations is solved algebraically by eliminating all but one variable from the set of equations and solving that equation for the unknown variable. The other variables are subsequently found by substitution. Applying the method of algebraic elimination to a system of two equations of the form

$$a_1x_1 + a_2x_2 = b_1 \tag{A-8}$$

$$a_3x_1 + a_4x_2 = b_2 \tag{A-9}$$

where x_1 and x_2 are the variables or unknowns; a_1, a_2, a_3, and a_4 are the respective coefficients; and b_1 and b_2 are constant terms, we obtain

$$x_1 = \frac{a_4b_1 - a_2b_2}{a_1a_4 - a_2a_3} \tag{A-10}$$

and

$$x_2 = \frac{a_1b_2 - a_3b_1}{a_1a_4 - a_2a_3} \tag{A-11}$$

Notice that the denominators of Eqs. A-10 and A-11 can be expressed in determinant form in accordance with Eq. A-2. By applying the determinant definition to the numerator, we can also write

$$a_4b_1 - a_2b_2 = \begin{vmatrix} b_1 & a_2 \\ b_2 & a_4 \end{vmatrix} \tag{A-12}$$

and

$$a_1b_2 - a_3b_1 = \begin{vmatrix} a_1 & b_1 \\ a_3 & b_2 \end{vmatrix} \tag{A-13}$$

Thus the solution of two linear equations may be written in the form

$$x_1 = \frac{\begin{vmatrix} b_1 & a_2 \\ b_2 & a_4 \end{vmatrix}}{\begin{vmatrix} a_1 & a_2 \\ a_3 & a_4 \end{vmatrix}} \qquad x_2 = \frac{\begin{vmatrix} a_1 & b_1 \\ a_3 & b_2 \end{vmatrix}}{\begin{vmatrix} a_1 & a_2 \\ a_3 & a_4 \end{vmatrix}} \tag{A-14}$$

Notice that the two denominators in Eq. A-14 are the same determinant that was formed from the coefficients of the variables x_1 and x_2. This determinant is called the *determinant of the system D.* For a finite solution, D cannot equal zero.

The determinant in the numerator of the value of either unknown may be obtained from the denominator determinant by replacing the column of coefficients of this unknown by the corresponding constant terms. These determinants are denoted by the symbols D_1 and D_2, respectively. Then

$$x_1 = \frac{D_1}{D} \qquad x_2 = \frac{D_2}{D} \tag{A-15}$$

EXAMPLE A-1

Solve the two equations

$$x_1 + 3x_2 = 20$$
$$2x_1 - 2x_2 = -5$$

Solution

$$D = \begin{vmatrix} 1 & 3 \\ 2 & -2 \end{vmatrix} = -2 - 6 = -8$$

$$D_1 = \begin{vmatrix} 20 & 3 \\ -5 & -2 \end{vmatrix} = -40 + 15 = -25$$

$$D_2 = \begin{vmatrix} 1 & 20 \\ 2 & -5 \end{vmatrix} = -5 - 40 = -45$$

$$x_1 = \frac{D_1}{D} = \frac{-25}{-8} = 3.125$$

$$x_2 = \frac{D_2}{D} = \frac{-45}{-8} = 5.625$$

In a similar manner, three linear equations of the form

$$a_1 x_1 + a_2 x_2 + a_3 x_3 = b_1 \tag{A-16}$$

$$a_4 x_1 + a_5 x_2 + a_6 x_3 = b_2 \tag{A-17}$$

$$a_7 x_1 + a_8 x_2 + a_9 x_3 = b_3 \tag{A-18}$$

can be solved simultaneously by determinants if one forms the determinant of the coefficients D and the determinants D_1, D_2, and D_3, where the respective determinants are formed when the column of coefficients of the respective unknown are replaced by the constant terms. Then

$$x_1 = \frac{D_1}{D} \qquad x_2 = \frac{D_2}{D} \qquad x_3 = \frac{D_3}{D} \tag{A-19}$$

EXAMPLE A-2

Solve the three equations

$$4x_1 - 3x_2 + x_3 = -5$$
$$3x_1 - x_2 - 2x_3 = 7$$
$$x_1 - 2x_2 - x_3 = 0$$

Solution

$$D = \begin{vmatrix} 4 & -3 & 1 \\ 3 & -1 & -2 \\ 1 & -2 & -1 \end{vmatrix} = 4 + 6 - 6 + 1 - 16 - 9 = -20$$

$$D_1 = \begin{vmatrix} -5 & -3 & 1 \\ 7 & -1 & -2 \\ 0 & -2 & -1 \end{vmatrix} = -5 + 0 - 14 + 0 + 20 - 21 = -20$$

$$D_2 = \begin{vmatrix} 4 & -5 & 1 \\ 3 & 7 & -2 \\ 1 & 0 & -1 \end{vmatrix} = -28 + 10 + 0 - 7 - 15 + 0 = -40$$

$$D_3 = \begin{vmatrix} 4 & -3 & -5 \\ 3 & -1 & 7 \\ 1 & -2 & 0 \end{vmatrix} = 0 - 21 + 30 - 5 + 56 + 0 = 60$$

$$x_1 = \frac{D_1}{D} = \frac{-20}{-20} = 1$$

$$x_2 = \frac{D_2}{D} = \frac{-40}{-20} = 2$$

$$x_3 = \frac{D_3}{D} = \frac{60}{-20} = -3$$

As was pointed out in Chapter 6, the application of Kirchhoff's laws to some circuits results in a number of simultaneous equations. In these equations the unknown currents, signified by subscripted Is, or the unknown voltages, signified by subscripted Vs, are the variables and replace the xs in the previous equations.

As was pointed out in Chapter 14, certain simultaneous circuit equations contain complex coefficients. The solution of these equations by determinants proceeds in the same manner as those with real coefficients.

EXAMPLE A-3

Solve the two equations

$$(10 + j0)x_1 + (-12 + j4)x_2 = \quad 20 + j30 = 36.05 \angle 56.3°$$
$$(-12 - j3)x_1 + \quad (10 + j0)x_2 = -50 + j0 \quad = 50 \angle 180°$$

Solution

$$D = \begin{vmatrix} 10 & -12 + j4 \\ -12 - j3 & 10 \end{vmatrix}$$
$$= 100 - (144 + j36 - j48 - j^2 12)$$
$$= 100 - 144 - j36 + j48 - 12 = -56 + j12$$

$$D_1 = \begin{vmatrix} 20 + j30 & -12 + j4 \\ -50 & 10 \end{vmatrix}$$
$$= 200 + j300 - 600 + j200$$
$$= -400 + j500$$

$$D_2 = \begin{vmatrix} 10 & 20 + j30 \\ -12 - j3 & -50 \end{vmatrix}$$
$$= -500 - (-240 - j60 - j360 - j^2 90)$$
$$= -500 + 240 + j60 + j360 + j^2 90 = -350 + j420$$

$$x_1 = \frac{D_1}{D} = \frac{-400 + j500}{-56 + j12} = \frac{100(-4 + j5)}{4(-14 + j3)}$$
$$= \frac{25(-4 + j5)(-14 - j3)}{205}$$
$$= 0.122(56 - j70 + j12 - j^2 15) = 0.122(71 - j58)$$
$$= 8.66 - j7.07 = 11.2 \angle -39.2°$$

$$x_2 = \frac{D_2}{D} = \frac{-350 + j420}{-56 + j12} = \frac{70(-5 + j6)}{4(-14 + j3)}$$

$$= \frac{70(-5 + j6)(-14 - j3)}{4(205)}$$

$$= 0.0855(70 - j84 + j15 - j^2 18) = 0.0855(88 - j69)$$

$$= 7.52 - j5.9 = 9.56 \angle -38.1°$$

The solution of simultaneous circuit equations is a tedious task, particularly if complex coefficients are present. Fortunately, the availability of personal computers or microcomputers allows one to write computer programs that eliminate some of the tedious work. In Section A.2 some simple computer programs are introduced as aids for solving electric circuit problems. Included are programs for the solution of simultaneous equations.

A.2 COMPUTER ANALYSIS

Although a computer can do very simple electric circuit calculations, such as finding the equivalent resistance of two parallel resistors, its great usefulness lies in its ability to relieve the analyst of more tedious tasks, such as solving simultaneous equations or converting phasors from rectangular to polar form. Furthermore, when properly instructed or programmed, the computer will perform these tasks without the human errors often introduced by the analyst.

The material that follows is a brief introduction to the use of the computer in electric-circuit problem solving. The accompanying programs are written in BASIC (Beginner's All-Purpose Symbolic Instruction Code), a language available with most personal computers.

One communicates with a computer in the BASIC language with statements that resemble mathematical equations. First, one is concerned with either *integer numbers*—that is, whole numbers without a decimal point—or *floating point numbers*—that is, numbers with a decimal point. Names are assigned to the various constants and variables in a problem, names that consist of one alphabetic character usually followed by an alphameric (alphabetic or numeric) character.

The various equations are formed by using the mathematical operations shown

TABLE A-1 BASIC MATHEMATICAL OPERATORS

Operator	Meaning
↑	Exponentiation (to the power)
*	Multiplication
/	Division
+	Addition
−	Subtraction
=	Replace the value of the quantity to the left of the equality sign with that from the right

TABLE A-2 BASIC RELATIONAL OPERATORS

Operator	Meaning
=	Equality
< >	Inequality
<	Less than
>	Greater than
< =	Less than or equal to
> =	Greater than or equal to

in Table A-1. The hierarchy, or *order of priority*, of mathematical operations is exponentiation, then multiplication or division, and then addition or subtraction. However, when parentheses are present, the operations within the parentheses are performed first. In addition, the BASIC language allows the use of the relational operators specified in Table A-2. These relational operators allow the programmer to test whether a statement is true or false. Based upon the true or false condition, the computer makes a decision to branch or not to branch to a different part of the program.

Although there are a few different versions or dialects of the BASIC language, most of them provide built-in mathematical functions such as those defined in Table A-3. Many of the built-in functions of Table A-3 are used in the programs given in this appendix.

Before using the following programs, the student should compare the operators and built-in functions in the given programs to those available on his or her microcomputer. By referring to the instruction manual for a particular computer, one can easily make any necessary changes, which should be minor, to the given programs.

Two Simultaneous Equations with Real Coefficients

This program allows one to solve two simultaneous equations. The program follows the determinant format of Appendix 1.

TABLE A-3 TYPICAL BUILT-IN BASIC FUNCTIONS

Function	Action
ABS(X)	Calculates the absolute value of the number X
ATN(X)	Calculates the angle, measured in radians, whose tangent is X
COS(X)	Calculates the value of the cosine of X, where X is an angle measured in radians
EXP(X)	Calculates the value of the constant e ($=2.718$) raised to the X power
INT(X)	Converts by truncation the number X to an integer
LOG(X)	Returns the natural logarithm of X
SIN(X)	Calculates the sine of an angle X, where X is in radians
SPC(X)	Prints X number of spaces
SQR(X)	Calculates the square root of a positive number X
TAB(X)	Moves the cursor or print head to position X
TAN(X)	Calculates the tangent of an angle X, where X is in radians

```
10   PRINT "PROG. FOR SOLUTION OF TWO"
20   PRINT "SIMULTANEOUS EQS. OF THE FORM"
30   PRINT "A1*X1 + A2*X2 = B1"
40   PRINT "A3*X1 + A4*X2 = B2"
45   PRINT "INPUT A1,A2,B1,A3,A4,B2"
50   INPUT A1,A2,B1,A3,A4,B2
60   DET = A1*A4-A2*A3
70   X1 = (A4*B1-A2*B2)/DET
80   X2 = (A1*B2-A3*B1)/DET
90   PRINT A1;A2;B1;A3;A4;B2
100  PRINT "VALUE OF X1 EQUALS"; X1
110  PRINT "VALUE OF X2 EQUALS"; X2
120  END
```

When the constants for Example A-1 are substituted into the program, the screen monitor or printout is

```
1  3  20  2  -2  -5
VALUE OF X1 EQUALS 3.125
VALUE OF X2 EQUALS 5.625
```

N Simultaneous Equations with Real Coefficients

This program allows one to solve N simultaneous equations. Unlike the program for the solution of two simultaneous equations, this program does not use determinants but a procedure called Gaussian elimination. In the program, the coefficients are entered as terms of a matrix. Hence, the coefficients are identified by a row number (J) and a column number (K), corresponding, respectively, to an equation number and a variable subscript.

```
10   PRINT "PROG FOR SOLVING N SIMULTANEOUS EQUATIONS"
12   PRINT "THE EQS ARE IN THE FORM"
14   PRINT "A(1,1)X1 +A(1,2)X2 +... = B1"
16   PRINT "A(2,1)X1 +A(2,2)X2 +... = B2"
18   PRINT "            ETC           "
20   INPUT "NUMBER OF EQS.";N
30   DIM A(N,N), B(N), X(N)
40   FOR J=1 TO N
50   FOR K=1 TO N
60   PRINT "ENTER COEFFICIENT OF ROW";J; "COLUMN"; K
70   INPUT A(J,K)
80   NEXT K
90   PRINT "ENTER CONSTANT B";J
100  INPUT B(J)
110  NEXT J
```

```
120  FOR K=1 TO N-1
130  I=K+1
140  L=K
150  IF ABS(A(I,K))>ABS(A(L,K)) THEN L=1
160  IF I<N THEN I=I+1:GOTO 150
170  IF L=K THEN 220
180  FOR J=K TO N:Q=A(K,J):A(K,J)=A(L,J)
190  A(L,J)=Q
200  NEXT
210  Q=B(K):B(K)=B(L):B(L)=Q
220  I=K+1
230  Q=A(I,K)/A(K,K):A(I,K)=0
240  FOR J=K+1 TO N:A(I,J)=A(I,J)-Q*A(K,J):NEXT
250  B(I)=B(I)-Q*B(K):IF I<N THEN I=I+1:GOTO 230
260  NEXT
270  X(N)=B(N)/A(N,N):FOR I=N-1 TO 1 STEP-1
280  Q=0:FOR J=I+1 TO N:Q=Q+A(I,J)*X(J)
290  X(I)=(B(I)-Q)/A(I,I):NEXT:NEXT
300  FOR I=1 TO N
310  PRINT "X";I;"=";X(I)
320  NEXT
330  END
```

If the program is used to solve the three simultaneous equations of Example A-2, one enters the values $A(1, 1) = 4$, $A(1, 2) = -3$, $A(1, 3) = 1$, $B(1) = -5$, etc. Upon running the program, one obtains the following data:

```
X  1=  1
X  2=  2
X  3=-3
```

Polar to Rectangular Conversion

This program takes the entered values of a magnitude M and angle A in degrees of a phasor in polar form and converts the phasor to rectangular form:

```
10   PRINT "PROG TO CONVERT POLAR TO RECT"
20   PRINT "M IS MAGNITUDE & A IS ANGLE"
30   PRINT "X IS REAL PART & Y IS IMAG PART"
40   INPUT "M EQUALS"; M
50   INPUT "A EQUALS"; A
60   A=A*3.1416/180
70   X=M*COS(A)
80   Y=M*SIN(A)
90   PRINT "X+JY=";X;"+J"; Y
100  END
```

Rectangular to Polar Conversion

This program takes the entered values of a phasor in rectangular form and converts them to the equivalent polar form. Steps within the program test the algebraic sign of the x and y terms, and from this test the computer determines the quadrant in which the phasor lies.

```
10   PRINT "PROG TO CONVERT RECT TO POLAR"
20   PRINT "X=REAL PART & Y=IMAG PART"
30   PRINT "M=MAGNITUDE & A=ANGLE"
40   INPUT "X EQUALS";X
50   INPUT "Y EQUALS";Y
60   M=SQR(X↑2+Y↑2)
70   IF X=0 GOTO 100
80   A=ATN(ABS(Y)/ABS(X))*180/3.1416
90   GOTO 110
100  A=90
110  IF X>=0 GOTO 170
120  IF Y>=0 GOTO 150
130  A=-180+A
140  GOTO 190
150  A=180-A
160  GOTO 190
170  IF Y>=0 GOTO 190
180  A=-A
190  PRINT "M=";M;"A=";A;"DEGREES"
200  END
```

Two Simultaneous Equations with Complex Coefficients

This program solves two simultaneous equations having complex coefficients. Although the program uses the determinant method of solution, many steps are required because the complex coefficients must be manipulated within the determinants. To aid in this manipulation, the polar to rectangular and rectangular to polar conversion programs are used as subroutines.

```
2    PRINT "PROG. TO SOLVE TWO SIM. EQS."
4    PRINT "WITH COMPLEX COEFF. OF FORM"
6    PRINT "A(1)*X1+A(2)*X2=B(1)"
8    PRINT "A(3)*X1+A(4)*X2=B(2)"
10   DIM RA(4),IA(4),MB(2),AB(2),MA(4),AA(4)
20   FOR I=1 TO 4
30   INPUT "REAL PART A(I),IMAG PART A(I)";RA(I),IA(I)
40   X=RA(I)
50   Y=IA(I)
60   GOSUB 660
70   MA(I)=M
80   AA(I)=A
```

```
90    NEXT I
100   FOR I=1 TO 2
110   INPUT "MAGNITUDE B(I),ANGLE B(I)";MB(I),AB(I)
120   NEXT I
130   M=MA(1)*MA(4)
140   A=AA(1)+AA(4)
150   GOSUB 800
160   RS=X
170   IS=Y
180   M=MA(2)*MA(3)
190   A=AA(2)+AA(3)
200   GOSUB 800
210   RT=X
220   IT=Y
230   X=RS-RT
240   Y=IS-IT
250   GOSUB 660
260   MD=M
270   AD=A
280   M=MB(1)*MA(4)
290   A=AB(1)+AA(4)
300   GOSUB 800
310   RS=X
320   IS=Y
330   M=MB(2)*MA(2)
340   A=AB(2)+AA(2)
350   GOSUB 800
360   RT=X
370   IT=Y
380   X=RS-RT
390   Y=IS-IT
400   GOSUB 660
410   M1=M
420   A1=A
430   M=MB(2)*MA(1)
440   A=AB(2)+AA(1)
450   GOSUB 800
460   RS=X
470   IS=Y
480   M=MB(1)*MA(3)
490   A=AB(1)+AA(3)
500   GOSUB 800
510   RT=X
520   IT=Y
530   X=RS-RT
540   Y=IS-IT
```

```
550  GOSUB  660
560  M2=M
570  A2=A
580  M=M1/MD
590  A=A1-AD
600  PRINT  "X1=";M;"AT  ANGLE  OF";A;"DEG"
610  M=M2/MD
620  A=A2-AD
630  PRINT  "X2=";M;"AT  ANGLE  OF";A;"DEG"
640  END
660  M=SQR(X↑2+Y↑2)
670  IF  X=0  GOTO  700
680  A=ATN(ABS(Y)/ABS(X))*180/3.1416
690  GOTO  710
700  A=90
710  IF  X>=0  GOTO  770
720  IF  Y>=0  GOTO  750
730  A=-180+A
740  GOTO  790
750  A=180-A
760  GOTO  790
770  IF  Y>0  GOTO  790
780  A=-A
790  RETURN
800  A=A*3.1416/180
810  X=M*COS(A)
820  Y=M*SIN(A)
830  RETURN
```

Because the ac circuit analysis problems in the text are specified with the impedances and admittances in rectangular form and the voltage and current sources in polar form, the program is written to prompt the analyst to specify the coefficients $A(I)$ in rectangular form and the sources $B(I)$ in polar form. The solutions appear in polar form.

For simplicity, the program given prints or displays only the answers. However, if one desires to check the constants that have been entered, a printing or display routine can be added prior to running the program:

```
593  FOR  I=1  TO  4
594  PRINT  "COEFFICIENT  A";I;"=";RA(I);"+J";IA(I)
595  NEXT  I
596  FOR  I=1  TO  2
597  PRINT  "CONSTANT  B";I;"=";MB(I);"AT  ANGLE  OF";
     AB(I);"DEG"
598  NEXT  I
```

As an example of the use of the program and the accompanying coefficient display routine, consider the two equations of Example A-3. When the constants are entered and the results displayed, the following display results:

```
COEFFICIENT A 1 =  10 +J  0
COEFFICIENT A 2 =-12 +J  4
COEFFICIENT A 3 =-12 +J-3
COEFFICIENT A 4 =  10 +J  0
CONSTANT B 1 = 36.05 AT ANGLE OF 56.3 DEG
CONSTANT B 2 = 50 AT ANGLE OF 180 DEG
X1 =  11.1789338 AT ANGLE OF -39.2416161 DEG
X2 = 9.54425053 AT ANGLE OF -38.097417 DEG
```

Nonsinusoidal Waveforms

As described in Chapter 16, nonsinusoidal waveforms consist of harmonically related sinusoids. The following program plots the waveform of a nonsinusoidal function composed of two harmonically related sine waves.

```
10    PRINT "PROG FOR SCROLL OF SUM OF"
20    PRINT "TWO HARMONIC WAVES OF FORM"
30    PRINT "A1*SIN(X) AND A2*SIN(N*X+B)"
40    INPUT "CONSTANTS A1,A2 ARE";A1,A2
50    INPUT "HARMONIC,N, IS";N
60    INPUT "PHASE ANGLE,B, IS (RAD)";B
66    OPEN3,4
68    CMD3
70    PRINT TAB(18) "-2.0" TAB(18) "0" TAB(18) "2.0"
80    PRINT TAB(40) "+"
90    FOR J=10 TO 85
100   IF J/10=INT(J/10) THEN PRINT "+";:GOTO 120
110   PRINT "-";
120   NEXT J
130   FOR X=0 TO 6.28 STEP .2
140   V1=A1*SIN(X)
150   V2=A2*SIN(N*X+B)
160   VT=(V1+V2)*10
170   PRINT TAB(VT+40) "*"
180   NEXT X
190   PRINT TAB(40) "+"
196   CLOSE3,4
198   CMD3
200   END
```

The program, as listed, scrolls the waveform vertically from an 80-column printer. Statements 66, 68, 196, and 198 are used to open and close the transmissions to the printer and are unique to the personal computer and printer used by the author. If these statements are left out, the program will scroll vertically on the monitor or

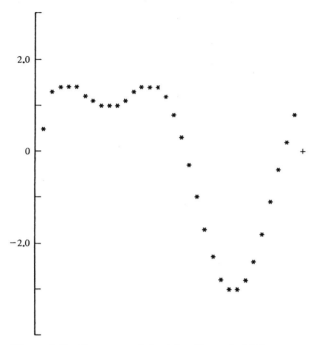

Figure A-1   Program printout for Example 16-2.

display. If an 80-column display format is not available, changes must also be made in the TAB expressions to limit the display width. Since scaling is provided by the multiplying factor of statement 160, that factor must also be changed for a column display less than 80 characters in width.

When the program is run with the constants $A1 = 2$, $A2 = 1$, $N = 2$, and $B = \pi/2 = 1.5708$ rad (from Example 16-2), the printout of Fig. A-1 is obtained.

By adding statements describing other harmonics, one can obtain a graphical representation of functions composed of three or more harmonically related sinusoids.

A.3 MATHEMATICAL SIGNS AND SYMBOLS; THE GREEK ALPHABET

Mathematical Signs and Symbols

$\pm$ ($\mp$)	plus or minus (minus or plus)
$<$	less than
$>$	greater than
$\leq$	less than or equal to

$\geq$	greater than or equal to
$\approx$	approximately equal to
$\propto$	proportional to
$\neq$	not equal to
∞	infinity
$\sqrt{}$	square root
$\angle$	angle
$\perp$	perpendicular or normal to
$\parallel$	parallel to
$\equiv$	equivalent to
B	(boldface) phasor or vector quantity
B	(italics) scalar quantity or magnitude
$\log$ or $\log_{10}$	common logarithm (base 10)
e	base (2.718) of the natural logarithm
$\sin$	sine
$\cos$	cosine
$\tan$	tangent
$\sin^{-1}$	angle whose sine is
$\cos^{-1}$	angle whose cosine is
$f(x)$	function of x
Δx	increment of x
$\sum$	summation of
dx	differential of x
$\dfrac{dy}{dx}$	derivative of y with respect to x
$\int$	integral of

Greek Alphabet

A α	alpha	I ι	iota	P ρ	rho
B β	beta	K κ	kappa	Σ σ	sigma
Γ γ	gamma	Λ λ	lambda	T τ	tau
Δ δ	delta	M μ	mu	Υ υ	upsilon
E ϵ	epsilon	N ν	nu	Φ ϕ	phi
Z ζ	zeta	Ξ ξ	xi	X χ	chi
H η	eta	O o	omicron	Ψ ψ	psi
Θ θ	theta	Π π	pi	Ω ω	omega

Glossary

absolute permeability (μ)   The magnetic specific conductivity per unit length and cross section of a material.

absolute permittivity (ϵ)   The ratio of flux density to electric-field intensity for a dielectric.

accuracy   The amount by which a measured value agrees or conforms to a standard or true value.

active network   A network containing voltage or current sources.

admittance (Y)   The reciprocal of impedance, measured in siemens.

admittance diagram   A right triangle relating conductance, susceptance, and admittance.

admittance parameters (y)   A set of two-port constants (measured in siemens) that relates the port currents to the port voltages.

alternating current (ac)   A current that periodically changes direction.

ammeter   The basic instrument used to measure electrical current in amperes.

ampere (A)   The SI unit of electric current. It equals 1 C of charge flowing past a point in 1 s.

antiresonance   A term that describes the parallel resonant condition.

apparent power   The product of voltage and current in an ac circuit.

autotransformer   A transformer having one continuous winding that is partially shared by both the primary and secondary circuits.

average value   The area under one cycle of a waveform divided by the period.

balance   The condition for which the detector or bridge current of a resistance or impedance bridge equals zero.

balanced delta   A delta configuration with equal impedances.

balanced system   A polyphase system having applied voltages of equal magnitudes and a load with equal phase impedances.

balanced wye   A wye configuration with equal impedances.

bandwidth   The spread of frequencies passed by a resonant circuit.

BASIC   A computer language whose name is an acronymn for Beginners All Purpose Symbolic Code.

battery   A collection of interconnected cells.

bilateral resistance   A resistance exhibiting the same voltage versus current relationship regardless of the current direction through the resistance.

bound electron   An electron tightly held in orbit about the nucleus of an atom.

branch   A single element or device connected between two nodes; often considered an element or string of elements connected between major nodes.

breakdown   The point at which substantial current begins to flow in an insulator.

capacitance (C)   The property of an electric circuit to oppose any change in voltage across the circuit.

capacitive reactance   The opposition that a capacitor offers to ac.

capacitor   A device specifically designed to have capacitance.

cell   A basic unit that converts chemical or solar energy into electrical energy.

cell capacity   The rating of a chemical cell that specifies for how long a time a certain level of current can be supplied.

charge   Electrical energy that is supplied to a cell in order to restore the active materials after a discharge.

circuit breaker   A manual switch that can open itself automatically under overload conditions.

circuit diagram   A drawing that shows schematically the interconnection of electric circuit components.

circular mil   The cross-sectional area of a wire having a 1 mil diameter.

coefficient of coupling   The ratio of mutual flux to primary flux in a coupled circuit.

combinational system   A polyphase system that contains a variety of delta, wye, and single-phase loads.

complex circuit   A circuit that contains two or more sources in different branches.

complex conjugate   The complex number formed when the sign of the imaginary part of a complex number is changed.

complex power   The product of phasor voltage and the conjugate of phasor current for an ac circuit.

conductance (G)   The reciprocal of resistance or a measure of the ease with which current flows in a material.

conductivity (σ)   The specific conductance or conductance per unit length and cross section of a material.

conductor   A material that allows appreciable current flow with only a small applied voltage.

conventional current   A current flow adopted by the IEEE in which current flow is defined in a direction opposite to electron flow.

conversion factor   An equality that relates a measurement in one measurement system to an equivalent in another system.

core loss   The combined hysteresis and eddy current loss for a magnetic circuit.

counter emf   An opposing emf generated in accordance with Lenz's law.

cycle   The smallest nonrepetitive portion of a periodic waveform.

decibel (dB)   A logarithmic unit used to relate the ratio of two powers.

delta network (Δ)   A network of elements arranged in the configuration of the Greek capital letter delta.

demagnetization curve   The second quadrant of a hysteresis loop.

dependent source   An electrical source whose electrical operation depends on a voltage or current in some other part of the circuit.

determinant   A square array of numbers representing the algebraic sum of a set of possible products.

dielectric constant   *See* relative permittivity.

dielectric strength   The voltage per unit thickness at which breakdown occurs.

direct current (dc)   An unchanging, unidirectional current.

discharge   To withdraw electrical energy from a cell.

discharge curve   A graph of terminal voltage versus time for a chemical cell or battery.

DMM   The abbreviation used to denote the digital multimeter.

domain   A large group of neighboring atoms (approximately 10^{15}) that have their magnetic spins aligned.

dot notation   Marks used to identify the terminals of coils having the same induced polarities.

double subscripted voltage   A notation in which the first subscript designates the point at which a voltage is measured with respect to a reference point, designated by a second subscript.

duty cycle   The ratio of "on time" to total period for repeating pulses.

DVM   The abbreviation for digital voltmeter.

eddy current   A circulating current due to induced emf within a magnetic material.

effective resistance   The total of all resistive effects for an ac circuit including ohmic resistance and core and radiation losses.

effective value   A value for a periodic waveform relating its heating effect to a constant or dc value.

efficiency (η)   The ratio of output energy to input energy.

electric charge   A term used to imply electrification or the presence of electrical potential energy.

electric circuit   A network that contains at least one closed path.

electric current (I)   The net flow of charge past a given point per unit time.

electric field   The region in which a charge exerts a force of attraction or repulsion on another charge.

electric field intensity ($\mathscr{E}$)   The electric force per unit charge at a particular point in space.

electric potential   The work required to move a unit charge in an electric field.

electromotive force (emf)   A potential difference generated by the conversion of other forms of energy into electrical energy.

electron   A negative polarity particle that orbits the nucleus of an atom.

engineering notation   A form of scientific notation in which a number is shown with a power of 10 that is divisible by 3 so that the number can be directly related to an SI prefix.

error   The difference between the measured and true values.

even function   A waveform that possesses symmetry about the vertical axis and is defined by $f(t) = f(-t)$.

farad (F)   The unit of capacitance.

Faraday's law   A mathematical statement relating an induced emf to the rate of change of magnetic flux.

filter   A circuit designed to separate different frequencies.

final conditions   The voltage and current values at the end of a transient period.

flux density   The number of flux lines per unit cross-sectional area.

free electron   An electron that is out of its valence state and free to move into the outer atomic shell of a neighboring atom.

frequency (f)   The number of cycles per unit time.

fringing   The leakage of flux lines outside the area designated for an electrical or magnetic field.

fundamental frequency   The lowest frequency component in the harmonic or Fourier expansion of a nonsinusoidal quantity.

fuse   A two-terminal device consisting of a wire or strip of fusible metal that melts when the current through it is excessive.

generalized Kirchhoff analysis   A circuit analysis method in which both KVL and KCL equations are written in order to solve for the branch currents of the circuit.

half-power frequency The frequency at which the voltage or current response for a frequency selective network is 0.707 of the maximum value.

harmonic An integer multiple of the fundamental frequency component of a nonsinusoidal quantity.

henry (H) The unit of inductance.

hertz (Hz) The SI unit for frequency.

hole An electron vacancy in an atomic structure.

hybrid parameters (h) A set of two-port constants with mixed units (ohms, siemens, unitless).

hysteresis loop The loop formed when a complete magnetization cycle is plotted as a B–H curve.

ideal transformer A transformer in which the winding resistances are negligible, the coupling coefficient is 1, the core losses are negligible, and a negligible amount of excitation current is needed.

impedance (Z) The total opposition, representing a combination of resistance and reactance, to the flow of ac.

impedance bridge An instrument using a bridge configuration for the determination of resistance, capacitance, and inductance.

impedance diagram A right triangle relating resistance, reactance, and impedance.

impedance parameters (z) A set of two-port constants with units of ohms that relate the port voltages to port currents.

independent source An electrical source whose electrical operation depends only on its own electrical characteristics.

inductance (L) The property of a circuit or coil to oppose a change in current through the circuit or coil.

inductive reactance The opposition that an inductor offers to ac.

inductor (coil) A device designed to have inductance.

inferred zero-resistance temperature The temperature at which the extrapolation of a resistance versus temperature curve intersects the temperature axis.

initial conditions The voltage and current values at the start of a transient period.

input impedance The ratio of voltage to current at a pair of network terminals considered to be an input.

insulator A material that allows very little current flow with the application of a voltage.

internal impedance The inherent series impedance within an ac source.

internal resistance The inherent internal resistance of an electrical energy source.

j operator A mathematical operator that rotates a phasor by 90° counterclockwise.

Kirchhoff's current law (KCL) A law that states that the algebraic sum of the currents entering and leaving a node is zero.

Kirchhoff's voltage law (KVL) A law that states that the algebraic sum of the voltage drops and rises around a closed loop is zero.

lagging A term used to describe a waveform that is delayed relative to another.

leading A term used to describe a waveform that is advanced relative to another.

leakage flux The flux that passes outside an area considered for the passage of flux.

leakage resistance The relatively high resistance representing the dielectric in a capacitor.

Lenz's law A statement that relates the direction of an induced emf to the changing flux creating that emf.

linear resistance A resistance that has a linear voltage versus current relationship.

line current The current that flows in a line connecting a single-phase source to a load or a polyphase source to a polyphase load.

line of force (flux line) The path along which a test charge moves in an electric field or a test magnetic pole moves in a magnetic field.

line voltage The voltage between two lines of either a single-phase or polyphase circuit.

load line The voltage–current characteristic of a device or group of devices used as a load.

loop Any closed path in an electric circuit.

loop analysis A circuit analysis method in which KVL equations are written in order to solve for loop currents of the circuit.

magnetic circuit The physical interpretation for a source of mmf connected to a magnetic path.

magnetic field A region in space where a force acts upon a magnetic body.

magnetic field intensity (*H*) The force per unit magnetic pole.

magnetic flux density (*B*) The magnetic flux per unit area.

magnetism A property associated with materials that attract iron and iron alloys.

magnetization curve A graph of flux density versus magnetic-field intensity for a magnetic material (*B–H* curve).

magnetomotive force (mmf) The force that causes magnetic flux to be established.

major node A connection point for three or more electrical elements.

maximum-power transfer theorem A theorem used in determining the load resistance or impedance that when connected to a two-terminal network will receive maximum power from the network.

measured value The magnitude of a measured quantity as indicated at the output of an instrument.

mesh A closed loop containing no other closed loop.

mutual inductance The property relating the changing of flux in one coil to an emf generated in a second coil.

network A connection of two or more circuit elements.

nodal analysis A circuit analysis method in which KCL equations are written in order to solve for node voltages of the circuit.

node A connection point between two or more electrical elements.

nonlinear resistance A resistance that has a voltage versus current relationship that is not linear.

Norton's theorem A theorem used in reducing a two-terminal network to an equivalent network consisting of a single current source and a parallel resistance or impedance.

odd function A waveform that possesses the mathematical relationship $f(t) = -f(-t)$.

ohm (Ω) The unit of measurement applied to resistance and impedance.

ohmmeter An instrument having a self-contained source and used for measuring resistance.

one-line equivalent circuit The single-phase circuit that represents one of the phases of a balanced three-phase circuit.

open circuit An electrical path that is not closed.

open-circuit impedance parameters The set of z parameters.

open-circuit test A laboratory test used to determine the core losses and excitation constants of a transformer.

open-circuit voltage The voltage between two terminals of a network when no current flows from the terminals.

operating point The coordinates of the point at which the source line and load line form a graphical solution.

oscilloscope An electronic instrument that uses a cathode-ray tube to display the waveform of a varying voltage or current.

parallel circuit A circuit with two common points so that the same voltage is across all elements.

passive network A network containing no sources.

period (*T*) The time interval between successive cycles of a repetitive waveform.

periodic waveform A waveform that repeats itself after given time intervals.

permeability (μ) A measure of the ease with which magnetic flux is established in a material. It is also called absolute permeability.

permeance ($\mathscr{P}$) The unit of magnetic conductance.

permittivity (ϵ) A measure of the ease with which electric flux is established in a dielectric. It is also called absolute permittivity.

phase angle The angular displacement of a sinusoid or vector quantity.

phase current The current that flows in a single-phase circuit or in one phase of a polyphase circuit.

phase difference The fractional part of a period by which two sinusoidal waveforms are separated.

phase sequence The order in which polyphase voltages pass through their respective maximum values.

phase voltage The voltage across one phase of a polyphase circuit.

phasor A rotating vector used to represent a sinusoidal quantity.

phasor diagram A two-dimensional plot showing the phase and magnitude relationships of phasors.

photoconductive cell A transducer used to convert changes in light intensity into resistance changes.

piezoelectric effect The ability of certain crystalline substances to transform mechanical strain into electric charge and vice versa.

pi network (π) A network of elements that are arranged in the configuration of the Greek letter pi (also called a delta network).

polar form The definition of a phasor by a magnitude and the angle it makes with a reference axis.

polygon method A graphical technique for adding phasors.

port A pair of network terminals.

potential difference The difference in electrical potential between two points in an electric field or electric circuit.

potentiometer A variable resistor with three terminals one of which connects to a movable contact and the remaining two to the resistor ends.

power The rate of doing work.

power factor The cosine of the angle between the voltage and current in an ac circuit.

power factor correction The addition of a capacitor to an inductive circuit in order to reduce the apparent power.

power supply A device that converts one type of electrical potential or current into another type.

power triangle The right triangle relating apparent power, real power, and reactive power.

precision The smallest unit used in specifying a measurement or a number.

primary cell A cell that is not designed to be charged.

primary winding The winding or coil of a transformer, or two coupled coils, to which electrical energy is applied.

pulse repetition rate The frequency of repetitive pulses.

quality factor (Q) A figure of merit for a coil or resonant circuit.

radian The SI unit of angular measure, approximately equal to 57.3°.

reactive power The power stored in a capacitive or inductive element.

reactive volt-ampere The unit of reactive power.

rectangular form The definition of a phasor by its projections on a coordinate set of axes.

reflected impedance The impedance appearing at the primary of a transformer as a direct result of an impedance across the secondary.

relative permeability (μ_r)   The ratio of the absolute permeability of a material to that for a vacuum.

relative permittivity (ϵ_r)   The ratio of absolute permittivity of a material to that for a vacuum.

reluctance ($\mathscr{R}$)   The opposition to the establishment of magnetic flux.

resistance (R)   A measure of the opposition to the flow of electric current.

resistivity (ρ)   The specific resistance or resistance per unit length and cross-sectional area of a material.

resistor   A device specifically designed to have resistance.

resonance   A condition for an ac circuit containing resistance and reactance in which the current and voltage are in phase.

right-hand rule   A rule in which the right hand relates the directions of flux and current for a current-carrying conductor or coil.

root-mean-square value   Effective value.

scalar   A quantity that has only a magnitude.

scientific notation   A format for writing very large or very small numbers as a number between 1 and 10 × an integral power of 10.

secondary cell   A cell that is designed to be charged after discharge.

secondary winding   The winding or coil of a transformer, or two coupled coils, to which electrical energy is applied.

semiconductor   A material with a conductance property between that of a conductor and an insulator.

sensitivity   A measure, in ohms per volt, of the loading effect produced by the connection of a voltmeter.

series circuit   A circuit having only one path for the current so that all elements have the same current.

short-circuit   An electrical network that is closed by a resistance of $0\ \Omega$.

short-circuit admittance parameters   The set of y parameters.

short-circuit current   The current flowing in a zero-resistance load connected between two terminals of a network.

short-circuit test   A laboratory test used for determining the transformer constants.

siemen (S)   The unit of measurement applied by conductance and admittance.

significant digit   Any digit that is necessary to define a value or quantity.

single-phase circuit   An electric circuit energized by a single ac source.

SI units   A system of measurement units proposed for international use in 1960 and adopted by the IEEE in 1965.

solar cell   A device designed to directly convert sunlight into electrical energy.

source line   The voltage–current characteristic of an active device used as an electrical energy source.

step-down transformer   A transformer whose secondary voltage is less than its primary voltage.

step-up transformer   A transformer whose secondary voltage is greater than the primary voltage.

strain gauge (gage)   A transducer used to convert linear-dimension changes into resistive changes.

stray capacitance   Capacitance that exists not by design but simply because two charged surfaces are close to each other.

superposition theorem   An analysis theorem that allows the effects of each independent source to be superimposed.

susceptance (B)   The imaginary part of the admittance.

switch   A device designed to make, break, or otherwise change electrical connections.

symmetrical network   A network of elements that can be divided into two symmetrical halves.

tee network (T)   A network of elements that are arranged in the configuration of the letter T (also called a Y network).

temperature coefficient of resistance (α)   The per unit change in resistance per change in temperature referred to a particular temperature.

thermistor   A two-terminal device designed to exhibit a change in resistance with a change in its body temperature.

Thevenin's theorem   A theorem used in reducing a two-terminal network to an equivalent network consisting of a single voltage source and a series resistance or impedance.

three-phase circuit   An electric circuit energized by three single-phase sources each out of phase with the other.

three-wattmeter method   A three-phase power measurement method utilizing three wattmeters.

time constant (τ)   The time required for an increasing exponential quantity to reach 63.2% of its final value or a decreasing exponential quantity to fall to 36.8% of its initial value.

transducer   A device that responds to one physical effect by producing another physical effect.

transformation ratio   The ratio of primary to secondary turns of a transformer.

transformer   The network formed by two magnetically coupled coils.

transformer bank   A group of similar single transformers connected in delta or wye configurations for three-phase operation.

transformer constants   The resistances and reactances used to describe electrically an equivalent circuit for a transformer.

transformer polarity   The relative directions of the induced voltages with respect to the terminals of the transformer windings.

transient state   A temporary or unsettled state generally associated with the opening and closing of switches.

true value   The actual magnitude of a quantity being measured.

two-port   A network having two distinct pairs of terminals.

two-wattmeter method   A three-phase power measurement method utilizing two wattmeters.

unbalanced system   A polyphase system having applied voltages that may not be equal in magnitude or a load with unequal phase impedances.

unilateral resistance   A resistance exhibiting a markedly different voltage versus current relationship with the reversal of current through the resistance.

vector quantity   A quantity that has both a magnitude and direction.

volt (V)   The SI unit of potential difference. It equals 1 J of energy in the movement of a 1 C charge.

voltage divider   A series combination of resistors chosen so that a single voltage source can supply one or more reduced voltages.

voltage regulation   The voltage variation of a source in going from no load to full load as a ratio of full-load voltage.

volt-ampere (VA)   The unit for apparent power.

voltmeter   A basic instrument used to measure electrical potential in volts.

VOM   The abbreviation for a multiple function instrument known as a volt-ohm-milliammeter.

watt-hour meter   An instrument for measuring energy.

wattmeter   An instrument that directly measures power.

waveform   A graph or plot of voltage or current, generally as a function of time.

Wheatstone bridge   An instrument that uses a bridge configuration to establish a null condition used in determining an unknown resistance.

wye (Y) network   A network of elements that are arranged in the configuration of the letter Y.

Answers to Odd-Numbered Problems

CHAPTER 1

1. **(a)** 75 000 μs.   **(b)** 75 000 000 ns.   **(c)** 0.0387 m.
     (d) 1.61×10^{-2} m².   **(e)** 1.45×10^{-2} m.   **(f)** 1.27 hp.
     (g) 24.58 m/s.

3. **(a)** 1.05 kΩ; 1.05×10^{-3} MΩ.   **(b)** 7.252 kW; 7.252×10^{-3} MW.
     (c) 75 km; 0.075 Mm.   **(d)** 7252 kW; 7.252 MW.

5. 1 psi = 6919 Pa.

7. **(a)** 2×10^9.   **(b)** 450×10^3.   **(c)** 100×10^0.   **(d)** 13.7×10^{-6}.
     (e) 543×10^{-15}.   **(f)** -8×10^{-6}.   **(g)** 27×10^6.   **(h)** 62.5×10^3.

9. 1.62×10^3 N.

11. 2.07 μC; 4.14 μC.

CHAPTER 2

1. 18.1 A.

3. 6.4×10^3 A.

5. 8.75 ms.

7. 10 C.

9. 1.67 V.

11. ≈ 4.25 years.

13. 60 Ah.

15. $\frac{1}{20}$ C.

17. **(a)** 33.6 Ah.   **(b)** 0.336 h.

19. 0.15 mm.

CHAPTER 3

1. $5.73 \times 10^{-4} \ \Omega$.

3. $2.26 \times 10^{-4} \ \Omega$.

5. **(a)** 625 cir mil.   **(b)** 1.27×10^{6} cir mil.
   (c) 7.4×10^{4} cir mil.   **(d)** 1.3×10^{4} cir mil.

7. 1.69 Ω; 2.25 Ω.

9. 10.5 $\Omega \cdot$cir mil/ft.

11. 13.1 Ω.

13. Use AWG 12.

15. **(a)** AWG 1/0.   **(b)** 0.517 Ω.

17. 24.1 Ω.

19. 0.00393 $\Omega/°C \cdot \Omega$; 0.00339 $\Omega/°C \cdot \Omega$.

21. 48.6°C.

23. **(a)** White, brown, orange, silver.   **(b)** Blue, red, brown.
    (c) Red, red, red, gold.   **(d)** Brown, gray, silver, gold.
    (e) Green, blue, silver, silver.   **(f)** Orange, white, yellow.

25. 250 mil.

27. 10 kΩ; 2.9 kΩ; 1 kΩ.

29. **(a)** 0.0021 S.   **(b)** 0.192 μS.   **(c)** 4 S.

31. 10^{6} S/m.

33. 105%; silver conducts more readily than copper.

CHAPTER 4

1. 146 V.

3. 1 MΩ.

5. 8.64 V.

7. 24–10 mA.

9. **(a)** 700 Ω.   **(b)** 80 Ω.

11. 1 Ω; 1 S.

13. 0.8 kWh.

15. 36.7 V.

17. 1.44 W; 50 Ω.

19. 0.577 mA.

21. 19.98 W.

23. 4.05×10^{6} J.

25. 3.24¢.

27. $144.10.
29. 17.2 hp.
31. 1.94 A.
33. 75%.
35. 246 W.

CHAPTER 5

1. **(a)** 7; 4. **(b)** 8; 5.
3. **(a)** 6 V. **(b)** -2 V.
5. **(a)** 7 A. **(b)** -5 A.
   (c) $I_1 = -18$ A; $I_2 = -7$ A. **(d)** $I_1 = -3$ A; $I_2 = 9$ A.
7. **(a)** $E = 21$ V; $V_3 = 4$ V. **(b)** $V_1 = 1.6$ V; $P_3 = 0.96$ W.
9. 0.218 mA; 2.62 mW; 4.36 V; 7.63 V.
11. 0.4 A; 37.5 Ω.
13. 14.98 kΩ.
15. 114.8 V.
17. 257 ft.
19. 100 A; 50 A; 20 A; 10 A; 5 A; 185 A; 0.54 Ω.
21. 600 kΩ.
23. 115 Ω; 1.04 A.
25. **(a)** 0.91 A; 10 A; 2.18 A. **(b)** 8.4 Ω. **(c)** 13.09 A.
27. 5.26 Ω.
29. **(a)** 22 Ω. **(b)** 3.23 Ω.
31. 1 V; 4 V.
33. 48.6 V.
35. 1.37 W.
37. 25 V.
39. 83.5%.
41. 9.2 Ω.
43. **(a)** 40 A; 3 Ω. **(b)** 250 V; 50 kΩ.
45. **(a)** 1 V; 500 Ω. **(b)** 14.4 V; 1.2 Ω. **(c)** 40 V; 8 Ω.
47. 3.69 V.
49. 12 Ω, 3 W; 4 Ω, 9 W; 2.67 Ω, 54 W.
51. 8.57 kΩ; 526 Ω; 1.11 kΩ; all less than 0.1 W.

CHAPTER 6

1. 4; 6.
3. 4 V.

   5. 6 A; 4 A; 2 A.

   7. 0.398 A; 0.289 A; 0.108 A.

   9. 0.565 A; 0.0088 A.

   11. 0.692 A.

   13. 0.333 A.

   15. 0.565 A; 0.0091 A.

   17. 5.33 A; 106.7 V.

   19. 1.333 V; 10 V; 2.993 V.

   21. 1.6 A.

   23. 3.92 A; 4.86 A; 5.23 A; 4.31 A.

   25. -6.06 V.

   27. 11.6–6.1 A.

   29. 7.33 V; 1933 Ω.

   31. **(a)** 12.6 V; 0.315 Ω.   **(b)** 40 A; 0.315 Ω.

   33. 3.79 mA; 1933 Ω.

   35. **(a)** 3.   **(b)** 2; 1.86 A.

   37. 12.12 kΩ; 0.876 W.

   39. **(a)** 16.7 V.   **(b)** 2.63 V.

   41. -13.6 V; -6.82 mA.

   43. **(a)** 9.5 kΩ; 19 kΩ; 7.6 kΩ.   **(b)** 50 Ω; 50 Ω; 50 Ω.

   45. **(a)** 4.59 Ω.   **(b)** 15 Ω.

CHAPTER 7

   1. 0.144 N.

   3. 72×10^3 N/C.

   5. **(a)** 4×10^4 N/C.   **(b)** 0.04 N.

   7. 55.7×10^{-12} F/m.

   9. 6 mC.

   11. **(a)** 0.354 pF.   **(b)** 1.77 pF.

   13. 4.

   15. 60.3 pF.

   17. 0.00667 μF; 66.7 V; 33.3 V.

   19. 100 μA.

   21. E/RC; if the capacitor voltage were to increase at the initial rate, it would equal the supply voltage E after one time constant.

   23. **(a)** 1.32 mA.   **(b)** 0.7 mA.

   25. 92.1 mS.

   27. **(a)** 6.97 μA.   **(b)** 10.1 V.

   29. 36.4 mS.

·**31.** **(a)** 81.1 V.   **(b)** 3.15 μA; 9.45 μA.

33. 1 mA; 20 V.

35. 40 V; 80 V.

37. 26.67 V; 7.33 kΩ; 3.64 mA.

39. 0.131 μF/mi.

41. 0.194 J.

CHAPTER 8

1. 1.01 N.

3. 0.9 T; 9000 G.

5. **(a)** 2000 At.   **(b)** 200 lines.   **(c)** 5.16 $\times$ 10^4 lines/in^2.

7. **(a)** Into *a.*   **(b)** Out of *a.*

9. 3.15 N.

11. 900 At/m.

13. **(a)** 375 At.   **(b)** 3125 At/m.

15. 2.65 $\times$ 10^6 At/Wb.

17. 7.54 $\times$ 10^{-4} Wb.

19. 191; 820; 1114.

21. 350 At.

23. 6.52 $\times$ 10^{-5} Wb.

25. 10 A.

27. 208 At.

29. 256 At.

31. 0.781 W.

33. **(a)** -3.77×10^{-5}.   **(b)** 1.2 $\times$ 10^{-4} Wb.

35. 0.375 T.

37. 4131 At.

CHAPTER 9

1. 7.5 V.

3. 32*t.*

5. 0.12 V.

7. 9.6 V.

9. 222 mH.

11. 66.6 mH.

13. **(a)** 48 mH.   **(b)** 0.75.   **(c)** 18 mH.

15. **(a)** 60.2 H.   **(b)** 35 H.

17. 7.8 H.

19. 5.71 A.

21. 0.026 s.

23. (a) 8.65 A.   (b) 2.71 V.

25. (a) i $= 4 + 6e^{-t/2}$ A, where t is in milliseconds.
    (b) 4.81 A.

27. (a) 1 A; 2 A.   (b) 20 V; 0 V.

29. 10.5 mA; -0.1 V.

31. 80.6 J.

CHAPTER 10

1. Construction

3. (a) 4.83 A.   (b) -4.33 A.

5. 75 mA.

7. 500 Hz, 2 ms; 1 kHz, 1 ms.

9. 50.9 mA.

11. (a) 5 μs; 200 kHz.   (b) 0.03 s; 33.3 Hz.
    (c) 70 ps; 14.3 GHz.   (d) 0.2 s; 5 Hz.

13. 12 μs; 83.3 kHz; 50%.

15. (a) 3 V.   (b) 500 kHz.
    (c) $v = -0.5 \sin(3.14 \times 10^6 \, t)$ V.

17. 1.98 V.

19. (a) 0.857 A; 1.69 A.   (b) 1 mA; 2.24 mA.

21. 20.3 A.

23. 60 V; 6 A.

25. 2 A; 2.828 A.

27. 13.3 Ω; 0.796 Ω.

29. 497 Hz.

31. 163 var.

33. 528 Ω; 8.8 kΩ.

35. (a) 10 kΩ.   (b) 1.59 H.   (c) 0 W.

37. (a) 100 V.   (b) 0.0265 H.

39. 2.05 MHz.

CHAPTER 11

1. (a) 84.8 $\angle\, 30°$ V.   (b) 35.4 $\angle\, 100°$ V.
    (c) 1.414 $\angle -78°$ mA.   (d) 7.07 $\angle\, 25°$ A.

3. **(a)** $25 \angle -138°$ A.   **(b)** $80 \angle 160°$ mA.   **(c)** $70.7 \angle -90°$ V.   **(d)** $35.4 \angle -60°$ V.

5. **(a)** $26.8 - j17.4$.   **(b)** $0.849 + j0.849$.   **(c)** $-11 + j19.1$.   **(d)** $-15.4 + j2.71$.
   (e) $-20 - j20$.   **(f)** $0.171 + j0.47$.   **(g)** $-69.3 - j40$.   **(h)** $99.6 + j8.72$.

7. **(a)** $1.41 \angle -135°$.   **(b)** $0.73 \angle -15.9°$.   **(c)** $11.3 \angle 45°$.   **(d)** $0.985 \angle 24°$.
   (e) $3.6 \angle -146°$.   **(f)** $156 \angle 33.7°$.   **(g)** $1.2 \angle 60°$.

9. **(a)** $100 \angle 126.9°$ V.   **(b)** $4.8 \angle 43.7°$ V.

11. **(a)** $60 \angle -25°$.   **(b)** $35 \angle -15°$.   **(c)** $220 \angle 165°$.   **(d)** $5 \angle 20°$.
   (e) $-j2$.   **(f)** $12 - j1$.   **(g)** $24 - j7$.   **(h)** $-0.5 - j1.4$.

13. 25.

15. **(a)** $0.056 - j0.09$.   **(b)** $0.08 - j0.04$.

17. **(a)** $63.6 \angle 45°$.   **(b)** $2.63 \angle -63.4°$.

CHAPTER 12

1. **(a)** $40 \angle 25°$ Ω.   **(b)** $0.06 \angle 30°$ A.
   (c) $i = 10 \sin(\omega t - 10°)$ mA.

3. **(a)** $v_1 = 147.5 \sin(\omega t + 66°)$ V.   **(b)** $i_3 = 25.2 \sin(\omega t + 41.8°)$ mA.

5. $2.78 \angle -72.3°$ A.

7. 60.5 V; 71.5 V.

9. 9.55 Ω; 48.9 μF.

11. **(a)** $10.6 \angle 58°$ V; $16.96 \angle -32°$ V.   **(b)** 22.5 W.

13. $V_R = 49.1$ V; $V_L = 196$ V; $V_C = 295$ V.

15. 64.8 MHz.

17. **(a)** $3.72 \angle 7.15°$ Ω.   **(b)** $2.96 \angle -53°$ Ω.

19. $14.9 \angle -70.6°$ A

21. $0.894 \angle -63.4°$ A; $0.588 \angle 61.93°$ A; $8.15 \angle 88.5°$ V.

23. **(a)** $31.3 \angle 20.8°$ V.   **(b)** $44 \angle -36.9°$ V.

25. $20 \angle 30°$ Ω.

27. $2.78 \angle -13.7°$ mA; $5.56 \angle 76.3°$ V.

29. $7.07 \angle -45°$ A; $50 \angle -90°$ V.

31. **(a)** $0.141 \angle 45°$ S.   **(b)** 11.6 Ω.   **(c)** $50 \angle -25°$ mS.

33. 0.725 lagging.

35. 180 kW.

37. **(a)** 200 W.   **(b)** 125 var (capacitive).
   (c) 236 VA.   **(d)** 0.85 leading.

39. **(a)** 124 W.   **(b)** 28.2 var (inductive).
   (c) 127 VA.   **(d)** 0.975 lagging.

41. **(a)** $935 \angle 10.5°$ VA.   **(b)** 919 W.   **(c)** 171 var.

43. 640 kvar.

45. **(a)** 1253 kvar.   **(b)** 1693 kVA.

47. (a) $60 \, \Omega$. **(b)** 0.1 W.
49. $7.1 \, \Omega$.

CHAPTER 13

1. 31.8 Hz.

3. 79.6 Hz; decrease f_c by increasing R or C; 0.00565 μF.

5. 57 dB.

7. 1 MHz.

9. (a) 3.27 pF. **(b)** 2 μA.

11. 1106.

13. (a) 50 MHz. **(b)** 3.82 MHz.
(c) 48.09 MHz; 51.91 MHz.

15. (a) 1 kHz. **(b)** 80 Hz. **(c)** 2 Hz. **(d)** 0.01 A; 0.4 A.

(e)

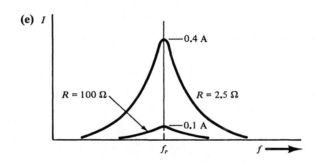

17. Yes; 1.59 MHz; 100.

19. (a) 1 MHz. **(b)** 6 mA.

21. (a) 64.9 Hz. **(b)** 42.9 mA.

23. (a) 7.9 Ω. **(b)** 12.64 kΩ.

25. (a) 5 kHz; 5 kHz; the filter passes 5 kHz with the response curve

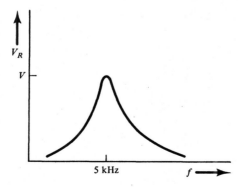

(b) 10 kHz; 10 kHz; the filter rejects 10 kHz with the response curve

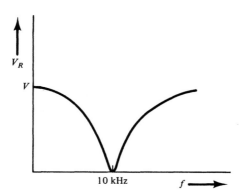

CHAPTER 14

1. (a) 5.66 $\angle$ 90° (measured b to a); 1.414 $\angle$ 45° kΩ.
   (b) 2 $\angle$ 53.1° A; 5 $\angle$ −53.1° Ω.

3. 4.47 $\angle$ −116.6° A; 10 $\angle$ 36.8° A.

5. 6.74 $\angle$ 43.2° A.

7. 70.7 $\angle$ 45° V.

9. (a) 15.9 $\angle$ 71.5° A. (b) 39.2 $\angle$ −61.4° A. (c) 1.65 $\angle$ −45.7° A.

11. 33.3 $\angle$ 6.9° V.

13. 18.9 $\angle$ 58.6° V.

15. 1.73 $\angle$ 108.4° A.

17. 4.47 $\angle$ −26.6° A; 3.61 $\angle$ 123.7° A; 2.24 $\angle$ 26.6° A.

19. (a) 7.81 $\angle$ 51.4° A; 8 $\angle$ 38.6° Ω.
    (b) 149 $\angle$ 29.7° mV; 2.48 $\angle$ −23.4° Ω.
    (c) 6.67 $\angle$ −90° A; 30 $\angle$ 90° Ω.

21. 16.6 $\angle$ 37.2° A.

23. 16.6 $\angle$ 33.7° V; 6.66 $\angle$ −56.3° Ω, 2.5 $\angle$ 90° A; 6.66 $\angle$ −56.3° Ω.

25. (a) 3.125 − j5 Ω.
    (b) 1290 W.

27. 11.25 nW

29. 1 Ω; 2 Ω; j2 Ω.

31. 0.053 $\angle$ −18.4 A; 0.053 $\angle$ −18.4° A; 0.105 $\angle$ −18.4° A.

33. 1.21 $\angle$ 1.7° A; 0.428 $\angle$ 9.8° A.

35. 19.6 mV: $\angle$ 191.3° phase shift.

37. 1.58 $\angle$ −18.43° Ω.

CHAPTER 15

1. (a) $208 \angle -150°$ V.   **(b)** $208 \angle 90°$ V.   **(c)** $208 \angle -30°$ V.

3.

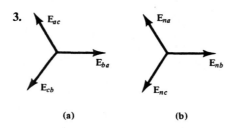

         (a)   (b)

5. 15.2 A.

7. $\mathbf{I}_{ab} = 10 \angle 0°$ A; $\mathbf{I}_{bc} = 10 \angle 120°$ A; $\mathbf{I}_{ca} = 10 \angle -120°$ A; $\mathbf{I}_a = 17.3 \angle 30°$ A; $\mathbf{I}_b =$ 17.3 $\angle 150°$ A; $\mathbf{I}_c = 17.3 \angle -90°$ A.

9. 6.64 A.

11. 41 A.

13. 7200 W; 9600 var.

15. 866 W; 1200 W; 1000 W; 3066 W.

17. 232 W; -50 W.

19. (a) 0.749.   **(b)** 18.65 kW; 57.55 kW.

21. $\mathbf{I}_a = 0$ A; $\mathbf{I}_b = 30 \angle -120°$ A; $\mathbf{I}_c = 30 \angle 60°$ A.

23. 2.4 A.

25. $\mathbf{I}_a = 1.65 \angle -49.1°$ A; $\mathbf{I}_b = 2.16 \angle 180°$ A; $\mathbf{I}_c = 1.65 \angle 49.1°$ A.

27. $65.2 \angle -162°$ A.

29. $\mathbf{I}_a = 13.1 \angle -18.5°$ A; $\mathbf{I}_b = 13.1 \angle -138.5°$ A; $\mathbf{I}_c = 13.1 \angle 101.5°$ A.

CHAPTER 16

1.

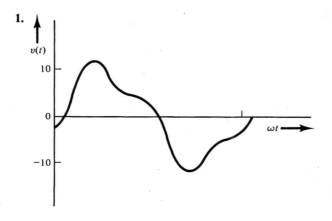

3.

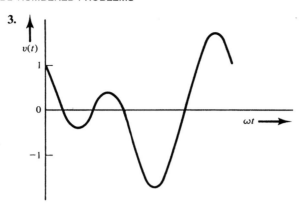

5. (a) 0.5 V; cosine only.
  (b) V/π; cosine only.
  (c) 6/π A; cosine only.
  (d) 0.833 A; sine and cosine.

7. (a) 101.3 V.
  (b) 684 W.

9. 5.67 V.

11. $i = 5 \sin(\omega t + 36.9°) + 2.89 \sin(3\omega t + 37.4°)$ A.

13. (a) $i = 54.4 \sin(377t - 36°) + 14.3 \sin(1131t + 11.8°)$ A.
  (b) $I = 39.8$ A.

CHAPTER 17

1. 2.86%.

3. 4.7–5.1 A.

5. 2×10^{-4} Ω.

7. $R_1 = 1.515$ Ω; $R_2 = 1.546$ Ω; $R_3 = 34.44$ Ω.

9. 99 990 Ω.

11. (a) 50 μA.
  (b) 50 kΩ; 200 kΩ; 1 MΩ; 5 MΩ.

13. 1.2%.

15. 1.32 MΩ.

17. 5.

19. 4 μF; 50 kΩ.

21. (a) 10 V; 15 V.
  (b) 0.5 ms; 2 kHz.

CHAPTER 18

1.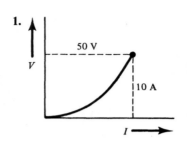

3.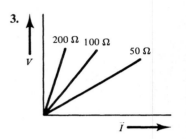

5. 140 V; 70 mA.

7. (a) 89 V; 0.76 A.   (b) 90 V; 0.73 A.

9. (a) 2.7 V.   (b) 11.6 mA.   (c) 1.15 V.

11. 5.5 V; 1.3 A; 1.05 A; 0.25 A.

13. (a) 6.8 mA.   (b) 8 mA.   (c) 2 V.

15. 26 A; 62 A.

17. 1.9 V; 305 mA; 120 mA.

19. 5.5 mA; 4.25 V.

CHAPTER 19

1. (a) 24.   (b) 5520 V.   (c) 13.8 kVA.

3. 24 V.

5. $\frac{1}{7}$.

7. (a) 121.3 V.   (b) 8.09 A.   (c) 0.755 A.   (d) 981 W.

9. Connections: b to d; c to e; a and f to load.

11. $R_e = 0.15\ \Omega$; $X_e = 0.9\ \Omega$; $Z'_L = 25\ \Omega$.

13. $R_e = 2.3\ \Omega$; $X_e = 4.2\ \Omega$.

15. (a) 96.8%.   (b) 97%.

17. 55 kVA.

19. (a) 7621/220V; each 20 kVA.   (b) 2.62 A; 157.5 A.

CHAPTER 20

1. (a) 5 kΩ; 5 kΩ; 5 kΩ; 5 kΩ.
   (b) 10.2 $\angle -78.7°$ kΩ; 10 $\angle -90°$ kΩ; 10 $\angle -90°$ kΩ; 10 $\angle -90°$ kΩ.

3. 1.71 V.

5. -0.798 V.

7. 1.98 kΩ; -9.9×10^{-3}; 9.9×10^{-3}; 1.05×10^{-4} S.

9. 520 Ω.

Index